X-Ray Fluorescence Spectroscopy and Chemometrics

Vivek Kumar Singh
Editor

X-Ray Fluorescence Spectroscopy and Chemometrics

Instrumentations, Techniques, and Applications

Springer

Editor
Vivek Kumar Singh
Department of Physics
University of Lucknow
Lucknow, Uttar Pradesh, India

ISBN 978-3-031-98374-0 ISBN 978-3-031-98375-7 (eBook)
https://doi.org/10.1007/978-3-031-98375-7

This Springer imprint is published by the registered company Springer Nature Switzerland AG
The registered company address is: Gewerbestrasse 11, 6330 Cham, Switzerland

To my parents, Dr. Markandeya Singh and Mrs. Chandratara Devi, who planted the seed of knowledge in my mind and lovingly nurtured it to flourish.

—Vivek Kumar Singh

Foreword

Dr. Singh has for more than 15 years applied different analytical spectroscopy methods to analyze different biological specimens. His results are published in international journals and as book chapters. In this book, Dr. Vivek Kumar Singh brought together a distinguished group of researchers with extensive expertise in *X-Ray Fluorescence Spectroscopy and Chemometrics: Instrumentations, Techniques, and Applications.* The result is a timely and comprehensive overview of the current state of the art in these analytical techniques—ranging from recent methodological advancements to a wide array of practical applications.

X-ray fluorescence, known for its non-destructive nature and inherent multi-elemental capabilities, is a highly versatile technique adaptable to numerous scientific disciplines. This book showcases several of these fields, offering diverse perspectives on how XRF is applied under varying analytical conditions.

By combining theoretical depth with real-world case studies, this collection provides a thorough and insightful exploration of modern XRF applications. It serves as a valuable resource for researchers, practitioners, and students seeking to deepen their understanding of this well-established and continually evolving analytical method.

June 2025

Professor Emeritus Johan Boman
Editor-in-Chief "X-Ray Spectrometry"
University of Gothenburg
Gothenburg, Sweden

Foreword by Revenko and Maltsev

X-Ray Fluorescence (XRF) spectrometry is a widely employed analytical method valued for its rapid, non-destructive analysis, and its ability to determine elemental compositions across various sample matrices. Its applications include environmental monitoring, materials characterization, metallurgy, archaeology, and quality assurance. Despite its versatility, the complexity of spectral data—often featuring overlapping peaks, matrix effects, and background noise—poses significant challenges for accurate quantitative analysis.

Chemometrics is an interdisciplinary science that employs methods commonly used in data science disciplines such as multivariate statistics, applied mathematics, and computer science to solve problems in chemistry, analytical chemistry, biochemistry, chemical engineering, biology, and medicine. The development of chemometric methods also contributes to the enhancement of analytical equipment and methodologies. The number of annual publications applying chemometric techniques to resolve various issues in XRF analysis is continually increasing. Since 1980, specialized journals dedicated to chemometrics have been published. These include, for example, the *Journal of Chemometrics*, *Chemometrics and Intelligent Laboratory Systems*, and the *Journal of Chemical Information and Modeling*. These journals focus on fundamental and methodological research in the field of chemometrics. Additionally, numerous practical applications of chemometric methods are published in journals related to analytical chemistry, such as *Spectrochimica Acta Part B: Atomic Spectroscopy*, *Analytical Chemistry*, *Journal of Analytical Atomic Spectrometry*, *Applied Spectroscopy*, and *Analytica Chimica Acta*.

Chemometrics, which integrates statistical, mathematical, and computational methods, has become an essential complement to XRF spectrometry. Techniques such as multivariate calibration, principal component analysis (PCA), partial least squares regression (PLS), and advanced machine learning algorithms have been increasingly adopted to enhance data interpretation, correct spectral interferences, and improve analytical accuracy. Recent studies demonstrate that these approaches significantly increase the sensitivity and robustness of XRF analysis, especially in complex matrices and at trace levels. The integration of chemometric techniques enables the correction of matrix effects, noise reduction, and feature extraction, broadening the applicability of XRF to new challenging contexts such as environmental pollutant detection and microanalysis. With the advent of machine learning and artificial intelligence, newer algorithms like support vector machines (SVM), random forests, and neural networks have shown superior performance in spectral classification and quantification tasks. Chemometric methods and models are used to address problems of qualitative and quantitative analysis in the field of chemometrics over the past 20 years and explores the possibilities for the development. Unfortunately, we could not find a textbook among publications on chemometrics that clearly outlines the fundamental principles (methods) of chemometrics along with examples of their application in XRF practice. The published book partially fills this gap.

This book "*X-Ray Fluorescence Spectroscopy and Chemometrics: Instrumentations, Techniques, and Applications*" edited by Dr. Vivek K. Singh provides a comprehensive overview of the fundamental principles and recent advances in WDXRF, EDXRF, TXRF, and SRXRF methods, as well as X-ray detectors and the application of chemometrics to XRF spectrometry. Emphasizing recent methodological developments, practical applications, and future directions, it aims to assist researchers in optimizing their analytical workflows through advanced chemometric strategies. Although the book offers only a limited number of examples demonstrating the integration of XRF and chemometrics, its approaches can be broadly extended to other similar samples and studies. In conclusion, it is important to highlight the diverse geographic contributions of the authors involved in this collective work. Participants include colleagues from India (who contributed to six chapters), Nigeria (four chapters), Turkey (three chapters), and Slovenia (two chapters). Representatives from seven countries—Algeria, Brazil, Egypt, Spain, Cuba, Portugal, and Russia—each contributed one chapter. Some chapters, such as chapters 1, 6, and 15, were prepared collaboratively by authors from multiple countries.

This book is designed as a valuable resource for graduate students, research scientists, and other users. We hope it will find a prominent place in the libraries of every laboratory engaged in X-ray fluorescence analysis.

June 2025

Anatoly G. Revenko
Leading Researcher, Center for
Geodynamics and Geochronology
Institute of the Earth's crust Russian
Academy of Sciences Siberian Branch
Irkutsk, Russia

Artem S. Maltsev
Senior Researcher, Center for
Geodynamics and Geochronology
Institute of the Earth's crust Russian
Academy of Sciences Siberian Branch
Irkutsk, Russia

Preface

X-ray fluorescence (XRF) spectroscopy is a valuable tool in environmental studies for identifying the elemental composition of soil, sediments, water, and air samples for pollutants and contaminants. It is used to monitor air quality, analyze heavy metals in water sources, and determine the elemental composition of various environmental samples. During the past decade, XRF spectrometry has gone through major changes in the field of environmental sciences. This book is a guide which provides an up-to-date review information of XRF spectrometry for the researchers working in the field of plant science, environmental science, environmental toxicology, and geological science. It covers the basic fundamental principles, latest instrumentation developments, data analysis tools and chemometrics in X-ray fluorescence for applications in different field of science and technology. Data analysis methods and chemometrics in XRF is an excellent approach to solve qualitative and quantitative problems in XRF analyses. This book includes the practical exploration of the use of advanced chemometric methods to XRF cases in environmental science. It provides a concise guide to solve commonly encountered problems involving XRF-based spectro-analytical techniques and their applications.

The chapters are contributed by independent groups from different countries, Algeria, Brazil, Cuba, Egypt, India, Ljubljana, Nigeria, Portugal, Russia, Spain, and Turkey. Four chapters are contributed by the members of editorial advisory board of the journal "X-Ray Spectrometry" from John Wiley & Sons, and Spectrochimica Acta B, from Elsevier.

The book consists of 17 chapters. Chapter 1 covers the introduction, latest developments and applications of XRF spectroscopy. Chapter 2 highlights the recent developments particularly in WDXRF and EDXRF spectroscopy. The instrumentation, analytical performance, and emerging applications have been covered thoroughly. Chapter 3 provides a detailed discussion on gas-filled detectors, scintillation and semiconductor detectors and associated imaging applications. Chapter 4 highlights the principles of SXRFI, its applications, and recent technological developments shaping its future in scientific research. Chapter 5 discusses the recent developments in TXRF spectroscopy and applications. Chapter 6 presents a comprehensive overview of a myriad of studies utilizing chemometrics and XRF to prove the worthiness of the combination of XRF with chemometrics which may open new interesting applications of XRF in many different fields in the near future. Chapter 7 describe the review of the current tools and future trends used to statistically process data previously collected through XRF devices in the field of environmental toxicology. This chapter equally addresses the main advantages and disadvantages of techniques. Finally, newer mathematical models whose preliminary results denote an auspicious future, namely, recurrent, deep, and convolution neural networks-based algorithms, are described here and properly framed in the environmental toxicology field. Chapter 8 discusses recent advancements in the application of XRF in plant biology with an emphasis on the element distribution in plant organs by benchtop μ-XRF systems. In addition, it also addresses the challenges in sample preparation, quantification methods, and the analysis and interpretation of complex data sets. Chapter 9 explores the application of XRF for analyzing vegetation tissues. Chapter 10 explores the application of XRF in agriculture, with a particular focus on its use for measuring contamination and the integration of chemometric tools to enhance the accuracy and interpretability of the results. Chapter 11 explores the principles and applications of XRF spectroscopy in food analysis, emphasizing its advantages and limitations. Chapter 12 intends to delineate the multivariate approaches being used in the field of XRF analysis for geological samples. Chapter 13 includes the technological developments and expanding applications TXRF. Chapter 14 highlights the potential application of XRF for detection of the toxicity of fluorine, chromium, arsenic, cadmium, mercury, lead, and uranium in drinking water. In continuation, Chapter 15 gives the overview on the different XRF methods for detection of chemical hazards in drinking water. Chapter 16 provides to inform the reader about the recent studies and applications about the detection of atmospheric chemicals by XRF along with possible future studies. Chapter 17, which is the last chapter of the book, explores the application of WDXRF in diagnosing nematode-induced damage in crops, demonstrating its potential to enhance food security while promoting environmentally friendly farming approaches.

I would like to take this opportunity to express our gratitude to all of the authors for their excellent contributions in this book. We hope the readers will enjoy this book "*X-Ray Fluorescence Spectroscopy and Chemometrics: Instrumentations, Techniques, and Applications*" and that it contributes to the continued instrumental developments of XRF and environmental applications. We also hope that it encourages and inspires the beginners to the field in exploring the multifaceted aspects of XRF.

Finally, the critical evaluations and recommendations by the reviewers for the applicability of the XRF methods to environmental samples will make this book a valuable asset for anyone employing or improving upon these techniques.

Lucknow, India
June 2025

Vivek Kumar Singh

Contents

1 X-Ray Fluorescence Spectroscopy: Introduction, Latest Developments and Applications

Neslihan Ekinci, Ahmet Furkan Ekinci, Yasser Rammah, Neha Sharma, and Vivek K. Singh

Abstract

One of the most popular and straightforward methods for non-destructive multi-element material analysis is X-ray fluorescence (XRF) spectroscopy. In the last few years, the method has advanced remarkably and shown to be helpful in several disciplines, including geology, environment, materials science, and archaeology. The technique has proven successful in industry in addition to research applications, particularly in preserving the purity of ultra-pure grade chemicals, reagents, and products. These practical applications of XRF provide a wealth of knowledge and understanding for scientists and industrial users. Due to recent developments in science and technology, we decided to compile this section as a resource that provides sufficient information for scientists and industrial users to design and set up modern XRF experiments for use in a wide range of practical applications. This compilation aims to keep you updated and aware of the latest advancements in XRF. We have also briefly discussed the application of machine learning and artificial intelligence which has the potential to expand the capabilities of XRF analysis.

N. Ekinci (✉)
Department of Physics, Atatürk University, Erzurum, Turkey
e-mail: nekinci@atauni.edu.tr

A. F. Ekinci
Defense Systems Division, Aselsan Joint Stock Company, Ankara, Turkey

Y. Rammah
Department of Physics, Faculty of Science, Menoufia University, Shebin El-Koom, Menoufia, Egypt

N. Sharma
Department of Applied Sciences & Humanities, KCC Institute of Technology and Management, Greater Noida,
Uttar Pradesh, India

V. K. Singh
Department of Physics, University of Lucknow, Lucknow, Uttar Pradesh, India

V. K. Singh (ed.), *X-Ray Fluorescence Spectroscopy and Chemometrics*,
https://doi.org/10.1007/978-3-031-98375-7_1

1.1 Introduction: X-Ray Fluorescence Spectroscopy

X-ray fluorescence (XRF) analysis is a widely used multi-element analysis technique with numerous practical applications, especially for non-destructive procedures. Over time, steady improvements have been observed in XRF, methodologically and instrumentally.

Recent technological advancements, such as tabletop instruments utilizing innovative low-power micro-focus tubes, new X-ray optics and detectors, and simplified access to synchrotron radiation, have revolutionized XRF. These advancements have enabled XRF to expand to include low Z elements and the acquisition of two- and three-dimensional data from a sample at a micrometer-scale. For example, developing portable equipment has facilitated more versatile use of XRF in various new contexts, including archaeometry and process control. These developments have significantly improved XRF's accuracy, speed, and versatility, making it an indispensable tool in a wide range of industries.

XRF spectroscopy is the most effective method for analyzing materials' qualitative and quantitative composition. There are two types of XRF methods: wavelength-dispersive XRF (WDXRF) and energy-dispersive XRF (EDXRF). Both methods have advantages and disadvantages. WDXRF offers a comprehensive, high-resolution examination of specific elements, especially in more complex matrices, while EDXRF is suitable for rapid and flexible analysis of various samples. So, both EDXRF and WDXRF are required for many different industries. In EDXRF, trends include miniaturization, development, and optimization of high-resolution room-temperature detectors and extension of the application range for determining light elements.

The status of WDXRF spectrometers is as follows: these spectrometers have gained popularity in recent years due to technological advancements that have increased their sensitivity, accuracy, and speed. The demand for elemental analysis with high sensitivity and reliability is one of the main reasons for the rise of the WDXRF market. Furthermore, stringent industry safety and quality assurance guidelines have significantly increased the demand for WDXRF spectrometers. So, the growth of WDXRF spectrometer trends is primarily driven by advancements in analytical technologies, raising awareness of quality control, and increasing demand from end-user industries. The key trends in the WDXRF market include incorporating automation and software solutions for advanced data processing and reporting, as well as developing portable and benchtop devices to meet a wide range of application needs.

Another significant trend is the ability of WDXRF spectrometers to provide accurate analysis of complex samples. However, these WDXRF trends are constrained by factors such as the high initial costs of the equipment, the need for skilled operators, and competition from alternative technologies such as EDXRF.

1.2 Principle and Instrumentation of EDXRF

As seen in Fig. 1.1, an EDXRF system comprises the following fundamental functional parts: a multichannel pulse height analyzer, sample chamber, Si(Li) detector, preamplifier, main amplifier, and an X-ray excitation source. The characteristics and capabilities of an EDXRF system vary depending on the electronics and computer upgrades.

Identifying characteristic lines in EDXRF is accomplished by utilizing detectors that directly measure photon energy. In its most basic form, photo absorption ejects an electron from an atom of the detector material. The loss of energy of this newly formed primary electron causes a shower of electron–ion pairs in a proportional counter, visual excitations in a scintillation counter, or showers of electron–hole pairs in a semiconductor detector. Unlike wavelength dispersion, which uses a crystal's Bragg reflecting characteristics to disperse X-rays at different reflection angles based on their wavelengths, the resulting detector signal is proportional to the original photon's energy. Although energy-dispersive detectors have lower energy resolution than wavelength-dispersive analyzers, they may detect many energies simultaneously. The most common EDXRF detector is the silicon semiconductor detector, which can achieve outstanding energy resolution.

The two alternative types of detectors discussed above, with their lower energy resolution, are limited to particular scenarios in which specific semiconductor properties are unacceptable. The germanium semiconductor detector is disadvantageous compared to traditional XRF because it can obscure other lines of interest due to escape peaks from powerful signals.

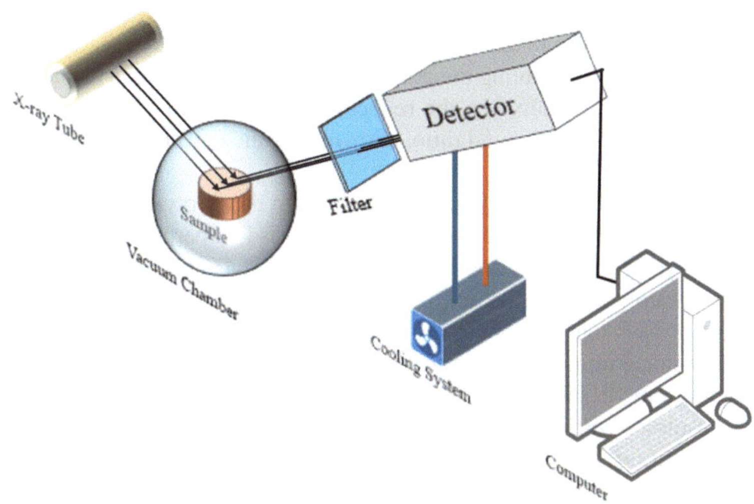

Fig. 1.1 The essential components of the EDXRF spectrometer

EDXRF spectrometers can be divided into spectrometers with 2D and 3D optics [3–7]. Both types have a source and an energy dispersive detector, but the difference is found in the X-ray optical path. For 2D spectrometers, the X-ray path is in one plane, so in 2 dimensions. For the 3D spectrometers, the path is not limited to one plane but involves 3 dimensions.

1.2.1 EDXRF Spectrometer with 2D Optics

EDXRF spectrometers with 2D optics position the X-ray tube, sample, and detector within the same plane. This is also known as direct excitation geometry which allows for simpler beam paths and direct interaction between the X-rays and the sample. 2D optics are generally used in benchtop EDXRF instruments [3–7]. This configuration facilitates a straightforward path for the X-ray beam to interact with the sample and for the emitted fluorescence to be detected. This configuration is simple and enables more compact instrument as compared 3D optics setup and are well-suited for analyzing various sample types, including liquids, powders, solids, and bulk materials. These spectrometers are widely used in industries such as metals and alloys manufacturing, petrochemicals, forensics, food analysis, and environmental analysis. Many benchtop EDXRF instruments utilize 2D optics, offering a compact and accessible solution for elemental analysis. Some 2D instruments, such as those using polarized X-rays, can achieve high sensitivity for specific elements like sulfur, phosphorus, and chlorine in petrochemicals.

1.2.2 EDXRF Spectrometers with 3D Optics

EDXRF spectrometers with 3D optics directs primary X-rays to a secondary target and then to the sample, enabling the complete elimination of primary radiation and resulting background noise when the reflections are at exactly 90° [3, 4, 7]. This enhances the peak-to-background ratio, allowing for the detection of low elemental concentrations.

This geometry involves a 3D arrangement where the X-ray tube, secondary target, and sample are in one plane, while the detector is in a perpendicular plane. The primary X-rays are first directed to a secondary target, and then to the sample, effectively creating a 90° reflection path. When the reflections are at exactly 90°, the primary X-rays that contribute to background noise are eliminated, improving the signal-to-noise ratio and allowing for more accurate analysis of lower elemental concentrations. EDXRF instruments with 3D optics are particularly useful for analyzing samples with complex matrices, where background reduction is crucial, and for detecting trace elements.

1.3 Principle and Instrumentation of WDXRF

The experimental setup of WDXRF is shown in Fig. 1.2, which consists of a (i) X-ray source; (ii) primary filters; (iii) collimators; (iv) analyzing crystals; (v) detectors, and (vi) electronics and data acquisition systems.

In WDXRF spectrometers the analyzing crystal acts as a monochromator that selectively transmits the characteristic X-rays of particular wavelength produced by the sample elements to the X-ray detector. When the emitted X-ray radiations of different wavelengths fall into the analyzing crystal at a certain angle, the diffraction pattern has been observed only for those radiations that satisfy Bragg's law (Fig. 1.3):

$$n.\lambda = 2d.\sin\theta$$

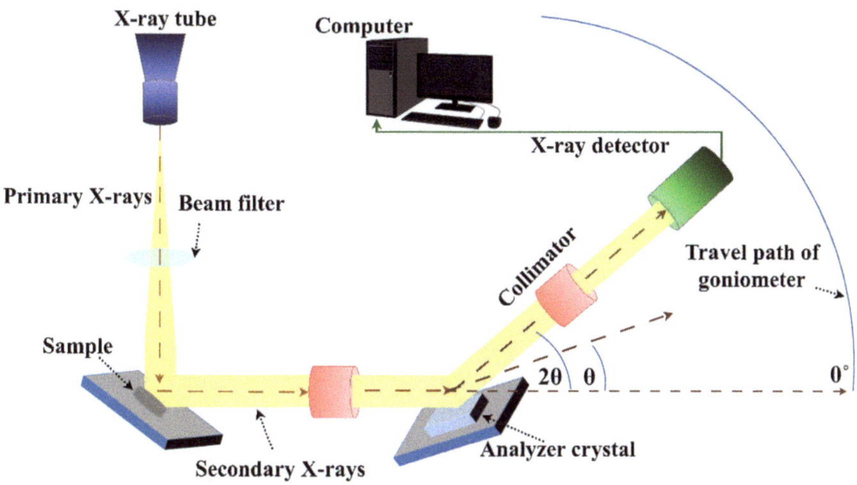

Fig. 1.2 The essential components of the WDXRF spectrometer

Fig. 1.3 Representation of Bragg's law

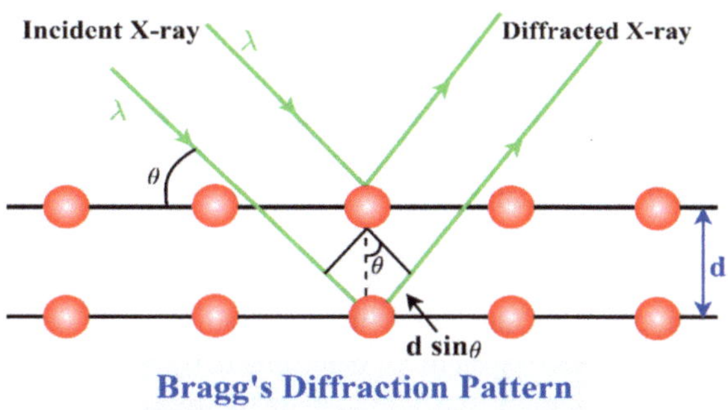

Table 1.1 List of some analyzing crystals generally used in WDXRF spectrometry system

S. no.	Crystal	Reflection plane	2d(nm)	Element range	Reflectivity
1	Lithium fluoride (LiF)	(420)	0.180	Ni-U	Average
2	Lithium fluoride (LiF)	(220)	0.284	V-U	High
3	Lithium fluoride (LiF)	(200)	0.402	K-U	Very high
4	Germanium (Ge)	(111)	0.653	P, S, Cl	High
5	Indium antimonide (InSb)	(111)	0.748	Si	Very high
6	Penthaerythritol (PET)	(002)	0.874	Cl–Al	High
7	Ammonium dihydrogen phosphate (ADP)	(101)	1.064	Mg	Low
8	Thallium (I) hydrogen phthalate (TIAP)	(1010)	2.575	Mg–O	Medium
9	Ethylene diamine-D-tartrate (EDDT)	(020)	0.880	Lighter elements	High
10	Potassium hydrophthalate (KHP)	(1010)	2.64	Mg, Na, F	High

where d denotes the interplanar distance between the lattice spacing of the crystal (*nm*), n is the order of reflection, and 'θ'is the diffraction angle.

The diffraction pattern may be obtained for different wavelengths by changing the value of 'θ' (angle of diffraction) which can be done by rotating the crystal with respect to the emitted X-ray radiation. Consequently, only X-rays with specified wavelengths will reach the detector. Some analyzing crystals that are generally used in WDXRF are listed in Table 1.1.

1.4 Comparison of EDXRF with WDXRF

1.4.1 Resolution

It describes the width of the spectral peaks. The lower the resolution, the easier to distinguish an elemental line from adjacent X-ray line intensities. The resolution of a WDXRF system is determined by the crystal and optical design, particularly collimation, spacing, and spatial reproducibility. The effective resolution of a WDXRF system can range from 20 to 5 eV. The effective resolution of a suitable

WDXRF system is 5 eV or less. In these systems, the resolution is not detector-dependent. The detector resolution determines the resolution of an EDXRF system. For example, these resolution values can be 150 eV or less for a liquid nitrogen-cooled Si(Li) detector, 150–220 eV for various solid-state detectors, or 600 eV or more for a gas-filled proportional counter. The advantage of WDXRF is its high resolution, which means less spectral overlap and lower background intensities. The advantage of EDXRF is the prohibitive cost of the WDXRF crystal and optics.

1.4.2 Spectral Overlaps

When two spectral lines overlap, it is challenging to measure them separately, so spectral deconvolution is required to determine the net intensities. When a very high resolution (low eV count) WDXRF instrument is used, most components' spectral aliasing corrections are not required. The gross intensities of each element can be calculated in a single acquisition. The EDXRF analyzer is designed to detect several components simultaneously. Some form of deconvolution technique must be applied to correct for spectral aliasing. Overlapping is no longer a problem with 150 eV resolution instruments. This is significant when compared to WDXRF. The advantage of WDXRF is that Spectral deconvolution algorithms introduce statistical counting errors for the aliasing correction of each element being corrected. This can result in a doubling or tripling of the error. Spectral aliasing becomes more problematic at lower resolutions.

1.4.3 Background

Background radiation is essential to determining detection limit, repeatability, and reproducibility.

A WDXRF instrument usually uses direct radiation flux. The background in the area of interest is directly related to the amount of continuous radiation, and the WDXRF spectrometer has the advantage due to its resolution. If a peak is one-tenth as wide, it is one-tenth as wide as the background.

An EDXRF instrument uses filters and targets to reduce the amount of continuous radiation in the area of interest.

1.4.4 Excitation Efficiency

This is the second primary factor determining detection limits, repeatability, and reproducibility. It is usually expressed in PPM per second (cps) or similar units. The relative excitation efficiency can be increased by exciting an X-ray source close to the absorption edge of the element of interest but with slightly higher energy. To excite characteristic (fluorescent) X-rays of an element in a sample, excitation of the sample with a source closer to the absorption edge of interest but with slightly higher energy can increase the relative excitation efficiency.

WDXRF generally uses direct, unmodified X-ray excitation, which contains a continuum of energies, most of which are not optimal for exciting the element of interest.

EDXRF uses filters to reduce the continuous spectrum and effectively increase the percentage of X-ray intensity above the absorption edge of the element of interest. It is also possible to provide a monochromatic line source using secondary targets to obtain the best excitation efficiency for the element line of interest.

1.4.5 Comparison of Advantages, Limitations and Recent Development Parameters of EDXRF and WDXRF

ED-XRF and WD-XRF are both X-ray fluorescence techniques used for elemental analysis, but they differ in how they separate the emitted X-rays. ED-XRF uses energy to separate the X-rays, while WD-XRF uses wavelengths. ED-XRF is known for its speed, simplicity, and portability, while WD-XRF excels in high-precision analysis and low detection limits. Table 1.2 briefly prsents the advanatges, limitations and recent develonpments and future prospects of the use of XRF technology.

In summary, the choice between ED-XRF and WD-XRF depends on the specific application and requirements. ED-XRF is ideal for quick, versatile, and cost-effective analysis, while WD-XRF excels in high-precision, detailed analysis, and low detection limits. Both techniques are continuously evolving, with new developments in instrumentation, data analysis, and application areas.

1.5 Current Interests in WDXRF and EDXRF Spectroscopy

Among the industries that EDXRF and WDXRF spectrometers are being used in the field of chemistry, forensics and environmental science, food and agriculture science, industries, life science research, material science research, semiconductor and nanotechnology, microbiology and diagnostic applications, minerals and mining petrochemicals, and pharma and biopharma industries.

1.5.1 Chemistry

Chemistry researchers are particularly interested in understanding material qualities, chemical mechanisms, and the physics underlying them.

In all fields of chemistry, EDXRF and WDXRF are valuable tools for examining material properties in both liquid and solid states. Innovations in EDXRF and WDXRF equipment are essential for advancing current research and achieving faster and more accurate results.

Table 1.2 Aadvantages, limitations, recent developments and future prospects of the use of XRF technology

Parameters	EDXRF	WDXRF
Advantages	• **Speed and Simplicity**: ED-XRF is faster and simpler to operate as compared to WD-XRF • **Versatility**: It can be used in various applications, from lab-based analysis to portable field use • **Simultaneous Multi-element Analysis**: Simultaneously analysis capability of multiple elements makes it efficient for screening	• **High Resolution**: Capability of superior spectral resolution allows for better separation of signals and accurate analysis of complex samples • **Low Detection Limits**: Lower detection limits for low-mass trace elements • **High Precision and Accuracy**: Capability of determining major elements with high precision and accuracy
Limitations	• **Lower Resolution**: ED-XRF generally has lower spectral resolution compared to WD-XRF • **Higher Detection Limits**: ED-XRF may have higher detection limits, particularly for light elements	• **Slower Analysis**: It is slower than ED-XRF, especially for analyzing a large number of elements • **Higher Cost**: More expensive due to the use of crystals and other optical components • **Limited Portability**: WD-XRF are typically larger and less portable than ED-XRF
Recent developments and future	• **Advanced Detectors**: Ongoing advancements in detector technology are improving resolution and sensitivity • **3D-EDXRF**: 3D-EDXRF with polarized X-ray beams enhances sensitivity and specificity for trace element analysis • **Miniaturization**: Miniaturized ED-XRF sensors are enabling integration into mobile platforms like autonomous drilling rigs and drones • **AI-Enhanced Analysis**: AI is being used to improve the speed and accuracy of ED-XRF analysis	• **Improved Crystals**: Synthetic thin-film multilayer crystals are being used to enhance sensitivity and resolution, especially for light element analysis • **Miniaturization**: Efforts are being made to miniaturize WD-XRF components, making them more portable • **AI-Driven Data Analysis**: AI is being used to improve data interpretation and automation in WD-XRF

1.5.2 Forensics and Environmental Science

EDXRF and WDXRF spectrometers contribute significantly to forensic and environmental science by identifying micropollutants, trace particles, and dangerous compounds. For example, identifying trace amounts of materials, hazardous chemicals, illicit substances, and explosives, as well as assisting in detecting fake paintings, significantly impacts human life.

1.5.3 Food Analysis and Agriculture

There are significant obstacles facing producers, consumers, governments, and brands due to the globalized food supply chain. Worldwide competition, the use of harmful substances in food ingredients and the strong demand for food quality control constantly increase the number of stringent regulations. EDXRF and WDXRF spectrometers protect the safety and quality of food and beverages, animal feed and agricultural products throughout the entire supply chain, from product creation to quality control in production and incoming inspections. EDXRF and WDXRF spectrometers with mobile, laboratory-based, and online systems are available to identify and quantify agents, substances, and trace elements.

1.5.4 Industrial

Today's industrial environment demands that EDXRF and WDXRF systems meet higher standards. Therefore, these analytical and quality control systems are not expected to be lower quality. The comprehensive product portfolio and innovative solutions for EDXRF and WDXRF systems must meet tomorrow's demands and help increase productivity and quality.

Precise measurements of materials and their components are required to test automotive and aerospace components under real-world conditions and help predict and prevent failure. Without sample preparation, EDXRF and WDXRF spectrometers provide spatially resolved, non-destructive material analysis (mapping) or layer thickness analysis down to trace levels.

In cement production, EDXRF and WDXRF systems are crucial in obtaining information about the chemical and mineralogical composition of raw materials and finished products, controlling and optimizing the process, and ensuring consistently high product quality. These systems provide the basis for efficient processes while ensuring cement quality and meeting international standards.

The purpose and properties of the industrial coating are of immense importance among the applications of EDXRF and WDXRF (Ekinci et al. 2001, 2006). EDXRF and WDXRF systems are used in surface refining, determining whether a product has a bioactive, photovoltaic, insular, or protective polymer coating, providing essential quality control and failure analysis.

In other words, these systems are beneficial in investigating coating properties, defects, and deterioration, especially in evaluating roughness, thickness, adhesion, homogeneity, and performance at the nanoscale and in determining the conditions under which engineers use coating.

Quality control is essential for maximizing efficiency in the highly competitive photovoltaic sector. Photovoltaic devices' efficiency, stability, and operational life have recently increased. To meet the global energy crisis, we can increase the overall optoelectronic performance of solar cells by better understanding their chemical composition, microstructure, flaws, and contaminants. This can be accomplished utilizing cutting-edge XRF (EDXRF-WDXRF).

Many semiconductor devices, including solar cells, now consist of several layers. XRF (ED XRF-WDXRF) enables simultaneous layer thickness and composition assessment to explore such layered systems in the final product.

Ceramics and glass are another industry segment. EDXRF and WDXRF systems provide comprehensive solutions for glass and ceramic analysis. We provide the necessary technologies, from chemical composition analysis of raw materials, intermediates, and finished products to spatial distribution of constituents for quality control and failure analysis of cooled samples.

Ceramics and refractors require a thorough elemental examination of raw materials, manufacturing processes, and finished products. For successful quality control and failure analysis of cooled samples, XRF analyzers provide non-destructive methods for performing spatially resolved assessments of texture and elemental distributions down to the ppm range. Typically, wavelength dispersive WDXRF is chosen because it can measure major and minor components and traces at lower concentrations. The WDXRF instruments used in this application are the S6 JAGUAR and S8 TIGER.

Different materials are mixed or dissolved in the raw glass to meet the usage requirements. XRF analysis can be performed with the EDXRF S2 PUMA or WDXRF S6 JAGUAR spectrometers for non-destructive analysis of the final glass or metallic coatings, depending on the elements and concentration ranges.

Using the wavelength dispersive spectrometer QUANTAX WDS, complex materials in the low energy or light element range can be determined better due to the excellent energy resolution and light element sensitivity.

Glasses have varying qualities depending on the application for which they will be utilized. Different mixing of various additives with raw glass can determine specific properties, such as mechanical strengths and electrical conductivities. XRF enables an exact assessment of the concentration of the components used to make a glass, allowing fine-tuning of its qualities.

One application of automated XRF is the separation of diverse types of glass for the glass recycling industry. Automated XRF instruments can accurately classify each type of glass prior to recycling.

XRF analytical systems also help the polymer industry efficiently produce high-quality plastics and develop new polymers. They are the ideal tools for simple quality control, process monitoring or research on complex materials. XRF spectrometers are also used to characterize polymers' multilayer composites, determining layer thickness and content uniformity.

XRF systems play an essential role in the research, development and characterization of new materials, raw material quality control, determination of the dynamics and kinetics of polymer reactions, and thermal stability and crystallization properties.

Life science requires the utmost accuracy in investigating various biological samples and molecular processes. Today, there are rapid and continuous changes in life science. XRF (EDXRF and WDXRF) has also found its place in this field as a new analytical instrument that allows these samples to be examined accurately with extremely high temporal and spatial resolution.

In addition, XRF analysis provides versatile, forward-thinking solutions for materials science research and testing, enabling companies to invest in process and product improvements and make continuous progress in this direction.

Due to industrial growth and technological changes, new products based on biomedical and nanotechnological materials, thin films, functional materials, carbons, or high-tech ceramics make our daily lives easier. Developing these new materials and nanostructures requires interdisciplinary collaboration in various research areas involving physics, biology, chemistry, and engineering elements.

XRF analytical systems help to produce usable products at the nanoscale. Our systems also ensure progress in efficient drug and additive production. XRF analysis offers solution-oriented approaches to any scientific problems encountered in geological sciences, mining, oil, and gas. By investigating all rocks, minerals, and materials from nanometer to planetary scale with accurate analyses, that is, by analyzing the raw materials that feed economies, it actively serves inclusive communities and humanity that support scientific discoveries.

1.6 Artificial Intelligence, Machine Learning, and Cloud Technologies: Tranforming XRF Analysis and Technology

AI, particularly machine learning (ML) and deep learning, is increasingly used to enhance EDXRF and WDXRF analysis by automating processes, improving data interpretation, and enabling real-time analysis. AI can streamline material analysis, increase accuracy, and reduce errors, ultimately boosting the efficiency and effectiveness of XRF techniques [8–11].

The application of ML models provides a better understanding of the past, as ML-enabled XRF can handle complex materials or detect trace elements [11]. XRF-ML integration has enabled the dating of French gilt bronzes, the determination of carbon in geological samples, and the identification of soil moisture levels [11].

Artificial Intelligence, Machine Learning, and Cloud Technologies are now tranforming the XRF Analysis.

A. **Automation and Streamlining**
   - **Automated Preprocessing**: AI algorithms can automatically preprocess raw XRF spectra, eliminating manual steps and reducing the time required for analysis.
   - **Feature Extraction**: AI can identify the most relevant features in the spectra, making it easier to extract meaningful information and reduce the need for manual feature selection.
   - **Data Reduction**: AI, like autoencoder neural networks, can reduce the dimensionality of XRF data, allowing for more efficient storage and processing.

B. **Improved Data Interpretation and Analysis**
   - **Pattern Recognition**: Deep learning models can learn complex patterns in XRF spectra, leading to more accurate and robust analysis.
   - **Real-Time Analysis**: AI can enable real-time analysis of XRF data, providing immediate feedback during the analytical process.
   - **Predictive Maintenance**: AI can be used to predict when XRF instruments may require maintenance, reducing downtime and improving reliability.

C. **Enhanced Accuracy and Reliability**
   - **Error Reduction**: AI algorithms can help identify and correct errors in XRF data, leading to more accurate results.
   - **Increased Reliability**: AI can enhance the reliability of XRF analysis by improving data interpretation and identifying potential issues.
   - **Improved Collaboration**: AI can facilitate improved collaboration between research groups and organizations by streamlining data processing and analysis.

D. **Specific Applications**
   - **Material Science**: AI can be used to analyze the elemental composition of various materials, including building materials, ores, and polymers.
   - **Environmental Monitoring**: AI can help assess pollution and contamination levels by analyzing XRF data from environmental samples.
   - **Cultural Heritage**: AI can aid in the analysis of materials from cultural heritage objects, helping to understand their composition and history.
   - **Medical Imaging**: AI can be used to enhance medical images, improving the accuracy and efficiency of diagnosis.

E. **Future Directions**
   - **Development of smaller and more powerful XRF instruments**: These instruments, coupled with AI and ML, will allow for more comprehensive material investigations. ML algorithms can also be used to create and implement calibrations. While traditional empirical modeling could create only material-specific calibrations, different ML models can be used in combination to improve these calibrations [11]. For example, ML-XRF analysis of archaeological ceramics can produce calibration curves with sufficient cross-validation for lighter elements like Si, Al, and Na, even when the instrument is working at higher voltage settings [11].
   - **Interdisciplinary approaches**: Combining expertise in XRF analysis, AI, and other fields will lead to significant advancements in material analysis technologies.
   - **AI-driven predictive maintenance and real-time analysis**: These capabilities will further enhance the sustainability and efficiency of XRF analysis.

1.7 Conclusion

The new developments in excitation and detection technologies render XRF benchtop spectrometers very robust. Since XRF's prominence in the geological field, EDXRF instruments are frequently used in geological samples alongside WDXRF instruments to quantify major and minor components. With the new generation databases developed, users can achieve alloy library expansion in these systems and can analyze more than 600 alloy types and more than 10,000 alloy materials simultaneously. The new parameter-fitting FP algorithm enables users to analyze metal material and element content accurately and with a low detection limit without changing mode. Both spectrometers do not damage the product during analysis, and the measurement can be conducted in extreme environmental conditions such as high and low temperatures, dust, darkness, and humidity. These spectrometers can be operated with a single button or remote control, which is very convenient. The device is also smaller and lighter, which is convenient for users to carry.

References

1. Ekinci, N., Orhan, E., Kurucu, Y., Sahin, Y.: Measurement of foil and cable thickness by a transmission method. Instrum. Sci. Technol. **29**(5), 415–421 (2001). https://doi.org/10.1081/CI-100107233
2. Ekinci, N., Orhan, E., Kurucu, Y., Sahin, Y.: Measurement of sample thickness by an intensity ratio method. Instrum. Sci. Technol. **29**(2), 145–151 (2006). https://doi.org/10.1081/CI-10010346
3. Singh, V.K., Kawai, J., Tripathi, D.K.: X-Ray Fluorescence in Biological Sciences: Principles, Instrumentation, and Applications. Wiley (2022)
4. Bauer, L.J., Mustafa, H.A., Zaslansky, P., Mantouvalou, I.: Chemical mapping of teeth in 2D and 3D: X-ray fluorescence reveals hidden details in dentine surrounding fillings. Acta Biomater. Biomater. **109**, 142–152 (2020)
5. Brouwer, P.: Theory of XRF. PANalytical BV, Almelo, Netherlands
6. Beckers, D., Kogan, V., Bolze, J., Lehmann, C.W., Schweim, H.G., Steffens, K.J., inventors; PANalytical BV, assignee. X-ray detection in packaging. United States patent US 7,756,248. 2010 Jul 13
7. Marguí, E., Zawisza, B., Sitko, R.: Trace and ultratrace analysis of liquid samples by X-ray fluorescence spectrometry. TrAC Trends Anal. Chem. **53**, 73–83 (2014)
8. Drake, B.L., Shugar, A., Kipnis, E.: Advances in portable X-ray fluorescence spectrometry instrumentation, application and interpretation. İn: Drake, B.L., MacDonald, B.L. (eds.) The Royal Society of Chemistry, ch. 17, pp. 490–534 (2022)
9. Andric, V., Kvascev, G., Cvetanovic, M., Stojanovic, S., Bacanin, N., Gajic-Kvascev, M.: Deep learning assisted XRF spectra classification. Sci. Rep. **14**(1) (2024). https://doi.org/10.1038/s41598-024-53988-z.
10. Zheng, X., Kankanallu, V.R., Lo, C.-A., Ajith Pattammattel, Chu, Y., Chen-Wiegart, K., Huang, X.: Deep learning enhanced super-resolution X-Ray fluorescence microscopy by a dual-branch network. Optica (2024). https://doi.org/10.1364/optica.503398.
11. Shugar, A.: Advancements in portable and lab-based XRF instrumentation for analysis in cultural heritage: a change in perspective. Micros. Microanal. (2021). https://doi.org/10.1017/s1431927621009077

Recent Developments in WDXRF and EDXRF: Instrumentation, Analytical Performance, and Emerging Applications

2

Arun Upmanyu

Abstract

This chapter investigates recent research-driven developments in wavelength dispersive X-ray fluorescence (WDXRF) and energy dispersive X-ray fluorescence (EDXRF), focusing on instrumentation, analytical performance, and emerging applications. In WDXRF, the study critically examines the evolution of detector systems including proportional counters and scintillation detectors alongside advancements in geometrical configurations such as tube-above setups, crystal technologies, improved signal-to-noise ratios, and advanced data processing algorithms, which collectively contribute to higher resolution and precision in multi-element analysis. The chapter also explores recent trends in detector architecture and system automation that are shaping the next generation of WDXRF instruments. In the context of EDXRF, this work evaluates ongoing innovations in compact and high-throughput systems, with particular focus on cutting-edge modalities such as total reflection XRF (TXRF), grazing emission XRF (GEXRF), and micro-XRF (μ-XRF). These techniques demonstrate extended analytical capabilities for trace, ultra-trace, and spatially resolved elemental analysis in complex matrices. This chapter provides a foundation for researchers seeking to understand the instrumental evolution of XRF and its implications for future analytical research and industrial integration.

A. Upmanyu (✉)
Chitkara University Institute of Engineering and Technology Chitkara University, Rajpura, Punjab, India
e-mail: arun.upmanyu@chitkara.edu.in

V. K. Singh (ed.), *X-Ray Fluorescence Spectroscopy and Chemometrics*,
https://doi.org/10.1007/978-3-031-98375-7_2

2.1 Introduction

X-ray fluorescence (XRF) is an analytical technique extensively utilized for determining the thickness, elemental composition of materials and quality control [1–3] This method operates on the principle of high-energy X-rays or gamma rays interacting with a sample, stimulating atoms within the material to emit secondary (fluorescent) X-rays. The photon-atom interactions involve in XRF include photon attenuation, photoionization, and scattering processes such as Rayleigh, Compton, and resonant Raman scattering. XRF analysis is primarily conducted using two spectrometers: (i) wavelength dispersive XRF (WDXRF) and (ii) energy dispersive XRF (EDXRF). Both spectrometers provide comparable insights into the intensity of emitted characteristic X-rays and scattered photons, but they differ significantly in their operational mechanisms and applications. The EDXRF spectrometer utilizes a semiconductor detector coupled with a multichannel analyzer (MCA) to separate radiation wavelengths based on their energies. In contrast, WDXRF spectrometer employs crystal diffraction to isolate X-rays of interest according to Bragg's law, $2d\sin\theta = n\lambda$, θ be the angle of diffraction. WDXRF systems analyze spectra at individual wavelengths, yielding higher intensities for specific photon energies but are limited to single-wavelength detection per scan. These distinctions highlight the complementary strengths of WDXRF for different research applications [4, 5] and offer tailored approaches to achieve precise elemental characterization [6, 7].

The instrumentation of WDXRF and EDXRF is critical for accurate and versatile across a wide range of scientific and industrial applications. In WDXRF, high spectral resolution is achieved through crystal diffraction using closely spaced X-ray wavelengths which offer superior accuracy in quantitative analysis. On the other hand, EDXRF offers rapid multi-element detection using semiconductor detectors, making it ideal for high-throughput and field applications. Additionally EDXRF caught interest of researcher due to its compact design and lower operational cost. Together, these instruments address diverse analytical requirements, from high-resolution analysis in materials research to fast and portable testing in environmental and industrial monitoring [8–11]. This chapter focuses on providing a comprehensive understanding of the instrumentation of WDXRF and EDXRF spectrometers. It delves into the principles, geometrical setups, and key components that define the functionality and performance of these systems. Additionally, recent advancements and trends in WDXRF and EDXRF technologies which enhance analytical capabilities and expand their applications in various scientific and industrial fields are discussed in the concluding section.

2.2 Instrumentation for WDXRF Spectrometer

The geometrical setup of a WDXRF spectrometer is illustrated in Fig. 2.1. The X-ray source, combined with a primary beam filter, ensures precise excitation of elements in the sample by selectively shaping the primary beam. A low-background mask changer, available in sizes such as 23, 28, and 34 mm, effectively blocks unwanted signals from the sample cup, preventing interference with the detector. A vacuum seal is incorporated between the sample chamber and the goniometer chamber, serving a critical function during sample loading. This seal remains closed during liquid sample measurements to prevent accidental spillage from damaging internal components and to conserve helium gas. It also allows the goniometer chamber to remain under vacuum while the sample chamber flushed with helium (for liquid samples) or re-evacuated.

A set of collimators is employed to enhance the resolution of the detected signals. The analyzer crystals, mounted on a crystal changer, disperse the multi-frequency fluorescence spectrum into specific wavelengths corresponding to

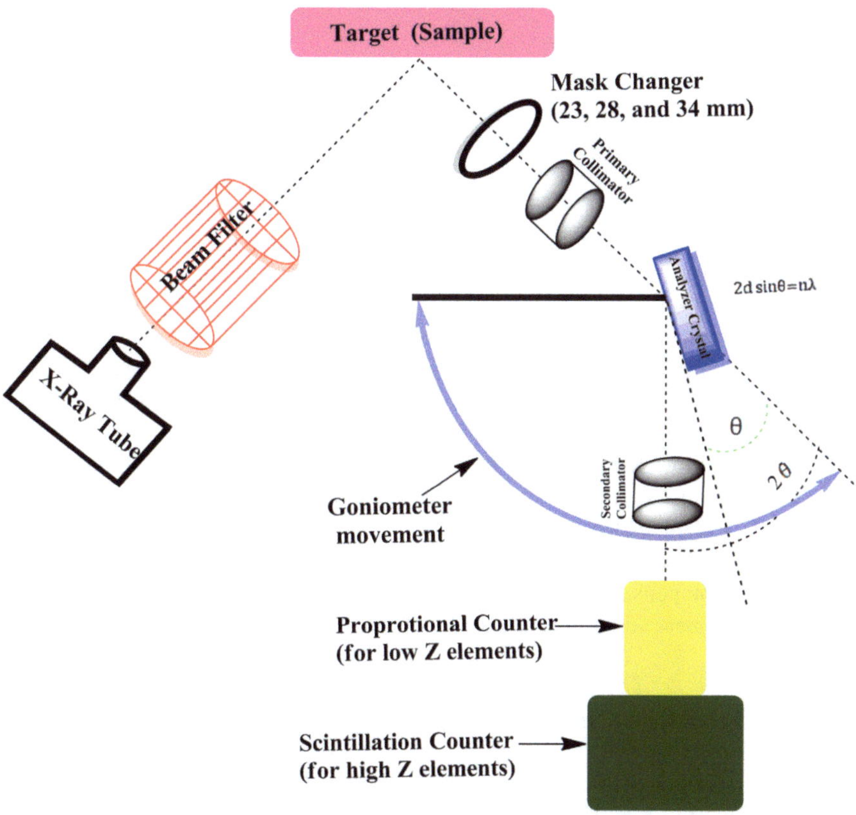

Fig. 2.1 Overview of the geometrical setup of WDXRF setup

Table 2.1 Different crystals used in WDXRF spectrometer

Crystal	Name	Element range	2d-value (nm)
LiF(420)	Lithium fluoride	$\geq$Co $K\beta_1$	0.1801
LiF(220)	Lithium fluoride	$\geq$V $K\alpha_1$	0.2848
LiF(200)	Lithium fluoride	$\geq$K $K\alpha_1$	0.4028
XS-400	Lithium fluoride	$\geq$K $K\alpha_1$	0.4028
Ge, XS-Ge-C	Germanium	P, S, Cl	0.653
InSb, XS-InSb-C	Indium antimonide	Si	0.7481
PET, XS-PET-C	Pentaerythite	Al–Ti	0.874
AdP	Ammonium dihydrogen phosphate	Mg	1.0648
TlAP	Thalliumhydrogenphthalate	F, Na	2.5760
XS-CEM	Multilayer	Al, Si, P, S	Trade secret
XS-55	Multilayer [W/Si]	O–Si (C)	5.5
XS-N	Multilayer [Ni/BN	N	11
XS-C	Multilayer [TiO2/C]	C	12
XS-B	Multilayer [La/B4C]	B (Be)	20

different elements, as detailed in The end window X-ray tube is the important component used in WDXRF setup. The configuration of end window X-ray tube used WDXRF spectrometer is shown in Table 2.1. This precise dispersion enables the spectrometer to achieve unparalleled sensitivity and exceptional resolution across a range of wavelengths. The system features a high-precision goniometer equipped with modern electronics and two independent stepper motors for separate θ and 2θ movements, ensuring accurate alignment during spectral scanning. Additionally, a robotic arm within the sample changer moves along the x- and y-axes, enabling fully automated and precise positioning of sample cups. The robotic arm, equipped with a magnetic grabber, ensures efficient and accurate sample handling, streamlining the analysis process. This integration of advanced components ensures reliable and high-quality elemental analysis.

The end window X-ray tube is the important component used in WDXRF setup. The configuration of end window X-ray tube used WDXRF spectrometer is shown in Fig. 2.2. It involves Rh-anode X-ray tube generator produce a power of 4 kW at a maximum high voltage of 60 kV and a maximum tube current of 170 mA. The temperature compensation is made by circulating the DI water inside the X ray tube. The complete setup is vacated to prevent the collisions with air molecules.

2.2.1 Detection System for WDXRF Spectrometer

The block diagram of the detection system utilized in WDXRF measurement is illustrated in Fig. 2.3. The detector, either a proportional counter or a scintillation counter, generates electronic pulses corresponding to the X-ray photons.

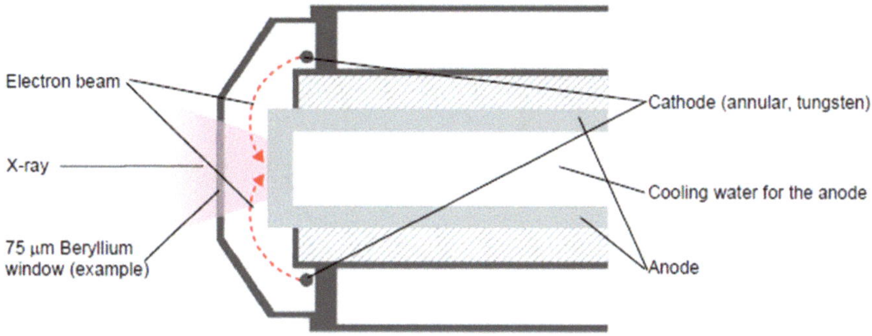

Fig. 2.2 Configuration of Rh anode end window X-ray tube [12]

These pulses are then processed electronically in the pulse processing unit. A concise explanation of each key component of the detection system is provided in subsequent sections. These two types of detectors are best suited for specific energy ranges. Within the spectrometer chamber, the flow counter operates at an angular range of 0 to 115°. Outside the chamber, the scintillation counter is positioned behind the flow counter, separated by a 0.1 mm aluminum foil, and functions within an angular range of 17 to 152°. Proportional and scintillation detectors can be employed either separately or together. Typically, a proportional counter is used for detecting light elements, while a scintillation counter is preferred for heavier elements. The flow counter, located inside the vacuum chamber, features an entrance window made of thin aluminum-coated foil, enabling optimal measurement of lighter elements, such as those ranging from Be to Na.

2.2.2 Proportional Counter

The gas proportional counter is designed as a cylindrical metallic tube with a thin counting wire positioned along its central axis, optimized for the measurement of light elements ranging from Be to Na. The tube is filled with P10 counting gas; a mixture of 90% argon and 10% methane, which flows at a constant rate to stabilize peak positions in the pulse height spectrum, located within the vacuum chamber, the flow counter features an entrance window made of thin aluminum-coated foil, available in thicknesses of either 0.6 or 0.3 µm, to enhance measurement precision. A high positive voltage (+U volts) is applied to the central wire, as illustrated in Fig. 2.4. The entrance window, mounted laterally on the tube, is designed to be transparent to X-rays, allowing photons to pass into the chamber. These photons ionize the gas atoms and molecules inside the tube, initiating the detection process.

Free electrons generated within the chamber drift toward the central anode wire, while the positive ions move toward the cathode (the tube housing). The number of electron–ion pairs formed is directly proportional to the energy of the incident X-ray photon. For instance, it takes approximately 0.03 keV of photon energy to

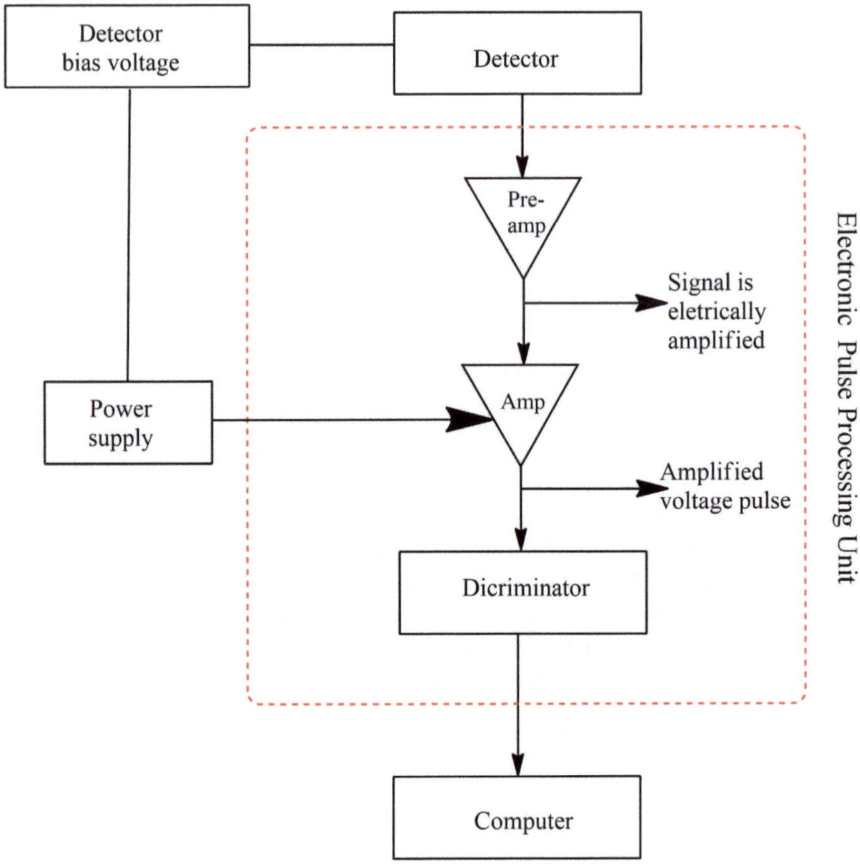

Fig. 2.3 Block diagram of the detection system used in WDXRF measurement

produce a single electron–ion pair. Accordingly, a molybdenum Kα photon with an energy of 17.5 keV generates around 583 electron–ion pairs, whereas a boron photon with an energy of 0.185 keV produces approximately 6 pairs.

2.2.3 Scintillation Counter

The scintillation counter, as illustrated in Fig. 2.5, consists of a sodium iodide (NaI) crystal doped with thallium for uniform distribution. The NaI(Tl) crystal facilitates the movement of primary electrons through an increasing electric field, leading to a cascade of secondary electrons and the generation of a measurable pulse, which is transmitted to the pre-amplifier. When an incident photon transfers its energy to the crystal, it produces a flash of light (scintillation), the intensity of which is directly proportional to the energy of the photon interacting

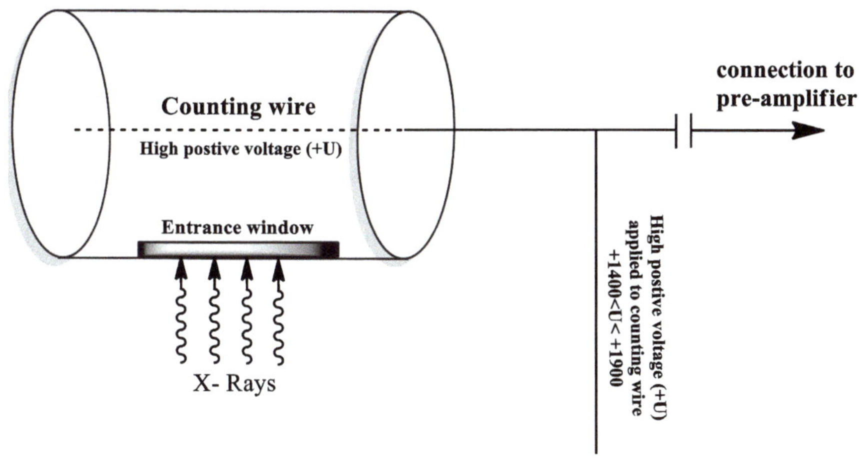

Fig. 2.4 Gas proportional counter used in WDXRF spectrometer

with the crystal. This flash of light ejects electrons upon striking the photocathode. Inside the photomultiplier tube, these ejected electrons are further accelerated and directed onto a series of dynodes, where they generate additional secondary electrons through a process of amplification. The resulting avalanche of electrons produces a measurable signal. The height of the pulse generated is proportional to the energy of the X-ray photon detected by the crystal. To reach the scintillation counter, the radiation must pass through a flow counter, a 0.1 mm thick vacuum-chamber sealing foil, and a 0.2 mm beryllium (Be) entrance window of the detector.

The scintillation counter is primarily used for detecting X-ray energies above approximately 4.5 keV (Cr-Kα_1), as lower-energy radiations are predominantly absorbed by the flow counter. In case of S8 TIGER spectrometer, the scintillation counter is positioned adjacent to the proportional counter within the spectrometer chamber, with a movable range between 17 and 152°. For the flow proportional

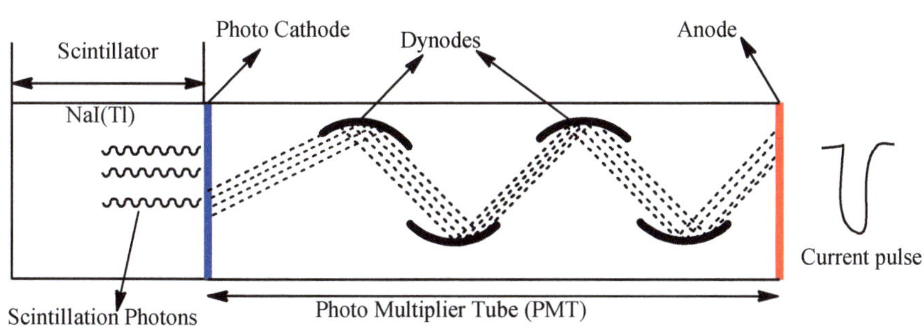

Fig. 2.5 Illustration of scintillation counter

counter, signals from the preamplifier are routed to the main amplifier. In contrast, for the scintillation counter, signals from the photomultiplier tube (PMT) are directly transmitted to the main amplifier. The preamplifier is also used to shape the signal before it undergoes further processing as voltage pulses in the main amplifier and discriminator. The discriminator applies a window to define lower and upper thresholds, recording only pulses with heights within this range. This mechanism suppresses higher-order peaks and interfering radiations, ensuring that only relevant pulses are counted while reducing background noise. The recorded pulses are further amplified electronically in the main amplifier, where the degree of amplification depends on the crystal used for detection. Additionally, the electronic amplification is synchronized with the goniometer's angular settings to ensure that peaks in the pulse height spectrum consistently appear at the correct positions. This synchronization is achieved by adjusting the main amplifier's amplification factor, which is proportional to sin θ.

2.3 Recent Trends in WDXRF

Since its inception, Wavelength Dispersive X-ray Fluorescence (WDXRF) has undergone significant sequential improvements. This section outlines the major advancements in its set-up.

2.3.1 Instrumentation Advancements in WDXRF

Recent years have witnessed significant instrumentation advancements in WDXRF systems, expanding their usability, precision, and efficiency. One major development is the evolution of portable XRF spectrometers, which now incorporate advanced detectors, improved battery life, and rugged designs—making them suitable for on-site applications in mining, environmental monitoring, and recycling. Additionally, the tube-above Additionally, the introduction of the tube-above configuration offers advantages in minimizing sample contamination and prolonging system selectivity, an essential factor in industries where continuous operations is critical [13]. The integration of automated sample preparation and analysis has further streamlined workflows, minimized manual errors and enhancing reproducibility in high-throughput environments. Complementing hardware improvements are software advancements that provide intuitive user interfaces, robust data analysis tools, real-time diagnostics, and customizable reporting features, significantly improving overall user experience and analytical confidence.

2.3.2 Advancements in Crystal Technology in WDXRF

Recent progress in crystal technology has significantly enhanced the performance and analytical capabilities of WDXRF systems. Synthetic thin-film multilayer crystals have gained prominence due to their superior sensitivity and spectral resolution, especially for light elements such as sodium, magnesium, and aluminum. These crystals enable more accurate peak discrimination, which is crucial for trace-level analysis in complex samples. Additionally, improved crystal stability has addressed limitations associated with traditional crystals, which often degrade under prolonged X-ray exposure or temperature fluctuations. Modern crystals now offer extended operational lifespans and consistent performance. Furthermore, ongoing crystal optimization efforts—including the refinement of materials, geometries, and coating techniques—have contributed to improved resolution, detection limits, and overall efficiency, solidifying the role of WDXRF in high-precision elemental analysis across scientific and industrial applications.

2.3.3 Advanced Data Processing in WDXRF

The analytical capabilities of WDXRF have been significantly enhanced through advancements in data processing techniques. One such innovation is the Fundamental Parameters (FP) method, which allows for rapid and accurate quantitative determination of both major and minor elements using minimal or even no calibration standards. This approach improves efficiency and versatility, especially when dealing with unknown or variable sample types. Additionally, matrix correction algorithms are being actively developed and refined to address matrix effects and inter-elemental interferences—common challenges in complex material analysis. These corrections contribute to more precise and reliable quantitative results across diverse sample matrices. Another breakthrough is the incorporation of elemental mapping capabilities into modern WDXRF systems. This feature enables spatially resolved analysis, providing detailed insight into sample heterogeneity, the distribution of trace elements, and the identification of inclusions or contaminants—proving especially valuable in materials science, geology, and quality control applications.

2.3.4 New Applications of WDXRF

With ongoing technological advancements, WDXRF has broadened its application scope across diverse scientific and industrial domains. One emerging area is light element analysis, where innovations in detectors and thin-window technology have enabled the reliable quantification of elements such as carbon, oxygen, nitrogen, and hydrogen—particularly important in coal analysis [14], combustion studies, and fuel characterization. In environmental monitoring [15, 16], WDXRF plays a crucial role in the detection of trace elements in soil, dust, and rock samples, aiding in the continuous assessment and remediation of contaminated sites. The technique is also extensively used in geological applications, offering rapid, non-destructive analysis of rocks, minerals, and sediments for both academic research and resource exploration. In materials science, WDXRF provides accurate compositional analysis of metals, alloys, ceramics, and advanced composites, supporting quality control and development of high-performance materials [17]. Furthermore, the method is finding increasing relevance in the pharmaceutical and healthcare sectors for analyzing raw materials, detecting elemental impurities, and ensuring compliance with regulatory standards such as ICH Q3D [17].

2.4 Instrumentation for EDXRF Spectrometer

The block diagram of EDXRF Setup is shown in Fig. 2.6. A brief description of the essential parts of the detection system is given below as.

2.4.1 Detectors

X-ray detection in EDXRF systems is carried out using detectors namely Low-Energy Germanium (LEGe), Lithium-drifted Silicon [Si(Li)], and Peltier-cooled Silicon Drift Detectors (SDD). Among these, the LEGe detector offers two key advantages: enhanced energy efficiency across a wide range of energies (from 1 keV to several hundred keV) and superior energy resolution compared to the Si(Li) detector. This performance difference arises because the mean energy required to generate an electron–hole pair is 2.97 eV in germanium, compared to 3.76 eV in silicon. The detailed specifications of the detector are summarized in Table 2.2.

In a solid-state detector, the absorption of a photon generates a current pulse at the input of the preamplifier. A charge-sensitive loop, consisting of a field-effect transistor (FET) input operational amplifier with capacitive feedback, integrates each of these current pulses. Both the detector and the FET are mounted on a copper rod connected to a liquid nitrogen reservoir within the cryostat. Cooling the detector minimizes the thermal generation of charge carriers, significantly reducing noise. The detector operates at liquid nitrogen temperature (77 K) to mitigate noise caused by excessive diode leakage currents, which occur at elevated temperatures. The LEGe detector demonstrates superior efficiency with an energy resolution of approximately 160 eV at 5.895 keV. Although LEGe detectors can be safely kept at room temperature for short periods without bias voltage, maintaining optimal performance typically requires that the detector remain continuously cooled.

2.4.2 Detector Bias Supply

The recommended bias voltage is supplied to the detector carefully. In case of accidental detector warm up, the damage to the detector is prevented by automatically disabling the high voltage power supply with help of a bias inhibiting circuit which is shown in Fig. 2.7. The warmup with bias applied to a detector can lead to FET damage. The internal temperature of the detector is monitored

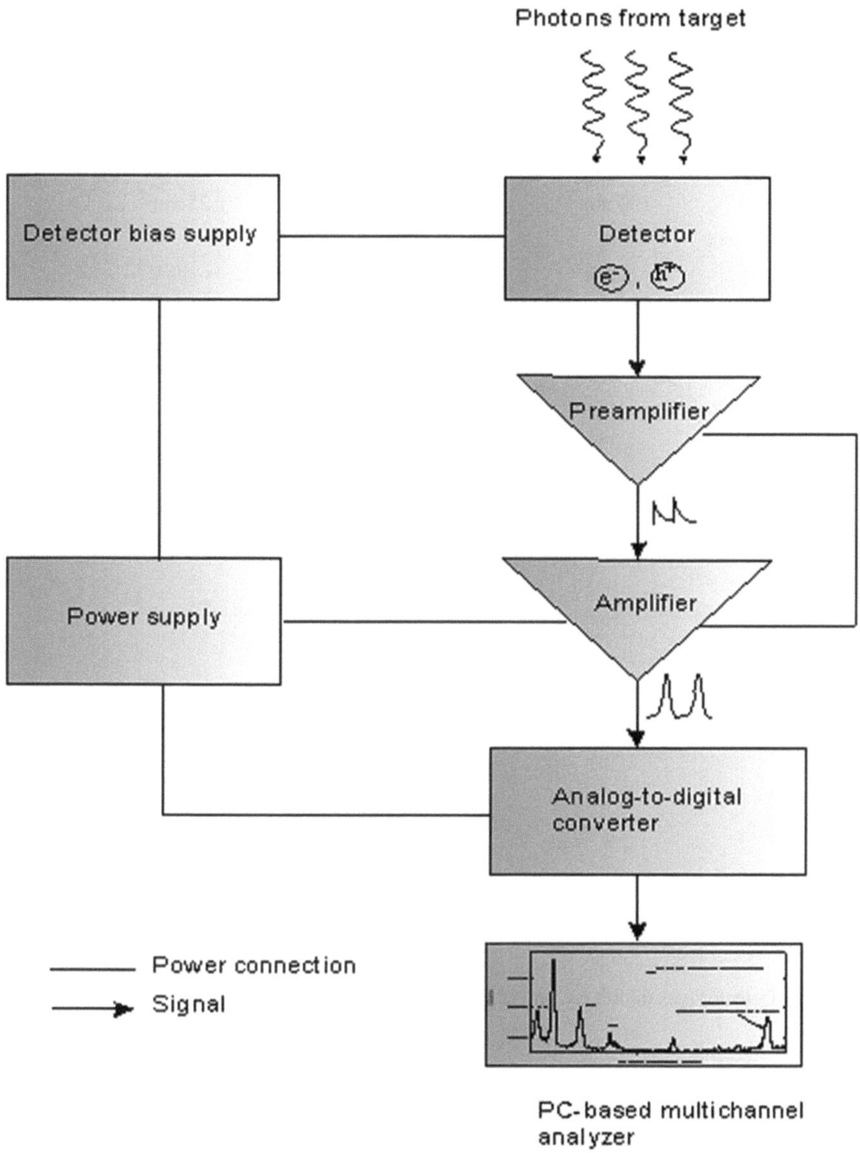

Fig. 2.6 Block diagram of the EDXRF setup [12]

with a Platinum resistance temperature detector (PRTD), amplifier and preamplifier power supply. The recommended bias voltage is applied to the detector using 0–6 kV high voltage power supply (Model 3106D, Canberra, US). This power supply is suitable for all types of detectors requiring current up to 300 μA and bias of 6 kV. The output voltage can be regulated gradually from ± 30 to ± 6000 V

Table 2.2 Specifications of the various semiconductor detectors used in present work

Parameters	LEGe detector	Si(Li) detector	SDD
Company	Canberra, US	Canberra, US	Amptex, Inc., US
Window (Be)	8 μm	8 μm	25 μm
Active thickness	10 mm	5.5 mm	500 mm
Cross section	100 mm^2	100 mm^2	25 mm^2
Window distance	5 mm	5mm	
Biasing	−500 V	−500 V	−50 V
Energy resolution	150 eV at 5.89 keV	180 eV at 5.89 keV	125 eV at 5.89 keV
Coolant	Liquid N_2	Liquid N_2	Peltier cooled

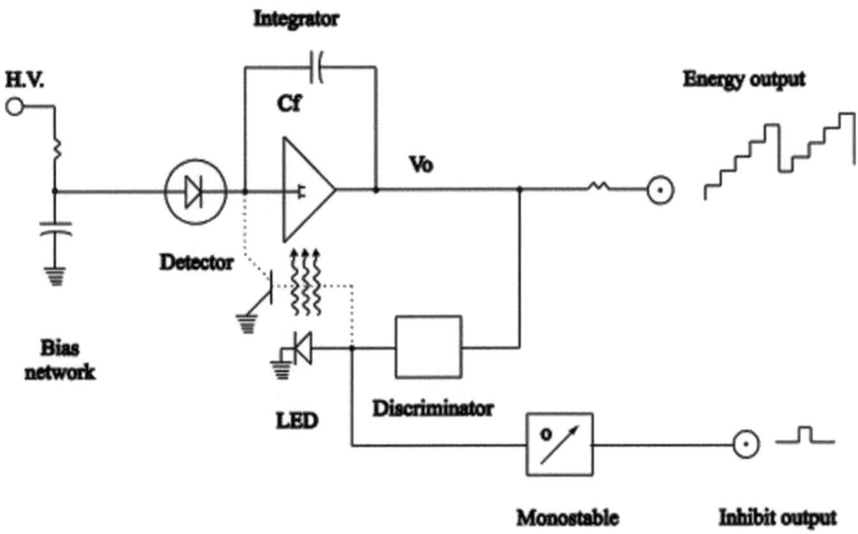

Fig. 2.7 Detector bias inhibit circuit [12]

dc. A secondary output having a range of ±3 V to ±600 V is also available for the low voltage detectors. With the help of a high voltage filter network, ripples from power supply and interference picked up by the cable between preamplifier and high voltage supply are eliminated from bias supply. This charge-sensitive circuit includes a high voltage filter network comprising of *RC* filter stages. The test input is coupled into the preamplifier through the capacitance of the detector element by means of a signal imposed on the ground side of the second filter capacitor. The signal from the preamplifier of the detector fed to spectroscopy amplifier and the amplified signal is further passed to the multichannel analyzer for processing of pulse height analysis. A portable bin/power supply (Model 1000, CANBERRA, US) is generally used as standards nuclear instrument module (NIM) for the high voltage power.

2.4.3 Pulsed Optical (PO) Reset Preamplifier

In conventional charge-sensitive preamplifiers, commonly used in spectroscopy applications, the charge generated by the detector is collected on a capacitor. This collected charge is then removed by an RC feedback circuit, as shown in Fig. 2.8, resulting in an exponential tail pulse. The characteristic time constant of this circuit is typically in the range of 40–50 µs. The output voltage of the preamplifier is directly proportional to the charge fed into its input, and it remains unaffected by the detector's capacitance or any stray capacitance from the cables. This design allows for the generation of large output pulse heights. As a result, the total charge extracted from the detector is forced to accumulate on the feedback capacitor C_f which then generates a voltage pulse at the output. The amplitude of this pulse is determined by the relationship outlined in Eq. (2.1).

$$V = \frac{E}{\varepsilon}\frac{q_e}{C_f}\exp\left(-\frac{t}{RC}\right) \tag{2.1}$$

The charge sensitive stage is an operational amplifier operating with capacitive feedback. The preamplifier transfer function is $V_o = Q_i/C_f$. The function of the feedback resistor (R) is to remove the charge slowly from C_f after arrival of each pulse. Thus, a reset charge restoration method is employed to discharge the integrator.

In RC preamplifiers, the feedback resistor is a significant source of noise. For low-energy detectors, where electronic noise can limit resolution, eliminating the feedback resistor is advantageous. To achieve this, the resistor is replaced with optical coupling. In the case of a pulsed feedback charge-sensitive preamplifier, the resistive feedback R_f is removed. Instead, the charge from each detected X-ray

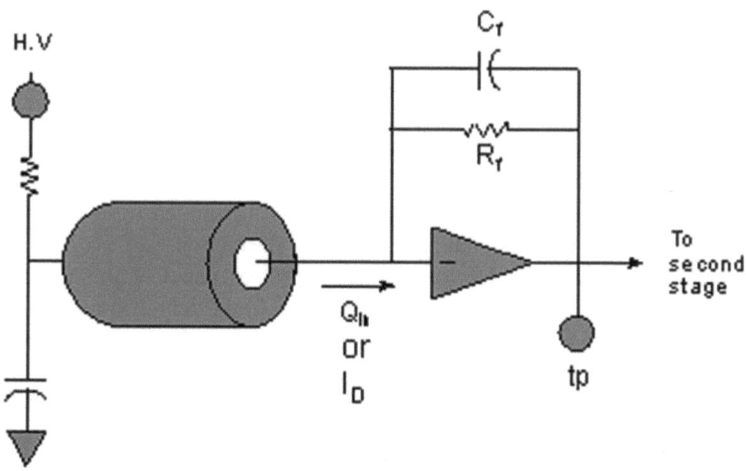

Fig. 2.8 RC feedback preamplifier [12]

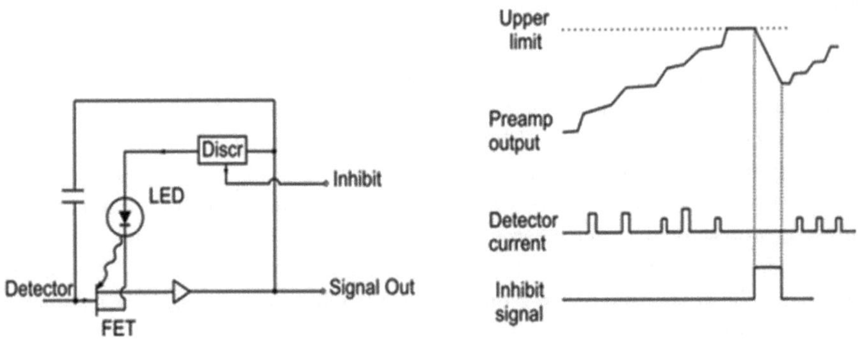

Fig. 2.9 Pulsed optical (PO) reset preamplifier [12]

photon accumulates on the feedback capacitor, resulting in a staircase pattern of pulses at the preamplifier output. As before, the amplitude of each pulse corresponds to the energy of the detected photon. However, this staircase pattern would eventually drive the preamplifier into saturation if left unchecked. To prevent this, a comparator, controlled by an LED, illuminates the gate-to-drain junction of the FET, discharging the feedback capacitor C_f before it reaches the operational limit. Once the output voltage returns to the initial level of the staircase, the LED is turned off, allowing the preamplifier to process another series of pulses. The comparator is set to trigger the LED when the output voltage reaches approximately 0 V and resets when it returns to about −4 V. Therefore, the normal operating output voltage range of the preamplifier is between −4 and 0 V. The ramp (or reset) rate depends on the detector leakage current and the energy deposition rate within the detector element, as depicted in Fig. 2.9.

When the preamplifier resets, its output swings negatively by about 4 V within a few microseconds. This large overload signal causes the associated shaping amplifier to temporarily stop functioning for several pulse widths, resulting in dead time and distorted signals. Additionally, spurious effects from light on the field-effect transistor (FET) exacerbate these issues. To address dead time loss and prevent the analysis of distorted signals, the preamplifier is equipped with a monostable multivibrator. This device generates a transistor-transistor logic (TTL) gating signal that can be used to disable the analog-to-digital converter (ADC), ensuring that only valid signals are processed. This gating signal is applied to the inhibit input for both rejection of invalid signals and live-time correction.

The preamplifier receives DC power from the rear panel of the spectroscopy amplifier, and the output pulses have a rise time of less than 200 ns and a decay time of $\leq$40 ns. A pulse of light from the LED is used to discharge the integrator. The spectroscopy amplifier serves two main functions: (i) amplifying the preamplifier pulses (which are typically in the millivolt range), and (ii) shaping the pulses into a convenient form for further processing. Unipolar pulse shaping is achieved using a differentiator and two integrators. The spectroscopy amplifier is designed to offer extremely low noise, a wide gain range, and excellent overload response for high-resolution spectroscopy. It accepts pulses with rise times of less than 650 ns and decay times greater than 40 μs. The amplifier has six selectable time constants for pulse shaping, with values of 0.5, 1, 2, 4, 8, and 12 μs, which can be chosen to optimize the balance between resolution and count rate. The amplifier has an input impedance of approximately 500 Ω and can accept either positive or negative pulses. The shaping time constant can be selected in steps of 0.5, 1, 2, 3, 6, and 10 μs. Selecting the appropriate shaping time constant is a balance between accommodating high-count rates (around 3000 counts per second) with a shorter time constant and achieving a better signal-to-noise ratio with a longer time constant. At low counting rates (below 100 counts per second), a 10 μs shaping time is optimal for a Si(Li) detector. For the LEGe detector, a spectroscopy amplifier (Model 2026) from CANBERRA (US) is used. This amplifier includes a variable pole-zero cancellation network that can be adjusted to match the preamplifier's output decay time (less than 40 ns), reducing undershoot after the first differentiator and improving the overload and count rate characteristics. Base-line restoration is employed to correct for shifts in the baseline of the unipolar signal, caused by uncanceled RC inter-stages in the amplifier. The DC output level of the amplifier is adjustable up to ±100 mV, providing both unipolar and bipolar output pulses with an output impedance of less than 1 Ω. The unipolar positive output pulse is fed into the Multi-Channel Analyzer (MCA, Model S100, Multiport II, CANBERRA, US) for pulse height analysis. The MCA, coupled with an ADC, memory, and display screen, digitizes the amplitude of the unipolar output pulses to obtain a processed energy spectrum of the X-ray photons incident on the detector. The MCA counts the number of events in each channel, generating the energy spectrum of the detected photons. A software program controls various functions such as data acquisition, spectrum display, and linear energy calibration. The output signal from the ADC is transmitted to the USB port of a PC, which runs the Gennie 2000 software (CANBERRA, US) for analysis and visualization.

2.4.4 Detection System for EDXRF Spectrometer Involving X-Ray Tube and Equipped with Silicon Drift Detector

Figure 2.10 illustrates the block diagram of an EDXRF detection system employing an X-ray tube. The core component is a Peltier-cooled Silicon Drift Detector (SDD) responsible for detecting emitted photons. The SDD's output is amplified by a charge-sensitive preamplifier and then processed by a Digital Pulse Processor (DP5). The DP5 optimizes the signal for analog-to-digital conversion (ADC) by adjusting gain, offset, and filtering. The digitized signal is further shaped by a digital pulse shaper. The SDD sensor, fabricated on a high-purity silicon wafer, interacts with incident photons, generating electron clouds proportional to the photon energy. A thermoelectric system cools the SDD and a field-effect transistor (FET) preamplifier converts the accumulated charge into a voltage signal.

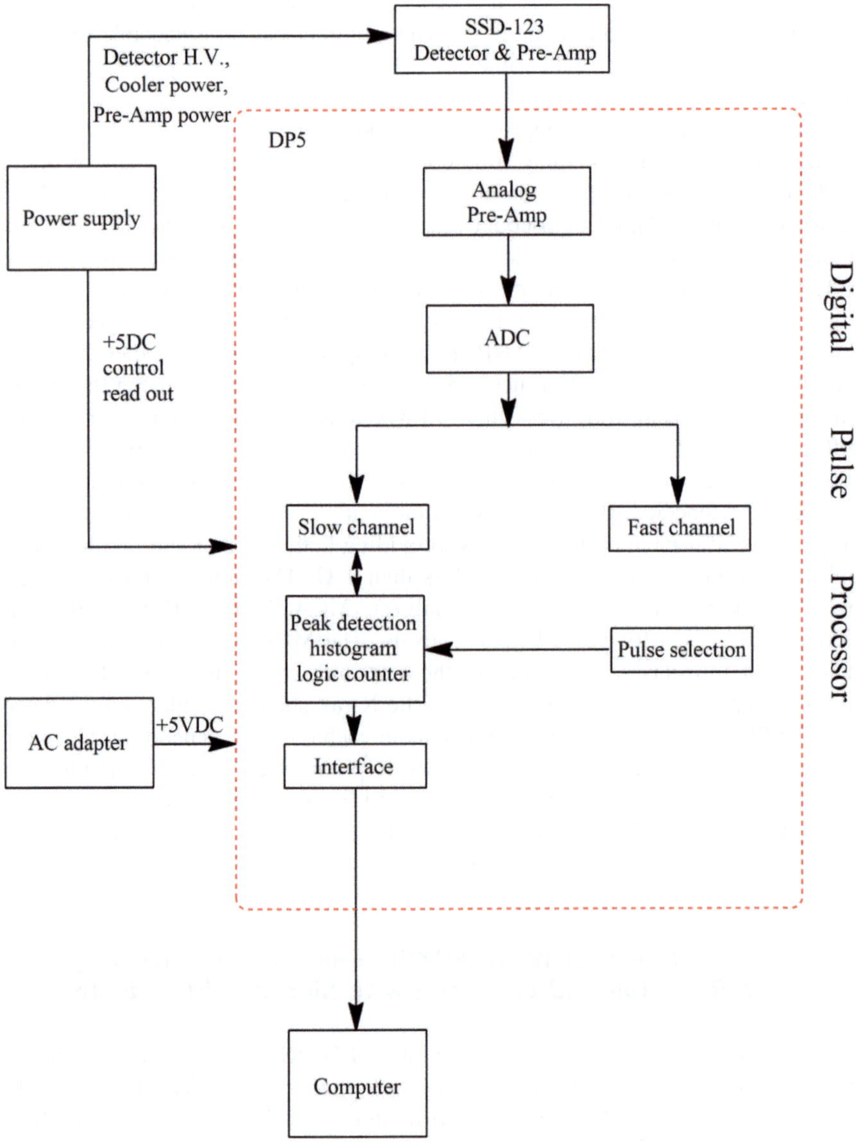

Fig. 2.10 Block diagram of the detection system equipped with SDD detector

Two primary sources contribute to charge accumulation: leakage current within the sensor material and charge generated by incident radiation. The preamplifier output is a voltage ramp with superimposed sharp steps, corresponding to charges from photon interactions. Incident photons generate electron–hole pairs in the silicon diode. Electrons are then drifted towards the anode, guided by concentric electrodes. The digital pulse shaper (DPP) optimizes the incoming signal using slow and fast channel techniques. The slow channel, with a longer shaping time constant, provides accurate pulse height measurements. The fast channel is optimized for count rate and pulse rise time measurements. Pulse selection logic ensures accurate pulse detection by rejecting pulses based on rise time and pile-up criteria. The DPP's primary output is an energy spectrum, displayed in a histogram memory. This memory functions like a traditional multi-channel analyzer (MCA), counting input pulses and rejected events. The DP5 system's interface, typically USB, Ethernet, or RS-232, enables data transmission to a computer and allows control of analog gain and pulse shaping time.

2.5 Recent Trends in EDXRF

2.5.1 Total Reflection X-Ray Fluorescence (TXRF)

The Total Reflection X-Ray Fluorescence (TXRF) technique is a specialized variant of the Energy Dispersive X-Ray Fluorescence (EDXRF) method and incorporates all the key features of conventional X-Ray Fluorescence (XRF). In TXRF, incident photons strike the sample at a glancing angle, resulting in total reflection. For X-rays, the refractive index of all materials is less than 1, enabling external total reflection when the incident angle is below the critical angle $\left[\varphi_c = (1.65/E_{cut})(Z\rho/M)^{\frac{1}{2}}\right]$, typically falls within the milliradian range and depends on the energy of the incident photons. At angles above this critical threshold, x-rays are absorbed by the sample instead of being reflected. To optimize the primary x-ray photons before they reach the sample reflector, various optical elements such as cutoff reflectors, multilayer filters, or natural crystals are employed. Among these, the cutoff reflector is preferred due to its availability and cost-effectiveness compared to other monochromatizing devices. The use of glancing angles in TXRF enhances surface sensitivity, distinguishing it from EDXRF, which is primarily a bulk-sensitive technique.

2.5.2 Grazing Emission X-Ray Fluorescence (GEXRF)

A related technique, grazing emission X-ray fluorescence (GEXRF), involves directing the primary X-ray beam almost perpendicularly to the sample, while the fluorescent radiation is measured as a function of the grazing exit angle. Energy-dispersive GEXRF is particularly effective for surface layer analysis, while wavelength dispersive GEXRF is more suited for examining depth variations in structured samples.

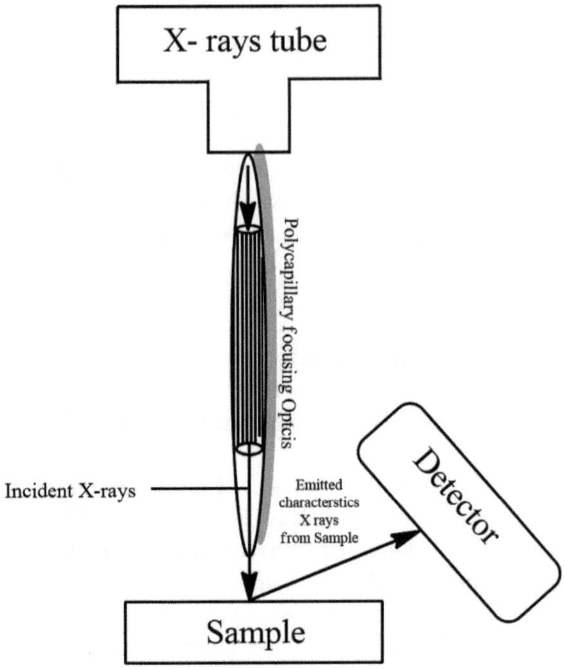

Fig. 2.11 Micro-XRF instrumentation using a polycapillary focusing optic

2.5.3 Micro-XRF (MXRF)

The micro-XRF (μXRF) technique represents a promising future approach for material analysis due to its minimal sample interaction and versatile capabilities. This method requires a concentrated photon beam focused to a micron-scale spot. Advances in x-ray optics, incorporating x-ray tube or synchrotron radiation sources, have facilitated the development of x-ray fluorescence microprobes. These optics often utilize total reflection or the diffraction properties of specialized gratings. Modern monocapillary and polycapillary x-ray optics enable the production of microbeams with diameters as small as 10–20 μm, allowing for compact systems capable of elemental microanalysis detectors), and advanced software for faster data analysis. The polycapillary X-ray optics incorporated with X-ray tube (XRT) to achieve the microbeam are depicted in Figs. 2.11, 2.12 and 2.13.

The combination of high-intensity X-ray beams and low background noise significantly improves signal detection in μXRF systems. Recent advancements in X-ray optics have led to groundbreaking capabilities. Modern μXRF instruments often feature dual optics for flexible micro and macro analysis, along with advanced detectors. Polycapillary optics offer an alternative approach to generate small focal spots with high X-ray flux, overcoming the inverse square law limitation. This enables the development of compact and low-power μXRF

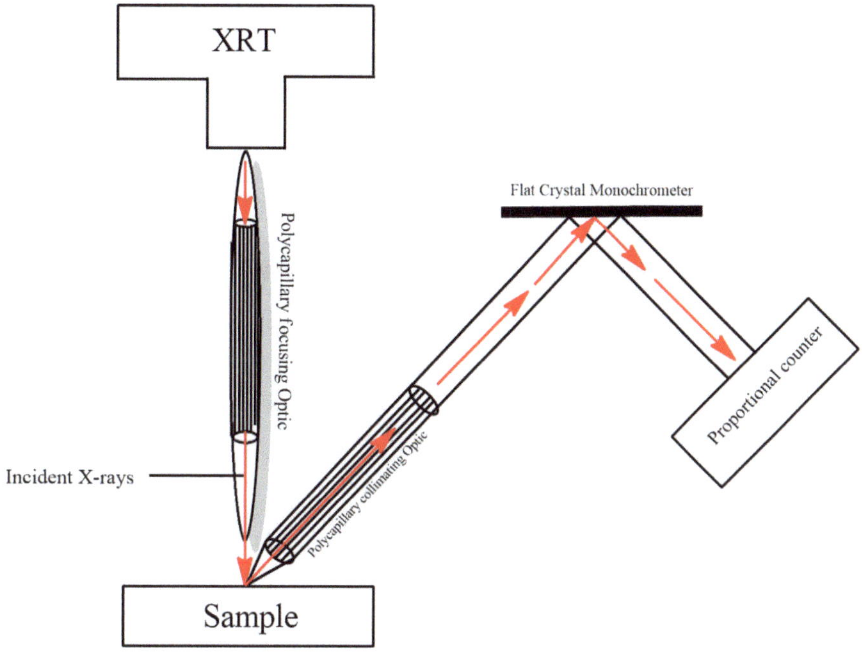

Fig. 2.12 Micro-XRF instrumentation using a polycarpellary focusing optic and flat crystal monochrometer

systems for in-line industrial applications and portable instruments. μXRF with X-ray optics has proven effective in various applications, including small feature evaluation, elemental mapping, film and plating thickness measurement, micro-contamination detection, multi-layer coating analysis, small particle analysis, and forensic investigations [8–10, 18–20].

2.5.4 High-Efficiency ED-XRF (SPECTRO XEPOS)

SPECTRO XEPOS represents a new generation of ED-XRF instruments offering superior analytical performance. The SPECTRO XEPOS is designed with enhanced excitation and detection technologies that dramatically increase sensitivity and reduce measurement time. These improvements enable the system to detect lower concentrations of elements with high precision, making it particularly suitable for trace element analysis. Unlike older ED-XRF models, which often required lengthy calibration processes and struggled with complex matrices, the SPECTRO XEPOS employs advanced software algorithms and adaptive excitation technology. This allows it to effectively analyse samples with variable compositions, including thin layers, powders, liquids, and heterogeneous materials, with minimal sample preparation. Moreover, high-efficiency ED-XRF

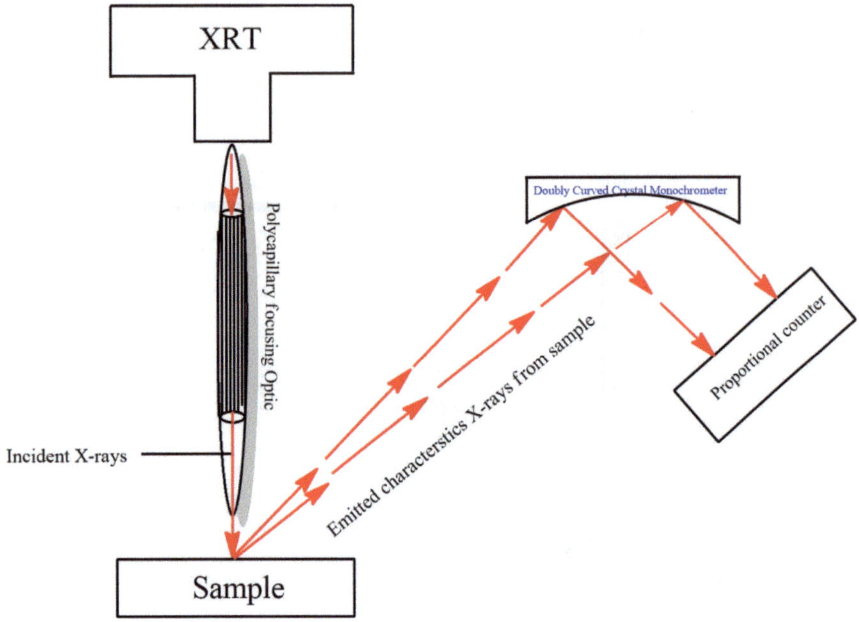

Fig. 2.13 Micro XRF arrangement using polycapillary focusing optic and doubly crystal monochromator

systems like XEPOS incorporate features such as automatic matrix correction and intelligent calibration routines. These enhancements simplify operation and allow for non-destructive, multi-elemental analysis across a wide concentration range. The ability to perform reliable analysis without extensive calibration or matrix-matching standards is a game-changer for industries such as environmental monitoring, metallurgy, mining, and petrochemicals. In summary, high-efficiency ED-XRF systems are revolutionizing the field of elemental analysis. By combining the speed and versatility of ED-XRF with near-WDXRF accuracy, instruments like the SPECTRO XEPOS are becoming essential tools for laboratories requiring cost-effective, high-performance solutions for complex analytical challenges.

2.5.5 Dual-Excitation XRF Systems

Traditional XRF systems generally use a single excitation source either an X-ray tube or radioactive isotope, to generate characteristic X-rays from the elements present in a sample. These setups are not much effective to analyse the light elements like sodium (Na), magnesium (Mg), and aluminum (Al). To overcome these limitations, dual-excitation XRF systems which utilize two distinct X-ray sources to enhance the detection capabilities of the instrument is most promote option. These sources can be differ in energy output, filtration, or geometry but are strategically combined to optimize excitation conditions for both light and heavy elements in the same analysis. Dual-excitation systems are particularly valuable in complex or heterogeneous samples where the co-existence of a wide range of elements poses a challenge to traditional single-source instruments. Such applications are common in fields like environmental analysis, metallurgy, mining, cement production, forensics, and archaeology, where accurate multi-element analysis is crucial. In these systems, the data from both excitation sources are either acquired simultaneously or sequentially and then integrated using advanced software algorithms to produce a comprehensive elemental profile. This dual approach not only increases detection limits for elements that are otherwise difficult to analyze but also improves precision and accuracy by reducing spectral overlaps and enhancing peak-to-background ratios. As the demand for more accurate, rapid, and versatile elemental analysis grows across various industries, dual-excitation XRF systems are emerging as a powerful solution. Their ability to provide detailed and reliable results for both light and heavy elements makes them a valuable innovation in modern analytical science.

2.5.6 Portable and Handheld-XRF

Portable and handheld X-ray fluorescence (XRF) analyzers have revolutionized on-site elemental analysis by enabling rapid, non-destructive testing in real time. While the technology itself is not new, recent advancements have significantly enhanced the sensitivity, versatility, and user-friendliness of these devices, making them indispensable for field applications. Modern portable XRF systems now incorporate high-resolution silicon drift detectors (SDDs), optimized X-ray tubes, and advanced software algorithms, allowing for lower detection limits (often in the ppm range) and improved accuracy for both light and heavy elements. These devices are particularly valuable in geology and mining, where they facilitate instant ore grade estimation, mineral exploration, and mine site remediation by providing immediate elemental composition data without the need for laboratory samples. In environmental monitoring, handheld XRF units are used to detect toxic heavy metals (e.g., lead, arsenic, and cadmium) in soil, water, and industrial waste, supporting regulatory compliance and pollution control efforts. Additionally, their compact design make them ideal for use in remote or challenging environments, such as archaeological sites or recycling facilities. With the integration of wireless connectivity and cloud-based data management, field technicians can now stream results in real time for further analysis and decision-making. As portable XRF technology continues to evolve, its applications are expanding into new domains, including consumer product safety, forensics, and industrial quality control, solidifying its role as a critical tool for rapid, on-the-spot elemental analysis.

2.5.7 Integrated XRF-XRD Systems

Integrated XRF-XRD systems represent a significant advancement in analytical instrumentation, combining X-ray fluorescence (XRF) for elemental analysis with X-ray diffraction (XRD) for structural characterization in a single measurement. These hybrid systems enable simultaneous determination of chemical composition and crystalline phase distribution, offering comprehensive material characterization for complex samples. By integrating both techniques, they overcome limitations of standalone instruments, providing correlated elemental and structural data that enhances detection sensitivity and phase identification accuracy. Such systems are particularly valuable in geology for mineral phase analysis, in materials science for studying catalysts and composites, and in pharmaceuticals for polymorph screening. Recent technological improvements, including co-axial optics, high-speed detectors, and advanced data processing software, have made these instruments more efficient and user-friendly. As the demand for multimodal analysis grows, integrated XRF-XRD systems are becoming indispensable tools for research and industrial applications where understanding both composition and structure is critical. Future developments are expected to focus on miniaturization, automation, and AI-enhanced data interpretation to further expand their utility across scientific disciplines.

References

1. Salvador, M.J., Lopes, G.N., Nascimento Filho, V.F., Zucchi, O.L.A.D.: Quality control of commercial tea by x-ray fluorescence. X-Ray Spectrom.: Int. J. **31**, 141–144 (2002)
2. Upmanyu, A.K., Kapil, A., Mehta, D., Kumar, S.: Thickness measurement of low-Z films fabricated on thick substrate using EDXRF technique. Vacuum **183**, 109852 (2021)
3. Singh, G., Singh, G., Rani, N., et al.: Contribution of flyash from coal-fired thermal power plants to uranium contamination of ground water. J. Radioanal. Nucl. Chem. **318**, 857–863 (2018). https://doi.org/10.1007/S10967-018-6079-7
4. Zhan, X.: Application of polarized EDXRF in geochemical sample analysis and comparison with WDXRF. X-Ray Spectrom. **34**, 207–212 (2005). https://doi.org/10.1002/XRS.794
5. Gazulla, M.F., Ventura, M.J., Orduña, M., et al.: Analysis of corrosion residues by WDXRF. X-Ray Spectrom. **46**, 271–276 (2017). https://doi.org/10.1002/XRS.2770
6. Kainth, H.S., Singh, T., Shahi, J.S., et al.: Instrumental detection limit and sensitivity of K and L shell X-ray emission lines of 17Cl, 37Rb, and 38Sr elements using PC-WDXRF spectrometer. X-Ray Spectrom. **47**, 352–358 (2018). https://doi.org/10.1002/XRS.2950
7. Kainth, H.S., Singh, T., Singh, G., et al.: Lab-scale wavelength dispersive x-ray fluorescence spectrometer and signal processing evaluation. X-Ray Fluoresc. Biol. Sci.: Princ., Instrum., Appl. 509–550 (2022). https://doi.org/10.1002/9781119645719.CH33
8. Weindorf, D., Bakr, N.: Agronomy YZ-A, 2014 Undefined Advances in Portable X-ray Fluorescence (PXRF) for Environmental, Pedological, and Agronomic Applications. Elsevier
9. Potts, P.J., Ellis, A.T., Kregsamer, P., et al.: Atomic spectrometry update. X-ray fluorescence spectrometry. J. Anal. At. Spectrom. **20**, 1124–1154 (2005)
10. Palmer, P.T., Jacobs, R., Baker, P.E., et al.: Use of field-portable XRF analyzers for rapid screening of toxic elements in FDA-regulated products. J. Agric. Food Chem. **57**, 2605–2613 (2009)

11. Kalnicky, D.J., Singhvi, R.: Field portable XRF analysis of environmental samples. J. Hazard. Mater. **83**, 93–122 (2001)
12. New Tube-Above WDXRF Spectrometer from Rigaku Features Advanced Guidance System and Automatic Application Setup | Technology Networks. https://www.technologynetworks.com/analysis/product-news/new-tubeabove-wdxrf-spectrometer-from-rigaku-features-advanced-guidance-system-and-automatic-application-setup-221112. Accessed 5 April 2025
13. Sverchkov, I., Matveeva, V.B.: MC-SAP, 2023 Undefined Determination of Carbon, Oxygen, Hydrogen and Nitrogen Content in Coals using WDXRF Scattering Spectra. Elsevier
14. An, J., Kim, K.H., Yoon, H.O., Seo, J.: Application of the wavelength dispersive X-ray fluorescence technique to determine soil fluorine with consideration of iron content in the matrix. Spectrochim. Acta Part B At. Spectrosc. **69**, 38–43 (2012). https://doi.org/10.1016/J.SAB.2012.02.006
15. Marguí, E., Queralt, I., de Almeida, E.: X-ray fluorescence spectrometry for environmental analysis: basic principles, instrumentation, applications and recent trends. Chemosphere **303**, 135006 (2022). https://doi.org/10.1016/J.CHEMOSPHERE.2022.135006
16. WDXRF | Wavelength Dispersive X-Ray Fluorescence Spectroscopy | Malvern Panalytical. https://www.malvernpanalytical.com/en/products/technology/xray-analysis/x-ray-fluorescence/wavelength-dispersive-x-ray-flourescence. Accessed 5 April 2025
17. Kaskes, P., Déhais T, de Graaff, S.J., et al.: Micro–X-ray fluorescence (μXRF) analysis of proximal impactites: High-resolution element mapping, digital image analysis, and quantifications (2021)
18. Wouters, B., Makarona, C., Nys, K., Claeys, P.: Characterization of archaeological metal remains in micromorphological thin sections using μXRF elemental mapping. Geoarchaeol. **32**, 311–318 (2017)
19. Chiti, M., Esposito, A., Gorghinian, A., et al.: Micro X-ray fluorescence analysis of Roman imperial coins from Nero. J. Archaeol. Sci. Rep. **37**, 102940 (2021)
20. Upmanyu, Arun.: Study of photon atom interaction processes in X RAY energy region and their applications in elemental analysis. Thesis, I. K. Gujral Punjab Technical University Punjab, India (2018)

X-Ray Detectors and Associated Imaging Applications: Emerging Trends and Challenges

3

Svetlana I. Shtel'makh

Abstract

This chapter provides a detailed discussion of gas-filled detectors (Geiger-Muller counters and gas-discharge proportional counters), scintillation and semiconductor detectors. The main characteristics (energy resolution, detection efficiency, and time resolution) of X-ray detectors are demonstrated. Gas proportional and scintillation detectors are widely used in modern wavelength-dispersive X-ray fluorescence (WDXRF) spectrometers. Semiconductor detectors have found wide application in modern energy-dispersive X-ray fluorescence (EDXRF) and total-reflection X-ray fluorescence (TXRF) spectrometers, as well as in EDXRF spectrometers, which are part of X-ray microprobe analysis systems and scanning electron microscopes. These microscopes are equipped with secondary electron (SE) detectors, backscattered electron (BSE) detectors, cathodoluminescence (CL) detectors in addition to semiconductor X-ray detectors. This is demonstrated using the example of a scanning electron microscope MIRA3 LMH produced by TESCAN, Czech Republic. Beam Deceleration Technology was implemented in these microscopes, which is designed to improve image quality at low accelerating voltages. Today, three large groups of semiconductor detectors used in X-ray imaging systems can be distinguished, which are the result of the development of different technological approaches to image formation (Coupled Charge Device-, Thin Film Transistors-, Complementary Metal Oxide Semiconductor-Technologies).

S. I. Shtel'makh (✉)
Institute of the Earth's Crust, Siberian Branch of the Russian Academy of Sciences, Irkutsk, Russia
e-mail: fotina78@gmail.com

V. K. Singh (ed.), *X-Ray Fluorescence Spectroscopy and Chemometrics*,
https://doi.org/10.1007/978-3-031-98375-7_3

3.1 Types of X-Ray Detectors

X-ray detector is the device for recording and determining the intensity of X-rays [1]. The principle of detection and determination of the X-ray radiation intensity is based on the absorption of an X-ray photon in the sensitive material of the detector and the conversion of the absorbed energy into a signal or image, which can be interpreted as the value associated with the number of absorbed photons, and sometimes with their energy and the coordinates of the absorption point [1].

The main difference between the different types of detectors is the destiny of the electrons released during the ionization process. The effects currently used to register X-ray radiation were discovered by Wilhelm C. Roentgen (1845–1923). Photochemical effect (the formation of a latent image on a photographic plate or film), electrical effect (gas ionization and ionization in a solid body), optical effect (scintillations) relate to such effects [2].

The photographic film or plate is the simplest type of detector. The photochemical action of X-rays reduces the amount of silver halide by converting it by ionization into free silver in this case. The total intensity of the X-ray radiation will be determined by the number of formed silver atoms. This number of atoms will be established after the photographic film development on the basis of the darkening [2].

In gas-filled detectors such as Geiger-Muller counters and different types of gas-discharge proportional counters free electrons are formed when exposed to X-rays. In this case, electrons are removed from the inert gas atom. This becomes possible in a gaseous medium due to the relatively large mean free path of the electron. Recombination is prevented by a large potential difference within the ionization volume, which also allows electrons to be collected at the anode. X-ray photons are converted into signals in the form of voltage pulses, which can be amplified and counted in the next step in these detectors [2].

The operating principle of scintillation and semiconductor detectors is based on the transitions of electrons in a solid body. The energy of the incident X-rays causes the valence band of electrons in a suitable scintillator to be elevated to a higher level in a scintillation counter. These photons are then converted into voltage pulses using a photomultiplier sensitive to this energy range. Electrons generated by X-ray radiation are transferred to the conduction band in a semiconductor. The corresponding change in the conductivity characteristics can be associated with the number of incident X-ray quanta [2].

Gas-discharge proportional counters, scintillation and semiconductor detectors are distinguished by very high efficiency of X-ray radiation registration. They allow to measure its intensity very quickly and with high reliability. The development of electronics has provided them with a number of advantages over the previously widely used photographic film and Geiger-Muller counters [2].

Gas-discharge proportional counters, scintillation and semiconductor detectors refer to detectors whose average pulse amplitude at the output is proportional to the energy of the absorbed photon [3]. Geiger-Muller counters, current ionization chambers, and photographic films are detectors that are not sensitive to quanta energy [3].

In addition to this, X-ray detectors can be classified according to the geometric feature of radiation registration. Detectors are divided into point detectors, which allow measuring the intensity of X-ray quanta at a point in space where the narrow receiving window of the counter is located, and position-sensitive detectors (position detectors (PSD)). PSD are capable of simultaneously measuring the intensity of the quantum flow and determining with high accuracy the spatial coordinates of the photon's incidence on the sensitive element of the detector [1]. Thus, a photographic plate or photographic film is a two-coordinate position-sensitive detector, since it allows one to determine two coordinates of the photographic reflex caused by the impact of a photon [1].

3.2 Main Characteristics of X-Ray Detectors

The quality of detectors is characterized by many technical parameters, the most important of which are presented in this section.

- **Detection Efficiency**: This parameter is also called the *absorption efficiency* or *quantum efficiency* of detector. It is the probability of registering particles or photons when they enter the working volume of the detector, which is defined as the ratio of the number of photons that reach the receiving window of the detector to the number of photons registered by the device. Registered photons were absorbed in the active volume of the detector and left a signals about themselves recorded by the measuring circuit or measuring medium (in the case of photographic film or plates with optical memory) [1].

Detection quantum efficiency (DQE) is an average measure of the efficiency and noise characteristics of the detector. It is determined by the formula from work [1]:

$$DQE = \frac{\left({}^{S_0}/_{N_0}\right)^2}{\left({}^{S_i}/_{N_i}\right)^2} \tag{3.1}$$

where S_0 is the value of the useful signal at the detector input, N_0 is noise magnitude at the detector input, S_i is the value of the useful signal at the exit from the detector, N_i is noise magnitude at the exit from the detector. Detectors always add their own noise during the process of recording and processing signals. Therefore, the value of DQE is less than 1. This quantity is associated with the probabilistic nature of the interaction of X-ray quanta with the detector and usually is in the range from 0.5 to 1.0 for most X-ray detectors [1].

Absorption of X-ray radiation on the way to the recording medium reduces the measured intensity. Besides, absorption is happening in the detector window

[2]. The window of modern X-ray detectors is made of beryllium due to its high transparency to X-ray radiation, mechanical strength, and corrosion resistance [4]. Modern trends in the development of X-ray equipment production are aimed at reducing the thickness of the window and simultaneously increasing the active surface area of detectors [4, 5].

When a parallel beam of X-ray radiation photons falls on the entrance window of a detector with a working volume thickness d and window thickness, d_0 detection efficiency ε_d is expressed by the following equation from work [6]:

$$d = B\left(1 - e^{-\mu d}\right)e^{-\mu_0 d_0} \tag{3.2}$$

where B is coefficient that takes into account the level of discrimination of the recording circuit; μ and μ_0 are total linear absorption coefficients of the material of the detector working volume and the input window.

As can be seen from formula (3.2), the efficiency of registration in the region of X-ray wavelengths is determined mainly by the thickness and absorption coefficient of the material of the working volume of the detector [2].

- **Detector Resolution Time**: It is the minimum time interval between the passage of two particles through the detector when they are registered separately, as separate events, when the detector signals do not overlap each other, and the detector can operate in the counting mode of individual quanta [1].
- **Dead Time of Detector**: This is the time during which the detector registers one particle and returns to its original state to register the next particle. This effect is associated with relaxation and recombination processes occurring inside the detector after the effective absorption of photon. Dead time is a measure of the detector's inertia. It limits the maximum radiation intensity that the detector can register [1]. So, dead time of Geiger-Muller counters is relatively large (more 1×10^{-4}) [7], while dead time of gas-filled proportional counters is 0.2×10^{-6} s [8].
- **Proportionality of Detector**: This parameter characterizes the proportionality between the useful electrical pulse of the detector and the energy of the photon that caused this pulse. Proportionality of detector closely related to energy resolution of detector and its dead time [1].
- **Current Sensitivity of Detector**: This parameter characterizes the minimum number of photons per unit of time that can be registered and recognized in comparison with the detector's own noise Current sensitivity of detector strongly depends on the signal-to-noise ratio and affects the accuracy of measurement intensity [1].
- **Detector Dynamics Range**: This is the characteristic that defines the lowest and highest intensities that a detector can measure in actual practice. Lowest intensity measured is determined by the detector's own background or noise. Intrinsic background not associated with X-ray radiation is caused either by cosmic radiation entering the detector, radioactive contamination, or electrical noise in the detector's electrical circuits. Lowest intensity measured varies from

several pulses per minute for proportional counters to several hundred pulses per minute for Geiger-Muller counters [1]. The maximum intensity at which the counter can still operate is determined by its dead time [1].

- **Spectral Resolution (Energy Resolution of Detector)**: This parameter determines the minimum difference δE in the energy of two photons at which the detector is still able to distinguish this difference. The detector cannot measure energy accurately due to the statistical nature of the conversion of photon energy into electrical signals. Energy is determined with some probability [1]. The energy losses ΔE for individual photons can be different when the detector is exposed to a uniform flow of monoenergetic quanta. This spread is described by a probability density function of the certain type [1]. As a result, we obtain some distribution with a maximum near their real energy when registering monoenergetic photons, instead of a sharp intensity peak with zero width on the graph of dependence on energy (wavelength). This distribution is close to the Gaussian (normal) distribution for a wide class of detectors [1]. The parameter characterizing the energy resolution of such detectors is the value $\delta_{1/2}$ equal to the ratio of the width of the distribution at half the height of its maximum to the value of the parameter corresponding to the maximum or center of the distribution on the energy or wavelength scale. This parameter can be either the energy of photons, or the wavelength of radiation, or the magnitude of the electrical pulse associated with the energy of the registered radiation. The ratio $\delta_{1/2}$ is called the relative resolution of detector and is one of the main characteristics of spectrometric instruments. The reciprocal $(\delta_{1/2})^{-1}$ of the relative resolution is called the spectral resolving power of detector [1].

Spectral resolution of detector is determined by the number of pairs formed n (pairs of ions in proportional counters, pairs of carriers in semiconductor counters, or photoelectrons in scintillation counters) [2]. The number of pairs formed n is determined by the formula from work [2]:

$$n = E\gamma/\epsilon \tag{3.3}$$

where E_γ is energy of the detected photon, ε is average energy of pair formation.

The process of carriers formation (photoelectrons) can represent a sequence of independent acts. Therefore, dispersion Δn is $\Delta n = \sqrt{n}$ [2]. Spectral resolution of detector (R) for a Gaussian distribution of pulses is determined by the formula from work [2]:

$$R = 2.36\sqrt{\epsilon/E\gamma} \tag{3.4}$$

As can be seen from formula (3.4), spectral resolution of the detector improves with increasing photon energy [2]. The values of average energy required to form a pair are minimal for semiconductor detectors. This defines their high spectral resolution [2].

The assumption of independence of acts of interaction is not always justified. In practice, the spectral resolution turns out to be slightly lower [2]. Spectral resolution of the detector (R) taking into account the Fano factor (F) is determined by the formula from work [2]:

$$R = 2.36\sqrt{F\epsilon/E\gamma}. \tag{3.5}$$

The value of the Fano factor is close to 1 for scintillation counters. It is 0.4 for proportional counters and 0.05–0.1 for semiconductor detectors [2].

Spectral resolution of detectors can be degraded by factors such as noise from measuring devices, the detector edge effects, and the influence of impurities in gases used to fill proportional counters [2].

- **Position Resolution (*For Position Detectors*)**: This is the accuracy with which the detector is able to localize the position of a particle in space. This characteristic shows the ability of the detector to distinguish adjacent signals in space as independent [1].
- **Temporal Stability of Detector**: This parameter determines the accuracy of X-ray structural and X-ray spectral analysis. The requirements for it in X-ray structural analysis are quite strict, since the intensities of diffraction reflections must be determined with a relative error not exceeding 1% [1].
- **Radiation Resistance of Detector**: It is a parameter that directly affects temporal stability, since radiation damage gradually reduces the sensitivity of the detector [1].

So, ionizing radiation acting over a long period of time causes significant disturbances in the crystal lattice of crystalline scintillators, as well as noticeable changes in the polymer matrix of plastic scintillators, which significantly changes the physical and mechanical properties of scintillation materials. As a result of these changes, radiation swelling of scintillators occurs, leading to the occurrence of radiation stresses, and, as a consequence, radiation degradation of scintillation detectors [9].

3.3 Ionization Detectors

Ionization detectors (gas-discharge and semiconductor) constitute the largest class of detectors used to detect X-rays. X-rays cannot ionize molecules and atoms directly because they have no electrical charge, but they cause ionization of matter through secondary processes such as the photoeffect, the Compton effect, and the formation of electron–positron pairs [1]. The photoeffect is the result of the interaction of an X-ray photon with one of the inner shells of an atom of the active medium of the detector [1]. The probability of the photoeffect decreases rapidly with increasing quanta energy (approximately proportionally $1/(Eq)^3$ in the energy range Eq below 100 keV) and is very dependent on the atomic number

(Z) (increases with increasing atomic number as Z^5). That is why materials with high Z are used to made detectors [1].

A large group of ionization detectors used in X-ray equipment are gas-discharge detectors. The main types of gas-discharge detectors are ionization chambers, gas-discharge proportional detectors and Geiger-Muller counters [1].

3.3.1 Gas-Discharge Proportional Detectors

These detectors are cylindrical metal tubes with an anode thin filament made from tungsten. It locate along the axis secured in insulators on both bases of the cylinder [3]. The walls serve as a cathode [3]. The tube is filled with gas, and there is a window in its side wall through which X-rays enter the counter chamber [3]. Argon is used very often [3]. X-ray photons entering the counter are absorbed by the gas due to the photoelectric effect. Electrons torn from the inner shells of gas atoms cause the emission of photoelectrons, Auger electrons and secondary X-ray (fluorescent) radiation, which can also be absorbed by gas atoms, generating additional electrons. The formation of one argon ion pair requires an average of 26.4 eV of energy [3].

Electrons move toward the anode (filament), and positive ions move toward the cathode (body) under the influence of the accelerating potential. Electrons are accelerated by the electric field on their way and repeatedly ionize gas atoms. The ratio of the number of initially generated electrons to the number of electrons reaching the anode (filament) is called the gas gain coefficient, which is usually equal to 10^4–10^5 [3]. Gas amplification in gas-filled detectors is due to the fact that photoelectrons arising during primary ionization at the moment of absorption of an X-ray quantum are accelerated by the electric field of the detector on the way to the collector electrode so much that their energy becomes sufficient to ionize other atoms in the sensitive medium [1]. In turn, the secondary ionization electrons are accelerated and ionize other molecules. It leads to the formation of the impact ionization avalanche near the collector and the increase of the electrical pulse. Therefore, the number of charge carriers falling on the collector can significantly exceed the number of charges created during singing ionization [1]. The value of the gas gain factor depends exponentially on the operating voltage on the detector. Therefore, the stability of the accelerating voltage should be no worse than $\pm 0.05\%$. To quench the pulse quickly, a small amount of quenching gas is added to the inert gas. The mixture of 90% argon and 10% methane is the most common [3]. The efficiency of the counter depends on the purity of the gas. Admixtures of water vapor and oxygen are unacceptable, since the ions of electronegative gases are low-mobility and easily recombine [3].

The flow-through proportional counter with similar construction is presented in Fig. 3.1.

There are two models of gas proportional counters used in modern wavelength-dispersive X-ray fluorescence (WDXRF) spectrometers for long-wavelength X-ray radiation, such as the flow counter (FC) and the sealed proportional counter (PC).

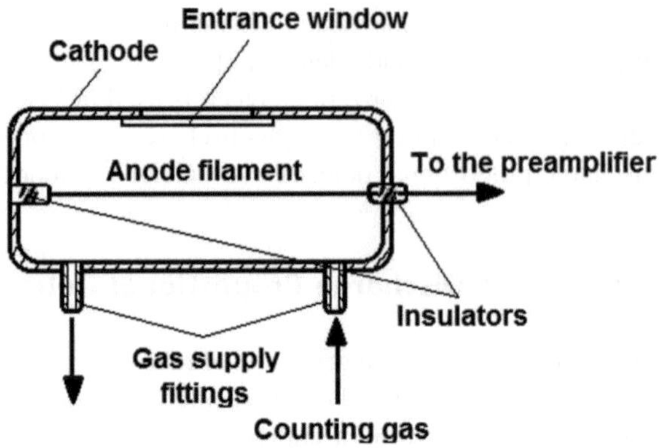

Fig. 3.1 Schematic diagram of the flow-through proportional counter

The FC is connected to a continuous supply of counting gas (Ar + 10% CH_4) and has the advantage of being able to be equipped with a very thin window (≤10.6 μm). Therefore, the FC is also suitable for measuring the very light elements and is very stable. The PC has a closed volume and requires a thick window normally made of beryllium. The absorption in this "thick" beryllium window prevents the measurement of the very light elements (from Be to Na) [10]. The vacuum-tight beryllium with a thickness of about 200 μm was used for windows in PC [3]. However, today success achieved in developing PC that are just as sensitive to the very light elements (from Be to Na) as flow counters due to the use of highly transparent organic materials [10].

X-rays cause secondary X-ray radiation of the gas in the process of their absorption in the working gas of the detector. The ionizations by secondary photons are lost if this secondary radiation leaves the working volume of the detector without being absorbed [3]. Furthermore, along with the main peak, the escape peak appears, which displaced from the main peak on the amount of lost energy of the working gas photons (Fig. 3.2).

Amplitude spectrum (Fig. 3.2) was obtained using WDXRF spectrometer S8 TIGER produced by Bruker AXS, Germany and counting gas (Ar + 10% CH_4). Detailed information about this spectrometer features can be found elsewhere [11].

It is necessary to take into account the influence of the escape peak during X-ray fluorescence analysis, because it can lead to unwanted overlap on the analytical line. Gas-discharge proportional detectors are used for quantitative determination of elements concentrations with atomic number $Z \leq 23$.

The total duration of the electrical pulse at the output of the proportional detector is quite large and is approximately 1×10^{-6} s due to the low-mobility ion component. However, using only the electronic component of the pulses with the help of the corresponding formation circuit gives the opportunity to receive pulses with the duration of about fractions of 1×10^{-6} s at the output of the detection unit [3].

The number of ion–electron pairs formed in the chamber of the gas-discharge detector upon absorption of an X-ray photon depends on the energy of the photon

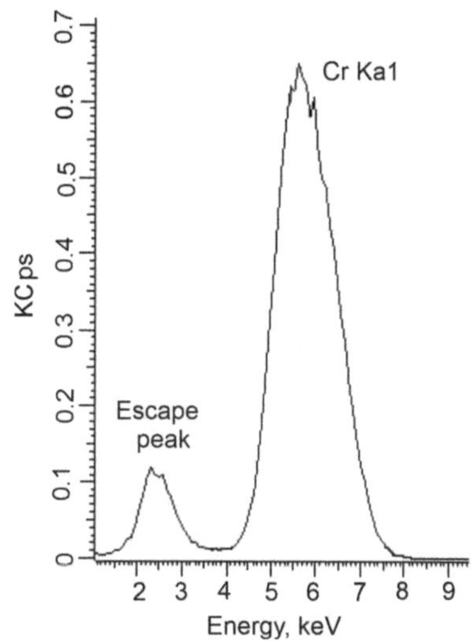

Fig. 3.2 Amplitude spectrum of the flow proportional counter (CrK_{a1}-line)

and the composition of the gas mixture [1]. The magnitude and nature of the electrical pulse in the detector circuit depends on the strength of the electric field in the detector at a constant energy of the detected X-ray quanta [1].

There is no fundamental difference in the design of ionization chambers, gas-discharge proportional detectors and Geiger-Muller counters, but they differ in the magnitude of the operating voltage (electric field in the gas volume), which determines their properties [1].

Geiger-Muller counter is the oldest X-ray detector. This counter operates at voltages that ensure an avalanche electrical breakdown of the medium when free charges arise in it. The gas ionization occurs when a charged particle passes through Geiger-Muller counter. Resulting electrons move towards the anode and enter the region of the strong electric field. They accelerate and ionize the gas molecules themselves. It leads to the corona discharge and current through the counter. The signal amplitude reaches several volts and can be easily measured. Geiger-Muller counter registers only the fact of a particle or photon hitting it but does not allow to measure their energy. This counter operates at voltages of 1000 V and higher, which is significantly higher than proportional counters [1]. The ionizing X-ray photon enter the counter and causes an electrical discharge between its electrodes, which creates a pulse of several tens of volts regardless of the energy of the absorbed photon. Each such discharge corresponds to one absorbed X-ray quantum or one charged particle. However, the avalanche process in it turns out to be very long due to the high voltage value on the counter. The avalanche lasts more than 1×10^{-4} s, and the subsequent relaxation of the electric field in the counter, determined by the accumulation of positive ions on the cathode, reaches 1.5×10^{-4}–3×10^{-4} s after the absorption of an ionizing particle or photon. During this period the counter is insensitive to other quanta, and this period is the dead time of the counter. At present, Geiger-Muller counters are not used in X-ray diffraction analysis due to the large dead time [1].

In the proportional counter in contrast to the Geiger-Muller counter, when a gas is ionized by a passing charged particle or photon, the corona discharge does not occur due to the selection of the supply voltage and the composition of the gas mixture. The proportional counter allows not only to register the fact of the hit but also to determine the energy of registered particles in contrast to the Geiger-Muller counter [1].

Proportional gas detectors have low noise (10–15 pulses/min). The efficiency of the proportional counter filled with argon usually exceeds 80% for X-ray quanta with the energy in the range of 10–20 keV. It is necessary to fill the counter with xenon in order to obtain high detection efficiency of X-ray radiation with energies above 20 keV. The values of the maximum counting rate are 10^5–10^6 c^{-1} usually at the values of the gas gain coefficients 10^4–10^5 for the gas-discharge proportional counter [1].

The main focus in the development of these detectors is on increasing the resources and reliability in operation. The radiation resource is characterized by the number of X-ray quanta that the detector can register without a noticeable change in its operating characteristics. The energy resolution of gas proportional detectors deteriorates over time during operation [2]. Research by a number of authors has shown that precipitation and polymerization of organic compounds occur at the anode [2, 12]. In this case, the shape of the anode filament is asymmetrically distorted [2].

The ionization chamber is filled with a gas mixture and has an anode and a cathode, as well as the proportional counter and the Geiger-Muller counter. The supply voltage in the ionization chamber is lower compared to the proportional counter. Increase of ionization does not occur in the ionization chamber [1]. The electron component of the current pulse alone or the electron and ion components are used depending on the experimental requirements to measure the particles energy. Ionization chambers are gas-discharge detectors operating at voltages in the saturation range. Ionization chambers are widely used at synchrotron radiation sources because they are the main monitor of the intensity of the primary radiation beams in almost all X-ray experiments. These devices are typically used as integrating detectors to measure the flow of photons in a beam, rather than to record individual quanta. The ability to pass powerful photon fluxes through itself with almost no attenuation and to measure their intensity at flux levels of 10^{12} photons/sec and higher is the main advantage of ionization chambers ensuring their wide application in all sources of synchrotron radiation. This ability is far beyond the limiting count rate of other detectors. Almost not weakened the rays that pass through the ionization chamber can be used for diffraction or spectral experiments [1].

One of the modifications of the gas-discharge proportional detector is the gas electroluminescent counter. The formation of a useful signal is carried out by registering network radiation using a photomultiplier tube, which occurs in the working gas during the process of its excitation and ionization [3].

Another type of gas-discharge detector is the gas position-sensitive detector (PSD). The high-resistance filament is used as an anode in the PSD. The registered signal is taken from both sides of the anode. The amplitude and shape of both output pulses are determined by the distances from the filament section corresponding to the primary ionization point to its ends. Subsequent processing of these signals allows determining both the energy of the absorbed photon and the linear coordinate of the point of its entry into the detector. The spatial resolution of the PSD reaches 50–100 μm [3].

PSDs can be divided into one-dimensional (linear), in which the coordinate of the photon hitting on the line of the detector's receiving window is determined, and two-dimensional detectors, which determine the coordinates of the registered particle on the surface [1]. PSDs can be either counters operating in the direct quantum counting mode or integrating analog detectors, in which the electrical signal is a function of the flow rate of the registered photons. The ability to continuously digitize electrical signals and process them by a computer directly during the measurement process is an advantage of electronic PSDs, as well as electronic point detectors [1]. Gas-discharge linear position-sensitive detectors (LPSDs) can be made straight or curved along a radius (focused on the sample). In the latter case, X-rays from the test sample always fall on the detector perpendicularly, and there is no parallax between the directions of the X-rays and the avalanches caused by them in the detector. Due to this, the spatial resolution and accuracy of determining the coordinates of reflections are improved in the X-ray diffraction pattern [1].

The technology for manufacturing proportional curved LPSDs with one extended anode, which allows for the high-precision manufacture of detectors of virtually any curvature, is described in detail in the work [13]. The anode is made of thin steel roll, about 40 μm thick, which can be bent to almost any radius in these detectors. The robust curved detector chamber has an entrance window covered with 25 μm thick aluminized Mylar film or beryllium foil [1]. Copper cathode electrodes connected to each other are located on the front and rear walls of the chamber. The anode is located between the cathodes in such a way that the working gap between the tip of the anode and the cathode surfaces is about 8 mm. The front cathode is a solid copper strip curved along the inner surface of the chamber body coaxially with the anode. The rear cathode is made in the form of a printed circuit with copper vertical strips 2 mm wide, spaced approximately at the same distance from each other. Decoding of the avalanche coordinate is carried out using an electronic delay line circuit located outside the chamber in this detector. The position of the center of gravity of the charge induced on the cathode during the formation of avalanches is determined by reading through the delay line. The signals from each end of the delay line are processed by preamplifiers, discriminators. They are converted by an amplitude-time converter and allow to determine the coordinates of the electron avalanche with a resolution of no worse than 200 μm. The detector chamber is filled with a mixture of inert gas and ethane under pressure, which is renewed in flow mode to reduce the static charge accumulated during ionization [1].

The company INEL (France) serially produces several models of detectors of this type, differing in focusing radius and receiving aperture size. These detectors operate with gas mixtures of argon-ethane or krypton-ethane under a pressure of 6.5 atm. The maximum count rate in one pixel is limited to 7×10^3 photons/sec or an integral intensity of 2×10^5 photons/sec for the entire detector aperture due to decoding of coordinates using the delay line [1]. The detectors are equipped with the necessary electronic devices for normal operation, such as gas flow and pressure controllers, a high-voltage source, preamplifiers, amplifier-discriminators, a coordinate processor, an intensimeter, a multichannel analyzer and the corresponding software controlling the detector calibration, data collection and its primary processing, issuance of an analysis protocol, and graphical presentation of the data [1]. The considered one-dimensional proportional detectors are widely used as part of X-ray diffractometers [14, 15]. These detectors are single-anode chambers. Their common disadvantage is the low recording speed in each individual channel, which is associated with the collection of signals from all channels from practically the same electrode [1].

One-dimensional X-ray detectors resistant to powerful radiation with a high recording speed based on multielectrode proportional chambers are widely used at synchrotron stations [16]. The main part of coordinate gas-discharge detectors with a high recording speed is the chamber with the drift zone covering the absorbing volume of the detector [16]. The entire gas volume opposite the receiving window of the chamber represents the zone of drift of ions and electrons of primary ionization to the collector electrodes in such detectors. The volume of this zone is strictly limited by the electric field of the cathode grids. Its size in the vertical direction is comparable to the height of the receiving window (5–10 mm), while the depth in the direction of the X-rays can be from several tens of mm to tens of centimeters in order to provide the high probability of quanta absorption. The zone of electron avalanches of gas amplification with anode wires is located outside the receiving window. Therefore, X-rays do not reach the anode wires and do not damage them. The anode consists of a multitude of stretched wires that form the anode plane. They can be located either along the length of the receiving window or along the direction of the incident rays [1]. The cathode is made in the form of a printed circuit with narrow strips of electrodes directed along the path of the X-rays (by the fan focused on the sample) in order to eliminate parallax errors in determining the coordinates of photons [1].

Electrons produced by the photoelectric effect drift through the field-forming cathode wire grid to the anode wires, where avalanche amplification occurs. The avalanche coordinates are measured from the charge distribution induced on the signal cathode strips. The number of the strip with the maximum amplitude, the amplitudes from two adjacent strips $A(n - l)$, $A(n + l)$ and the time signal from the external timer are transmitted to the selected events processor. Based on this data, the processor determines coordinates and energy of the photon [17].

3.3.2 Semiconductor Detectors

Semiconductor detectors belong to the group of solid-state ionization chambers according to the principle of converting ionizing radiation into a detectable electrical analogue, since they operate due to ionization in a solid semiconductor medium [1]. These detectors are manufactured on the basis of one, two or three-element semiconductor compounds [18]. Modern semiconductor detectors are created in most cases on detector crystals operating at room and low (up to −50 °C) temperatures. These crystals include silicon (Si) [2, 18], germanium (Ge) [2], mercuric iodide (HgI_2) [2], gallium-arsenide (GaAs) and gallium-arsenide compensated of Cr (GaAs: Cr), cadmium telluride (CdTe) and cadmium-zinc telluride (CdZnTe) [18].

The operation of semiconductor detectors is based on two physical processes, such as the formation of electron–hole pairs as the result of the absorption of incident X-ray quanta by the solid body and the collection of the formed carriers in an electric field. The generated charge is proportional to the absorbed energy of the specific quantum. It became the basis for the creation of energy-dispersive X-ray fluorescence spectrometers (EDXRF) [2]. Semiconductor detectors have found wide application in modern total-reflection X-ray fluorescence (TXRF) spectrometers, as well as in EDXRF spectrometers, which are part of X-ray microprobe analysis systems and scanning electron microscopes.

Besides, semiconductor detectors are used in devices for monitoring radioactive contamination of the external environment, in dosimetry, and in devices for monitoring and diagnosing the operation of nuclear reactors [18].

The sensitive element of semiconductor electron–hole detectors is the layer of (p-n) junction artificially created in a semiconductor crystal, which is depleted of charge carriers and has the resistance on several orders of magnitude higher than the rest of the crystal volume. Thickness of (p-n) junction is less than 1 μm. X-ray photons fly through this area without having time to fully expend their energy. The sensitive element (the layer of (p-n) junction) has very low detection efficiency [1]. The bias voltage is applied to (p-n) junction in the direction opposite to the direction of electron conduction in order to increase the thickness of the active region. The bias voltage repels charge carriers from it and increases the thickness of the depleted layer by several tens of times. The rest of the crystal volume is insensitive to ionizing radiation. It is the dead layer of the detector forming the input window of the semiconductor detector. Semiconductor detector consists of the semiconductor layer with metal electrodes deposited on both sides, to which the bias voltage is applied. Nonequilibrium charge carriers (electrons and holes) are formed in the semiconductor in the result of ionization when a particle or ϒ-quantum hits it. Electrons and holes move towards the electrodes under the influence of an electric field. As a result, a current pulse is generated in the electrical circuit connected to the detector, which is converted by the external circuit into a voltage pulse, the amplitude of which is proportional to the energy release of a particle or photon in the semiconductor [1].

The (p-n) junction is created by depositing a metal on the surface of a semiconductor by evaporation in a vacuum in the surface-barrier semiconductor detectors. Complete depletion of the semiconductor crystal by carriers can be ensured and detectors can be obtained with a sensitive area equal to the entire wafer thickness of the semiconductor up to 2–3 mm at high values of semiconductor resistance and when the high bias voltage is applied to the resulting (p-n) junction (opposite to the direction of conduction). The (p-n) junction is created by diffuse introduction of donor-impurity atoms for the semiconductor with p-type conductivity and acceptor-impurity atoms for n-type semiconductor. The (p-n) junction is created by the introduction of impurity atoms into a crystal when it is irradiated with an ion beam. Boron is incorporated into n-type semiconductor, phosphorus is incorporated into p-type semiconductor in most cases [1].

Wafers of silicon detectors wafers are produced by cutting high-purity silicon single crystals, which are ground, etched and polished to obtain a surface without damaging the layer. The processed wafers are thoroughly cleaned by chemical and plasma methods. Mixtures of strong oxidizers (HNO_3, H_2SO_4, and H_2O_2) and water solution of NH_3 are used for chemical cleaning [18]. The layer of SiO_2 approximately 0.5 μm thick is created by thermal oxidation in dry oxygen on a carefully cleaned plane of the plate surface of monocrystalline silicon with n-type electrical conductivity. Photolithographic processing of this layer is performed. The layer of photoresist sensitive to ultraviolet radiation is applied to the oxidized silicon surface. The plate with the dried layer of photoresist is placed under a photomask (a plane-parallel quartz plate with a pattern), which is transparent to ultraviolet radiation in specified places. The photoresist in those places under which the SiO_2 layer should be preserved is polymerized (tanned) after exposure to ultraviolet light. The photoresist is removed from the rest of the plate, the exposed SiO_2 layer is removed by etching, after which the remaining photoresist is removed [18]. Ion implantation of boron (an acceptor impurity) is carried out in areas where there is no oxide film to create a region with p-type electrical conductivity in the material of the original plate. The ions energy used is in the range from 15 to 150 keV depending on the required depth of the (n-p) junction. Thermal distillation of the impurity and annealing of the plate are carried out at the following stages. The entire technological process is based on the use of temperatures not exceeding 1050–1100 °C [18].

Arsenic ions are introduced from the back side of the plate to obtain a thin layer with an increased concentration of the electron impurity n^+ for the purpose of subsequent metallization of this layer by aluminum to obtain the ohmic contact [18].

Contact windows are formed to create contacts on the surface of the wafer (in the masking layer of SiO_2), through which metal is then sprayed. Contact pads on the periphery and connecting tracks between the pads and windows are formed in this process. The metal is burned into the wafers at 400–450 °C in the hydrogen atmosphere in the next stage [18]. SiO_2 layers are produced by pyrolytic or chemical cooling. SiO_2 is the most common material for forming the detector's protective layer [18].

One of the main macroparameters characterizing the quality of the detector is the dark current or leakage current of the detector under reverse bias. The gettering process is widely used to reduce dark current [18, 19]. It consists of creating the region where the solubility of polluting, rapidly diffusing, recombination-active impurities (Au, Cu, and Fe) is much higher than in other regions [18]. Diffusion of phosphorus with high concentration is used to create this region in most cases [18, 19]. The dark current of latest detector samples does not exceed 1–2 nA/cm^2 [18].

The significant increase in the sensitive region in diffusion-drift semiconductor detectors was achieved by compensating the conductivity of the source material to its own conductivity using the drift of Li atoms to the (p-n) junction zone. The compensation is that Li atoms are introduced into the semiconductor using the diffusion saturation method, creating a layer of electron-capturing centers near the electron–hole junction. In this way, it is possible to obtain silicon and germanium crystals with the required characteristics and a large thickness of the sensitive layer, capable of completely absorbing X-ray radiation. Detectors created on their basis are called lithium-drift detectors (Si(Li) and Ge(Li)). These detectors are widely used in X-ray spectrometry [1].

Semiconductor detectors have important distinctive properties, such as the low energy for the formation of an electron–hole pair, the higher stopping power compared to gas-filled devices, the ability to register particles penetrating to different depths detector, the ability to obtain the entire spectrum simultaneously, which provides the higher speed of collecting statistical data [18].

Besides, modern semiconductor detectors are characterized by a higher energy resolution than other types of detectors due to the low energy of formation of electron–hole pairs. The formation energy of an electron–hole pair is 3.6 eV for Si, while the formation energy of an electron–ion pair in a gas reaches several tens of eV (26 eV for Ar) [1]. So, energy resolution of semiconductor detectors is 150 eV for $MnK_{\alpha 1}$-line, energy resolution of proportional detectors is 1 keV for $MnK_{\alpha 1}$-line, and energy resolution of scintillation detectors is 3 keV for $MnK_{\alpha 1}$-line, respectively [2].

The characteristics related to the detector size and its energy resolution are considered in the work [20]. It was found that the background is formed partly due to the degradation of pulses in the peripheral region of the detector. The design with a guard ring has been proposed that can reduce the background by about 40 times [20]. Detectors with very low values of energy resolution are typically small in size with the working area of about 10 mm^2 [20]. Energy resolution of semiconductor detector worsens as its size increases, and geometric efficiency, on the contrary, increases [2].

Semiconductor detectors also have disadvantages despite a large number of advantages. Certain difficulties arise when working with lithium-drift detectors (Si(Li) and Ge(Li)). Li has the smallest atomic size of all metals and extremely high diffusion mobility in Ge and Si at room temperature. It is necessary to constantly keep the crystal at low temperatures, where the diffusion mobility of Li becomes small in order to avoid rapid degradation of lithium-drift detectors due to the movement of Li atoms from the detector active zone. Liquid nitrogen temperature is required for storage of Ge(Li)-detector, temperature ((−20) – (−10 °C)) is required for storage of Si(Li)-detector, but these detectors must be operated at liquid nitrogen temperature to reduce dark current [1].

The difficulties encountered in creating low temperatures prompted the development of detection units with thermoelectric cooling [2]. So, Si PIN-detector AXR 100 CR (Amtek, USA) with thermoelectric cooling was used in EDXRF spectrometer with a polarizer to determine concentrations of Mo, Nb, Zr, Y, Sr, and Rb in rocks [21]. The small size of the semiconductor element makes it possible to cool it by Peltier elements. Its input window area is 5 mm^2, the thickness of its working layer is 680 μm. The energy resolution of this detector is 175–180 eV for radiation of $MnK_{\alpha 1}$-line. The efficiency of recording hard X-ray radiation by the Si PIN-detector turned out to be significantly lower than for the S1 80,180 Si(Li)-detector due to the thinner working layer. This is especially important when registering radiation in the energy range exceeding 20 keV [21].

In addition to the development of detection units with thermoelectric cooling, the direction associated with the creation of materials that allow the operation of detectors at room temperature is developing intensively [2]. Detectors based on CdTe (CdZnTe), GaAs, HgI_2 have acceptable energy resolution at room temperature [22–26]. The best results in energy resolution at room temperature were obtained on low-sensitivity detectors based on CdZnTe and HgI_2 [22]. The energy resolution of detectors based on GaAs in the low-energy region is at the level of proportional counter resolution with an efficiency of 5–18%. Detectors based on GaAs differ from proportional counters in size and are more stable in operation. However, at present, the problem of uncooled semiconductor detectors for X-ray spectrometry is not solved. The main disadvantages of detectors based on CdTe are polarization effects and high leakage current determined by the quality of the crystal surface. Uncooled detectors based on GaAs are currently not widely used due to their low sensitivity in detecting X-ray radiation. This is due to limitations in the thickness ($\leq$300 μm) and area ($\leq$10 mm^2) of the sensitive area [22]. Besides, the difficulty in producing adequate quality thin crystals of HgI_2 is one of the reasons holding back mass production of detectors based on HgI_2 [2].

The significant disadvantage of semiconductor detectors is their high intrinsic noise, which is caused by the presence of intrinsic charge carriers even in the purest semiconductor [1, 2]. In addition, acoustic noises can register by the semiconductor detector as electrical impulses creating additional measurement errors [1]. The advantages of semiconductor detectors compared to other detectors include almost ideal proportionality of the particle energy release signal in a substance over a wide energy range, small thickness of the input window, insensitivity to magnetic fields, high energy resolution, and compactness. However, the implementation of these characteristics requires the use of complex electronic devices and, in most cases, cryogenic temperatures. This complicates the work with semiconductor detectors [1].

The production of silicon semiconductor diodes (pin-diodes) with thick-layer *i* depleted of charge carriers has become possible due to the development of semiconductor manufacturing technology. A typical pin-diode manufactured using the planar technology is the thin silicon wafer with several hundred microns thick. The planar technology was developed by Jean Hoerni (1924–1997) in 1959. Sides of pin-diode are heavily alloyed in various ways. One side has high n-type conductivity, the other side has high p-type conductivity. The semiconductor layer is located between these areas. This semiconductor layer is strongly or completely depleted of charge carriers with a conductivity close to the intrinsic conductivity of pure silicon (specific resistance up to 4000 Ohm · cm) [1].

Pin-diodes are semiconductor adjustable resistive diodes, which are widely used in modern electronic devices as high-frequency switches, and pin-diodes with an open plate surface are used as photodiodes. Such diodes have the *i*-area with the thickness that ranges from 3 to 30 μm. Ordinary pin-diodes are sensitive to X-rays and can be used as X-ray detectors. Modern pin-diodes with the thick *i*-layer (several hundred μm) have the high efficiency of X-ray detection. Pin-diodes with the very thick active layer (200–500 μm) are used for direct detection of hard X-rays. Such pin-diodes are diodes of the series XRA and XRB (Detection Technology, Finland). Pin-diodes of the series XRB are specially optimized for detecting intense beams of hard X-rays. The detection efficiency of these diodes depends on the thickness of the diode base layer *i* and reaches 100% at a plate thickness of 300 μm for photons with the energy of 10 keV and 1% for photons with the energy of 150 keV. The interaction of photons with the diode material occurs through the Compton effect at their energies exceeding 60 keV in most cases. The active region of the diode is in electronic equilibrium with the environment, so Compton recoil electrons are also detected. As the result, the overall detection efficiency at energies above 150 keV is almost constant over the wide range of photon energies. Thus, the pin-diode can be considered as a solid-state analogue of the ionization chamber [1].

Energy resolution of diodes of the series XRA is 200–250 eV for $MnK_{\alpha 1}$-line (Detection Technology, Finland). Depletion layer *i* thickness of these diodes is 500 μm [1].

Operational amplifiers of the JFET type (field-effect transistor with p–n junction) are used as a preamplifier when using diodes in pulse detectors. These amplifiers of the JFET type have the very short recovery time (1.5×10^{-6} s) [27]. At the same time, the dead time of the detector is practically independent of the counting rate [1]. The detector MOXTEK PF-2500 (MOXTEK, USA) based on the pin-diode with thermoelectric cooling is designed specifically for X-ray diffractometry. The area of the active region of this pin-diode is 2×12 mm^2, and its thickness is 300 μm. Energy resolution of the detector MOXTEK PF-2500 does not exceed 350 eV. It remains unchanged over the entire range of counting rates. Window thickness of this detector is 16 μm [1]. It is made of special foil DuraBeryllium (MOXTEK, USA) [28]. The main advantage of this detector compared to the Si(Li)-detector is its ability to operate at temperatures from 0 to 30 °C and provide the high counting rate (20,000 pulses per second) in the individual photon counting mode [1].

X-ray pin-diodes can be used successfully in synchrotron radiation beams due to their ability to operate in the direct photocurrent measurement mode with virtually no limitation on the registered photon flux with high proportionality between the current and flux intensity, weak dependence of detection efficiency on photon energy in the range from 3 to 30 keV, high radiation resistance and small dimensions [1].

However, detectors based on the pin-diodes have significantly worse energy resolution compared to silicon drift detectors (SDD) [29]. These detectors were fabricated by a team in Munich [30]. This team was involved in development of detectors for space research. They adopted the lithographic techniques and planar technology of the microelectronics industry for detector fabrication. However, to realize the concept of sideward depletion, an important modification was introduced, which consisted of the double-sided processing of silicon wafers. The SDD overcame the difficulties present with Si(Li) detectors. Recent SDDs with thermoelectric cooling to −20 °C collect X-rays at least two orders of magnitude faster, at slightly better energy resolution, and without the need for liquid nitrogen cooling [29].

Since their invention in 1983, SDDs have left deep traces in the field of radiation detection. Emilio Gatti (1926–2016) and Pavel Rehak (1945–2009) worked hard at the topic of how to move the signal electrons generated in a fully depleted silicon crystal to a point-like, low-capacitance readout node [31].

The integrated single-sided n-channel JFET is used in structures of SDDs of cylindrical form [29]. The platelet of n-type semiconductor material (channel) is clamped between two smaller platelets of p-type (the gates). The ends of the channel are designated the drain (D) and the source (S), and the two platelets of p-type material are connected together. Their terminal is named the gate (G) [29]. The double-sided polished high-resistivity n-type silicon wafer can be entirely depleted of mobile charge carriers if both wafer surfaces are covered with rectifying p+junctions [29]. The signal-electron collecting anode is in the center of the device. The p+drift rings (field strips) produce the graded electric field that forces the signal-electrons to the read node. The homogeneous radiation entrance window is at the back of the device where the p+back contact is located [29]. Only a very small ohmic n + contact is needed to apply reverse bias voltages and drain off all mobile charge carriers. With proper bias voltages the depletion occurs from both wafer surfaces until the detector volume is completely sensitive [29].

This type of detectors is characterized by the small intrinsic capacitance of the anode (up to 0.1 pF for the sensitive area of about 1 cm^2) [32] and the low noise level [29]. Integration of the transistor in the SDD reduced the total input capacitance dramatically, thus reducing the electronic noise and increasing the speed of signal processing [29]. Besides, SDDs are characterized by the best energy resolution compared to other X-ray detectors. For example, the SDD with the area of 7 mm^2 and the sensitive area depth of 0.3 mm has energy resolution of no more than 140 eV for $MnK_{\alpha 1}$-line at a temperature −10 °C (Amtek, USA) [18]. Today, the best guaranteed values are 121 eV for $MnK_{\alpha 1}$-line at shaping times around 1×10^{-6} s. allowing for very high count rates [29].

Today, SDDs are widely used in modern TXRF spectrometers (for example, spectrometer S2 PICOFOX (Bruker Nano, Germany) [33–35], as well as in EDXRF spectrometers, which can be used as analytical unit of X-ray microprobe analysis systems [29] and scanning electron microscopes (for example, the scanning electron microscope MIRA3 LMH (TESCAN, Czech Republic [36]). In addition, SDDs are widely used in mobile μ-XRF spectrometers (for example, mobile μ-XRF spectrometer ARTAX (Bruker, Germany) [37, 38]). Spectrometer ARTAX is designed for the study of art and archeology objects. It is equipped with the polycapillary lens providing a beam size of 75–100 μm. The design feature is that its measuring unit can be placed directly in front of or above the object. This makes it possible to measure samples of any size or complex geometry [38]. It should be noted that the traditional SDD for microanalysis has the cylindrical shape with an integrated n-channel junction field-effect transistor (JFET) in the center to minimize read node capacitance [29]. However, the microanalyzer JXA-8200 (JEOL Ltd., Japan) released in 2006 is equipped with EDXRF spectrometer with Si(Li) detector (energy resolution of 133 eV for the line of $MnK_{\alpha 1}$-line). Besides, this microanalyzer is equipped with five WDXRF spectrometers and detectors of secondary and backscattered electrons (SE and BSE detector) [36, 39].

High rates of count from the SDD can produce high-quality element distribution maps in surprisingly short times (5–10 min) compared to the Si(Li) detector (hours). Moreover, modern commercial X-ray spectrometer systems employ spectrum-imaging software that can collect an entire spectrum at each image pixel, producing a data cube (2D image with 1D spectrum at each pixel) [29].

3.4 Scintillation Detectors

Scintillation detectors are used to register hard X-ray radiation (wavelength does not exceed 0.23 nm). Phosphor materials act as the working fluid in them. The conversion of the energy of the registered photons occurs in two stages. Primary ionization causes a flash (scintillation) in the phosphor material in the visible range of wavelengths. The flash energy is converted into an electrical pulse using a photomultiplier tube in the next step. The plate of sodium iodide monocrystal activated by thallium (NaI(Tl)) serves as phosphor material. The crystal (scintillator) is packaged in a sealed aluminum container with a beryllium input window and a glass output window [3]. NaI(Tl) is X-ray sensitive crystal (scintillator) [1, 3]. The X-ray quantum hits a crystal (scintillator) and tears the fast photo-electron from the atoms. This photo-electron moves in the crystal and ionizes its atoms. During relaxation, excited atoms are illuminated, emitting photons of ultraviolet radiation when used NaI(Tl). The flash of such radiation consists of several tens or hundreds of photons per absorbed quantum of X-ray radiation. The conversion of a photon flash into an electronic signal occurs in an electron photomultiplier, which has a photosensitive cathode and a series of successive electrodes that ensure the multiplication of photoelectrons obtained on the photocathode [1].

The scintillation detector with similar construction is presented in Fig. 3.3.

The signal from the registration of an X-ray quantum generated by the detector goes from the preamplifier to a wideband amplifier and to an amplitude discriminator-pulse analyzer at the next stage. The amplitude analyzer processes pulses coming from a wideband amplifier. The scintillation detector has a certain spectral resolution allowing to separate different harmonics of the registered quanta and low intrinsic noise not exceeding 10 pulses per 100 s. This detector allows for correct measurement of flows up to 10^5 photons/sec without violating the linearity of the conversion of quantum energy into an electrical pulse [1].

The proportionality of the transformation of quantum energy into an electrical signal is performed less accurately in scintillation detectors compared to gas-discharge proportional counters. The reason lies in the dependence of the conversion efficiency in the crystal (scintillator) on the degree of its ionization by X-ray radiation. This dependence for NaI(Tl) is negligibly small in the energy range from 10 to 10^3 keV. Crystal (scintillator) NaI(Tl) with a thickness of 0.5 mm is capable of absorbing about 96% of X-ray radiation, and this crystal with a thickness of 1 mm is capable of completely absorbing of X-ray radiation. The duration of the scintillation flash in these crystals is 25×10^{-8} s, so detectors based on them have the good dynamic range and the small dead time (1×10^{-6} s), which is determined mainly by the speed of signal processing by the electronic recording system. Scintillation counters are capable of recording intensities up to 5×10^4 pulses/sec without noticeable miscalculations [1].

Scintillation detectors are widely used in modern wavelength-dispersive X-ray fluorescence (WDXRF) spectrometers. Besides, detectors with scintillators are used as detectors of secondary and backscattered electrons (SE and BSE detectors) in the microanalyzers and modern scanning electron microscopes.

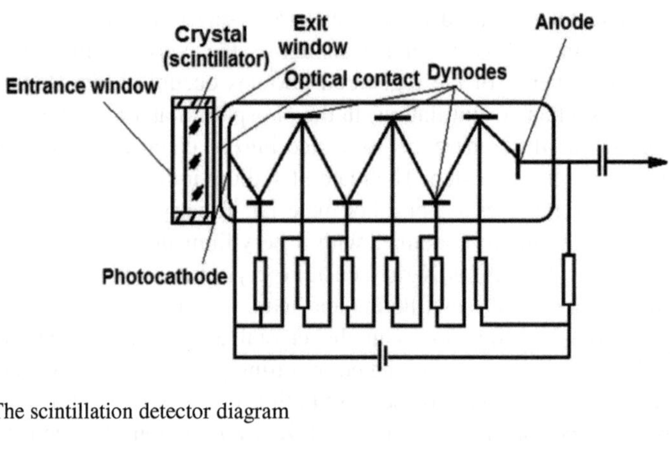

Fig. 3.3 The scintillation detector diagram

3.5 Detectors Used in Scanning Electron Microscopy: Acquiring Images

Scanning electron microscopy (SEM) is a technique that allows obtaining extensive information from an image created by the interaction between the sample being examined and an electron probe [40, 41]. The scanning electron microscope is based on the use of a preformed thin electron beam (probe), the position of which is controlled by electromagnetic fields. Scanning is similar in many ways to the scanning process in television picture tubes. The studied sample is scanned by a focused electron beam of medium energy under high vacuum conditions. Depending on the signal recording mechanism, several operating modes of scanning electron microscopy are distinguished, such as secondary electrons, backscattered electrons, cathodoluminescence, and X-ray radiation. The developed methods allow us to study not only the properties of the sample surface, but also to obtain and visualize information about surface structures, as well as to perform elemental analysis on micro-sections of the studied object [40, 41]. Resolution of scanning electron microscopes is significantly better compared to resolution of ordinary laboratory X-ray microscopes [40].

The scanning electron microscope MIRA3 LMH (TESCAN, Czech Republic) is equipped with an AztecLive Advanced Ultim Max 40 energy dispersive microanalysis system with a nitrogen-free detector (Oxford Instruments Analytical Ltd., England). This system have Silicon drift detection element with thermoelectric cooling (cascade of Peltier elements). The area of the detecting element is 40 mm^2. Energy resolution of 124 eV for the line of $MnK_{\alpha 1}$-line at a counting rate of 130,000 pulses/sec [36]. In addition, this scanning electron microscope are equipped with a secondary electron detector to obtain topographic contrast images (SE detector) with resolution of 1–2 nm at 30 kV and a motorized retractable backscattered electron detector (mR-BSE detector) to obtain compositional contrast images with 0.1Z atomic number resolution, and a motorized retractable detector for recording monochromatic cathodoluminescence (CL detector) in an extended wavelength range from 185 to 850 nm [36]. This scanning electron microscope have High Brightness Field Emission Schottky Cathode and is characterized by an increase from 2x to 1,000,000x. Schottky Cathode is the emitter, which is the single tungsten crystal coated with the layer of zirconium oxide (ZrO). It has the tip with a radius of curvature of several hundred nanometers. The ZrO coating significantly reduces the electron work function. Therefore, the large emission current can be obtained at a relatively low cathode temperature (about 1800 K). The negative voltage is applied to the pickoff electrode in order to screen the thermoelectrons emanating from the emitter [42].

The size of the elements distribution maps in the selected areas of the studied samples can be in 64, 128, 256, 512, 1024, 2048 and 4096 pixels, respectively. The accelerating voltage can vary from 200 V to 30 kV. The maximum scanning speed is 20 ns/pixel, and the minimum scanning speed is 10 ms/pixel. Software complex Aztec of this scanning electron microscope includes a control computer with an

IPS matrix (In-Plane Switching), as well as a LCD monitor with a diagonal of 27 inches and a resolution of 2560 × 1440 pixels [36].

- **SE Detectors (TESCAN)**: SE detectors represent the Everhart–Thornley secondary electron detector allowing selective identification of electrons with energies less than 50 eV. The Everhart–Thornley detector has a collector (Faraday cup), to which a positive potential (+250 to +300 V) is applied, due to which the trajectories of secondary electrons are curved, and they enter it. Inside the collector, the electrons fall on a scintillator, which is excited by applying a thin translucent metal layer to its outer surface and applying a positive voltage of about +12 kV accelerating low-energy electrons [43]. Additionally, the scintillator is surrounded by a Faraday cup to avoid, if possible, the influence of this potential on the incident beam electrons. By changing the voltage on the Faraday cup, it is possible to completely eliminate the influence of secondary electrons on the generated signal or achieve their most efficient use [44]. The light radiation generated by the scintillator material falls on the photocathode via a light guide, is converted into a photocurrent and then amplified by a photomultiplier, where it is converted into an electrical signal. The power of this signal and, consequently, the brightness of the corresponding point on the screen is determined by the topographic contrast [43]. Since the energy of secondary electrons is low, the electrostatic field between the sample and the collector grid strongly deflects their trajectories from the initial ones (after leaving the sample surface) towards the collector. This allows us to obtain an image of the sample areas located outside the direct movement of electrons to the collector (shadowless image) [44]. Elastically scattered electrons, which have much greater energy, change their trajectory little under the action of the same electrostatic field, moving along a straight line toward the collector. This causes a characteristic shadow image of the surface relief of the sample under study. Electrons scattered from areas closed from the collector by some obstacle do not reach the collector [44].
- **BSE Detectors (TESCAN)**: BSE detectors represent the detectors on based of synthetic highly sensitive YAG crystal (yttrium aluminium garnet) [45]. The ring-shaped YAG monocrystal with a conductive coating is placed on the optical axis of the microscope under the objective in order to collect backscattered electrons effectively. Backscattered electrons have high energy and excite the scintillator atoms without additional acceleration causing the generation of photons in the optical range. These photons are collected by the optical system, they are transmitted via a light guide to the photoelectron multiplier at the next stage.

The emission of reflected electrons is related to the average atomic number of the chemical element of the sample, so the image carries information about the distribution of various elements in the sample. Thus, on the surface of the sample, an area with a higher average atomic number reflects a greater number of electrons and appears lighter on the screen relative to other areas. The resulting contrast is called compositional [43].

The images in mode of backscattered electrons and element distribution maps were obtained when studying of the thin section of calcite marble using scanning electron microscope MIRA3 LMH TESCAN (A.P. Vinogradov Institute of Geochemistry, Siberian Branch of the Russian Academy of Sciences, Irkutsk, Russia). To remove the electrostatic charge, the thin section of calcite marble was sprayed with a thin carbon film on a Q150R ES vacuum unit (Quorum, Technologies). The thickness of this film was 20–30 nm. EDS-analysis was performed by collecting spectra by points on separate areas of grains. Basically, these points were located in the central part of mineral phases. The analysis areas were selected in the backscattered electron mode. The analysis was carried out at magnifications in the range from 595x to 21,970x. Inclusions of barite grains ($BaSO_4$) were identified in this calcite marble (Fig. 3.4).

White (bright) areas are barite grains. The analysis was performed using accelerating voltage of 20 kV, absorbed current of 4.6 nA, and beam diameter of 100 nm.

In order to confirm the presence of barite in the surrounding calcite matrix, X-ray mapping of the characteristic radiation of the determined elements was carried out in addition to the EDS- analysis (Fig. 3.5).

X-ray mapping was carried out using resolution 1024 pixels, accelerating voltage of 20 kV, absorbed current of 4.6 nA, and beam diameter of 100 nm. The mapping time for one element did not exceed 2 min.

- **CL Detectors (TESCAN)**: CL detectors are cathodoluminescence detectors. Cathodoluminescence is a type of luminescence that is excited by a primary

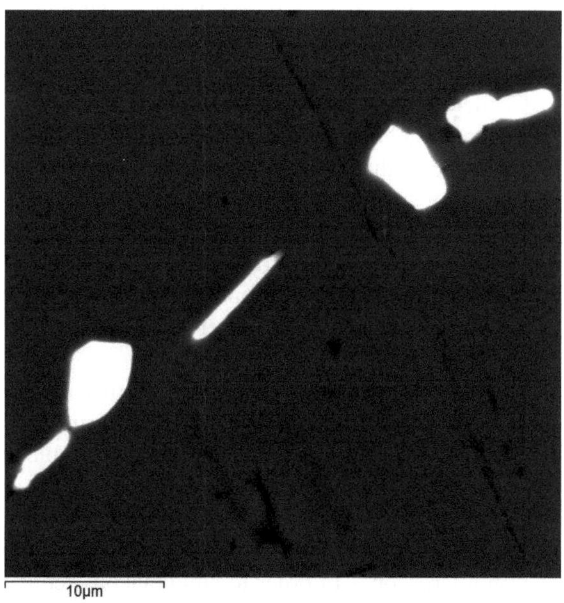

Fig. 3.4 The image in mode of backscattered electrons of barite grains ($BaSO_4$) in calcite marble

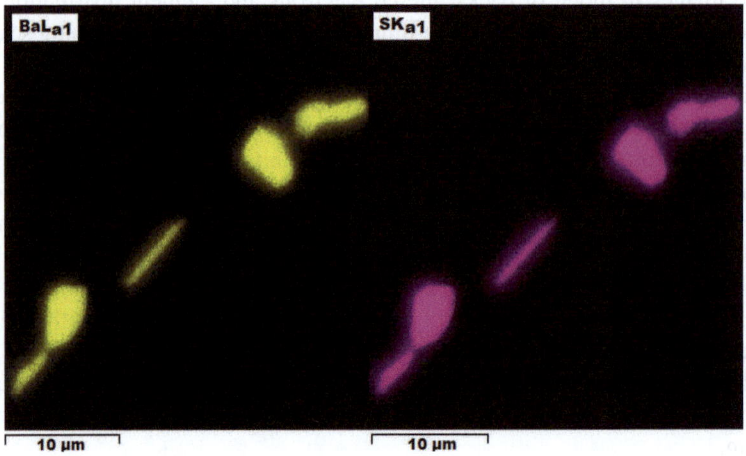

Fig. 3.5 Ba and S distribution maps in barite grains in calcite marble

electron beam and occurs as a result of radiative recombination of electron–hole pairs or internal transitions in phosphors. A number of materials generate visible spectrum radiation when irradiated with high-energy electrons. Such materials include semiconductors and many minerals (quartz, enstatite, alkali feldspars, plagioclases) [46].

Panchromatic CL detector allow to obtain black and white images, in which the brightness of the areas is proportional to the total intensity of CL-radiation from this area. Color CL detectors preserve the real color gamut of the emitted area [46].

CL detectors are interesting for studying materials because CL images potentially contain a huge amount of information (about micro-impurities, about crystal lattice defects) that is inaccessible by other methods. This sensitivity of the CL signal to the smallest inhomogeneities of matter is explained by the fact that the outer valence electron shells participate in the generation of the CL- radiation, and the inner electron shells participate in the generation of the X-ray radiation [44]. The energy of valence electrons depends on the defects of the crystal lattice and on impurity atoms, while the inner shells are not subject to change due to chemical bonds [46].

Excellent sensitivity of new CL detectors maintains high signal to noise ratio even at low accelerating voltages, which minimizes sample radiation damage. The small electron-specimen interaction volume enables the acquisition of images with the outstanding resolution [47].

- **Beam Deceleration Technology**: This is implemented in modern scanning electron microscopes in order to simultaneous reduce of electron beam energy and improve resolution and contrast [48]. Scanning electron microscopes (TESCAN) are equipped with High Brightness Field Emission Schottky Cathode and an immersion lens [36]. These elements are required for Low voltage microscopy [48]. The phenomenon of Schottky thermal emission (evaporation of electrons into a vacuum) is used in a Schottky cathode gun. This phenomenon occurs when heated conductors are placed in a high-voltage electric field. The advantage of this gun is the stability of the electron beam current, since the emitter placed in ultra-high vacuum conditions with a pressure of 10^{-7} Pa is heated to the high temperature without gas absorption [42].

The sharply focused electron beam is used in scanning electron microscopes to probe the sample surface. Such a probe is formed by electrons, the energy of which is determined by the electron potential of the electron source (accelerating voltage). Electrons leave the column with an energy corresponding to the accelerating voltage, which is usually within a range of 200 V to 30 kV for most scanning electron microscopes. The operator can choose appropriate energy of the electron beam for research work [48]. A high negative bias voltage is applied to sample, and the electrons leaving the column are slowed down afore they reach the sample. "Landing energy" (touchdown energy) describe the energy of the slowed down electrons when they hit the sample. Value of this parameter can be calculated by the formula from work [48]:

$$LE = -e(HV - SB) \tag{3.6}$$

where LE is touchdown energy of the probing electrons (eV), e is electron charge, HV is the accelerating voltage (negative), and SB is sample bias voltage (negative).

The beam deceleration application has three important advantages for imaging with the scanning electron microscope. First advantage is that electrostatic lens (cathode lens) created by an electric field has a great influence on the electron optical properties of the system [48]. Second advantage is efficiency improvement of signal collection due to higher energy of the signal electrons when they reach the detectors, as well as their changed trajectories [48]. Third advantage is that there is no requirement for a separate system in order to implement of imaging with very low touchdown energies [48].

The primary electron beam becomes even more focused during deceleration, because this electron beam is already being focused by the objective lens as it leaves the electron optical column. Besides, the slowing down electric field performs as collecting lens with additional immersion (cathode lens) [48].

The cathode lens strength for a specified geometrical configuration is characterized by only the "immersion ratio" (k), which is the ratio between primary beam energy and touchdown energy. This parameter can be calculated by the formula from work [48]:

$$k = -eHV/LE = HV/(HV - SB) \tag{3.7}$$

This lens becomes the stronger with increasing of immersion ratio value. Spherical and chromatic aberrations are very important properties of cathode lens in terms of resolution [48]. It should be noted that values of coefficients of spherical and chromatic aberrations decrease significantly when immersion ratio value increases. The coefficient of chromatic aberration has a great influence on resolution at low touchdown energies. It is practically inversely proportional to the immersion ratio. It results to limited equalization of the beam diameter degradation at low beam energy observed in the cases when beam slowing down is not applied [48].

In order to implement of beam deceleration mode several requirements must be fulfilled. The user safety is one of the most important requirements. The upper part of the microscope stage must be well insulated and connected to a high voltage source to apply several kilovolts to the sample [48]. Besides, the microscope stage should not be charged when the operator is in interaction with the sample. It must be guaranteed by the safety interlock system. In addition, the biased sample touches to different parts of the chamber should not pose any danger to the operator and the system. The sample holder at the upper part of the sample manipulator must be well insulated to withstand a high voltage of several kilovolts. This is the difficult task, because there is requirement for the applied voltage. It must be very stable [48]. The implementation of beam deceleration mode is totally incompatible with low vacuum conditions. Fluctuation of the sample voltage and loss of image quality occur due to the appearance of different micro-discharges and leakage currents. Besides, the microscope stage isolation is another difficult task for magnetic immersion lens columns due to the generation of the strong crossed electrical and magnetic fields. These fields are capable of creating of conditions for discharge in residual gases even in high vacuum mode in the chamber at short operating distances. The integration of the beam deceleration mode into the software controlling the optical elements of the electron column presents certain difficulties. The cathode lens is an additional flexible lens that operates at different combination of high voltages, sample bias voltages, and operating distances. Therefore, it is necessary to ensure that the operating distance reading and magnification precision are not degraded for different combinations [48]. In addition, there is the need to develop the system of calibration and alignment procedures. Nevertheless, improvement of the system reliability is an additional benefit of using of beam deceleration. So, the column is operated in mode of the increased high voltage in order to achieve certain touchdown energy with beam deceleration. Mode of the increased high voltage provides lower sensitivity to the external magnetic field and greater resistance to charging of different parts of column and the aperture's quality [48].

Beam deceleration is useful for many different types of scanning electron microscope systems to improve resolution and contrast, and the detection limit. Besides, this technique can be used to improve usability at low touchdown energy. Beam deceleration provides the best capabilities in low-energy imaging for immersion lens systems. In addition, the productivity gains for low-energy imaging can be proportionately higher. This significantly expands the imaging productivity and flexibility of these systems for scanning electron microscopes based on field emission gun (FEG) and thermionic emitters with regular lenses [48].

3.6 Detectors in X-Ray Imaging Systems

Various digital detectors can be divided according to the detection type into a group of indirect conversion detectors and a group of direct conversion detectors. Scintillators are used as an intermediate link in indirect conversion detectors. The scintillator absorbs high-energy quanta and emits in the low-energy, visible range. The properties of the scintillator largely determine the properties of the detecting system as a whole. X-ray quanta are recorded directly in direct conversion detectors [49].

Flat panel (linear) detectors with a "scintillator-semiconductor" structure are classified as indirect conversion detectors. Amorphous silicon (aSi) panels are most commonly used as indirect conversion flat panel detectors. The most common scintillators are made from gadolinium oxysulfide (Gd_2O_2S) and cesium iodide (CsI) in these detectors [49].

X-ray quanta first interact with a scintillator to form photons of light, and the light is converted either directly into an electrical signal, as in solid-state flat-panel detectors, or into a stream of electrons in the X-ray electron-optical converter, which creates a visible image on the output phosphor screen [49]. X-ray electron-optical converters with the electron-optical image wrapping system and X-ray electron-optical converters with the electrostatic direct transfer system constitute the large class of primary converters for various types of X-ray image intensifiers for medical and industrial purposes [18]. Their operating principle is based on the primary conversion of the X-ray radiation flow in the phosphor or scintillator layer of the device's input window into an adequate light flow, which is transformed by the X-ray electron-optical converter photocathode into a corresponding flow of photoelectrons. This flow of photoelectrons is accelerated by the electrostatic field of the electron-optical system. Both types of X-ray electron-optical converters are electric vacuum devices. The operating principle of the X-ray electron-optical converters with an electrostatic direct transfer system and with image wrapping is identical, but the X-ray electron-optical converters with image wrapping contain a number of additional electrodes that form an electron-optical system, which is an electrostatic lens for compressing the original electron image and flipping (image reversal) it on the output window. X-ray electron-optical converters are produced by many companies (for example, Siemens (Germany), Thomson (France), Toshiba (Japan)) [18].

Early experiments with aSi showed high noise levels in the above mentioned panels. Advances in technology, coupled with software capabilities that allow averaging over many frames, have dramatically improved the signal-to-noise ratio. The image quality now produced by these panels exceeds that of memory wafers and approaches that of amorphous selenium (aSe) panels. Moreover, aSi panels are less sensitive to environmental conditions. This makes them suitable for use under field conditions and in uncontrolled applications [49].

The sensitivity increases with increasing scintillator bulk, but the resolution decreases due to scattering of visible radiation quanta in the scintillator material. The bulk of the scintillation layer is a kind of compromise between sensitivity and spatial resolution of the detector. An increase in the bulk of the scintillator is required with an increase in the energy of X-ray quanta, which is possible up to the certain limit [49].

Gd_2O_2S is used at relatively low X-ray radiation energy (up to 80 keV). CsI used for higher energies of X-ray radiation (up to 300 keV). The peculiarity of this scintillator lies in its needle-shaped structure. Light quanta move along the needles practically without scattering, as in optical fiber. This feature of CsI crystals allows to significantly increase the bulk of the scintillation layer without loss of the detector's energy resolution [49].

Flat panel (linear) detectors on the basis of amorphous semiconductors (aSi, aSe), monocrystalline semiconductors (Si, GaAs, InP, CdZnTe, CdTe), and polycrystalline semiconductors (HgI_2, PbI_2) are classified as direct conversion detectors [49].

Electron–hole pairs are generated immediately upon impact of an X-ray quantum in the bulk of a semiconductor in direct conversion detectors. Electron current is generated by the applied voltage, which can be amplified and converted into an image by the reading electronics [49].

Panels on the basis of amorphous selenium (aSe) and monocrystalline cadmium telluride (CdTe) are used in most cases. In the first case, an assembly of thin-film transistors (TFT) is coated with aSe. This allows to convert the X-rays radiation directly into a digital signal without the use of scintillators. The image quality reaches the quality typical of medium-grain film due to the absence of scattering influence and an optimal signal-to-noise ratio. The limitation of the selenium detector is its narrow operating temperature range. To avoid destruction of the selenium layer, the detector must be maintained in the temperature range of 5–30 °C both for operation and during storage. In addition, selenium is prone to ghost imaging at high energies (>180 keV). All these limitations make it possible to use aSe panels in very specific applications where the operating conditions required for these panels can be strictly controlled [49].

The CdTe-based direct conversion detector system is free from a number of disadvantages of the aSe-based system. It can operate in a wider temperature range at energies of X-ray quanta up to 300 keV and has greater sensitivity [49].

3.6.1 Coupled Charge Device, Thin Film Transistors and Complementary Metal Oxide Semiconductor Technologies for Image Acquisition

3.6.1.1 Coupled Charge Device-Technology (CCD)

Willard S. Boyle (1924–2011) and George E. Smith invented the charge coupled device (CCD) [50] for imaging and signal processing purposes in 1969. The transfer of signal charges in a CCD was performed by rapidly changing voltages of isolated shift electrodes on the surface of the CCD to create time-variant potential wells inside the semiconductor [29].

CCD (coupled charge device) technology is one of the leading technologies for obtaining high-quality X-ray images [49, 51]. The signal is read by transporting charge packets along the silicon substrate in CCD-panels, unlike matrix panels such as TFT (Thin Film Transistors) and CMOS (Complementary Metal Oxide Semiconductor). The advantage of such panels is their low noise and high sensitivity. Fiber-optic buffer inserts are usually used in CCD panels. These inserts allow increasing the area of the panel's receiving window and simultaneously serve as radiation protection. CCD-panels can only operate at X-ray quantum energies not exceeding 50 keV, even with such additional shielding. Their production requires special conditions, as in the case of amorphous silicon [49]. It should be noted that CCD-panels are used in conjunction with X-ray electron-optical converters [18] in order to provide the good signal-to-noise ratio and the high spatial resolution [52].

The thin layer of dielectric is located on a semiconductor surface (for example, silicon oxide), on which strips of conducting electrodes (made of metal or polycrystalline silicon) are located in the CCD-matrix. These electrodes form the regular matrix system. The operating principle of the CCD-matrix is based on the generation, storage and directed transfer of charge packets in potential wells [53] formed in the subsurface layer of the semiconductor when external electrical voltages are applied to the electrodes.

CCD detectors are susceptible to visible photons, X-rays, and electrons. Above mentioned of sources of incident radiation can create a sufficient number of electron–hole pairs. Required energy is 3.6 eV in order to create an electron–hole pair in silicon, while the energy released by electrons used under conditions of standard imaging in electron microscope (100 keV or greater) is more than sufficient to produce a quite high signal [54].

Setting up the appropriate voltages on the gates is one of the main functions of the control electronics of the CCD detector. Such setting up provides the image to be integrated during the exposure part of a cycle. Besides, it allows to read out and to measure the accumulated charge signal from the individual pixels at the end of the exposure. There is the need for the preset sequence of clock pulses depending on the number of phases, the number of pixels per row and the number of columns for the readout the image from this detector [55]. CCD manufacturers typically provide recommendations relatively the size and shape of the clock pulses, which differ for different devices. The special low-noise circuits are necessary to use for measurement of the charge, which was transferred to the output sense node of the

CCD detector. These low-noise circuits are capable to amplifying the signal and minimizing the noise [54].

Direct detection of electrons is not yet available as the practical option for most applications. The signal created by electron with energy of even 100 keV is so large that it will be able to fill a significant part of the pixel well capacity (taking into account that all the energy is deposited in a single pixel) and will reduce the dynamic range of the measurements to very low levels in CCD detectors. It should be noted that the silicon device use as the counter of photon in hybrid pixel detectors, which record one count for each incident electron rather than store the charge in analog form in contrast to CCD detectors. The level of radiation damage can be very high due to the energy released by electrons in the front surface, which contains the polysilicon gates used to supply the voltages required for the readout process. Investigations of the damage to CCD detectors by photons of different energies and gamma rays show that the charge transfer efficiency is reduced, and the lifetime of the device can be reduced by only a few hours due to charge trapping at the interface between silicon and silicon dioxide [56]. CCD detectors with back-side illuminated can receive somewhat less radiation damage, because the polysilicon gates are on the reverse side. Before reaching the gates, the electrons must first pass through the bulk of silicon. However, the degree of damage would still be very high from electrons which cross the CCD thickness. A number of X-ray quanta that occur further up in the column of the microscope will be registered as very high (and false) events in the CCD. This problem is largely solved in CCD detectors where scintillators (phosphors) are applied. Scintillators (phosphors) convert the incident energy of the electrons into photons. Most X-ray photons are absorbed by the fiber-optics assembly that connects the phosphor to the CCD [54].

The main advantage of pixel detectors is the possibility to obtain of better spatial resolution in contrast to CCD detectors. In addition, pixel detectors are characterized by much faster readout times. The dynamic range is determined by the memory size in digital detectors, not by the pixel cell size as in CCD detectors [54]. Authors of the work [57], who carried out research using an 8¬8 pixel detector developed originally for protein crystallography, believe that problems associated with X-ray induced noise signals should be less problematic, because they can only be recorded as a single count. It was found that the charge distribution between adjacent pixels (the size of which were 150 μm^2) becomes significant only at energies above 200 keV due to the greater spread in the generated charge by the falling electron, as predicted by Monte Carlo simulations [54].

3.6.1.2 Thin Film Transistors Technology (TFT)

Since TFT belongs to a category of MOSFET (field effect transistor with the "metal-oxide semiconductor" structure), the early development history of TFT is also a part of the history of MOSFET. In 1957, John T. Wallmark (1919–2007) invented the thin film MOSFET where the materials of gate dielectric were made from germanium oxide. In 1962, Paul K. Weimer (1914–2005) developed the world's first TFT. The material in the channel of the very first TFT is a thin film made of polycrystalline CdS [58, 59]. In 1979, Peter G. Le Comber with employees manufactured the first hydrogenated amorphous silicon TFT [59].

TFT-technology is used mainly in panels based on amorphous silicon or selenium. Indirect conversion detector panels are typical photodiodes assemblies on an active TFT matrix. This technology is used in modern televisions, monitors, phone displays and other devices. The detector layer is located directly on this matrix for direct conversion detectors. The TFT transistor matrix is a fairly simple electronic switch structure [49, 60].

The capacitance of each light-emitting diode is initially charged to a voltage (U+). When exposed to light, it is discharged (the more light, the less residual charge). When reading the TFT, the transistor opens and the capacitance is charged again through the readout amplifier to a value (U+). The charge value is measured by the readout amplifier. It is fed to the analog-to-digital converter in the form of voltage (U) [49, 60].

The panels can be very large (40×40 cm^2). They do not have a very high resolution (typical pixel size is 127 μm). A distinctive feature of amorphous silicon is its very high radiation resistance, which allows such panels to be used at high energies of X-ray quanta more than 10 MeV [49].

There are constant efforts in searching for new types of TFTs to improve performance or to lower the production costs of Si-based TFTs [58]. Authors of work [61] provide materials on the recent advancements in Sn Oxide-Based Thin-Film Transistors for Large-Area electronics. Besides, the source of the dual conductivity of Sn oxide is addressed, which will be helpful in designing complementary metal oxide semiconductor devices [61].

3.6.1.3 Complementary Metal Oxide Semiconductor Technology (CMOS)

Patent for the invention of the metal–oxide–semiconductor field-effect transistor of John Bardeen (1908–1991) and inversion layer concept form the basis of CMOS-technology [62]. This technology is the matrix assembly, like TFT-technology. CMOS-panels are made from crystalline silicon. The main advantage of CMOS-technology over CCD- and TFT- technologies is the use of a well-developed industrial base for the production of integrated circuits. The noise level of CMOS panels has approached the level of CCD-panels, the dynamic range is higher by several times. The highest spatial resolution (pixel size does not exceed 10 μm) is achieved on CMOS-panels [49]. CMOS panels have a higher image formation speed compared to TFT-detectors. At the same time, the number of detecting elements in large-size matrices is comparable.

Advantages of the CMOS imagers include the higher readout speed and lower noise due to the much higher electrical charge mobility in crystalline silicon compared to amorphous silicon. The lower noise provides a wider dynamic range in CMOS even when the total pixel storage capacitance is the same as in a-Si. In addition, because the noise floor is lower, the low dose Detective Quantum Efficiency (DQE) is significantly higher and the X-ray detection is quantum noise limited down to very low dose levels. It is very important when using CMOS-panels in computer tomography units in medicine [63].

However, creating the same large detecting elements as in TFT-technology was difficult. Therefore, CMOS-panels was inferior in the maximum size of the detecting area [49], but CMOS-technology is continuously evolving.

While aSi TFT-technology is versatile and manufacturable in large areas, this technology is characterized by the high read noise from combined pixel and line noise sources (more than 1000 electrons) [64, 65], pixel-pitch limitations, and resistance to lowering of manufacturing costs. CCD technology, while offering excellent low-noise performance, does not economically scale to larger imaging formats [65]. The very large area (VLA) CMOS active-pixel-sensor (APS) technology was developed for digital radiography applications to eliminate the limitations of currently established technologies. The large active areas of the detector required to obtain a wide-format X-ray image are formed by the active areas of several VLA CMOS sensors connected by their edges to form a large sensor panel. As a result, the connection is obtained in the form of a tiled structure [65]. The three faces of the active area of the CMOS-sensor with active pixel can be mated with the faces of similar VLA CMOS sensors to create a large X-ray detector panel due to the tiled structure. The pixel pitch is 96 μm, the active area size is 49.1 × 98.3 mm. The signal processing circuit integrated on the CMOS-crystal includes bias circuits, control logic, clock pulse generators, an address line register, amplifiers, and a vertical line register. Connecting several sensors into one large detector requires placing the address line register in the center of the matrix, where it occupies an active area equal to 1.5 columns. Pixels adjacent to the address register are scaled to maintain a whole pixel interval. The active pixel area is located as close as possible to the chip edge (scribe channel) on each of the three sides of the sensor. The gap between the outermost pixels of adjacent sensors usually occupies 1.0 to 1.5 pixels, which can be compensated for by scaling the outermost pixels to a narrower width [65].

Fiber-optic faceplate are used in X-ray detector panels assembled from several VLA CMOS sensors connected to each other by the edges of the active areas. The fiber-optic faceplate is the effective absorber of X-ray radiation. They protects the CMOS-sensor from direct absorption of X-ray photons, reducing noise levels and increasing quantum efficiency at high spatial frequencies [65].

The active-pixel CMOS-sensor operates in the visible wavelength range. The value of the DQE is $47\% \pm 3\%$ at 550 nm wavelength. The depth of the p–n junction and depletion layer of the CMOS-photodiode is sufficient to capture the significant amount of carrier electrons generated by 550 nm wavelength photons emitted by the scintillator. The value of the DQE also depends on the internal structure of the photodiode, which limits the area of the active pixel. Since the open aperture of the pixel is more than 85% of the total area, there is no need to use microlenses [65].

Large-tiled CMOS-radiographic panels can compete in performance with passive-pixel arrays, aSi thin-film-transistor panels, and phosphor-panel technologies [65]. Although, CMOS sensors have become the key technology in low-cost consumer camera products, CMOS APS technology is also suited for manufacture of large-format imagers used to construct radiographic-detector panels. Large-area CMOS radiographic sensors combine a large full well over 3.5 million electron with low read noise less than 300 electron to provide wide dynamic range and improved signal-to-noise ratio under demanding radiographic imaging conditions. Applications include nondestructive testing, scientific imaging, security screening, and medical radiography [65].

Hybrid pixel detectors made by direct converting high-Z semi-insulating single crystalline material coupled to complementary-metal-oxide semiconductor (CMOS) readout electronics are described in detail in the work [66]. The main advantage of these detectors is that the direct conversion exterminates all the problems of spatial localization related to light diffusion, energy resolution, is far superior from the combination of scintillation crystals and photomultipliers, and lithography can be used to pattern electrodes with very fine pitch. Authors of work [66] are developing 2-D pixel CMOS ASICs (application specific integrated circuits), connect them to pixilated CdTe crystals with the flip chip and bump bonding method and characterize the hybrids. The detectors are targeting Compton scatter imaging and they can be used for coded aperture imaging too. Hybridization using CMOS-technology can overcome the limit put on pixel circuit complexity by the use of thin film transistors (TFT) in large flat panels. Hybrid active pixel sensors are used in dental imaging, industrial computed tomography (CT) and other applications [66].

The readout channel for applications to X-ray diffraction imaging at free electron lasers (FELs) was developed with using a 65 nm CMOS-technology [67]. This readout channel includes a charge preamplifier with dynamic signal compression, trapezoidal shaper and a 10 bit time-inter leaved SAR ADC (successive approximation register (SAR) based analog-to-digital converter (ADC)). Measurements performed on the single blocks and on the full readout channel provided some very encouraging results. The principle of signal compression based on the non-linear features of MOS-capacitors (capacitors with the "metal–oxide–semiconductor" structure) was proven. Operation in a time-inter leaved fashion of a low power, small area10 bit SAR ADC for in-pixel data conversion was also demonstrated [67].

The obvious superiority of the CMOS-detectors and the continuous development of the CMOS-technology allow us to assume their widest application in the future in the field of scientific research, in industry, and in medicine.

Acknowledgements The author gratefully acknowledge Anatoliy Grigoryevich Revenko DSc (Engineering Science) for the help and support provided. I thank Olga Yuryevna Belozerova PhD (Chemical Sciences) and Maria Nikolaevna Rubtsova for helping in research using the scanning electron microscope MIRA3 LMH TESCAN.

The equipment of the Center for Collective Use "Geodynamics and Geochronology" of the Institute of the Earth's Crust, SB RAS was used in this work within the framework of the grant No.

075-15-2021-682. In addition, equipment of the Center for Collective Use "Isotope-Geochemical Research" of the A. P. Vinogradov Institute of Geochemistry, SB RAS was used.

References

1. Fetisov, G.V.: Synchrotron Radiation. Methods of Studying the Structure of Substances. Moscow: FIZMATLIT (in Russian) (2007)
2. Revenko, A.G.: X-ray Fluorescence Analysis of natural materials. Nauka, Novosibirsk (in Russian) (1994)
3. Afonin, V.P., Komyak, N.I., Nikolaev, V.P., Plotnikov, R.I.: X-ray Fluorescence Analysis. Nauka, Novosibirsk (in Russian) (1991)
4. Beckhoff, B., Kanngier, B., Langhoff, N., Wedell, R., Wolff, H.: Handbook of Practical X-ray Fluorescence Analysis. Springer, Berlin Heidelberg, Berlin (2006)
5. Schlosser, D.M., Lechner, P., Lutz, G., Niculae, A., Soltau, H., Strüder, L., Eckhardt, R., Hermenau, K., Schaller, G., Schopper, F., Jaritschin, O., Liebel, A., Simsek, A., Fiorini, C., Longoni, A.: Expanding the detection efficiency of silicon drift detectors. Nucl. Instrum. Methods Phys. Res. **624**, 270–276 (2010)
6. Mamikonyan, S.V.: Equipment and methods of fluorescent X-ray radiometric analysis. Atomizdat (in Russian) (1977)
7. Boyko, VI, Zherin, II, Karataev, V.D., Nedbaylo, YuV, Silaev, M.E.: Methods and devices for measuring nuclear and other radioactive materials. Publishing House of Tomsk Polytechnic University (in Russian) (2011)
8. Thompson, A.C., Vaughan, D.: X-Ray Data Booklet, 2nd edn. University of California Berkeley, LBNL (2001)
9. Zolochevskiy, A.A., Goncharova, G.V., Borodenko, Y.A., Kozmin, Y.S.: Degradation modeling for scintillation detectors. J. National Tech. Univ. Kharkov Polytechnic Institute **9**: 17–28 (in Russian) (2008)
10. Schlotz, R., Uhlig, S.: Introduction to X-ray Fluorescence Analysis. Bruker AXS GmbH, Karlruhe (2006)
11. Service Manual: S8 TIGER XRF Spectrometer. Bruker AXS, Berlin (2007)
12. Alekseev, G.D., Kruglov, V.V., Khazins, D.M.: Self-quenching streamer (SQS) discharge in a wire chamber. Phys. Element. Particles Atomic Nuclei. **13**, 703–748 (in Russian) (1982)
13. Ballon, J., Comparat, V., Pouxe, J.: The blade chamber: a solution for curved gaseous detectors. Nucl. Instr. Method **217**, 213–216 (1983)
14. Stahl, K., Thomasson, R.: Using CPS120 (curved position-sensitive detector covering 120°) powder diffraction data in Rietveld analysis. The dehydration process in the zeolite thomsonite. J. Appl. Cryst. **25**, 251–258 (1992)
15. Masson, O., Boulle, A., Guinebretière, R., Lecomte, A., Dauger, A.: On the use of one-dimensional position sensitive detector for x-ray diffraction reciprocal space mapping: data quality and limitations. Rev. Sci. Instrum. **76**(6) 063912, 1–7 (2005)
16. Aulchenko, V.M., Baru, S.E., Dubrovin, M.S., Savinov, G.A., Shekhtman, L.I., Titov, V.M., Velikzhanin, Yu.S., Ancharov, A.I., Gaponov, Yu.A., Evdokov, O.V., Tolochko, B.P.: One- and two-coordinate detectors in BINP. J. Synchrotron Radiat. **5**(3), 263–267 (1998)
17. Aulchenko, V.M., Baru, S.E., Dubrovin, M.S., Titov, V.M., Velikzhanin, J.S., Usov, J.V.: The OD-3 fast one-coordinate X-ray detector. Nucl. Instr. Meth. Phys. Res. A. **367**, 79–82 (1995)
18. Kornev, E.A., Lelyukhin, A.S.: Microstructural Detectors of X-ray Information-Measuring Systems. Publishing House of Orenburg State University (in Russian) (2018)
19. Chistokhin, I.B., Fritzler, K.B.: Effect of conditions of getter formation in high-resistance silicon on the characteristics of PIN photodiodes. Lett. JTPh. **46**(21), 11–13 (2020) (in Russian)
20. Goulding, F.S., Jaklevich, J.M., Jarrett, B.V., Landis, D.A.: Detector background and sensitivity of semiconductor X-ray fluorescence spectrometers. Adv. X-ray Anal. **15**, 470–482 (1972)

21. Shtel'makh, S.I., Revenko, A.G.: Determination of Mo, Nb, Zr, Y, Sr, and Rb contents in rocks using X-Ray fluorescence analysis. In: Proceedings of Conference on X-Ray Analysis, UlaanBaatar, Mongolia, pp. 89–96 (in Russian) (2006)
22. Zaletin, V.M.: Development of semiconductor detectors based on wide-band materials. At. Energ. **97**(5), 362–370 (in Russian) (2004)
23. Matveev, O.A.: Cadmium telluride and nuclear radiation detectors based on it. Abstract of dis. for the Degree of Doctor of Technical Sciences. Leningrad, PhTI of the Academy of Sciences named after A.F. Ioffe (in Russian) (1977)
24. Zaletin, V.M., Mel'nikov, A.A., Nozhkina, I.M., Fomin, V.I.: Uncooled semiconductor detectors based on crystals of wide-gap compounds grown from the gas phase. At. Energ. **86**(5), 391–397 (in Russian) (1999)
25. Hitomi, K., Muroi, O., Shoji, T., Suehiro, T., Hiratate, Y.: Room temperature X- and gamma-ray detectors using thallium bromide crystals. Nucl. Instrum. Meth. A. **436**, 160–164 (1999)
26. Khusainov, A., Arlt, R., Siffert, P.: Performance of a high resolution CdTe and CdZnTe p-i-n-detectors. Nucl. Instrum. Meth. A. **380**, 245–251 (1996)
27. Bogart, T.F., Beasley, Jr. S., Rico, G.: Electronic devices and circuits, sixth edition, Prentice Hall International Inc. (2004)
28. Crichton, R.: DuraBeryllium plus X-ray windows: advantages of use in aggressive chemical environments. Analytics **2**(9), 60–65 (in Russian) (2013)
29. Strüder, L., Niculae, A., Holl, P., Soltau, H.: Development of the silicon drift detector for electron microscopy applications. Micros. Today **28**(5), 46–53 (2020)
30. Lechner, P., Eckbauer, S., Hartmann, R., Krisch, S., Hauff, D., Richter, R., Soltau, H., Strüder, L., Fiorini, C., Gatti, E., Longoni, A., Sampietro, M.: Silicon drift detectors for high resolution room temperature X-ray spectroscopy Nucl. Instrum. Meth. A **377**, 346–351 (1996)
31. Gatti, E., Rehak, P.: Semiconductor drift chamber – an application of a novel charge transport scheme. Nucl. Instrum. Methods **225**, 608–614 (1984)
32. Chilingarov, A.G.: Coordinate semiconductor detectors in elementary particle physics. Phys. Element. Particl. Atomic Nuclei. **23**(3), 785–837 (in Russian) (1992)
33. Maltsev, A.S., Pashkova, G.V.: Application of Total-Reflection X-Ray Fluorescence Spectrometry (TXRF) to geological objects: experience of the TXRF laboratory, centre for geodynamics and geochronology. Geodynam. Tectonophysics. **13**(2), 0601 (2022) (in Russian)
34. Chubarov, V.M., Cherkashina, T.Y., Maltsev, A.S., Chuparina, E.V., Amosova, A.A., Prosekin, S.N.: Investigation of soils and pine needles using WDXRF and TXRF techniques for assessment of the environmental pollution of Shelekhov District, Eastern Siberia, by the Aluminum Industry and Heat Power Engineering. Agronomy **12**(2), 454 (2022). https://doi.org/10.3390/agronomy12020454
35. Shtel'makh, S.I., Zhilicheva, A.N., Vasil'eva, I.E.: Determination of macro- and microelements concentrations by wavelength-dispersive X-ray fluorescence spectrometry in the objects of the technogenic ecosystem. X-Ray Spectrom 1 (2024). https://doi.org/10.1002/xrs.3443
36. Skuzovatov, S.Y., Belozerova, O.Y., Vasil'eva, I.E., Zarubina, O.V., Kaneva, E.V., Sokolnikova, Y.V., Chubarov, V.M., Shabanova, E.V.: Centre of Isotopic and Geochemical Research (IGC SB RAS): current state of micro- and macroanalysis. Geodynam. Tectonophy. **13**(2), 0585 (2022). https://doi.org/10.5800/GT-2022-13-2-0585 (in Russian)
37. Bronk, H., Röhrs, S., Bjeoumikhov, A., Langhoff, N., Schmalz, J., Wedell, R., Gorny, H.E., Herold, A., Waldschläger, U.: ArtTAX – a new mobile spectrometer for energy-dispersive micro X-ray fluorescence spectrometry on art and archaeological objects. Fresenius J. Anal. Chem. **371**, 307–316 (2001)
38. Revenko, A.G., Revenko, V.A.: Application of X-ray spectral analysis method for studying cultural heritage materials. Methods Objects Chem. Anal. **2**(1), 4–29 (2007) (in Russian)
39. Shtel'makh, S.I., Pavlova, L.A., Ryashchenko, T.G.: Features of the mineral composition of the quaternary sediments of the development areas of the erosion processes on the territory of the Tunkinsk's depression. J. Geol. Res. Eng. **7**: 113–122 (2019)

40. Konnov, N.P., Volkov, Yu.P., Kuznetsov, O.S.: Modern technologies for examination of three-dimensional (3-D) ultra-fine structure and visualization of microorganisms. Prob. Esp. Dangerous Infect. **112**, 29–34 (in Russian) (2012)
41. Goldstein, J., Joy, D., Lifshin, E., Newbury, D., Fiori, C., Echlin, P.: Scanning Electron Microscopy and X-ray Microanalysis, vol. 1. Mir, Moscow (in Russian) (1984)
42. Lyapin, A.: Visualization of fine structures – modern technologies. Electronics **6**(00146), 146–152 (in Russian) (2015)
43. Zyablov, A.N.: Analytical probe microscopy. Voronezh: Publishing and Printing Center "Scientific Book" (in Russian) (2018)
44. Williams, D.B., Barry Carter, C.: Transmission electron microscopy. A Textbook for Materials Science. Second edition, Springer (2009)
45. Ishchenko, A.V., Cherepanov, A.N., Shul'gin, B.V., Vainstein, I.A., Vokhmintsev, A.S.: Low-dimensional detector materials for radiation physics devices. Ural Federal University, Ekaterinburg (in Russian) (2014)
46. Nesterina, E.M., Lukashova, M.V.: Latest instrumental techniques based on the use of a scanning electron microscope for mineralogical analysis: applications in forensic soil analysis. Theor. Practice Forensic Sci. **36**(4), 80–88 (2014)
47. Jiruse, J., Kolosova, J., Leichmann, J.: Cathodoluminescence detectors suitable for easy and fast analysis in geosciences. Microsc. Microanal. **19**(S2), 1346–1347 (2013)
48. Phifer, D., Tuma, L., Vystavel, T., Wandrol, P., Young, R.J.: Improving SEM imaging performance using beam deceleration. Microsc. Today **17**(4), 40–49 (2009)
49. Maiorov, A.A.: Digital technologies in the radiation inspection. In the world of non-destructive testing. **3**(45), 5–12 (in Russian) (2009)
50. Boyle, W.S., Smith, G.E.: Charge coupled semiconductor devices. Bell Syst. Tech. J. **49**, 587–593 (1970)
51. Tompsett, M.F., Amelio, G.F., Bertram, W.J., Buckley, R.R., McNamara, W.J., Mikkelsen, J.C., Sealer DA, D.A.: Charge-coupled imaging devices: experimental results. IEEE Trans. Electron Devices **18**(11), 992–996 (1971)
52. Masschaele, B.C., Cnudde, V., Dierick, M., Jacobs, P., Hoorebeke LVan, Vlassenbroeck J.: UGCT: New X-ray radiography and tomography facility. Nucl. Instrum. Methods Phys. Res. A **580**, 266–269 (2007)
53. Parshakov, A.N.: Lecture course on quantum physics. Publishing House of Perm State Technical University, Perm (in Russian) (2006)
54. Faruqi, A.R., Subramaniam, S.: CCD detectors in high-resolution biological electron microscopy. Q. Rev. Biophys. **33**(1), 1–27 (2000)
55. McLean, I.S.: Electronic and Computer-Aided Astronomy. Wiley (1989)
56. Roberts, P.T.E., Chapman, J.N., MacLeod, A.M.: A CCD-based recording system for CTEM. Ultramicroscopy **8**, 385–396 (1982)
57. Fan, G.Y., Datte, P., Beuville, E., Beche, J., Millaud, K., Downing, F., Burkard, M., Ellisman, M.H., Xuong, N.: ASIC based event-driven 2D digital electron counter for TEM imaging. Ultramicroscopy **70**, 107–113 (1998)
58. Kuo, Y.: Thin film transistor technology – past, present, and future. Electrochem. Soc. Interf. **22**(1), 55–61 (2013)
59. Lin, X.: Thin film transistor - basic principles and commercial status. Highlights Sci. Eng. Technol. **27**, 428–435 (2022)
60. Brotherton, S.D.: Introduction to Thin Film Transistors. Springer International Publishing, Switzerland (2013)
61. Jenifer, K., Arulkumar, S., Parthiban, S., Kwon, J.Y.: A review on the recent advancements in tin oxide-based thin-film transistors for large-area electronics. J. Electron. Mater. **49**(12), 7098–7111 (2020)
62. Howard, R.H.: John Bardeen and transistor physics. In: Proceedings of AIP Conference 550, Gaithersburg, Maryland, USA, pp. 3–32 (2001)

63. Zentai, G.: Comparison of CMOS and a-Si flat panel imagers for X-ray imaging. In: Proceedings of IEEE International Conference on Imaging Systems and Techniques (IST), pp. 194–200 (2011)
64. Weisfield, R.L., Bennett, N.R.: Electronic noise analysis of a 127 μm pixel TFT/photodiode array. Proc. SPIE Int. Soc. Opt. Eng. **10**(1117/12), 430955 (2001)
65. Farrier, M., Achterkirchen, T.G., Weckler, G.P., Mrozack, A.: Very large area CMOS active-pixel sensor for digital radiography. Photonics **38**(2), 88–102 (2013)
66. Hatzistratis, D., Theodoratos, G., Zografos, V., Kazas I., Loukas, D., Lambropoulos, C.P.: Hybrid pixel detectors for gamma/X-ray imaging. J. Phys. Conf. Series 637. https://doi.org/10.1088/1742-6596/637/1/012015
67. Ratti, L., Comotti, D., Fabris, L., Grassi, M., Lodola, L., Malcovati, P., Manghisoni, M., Reb, V., Traversi, G., Vacchi, C., Rizzo, G., Batignani, G., Bettarini, S., Casarosa, G., Forti, F., Giorgi, M., Morsani, F., Paladino, A., Paoloni, E., Pancheri, L., DallaBetta, G.F., Mendicino, R., Verzellesi, G., Xue, H., Benkechkache, M.A.: A 2D imager for X-ray FELs with a 65 nm CMOS readout based on per-pixel signal compression and 10 bit A/D conversion. Nucl. Instr. Meth. Phys. Res. A. **831**, 301–308 (2016)

4 Synchrotron X-Ray Fluorescence Imaging and Recent Developments

Katarina Vogel-Mikuš

Abstract

Synchrotron X-ray fluorescence imaging (SXRFI) is an advanced analytical technique that enables high-resolution, non-destructive imaging of element distribution in various materials. Recent advancements in SXRFI have significantly enhanced its capabilities, including improved spatial resolution through nanofocused beams, increased sensitivity for detecting trace elements, and the integration of multimodal imaging approaches. The application of machine learning and artificial intelligence in data processing is revolutionizing the analysis of SXRFI data, facilitating high-throughput studies and the extraction of quantitative information from complex datasets. This chapter provides an overview of the principles of SXRFI, its applications, and recent technological developments shaping its future in scientific research. After a short introduction to synchrotron radiation, facilities, and X-ray fluorescence techniques, the main principles and breakthroughs in the development of SXRFI are presented with some case studies from environmental research, biology, medicine and cultural heritage. An overview of the world's synchrotron beamlines offering SXRFI and Synchrotron X-ray fluorescence computer tomography (SXFCT) in 2024 is also provided to make it easier for users to choose the beamline that best suits their needs.

K. Vogel-Mikuš (✉)
Biotechnical Faculty, Department of Biology, University of Ljubljana, Ljubljana, Slovenia
e-mail: Katarina.VogelMikus@bf.uni-lj.si

Jozef Stefan Institute, Ljubljana, Slovenia

V. K. Singh (ed.), *X-Ray Fluorescence Spectroscopy and Chemometrics*,
https://doi.org/10.1007/978-3-031-98375-7_4

4.1 Introduction

Synchrotron X-ray fluorescence imaging (SXRFI) is an advanced analytical technique that enables high-resolution, non-destructive element distribution mapping in various materials. By utilizing the intense focused monochromatic X-ray beam generated by a synchrotron, SXRFI provides detailed spatial distributions of elements at micro- and nanoscale, making it invaluable in scientific fields such as material science, environmental science, biology, medicine, and cultural heritage.

Recent advancements in SXRFI have significantly enhanced its capabilities, including improved spatial resolution through nanofocused beams, increased sensitivity for detecting trace elements, and the integration of multimodal imaging approaches. The development of three-dimensional imaging techniques, such as X-ray fluorescence computed tomography (XFCT), has expanded the scope of SXRFI, allowing researchers to reconstruct complex chemical information in three dimensions and combine it with 3D structural information obtained with CT. XFCT is especially useful in life science studies, as it gives insights into the structure, physiological functions, and pathological processes of organisms without a need for demanding sample preparation.

Additionally, in situ and *in operando* studies have been made possible, enabling real-time observation of dynamic processes under controlled environmental conditions. The application of machine learning and artificial intelligence in data processing is revolutionizing the analysis of SXRFI data, facilitating high-throughput studies and the extraction of quantitative information from complex datasets.

This chapter provides an overview of the principles of SXRFI, its applications, and recent technological developments shaping its future in scientific research. After a short introduction to synchrotron radiation, facilities, and XRF techniques, the main principles and breakthroughs in the development of SXRFI are presented with some case studies from environmental research, biology, medicine and cultural heritage. An overview of the world's synchrotron beamlines offering SXRFI and SXFCT in 2024 is also provided to make it easier for users to choose the beamline that best suits their needs.

4.2 Synchrotron Light and Facilities

4.2.1 The Discovery of Synchrotron Radiation

The theoretical understanding of synchrotron radiation began in 1873 with Maxwell, who pointed out that changing charge densities and electric currents would produce electromagnetic fields that would radiate outward, headed for infinity. In 1887, Heinrich Hertz, a student of Helmholz, demonstrated such waves and thus laid the foundations for the theory of synchrotron radiation [1]. In 1898 Liénard first pointed out that an electric charge moving in a circular path should radiate energy and calculated the rate of radiation from the centripetal acceleration of an electron. The theory was subsequently extended by Schott, who received the Adams Prize in 1908 at Cambridge University for his essay »The Radiation from Electric Systems or Ions in Accelerated Motion and The Mechanical Reactions on their Motion which Arise from it«. Schott, attempting to provide the background for an electron theory of matter, calculated the amount and the angular distribution of radiation from relativistic electrons grouped in various ways in orbits of proposed atomic models [2].

Synchrotron light was detected and visualized only when it was possible to accelerate electrons at high (MeV) energies. This happened on April 24, 1947, at the General Electric Research Laboratory in Schenectady, New York, using a 70 MeV synchrotron. Blewet described this event as follows:»The machine was in operation and some components were being pushed to rather high operating levels. Some sparking had occurred, and a large mirror had been set up at the end of the shielding wall whose function was the protection of the control area from the synchrotron's intense output of X-rays. With this mirror, it was possible to view the vacuum chamber, which was transparent, not opaque like the one in the betatron. Floyd Haber, a machinist working with the synchrotron team, was detailed to watch the mirror. He signalled immediately to turn the machine off because he saw an arc. More careful observation made it clear to Pollock and Langmuir, who were operating the machine, that this was indeed the predicted radiation« [1]. Although synchrotron radiation had been unknowingly observed in astronomical phenomena for centuries, its laboratory detection marked a significant milestone in science [2].

4.2.2 Properties of Synchrotron Light

The main properties of synchrotron radiation are the following: high intensity; vast and continuous spectral range from infrared up to the hard X-ray region; natural narrow angular collimation; high degree of polarization; pulsed time structure; high brightness of the source due to small cross-section of the electron beam and a high degree of collimation of the radiation; ultra-high vacuum environment and high beam stability; all properties quantitatively evaluable [3, 4].

Synchrotron light, by its properties, is invaluable for various research fields and technology since it offers significant advantages over conventional light sources, particularly in terms of brightness. Third-generation synchrotron sources provide a trillion-fold increase in brightness compared to laboratory X-ray sources, allowing for improved spatial and temporal resolution in X-ray scattering and spectroscopy [5]. In the mid-infrared range, synchrotron sources demonstrate at least two orders of magnitude greater brightness than traditional globar sources, making them ideal for FTIR micro spectroscopic studies [6].

4.2.3 Modern Synchrotron Facilities

The basic principle of synchrotron operation involves the electrons or positrons that enter the storage ring after they have been accelerated by a linear accelerator until their energy reaches several millions of electron volts (MeV) and then by a booster ring that gives them a boost in energy from mega to giga electron volts (GeV); at that point, they are transferred to the final circular accelerator, where if needed, they can be further accelerated to higher energies by the radio frequency (RF) electric fields. When the electrons or positrons in the storage ring reach the expected energy, they are in a quasi-stationary state; forced to follow circular paths by the magnetic field of insertion devices (e.g. bending magnets, wigglers, undulators). During each turn they lose part of their energy, emitting synchrotron radiation (Fig. 4.1) [3].

Synchrotron radiation facilities have evolved through four generations. The first-generation synchrotron facilities were originally constructed after the Second World War as particle accelerators for the high-energy physics community to conduct experiments on the fundamental properties of matter. Soon after their commissioning, it was realized that large quantities of electromagnetic radiation, including X-rays, were produced. First-generation sources were, therefore, parasitic operations, exemplified by SPEAR at the Stanford Linear Accelerator (SLAC). Physicists were pleased (at least in principle) to have customers who paid for the waste radiation, and the rest of the scientists were happy to have access to this unique source to conduct various experiments of material characterization [1, 3, 7].

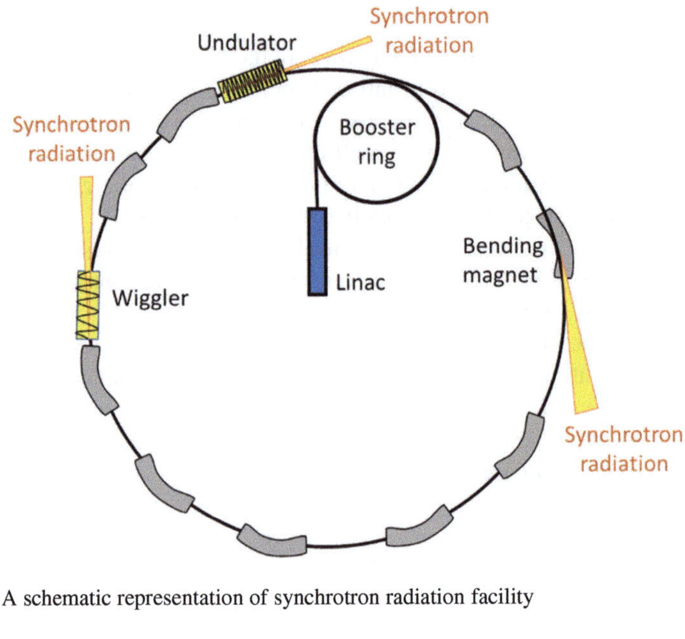

Fig. 4.1 A schematic representation of synchrotron radiation facility

Soon, however, demand for synchrotron light-based techniques outstripped availability at particle physics sites and dedicated facilities - the second-generation synchrotrons, such as the national synchrotron light source (NSLS) and the photon factory, were built to provide dozens of beamlines for use worldwide. These facilities provided significant improvement in brilliance and emittance, as well as in access, continuity and reliability of operation [7].

The third-generation sources combined many incremental advances to provide a one-to-three-order-of-magnitude improvement in performance. The facilities are, for example, nowadays represented by the European Synchrotron Radiation Facility (ESRF), Deutsche Synchrotron (DESY), the Advanced Photon Source (APS) and others [7]. In 2024, IAEA listed 60 synchrotron facilities all over the world (51 working and 9 planned) (https://nucleus.iaea.org/sites/accelerators/Pages/Interactive-Map-of-Accelerators.aspx).

The first fourth generation high energy synchrotron was opened in 2020, after a 20-month shutdown of ESRF.»ESRF-EBS (Extremely Brilliant Source) is characterized by the ring-shaped machine, 844 m in circumference, which generates X-ray beams 100 times brighter than its predecessor and 10 trillion times brighter than medical X-rays. This intense X-ray beam hails a new era for science to understand the complexity of materials and living matter at the nanometric level. ESRF-EBS will contribute to tackling global challenges in key areas such as health, environment, energy and new industrial materials and to unveiling hidden secrets of our natural and cultural heritage through the non-destructive investigation of precious artefacts and palaeontological treasures. A shining example of international cooperation, EBS has been funded by 22 countries joining forces to construct this innovative and world-unique research infrastructure with an investment of 150 million euros over 2015–2022, lighting the way for more than a dozen projects worldwide « (https://www.esrf.fr/home/news/general/content-news/general/opening-of-esrf-ebs-a-new-generation-of-synchrotron.html).

4.3 X-Ray Fluorescence

4.3.1 History of X-Ray Fluorescence

The history of XRF dates back to the accidental discovery of X-rays in 1895 by German physicist Wilhelm Conrad Roentgen. While studying cathode rays in a high-voltage, gaseous-discharge tube, Roentgen observed that even though the experimental tube was encased in a black cardboard box, a barium-platinocyanide screen, which was lying adjacent to the experiment, emitted fluorescent light whenever the tube was in operation. This was possible because the energies of X-ray photons are of the same order of magnitude as the binding energies of the inner-shell (K, L) electrons, and therefore, they can be used to excite and/ or probe these atomic levels [8].

4.3.2 Basic Principles

X-rays are electromagnetic waves spanning from about 80 nm (about 15 eV in photon energy) down to about 0.001 nm (about 1.2 MeV), overlapping to some extent with the spectral region of γ-rays. X-rays are generated by electrons slowed down in the outer field of the atomic nucleus or by transitions between bound states of electrons in the electronic shells of an atom, while γ-rays (usually above 1 MeV) are generated in nuclear processes.

X-ray fluorescence is induced by the excitation (ionization) of atoms in the tightly bound inner K (for the elements $20 < Z < 50$) and L ($Z > 50$) atomic shells with energies that must exceed the binding energies of the K and L electrons, through the process called the photoelectric effect. In the photoelectric absorption process, an X-ray photon from the X-ray source is completely absorbed by an atom. An electron (called a photoelectron) is ejected from one of the inner (K or L) shells. Part of the excitation photon energy is used to overcome the binding energy of the electron and the rest is transferred to the electron in the form of kinetic energy. After the interaction with an X-ray photon, the atom (in fact ion) is left in a highly excited state since a vacancy has been created in one of its inner (K or L) shells. This atom almost immediately returns to a more stable electronic configuration in the process called relaxation or deexcitation, where electrons from outer shells (L, M or N) fill the vacancy created in the K or L shells. Since during relaxation, the electrons pass from a higher to a lower energy state, the difference between the energies can be emitted in the form of characteristic X-ray photons (K, L, or M), or the energy can be absorbed by an electron in one of the outer shells (L, M, or N), which is then ejected from the atom as an Auger electron. Characteristic K, L and M X-ray photons are typical of particular elements, and this provides the basis for distinguishing between elements in X-ray spectrometry [8].

4.4 X-Ray Fluorescence Analytical Techniques

The previously described processes represent the fundamentals of various X-ray-based fluorescence analytical techniques, ranging from standard energy dispersive X-ray fluorescence analysis (EDXRF), total reflection XRF (TXRF), grazing incidence XRF (GIXRF), and grazing emission XRF (GEXRF), mainly dedicated to »bulk« material analyses, to highly sophisticated particle-induced and synchrotron-based X-ray fluorescence techniques such as, for example, micro-particle induced X-ray emission (PIXE), SXRFI [9–14] and XFCT [15], enabling spatial 2D or 3D localization of elements in various samples. In the past, the main restriction of XRF techniques was low sensitivity for light elements ($n < 10$) due to low fluorescence yields, but nowadays, with the development of XRF detector systems on one site and high brilliance X-ray sources on the other site, limits of detection are pushing down with the potential to detect single-atoms in the future [16] The main advantage of the so-called classical techniques such as

EDXRF is that simple sample preparation enables rapid insights into the chemical composition of various materials, even plastics [17].

4.4.1 Energy Dispersive X-Ray Fluorescence Analysis

In EDXRF, the dried, milled and homogenized samples (e.g. rocks, soil, plants, food) are pressed into pellets and excited by an X-ray tube or radioisotope source. The fluorescence X-rays are in modern spectrometers usually detected by silicon drift diode detectors that are coupled to multichannel analyzers [10]. Transportable EDXRF spectrometers are also available to perform robust material characterization in the field or in the case of large objects that cannot be transferred to laboratories (e.g. cultural heritage specimens) [18]. EDXRF is suitable for elemental fingerprinting of food items, where combined with chemometric analysis can be used to trace food geographical origin [19–23]. EDXRF is also especially suitable for determining alloy composition [24].

4.4.2 Total Reflection XRF

The basic fundamentals of TXRF are similar to those of EDXRF, although with different excitation modes. In TXRF systems, the samples are deposited as liquid drops (e.g. 10 mL) on an optically smooth, usually quartz, substrate and dried. They are then excited by a well-collimated X-ray beam (from an X-ray tube or synchrotron source) at an angle smaller than the angle of total reflection for the substrate (<1.8 mrad for quartz). In this case, the majority of the incident X-ray radiation is totally reflected from the quartz surface, and only a minor part of it is absorbed by the deposited sample to excite fluorescence. The penetration of the incident X-ray beam into the reflecting material is drastically reduced under these conditions, and the scattered and fluorescence radiation contributed by the carrier in this geometry is thus negligible. Consequently, the background radiation due to scattering on a small sample is very low, significantly increasing the sensitivity of TXRF when compared to standard XRF [10].

4.4.3 Grazing Incidence XRF

GIXRF is a powerful synchrotron light-based technique for non-destructive characterization of thin films, surfaces, and dopant profiles. It combines features of X-ray reflectivity (XRR) and XRF, allowing for depth profiling and analysis of layer thickness, interface roughness, and interlayer formation. Accurate modelling of GIXRF data requires precise knowledge of instrumental parameters, such as the incident beam profile and detector aperture, which can be determined using calibration samples. GIXRF is closely related to grazing incidence X-ray diffraction

(GIXD), which is used to study the structural parameters of surfaces, interfaces, and thin films by optimizing the diffraction intensity from the surface region [25].

4.4.4 Grazing Emission XRF

GEXRF is a powerful technique for non-destructive, depth-resolved elemental analysis of nanoscale materials. By varying the grazing emission angle, GEXRF can probe depths from a few to several hundred nanometers, making it suitable for surface studies and depth profiling. When combined with X-ray absorption spectroscopy, GEXRF also enables depth-resolved chemical speciation measurements with nanometer-scale accuracy [26].

In this chapter, however, we will focus on synchrotron-based 2D and 3D SXRFI and its applications.

4.5 Synchrotron X-Ray Fluorescence Imaging (SXRFI)

4.5.1 Developments of SXRFI

Since the discovery of X-rays by Wilhelm Conrad Röntgen in 1895, various modalities of X-ray imaging have been developed, which all share the common feature of providing insights into the structural and with XRF also the chemical composition of various materials in a non-invasive manner. After the confirmation of the diffraction with an alternative method performed by William Henry Bragg and William Lawrence Bragg in 1913, two new research fields were born: X-ray diffraction and X-ray spectrometry [27].

X-ray diffraction has since then developed into one of the most important methods for deciphering the three-dimensional structure of proteins, information which is essential to unravel the molecular function of proteins. Nevertheless, several obstacles are still associated with the structural determination of proteins, such as the necessity to obtain large and well-diffracting crystals and the instability of purified proteins, which can be overcome by new techniques and strategies developed over the past decade [28].

In addition to X-ray diffraction, X-ray spectrometry has become essential for obtaining bulk, and with the development of X-ray optics, also spatial chemical information of various materials. The first synchrotron (XRF) beamline in the world was developed at the Stanford Synchrotron Radiation Laboratory (SSRL), which is part of the SLAC National Accelerator Laboratory in California, USA. Already at that time, they have managed to focus the beam down to 450 um to analyze the elemental composition of human hair [29]. Development of synchrotron facilities, focusing optics devices, detector systems and data processing in the past 45 years led to the nano-imaging era of SRXRFI. In the next subchapters, some major breakthroughs are presented.

4.5.2 Increased Brilliance of the Synchrotron Light Sources

The major breakthrough in micro and nano SRXRFI was achieved with the development of 3rd and 4th-generation synchrotron light sources, which provide high flux, brilliance and high coherence of the beam.

As reported by Raimondi in the case of ESRF, France, in 2009, an ambitious upgrade of the facility was launched, which included technological developments on the accelerator, resulting in a brighter source and the possibility of a smaller, more focused beam at the sample position. During 2005–2012, new concepts for the 4th-generation circular accelerators were emerging, dwelling on the concept of multiple bend achromat lattice (MBA), leading to the possibility of building a new kind of storage ring with a much lower equilibrium horizontal emittance, which would approach the diffraction limit at 0.1 nm wavelength. This is the point where reducing the emittance of the electron beam does not reduce the emittance of the photon beam any further. This would enable an increase in the brilliance and coherence of the photon beam by two orders of magnitude. Key areas of research that would benefit from such a project included coherent diffraction imaging and microscopy, spectroscopy with nanometer spatial resolution, and time-resolved studies in the millisecond to sub-nanosecond scale. In 2012 a new lattice concept based on the MBA was developed at ESRF: the hybrid multibend achromat (HMBA) or Raimondi lattice, which is particularly well-adapted to high-energy synchrotron storage ring machines. Raimondi lattice was studied in its seven-bends version as the basis for a new storage ring for the ESRF with much improved horizontal emittance and thus increased brilliance and coherent fraction of the X ray beam ($\sim \times 100$). The ESRF governing bodies decided, in 2014, to launch the reconstruction of the ESRF storage ring based on the Raimondi lattice, aiming to transform the ESRF X-ray source into a new Extremely Brilliant Source (EBS) [30, 31].

ESRF-EBS was successfully commissioned in 2023, and the same concept will be utilized for most future 4 th generation storage ring light sources, such as the Advanced Photon Source upgrade at Argonne National Laboratory and the Advanced Light Source upgrade at Berkeley Lab in the US (APS-U and ALS-U); Spring 8–2 at the Japan Synchrotron Radiation Research Institute; the Shanghai Synchrotron Radiation Facility and the Beijing Synchrotron Radiation Facility in China (SSRF and IHEP) [30, 31].

4.5.3 Improvements in X-Ray Focusing Optics

The short wavelength of X-rays makes them an excellent choice for probing materials on the micro- and nanometer scale. With the development of nanomaterials in the past decades, there has been an increasing demand at SRXRFI facilities for better spatial resolution and the ability to resolve structural and chemical features at the nanometric scale.

The aim of micro- and nanofocusing optics is to produce a small focused beam size to get a very fine spatial probe while maintaining the highest X-ray flux on a small sample. Achieving nanometer-scale focused X-ray beam sizes puts great demands on the optical elements in an X-ray beamline—the optics must balance the requirements to de-magnify the electron beam X-ray source, to reduce the diffraction-limited focus size, and to minimize the contribution to the focus of aberrations in the optics while collecting the maximum X-ray flux into the focused beam [32].

Various nano-focusing optics, including Kirkpatrick-Baez (KB) and elliptical mirrors, Fresnel zone plates (ZP), compound refractive and kinoform lenses, and multilayer Laue lenses (MLL), are routinely used to achieve nano-focusing. The latest experiments utilizing KB mirrors have demonstrated focal spot size of 13 nm with the ZPs and MLLs being able to focus hard X-rays to 10 nm and below. When the focused beam size is reduced to below 100 nm, there is a need to develop a dedicated system tailored for a specific application which satisfies stringent mechanical, vibrational, and thermal characteristics to successfully enable nano-scale imaging. Multiple degrees of motion, including their positioning accuracy, resonance frequencies, and amount of heat being dissipated, have to be thoroughly considered in order to efficiently perform nano- focusing experiments, as reviewed in [33].

Various fabrication methods, including e-beam lithography, have been developed to create Fresnel zone plates (FZPs) for extreme ultraviolet and soft X-ray applications. Li et al. [34] demonstrated FZPs with 16 nm outermost zone width and 1.8 μm thickness, achieving a focus of 46×60 nm at 12 keV [34]. Using Kirkpatrick-Baez (KB) mirrors sub-15 nm spatial resolution has been achieved [33]. Multilayer KB mirrors with phase compensation have demonstrated focal spots as small as 26×17 nm [35].

Nanofocusing using refractive planar kinoform lenses and profiling of X-ray micro-focusing mirrors are two techniques being developed at Diamond Light Source for generating nanofocused beams and for the ability to change the focused beam size [32].

Improvements of X-ray focusing optics have also enabled the development of high-resolution X-ray imaging techniques, including full-field microscopy, which can be combined with computed tomography for non-destructive 3D and 4D nanoimaging [36].

Apart from the conventional SXRFI, different variations of XRF imaging techniques, X-ray fluorescence computed tomography (XFCT) and confocal XRF (CXRF) have been reported in recent years, the latter more typical of laboratory micro-XRF systems [37, 38]. Among these techniques, SXRFI has the best element sensitivity due to the absence of the bremsstrahlung background, while XFCT can provide a three-dimensional structural and elemental composition of materials. Full-field XRF imaging has an advantage in high-resolution imaging of large samples since it saves lots of time compared to classical scanning in SXRFI mode.

4.5.4 Improvements in XRF Detector Systems

Recent advancements in XRF detector system development have significantly improved sensitivity, spectral resolution and performance. Improvement of Silicon Drift Detectors (SDDs) has enabled the detection of low-energy X-rays (polymer window detectors) and higher counting rates [39]. Firmware-upgraded cadmium telluride (CdTe) detectors optimized for high X-ray flux operations have enhanced photon counting efficiency and reduced detection limits in XRF imaging systems [40]. A new collimated multichannel modular detection system using SDD arrays has demonstrated improved energy resolution and peak-to-background ratio [40].

The development of Maia detector has enabled high-throughput SRXRFI with large pixel counts, achieving this through a planar silicon detector array and event-mode data acquisition. It features 384 independent detector channels, allowing for cumulative count rates exceeding 10^6 counts s^{-1} with low pile-up probability [41]. To improve energy resolution and throughput, a new prototype replacing p-in-n diodes with silicon drift detectors (SDDs) was designed. The SDD-Maia detector maintains compatibility with existing electronics while offering enhanced performance [42].

These developments, along with advancements in X-ray optics and sources, have been applied across various XRF techniques, including EDXRF, WDXRF, TXRF, and XRF microscopy [43]. Improvements in vacuum systems, detector materials, and collimation have collectively contributed to enhanced accuracy, precision, and sensitivity in XRF analysis, enabling the detection of lower concentrations of elements and expanding the range of applications.

4.6 SXRFI Instrumentation and Methodology

4.6.1 SXRFI Beamlines

Typical micro or nano-XRF imaging synchrotron beamline consists of a monochromator, such as a silicon channel-cut or double-crystal design, to tune X-ray energies. Focusing optics, like zoneplate or Kirkpatrick-Baez mirrors, as mentioned above, can produce sub-micron-sized focal spots for high-resolution imaging [44] (Fig. 4.2). Besides XRF, many beamlines offer the possibility to scan the energy range, which is the basis for X-ray absorption spectrometry, namely X-ray near-edge structure (XANES).

The beam coming from the undulator is monochromatized by a crystal monochromator. I_0 counter enables the normalization of the incoming beam, which is important for quantitative XRF analysis and XANES. Kirkpatrick-Baez focusing mirrors can focus the beam down to a few 10 nm. The beam then hits the sample mounted on a 3D/rotating precision goniometer. X-ray fluorescence that is excited in the sample is detected by a silicon drift diode (SDD) detector. The transmitted beam is recorded by I_1 counter. Despite some general features of SXRFI beamlines, the unique construction of each synchrotron facility provides a large variety

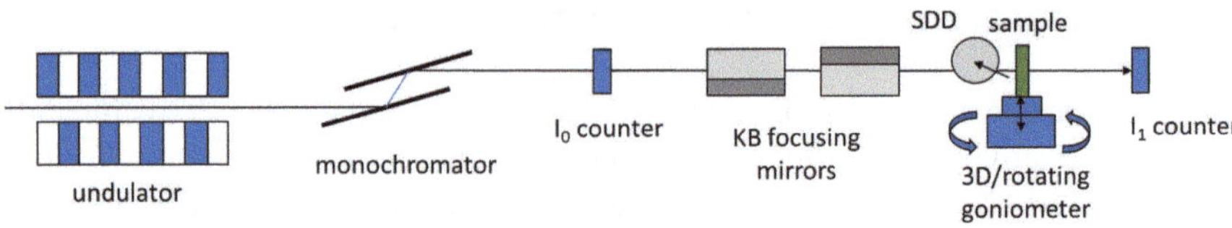

Fig. 4.2 Schematic presentation of an SXRFI beamline

of analytical setups. SXRFI beamlines thus feature multiple experimental stations and sample environments [45–48]. While this is a key benefit over lab-based systems, it also means considerable time needs to be spent on the planning of an experiment in order to choose a suitable beamline and maximize the output. Synchrotron radiation is highly customizable to individual needs. Various parameters like photon energy, photon flux, dwell time, pixel size, etc. have to be carefully selected to collect quality data and minimize image artefacts. Superior brilliance and extreme spatial resolution of synchrotron radiation allow the collection of large amounts of data in a rather short time frame. However, for optimal results of individual case studies, several scans with variable experimental setups and analytical settings are sometimes necessary [49].

Sometimes, it is difficult for users to choose a suitable beamline for a particular problem they want to address. Thus, in continuation of this chapter, an overview of world synchrotron facilities and their beamlines that provide SXRFI techniques can be found (Table 4.1).

4.6.2 Sample Preparation

SXRFI has found its application in various research fields, such as earth, planetary, and environmental sciences, as well as biology. In 2D SXRFI, normally, the sample inner structures are exposed by cutting [13], fracturing or polishing, depending on the material studied. In some cases, however, surface measurements are also performed, especially in the case of plant organs, where in situ element distribution at the whole leaf area can be determined [50].

The sample preparation of biological tissues is especially critical when it comes to being examined at micro- and nanometric scales. The final goal is to preserve spatial morphological and biochemical composition at cellular and sub-cellular levels to be as much as possible similar to that in in vivo stage. Rapid or flush freezing in cryogens such as propane or isopentane cooled by liquid nitrogen, followed by freeze-drying [14], or direct measurement of frozen-hydrated specimens is used at soft-X-ray beamlines with a vacuum sample environment, while at hard X-ray beamlines cryo jets [51] can be applied. In medical research, however, there is still a trend to use chemically fixed specimens, but there is a high probability of redistributing or washing out certain mobile biochemical species [52, 53]. Cryofixation of soft tissues, on the other hand, still remains challenging in the sense of preserving sample morphology [54].

Table 4.1 Beamlines with SXRF imaging mode (m−micro, n−nano)

SR facility	Beamline	Energy range (keV)	Beam size	Techniques	Sample environment	Research areas	Web page
ALS, Berkley lab California, USA	10.3.2 X-ray fluorescence microprobe (XFM)	2.1–14.0	1 μm	mXRF, mXAS	Ambient cryostage	Applied sciences, biological sciences, chemical sciences, energy sciences Earth & environmental sciences, materials sciences, physical sciences	https://als.lbl.gov/beamlines/10-3-2/ after ALS upgrade will be moved to 8.0
APS Illinois, USA	2-ID-D: microfluorescence and bionanoprobe	5.0–30.0	2 μm	mXRF, mXFCT, mXAS	Ambient cryostage	Chemistry, environmental science, life sciences, materials science	https://www.aps.anl.gov/Beamlines/Directory/Details?beamline_id=5
	2-ID-E: Hard X-ray microprobe	7.0–10.5	0.5 μm	mXRF	Ambient	Environmental science, life sciences, materials science	https://www.aps.anl.gov/Beamlines/Directory/Details?beamline_id=61
	13-ID-E: X-ray microprobe	2.4–28.0	1.5 μm	mXRF, mXAS, mXFCT	Ambient cryostage (He)	Earth, planetary, and environmental sciences	https://gsecars.uchicago.edu/gsecars-facility/sector-13-beamlines/13-ide/

(continued)

Table 4.1 (continued)

SR facility	Beamline	Energy range (keV)	Beam size	Techniques	Sample environment	Research areas	Web page
	19-ID-E: In-situ nanoprobe	4.8–30.0	20 nm	nXRF	Ambient	Catalysis, chemistry, environmental science, materials science	https://www.aps.anl.gov/Beamlines/Directory/Details?beamline_id=232
	25-ID-C: microprobe and advanced spectroscopy	5.0–20.0	2 µm	mXRF, mXAS confocal XRF	Ambient	Catalysis, chemistry, environmental science, materials science, physics	https://www.aps.anl.gov/Beamlines/Directory/Details?beamline_id=217
	25-ID-D,E: LERIX and time-resolved spectroscopy	5.0–20.0	2 µm	mXRF, mXAS	Ambient	Catalysis, chemistry, environmental science, materials science, physics	https://www.aps.anl.gov/Beamlines/Directory/Details?beamline_id=218
	26-ID-C: hard X-ray nanoprobe	7.0–12.0	20 nm	mXRF	Ambient	Materials science, physics	https://www.aps.anl.gov/Beamlines/Directory/Details?beamline_id=92

(continued)

Table 4.1 (continued)

SR facility	Beamline	Energy range (keV)	Beam size	Techniques	Sample environment	Research areas	Web page
	33-ID-C: ptychoProbe	5.0–30.0	5 nm	mXRF	Ambient	Catalysis, chemistry, environmental science, geoscience, materials science, physics	https://www.aps.anl.gov/Beamlines/Directory/Details?beamline_id=241
Australian synchrotron Clayton, Australia	XRF microscopy	4.1–27.0	Mili: 50–200 µm Micro: 1–10 µm	mXRF, mXFCT	Ambient	Chemistry, environmental science, life sciences, materials science	https://www.ansto.gov.au/facilities/australian-synchrotron/synchrotron-beamlines/x-ray-fluorescence-microscopy
BESSY II, Berlin, Germany	BAMline	4.5–60.0	1.5–200 µm	mXRF, ful-field mXRF, mXFCT, EXAFS	Ambient	Biology, corrosion, catalysis	https://www.helmholtz-berlin.de/pubbin/igama_output?modus=einzel&sprache=en&gid=2364&typoid=75136

(continued)

Table 4.1 (continued)

SR facility	Beamline	Energy range (keV)	Beam size	Techniques	Sample environment	Research areas	Web page
	mySpot	4.0–25.0	~5 µm	mXRF, mXAS, mXFCT	Ambient	Life sciences, environmental sciences	https://www.helmholtz-berlin.de/pubbin/igama_output?modus=einzel&sprache=en&gid=1697&typoid=75136
CLS, Saskatoon, Canada	BioXAS-imaging	5.0–21.0	Macro: 20–150 µm Micro: 5–10 µm	mXRF, mXAS	Ambient	Biological science Environmental science Cultural heritage	https://bioxas-imaging.lightsource.ca/
	VESPERS	6–30	2–4 µm	mXRF, mXAS	Ambient	Biological science Environmental science Cultural heritage	https://vespers.lightsource.ca/
NSLS II, New York, USA	3ID, hard X-ray nanoprobe	MLL: 12.0–18.0 ZP: 6.0–12.0	ML: 12 nm ZP: 40 nm	nXRF, nXAS, n-XFCT	Ambient in situ	Life sciences Environmental sciences Material sciences	https://www.bnl.gov/nsls2/beamlines/beamline.php?r=3-ID

(continued)

Table 4.1 (continued)

SR facility	Beamline	Energy range (keV)	Beam size	Techniques	Sample environment	Research areas	Web page
	4-BM, X-ray fluorescence microprobe	A: 2.1–21.0 B: 12.0–20.0	A: 5–10 μm B: 1–5 μm	mXRF, mXAS	Ambient	Molecular environmental science, plant and soil sciences, climate, earth and planetary sciences, life and biological sciences, materials and nanosciences, energy science	https://www.bnl.gov/nsls2/beamlines/beamline.php?r=4-BM
DESY, PETRA III Germany	P06	Micro: 5–23 (KB Mirrors) 15–90 (CRL) Nano: 7–10 keV (FZP) 10–30 keV (NFL)	Micro:350 nm (KB mirrors), 500 nm (CRL) Nano: 80 nm^2(NFL), 30 nm (FZP)	mXRF, mXAS, mXFCT nXRF, nXAS, nXFCT	Ambient cryo-jet	Applied sciences, biological sciences, chemical sciences, energy sciences, earth & environmental sciences, materials sciences, physical sciences	https://photon-science.desy.de/facilities/petra_iii/beamlines/p06_hard_x_ray_micro_nano_probe/index_eng.html

(continued)

Table 4.1 (continued)

SR facility	Beamline	Energy range (keV)	Beam size	Techniques	Sample environment	Research areas	Web page
Diamond, UK	I18	2.0–20.7	2 μm	mXRF, mXAS, mXFCT, mXAS-CT	He enclosure for low energy ambient vacuum cryostat	Geochemistry and space mineralogy, biochemistry in tumour and cancer treatment, analysis of agrochemicals, environmental and heritage research, catalysis structure and mechanistic pathways, ultrafast pump-probe XAS	https://www.diamond.ac.uk/Instruments/Spectroscopy/I18.html
	I08	0.25–4.4	1 μm	mXRF, mXANES	Vacuum ambient he cryostat	Earth, environmental science and geochemistry, biology, biotechnology and biogeochemistry, human health, nanotechnology, materials, energy and batteries, space and planetary science	https://www.diamond.ac.uk/Instruments/Imaging-and-Microscopy/I08.html

(continued)

Table 4.1 (continued)

SR facility	Beamline	Energy range (keV)	Beam size	Techniques	Sample environment	Research areas	Web page
	I14	5.0–23.0	50 nm	mXRF, mXANES	In situ environment	Material sciences earth & environmental sciences	https://www.diamond.ac.uk/Instruments/Imaging-and-Microscopy/I14.html
Elettra, Trieste, Italy	TwinMic	0.4–2.0	1 μm	mXRF	Vaccum	Biotechnology, nanotechnology, environmental science and geochemistry, clinical and medical applications, novel energy sources, biomaterial, cultural heritage and archeometry	https://www.elettra.eu/elettra-beamlines/twinmic.html
	XRF	2.0–14.0	50 μm	mXRF, mXANES	Vacuum (UHVC)	Materials and surfaces, cultural heritage, archaeology and forensics, biology, food safety, environmental and earth sciences	https://www.elettra.eu/lightsources/elettra/elettra-beamlines/microfluorescence/x-ray-fluorescence.html

(continued)

Table 4.1 (continued)

SR facility	Beamline	Energy range (keV)	Beam size	Techniques	Sample environment	Research areas	Web page
HEPS, Beijing, China	ID19–NAMI (under construction)	5.0–23.0	10 nm In situ: 40–50 nm	n-XRF, N-XANES	Ambient	Materials sciences	http://english.ihep.cas.cn/heps/fa/bl/202112/t20211222_295096.html
	ID21-Hard X-ray imaging beamline (under construction)	10–300	0.1–300 μm	Full field XRF	Ambient	Materials sciences	http://english.ihep.cas.cn/heps/fa/bl/202112/t20211222_295089.html
ESRF, Grenoble, France	ID13	7.0–30.0	100–20 μm	mXRF	Low temperature (90 K, liquid-nitrogen-based cryostream, microbranch only)	Life sciences, materials and engineering, environmental sciences, earth and planetary sciences, chemistry, physics, cultural heritage, medicine	https://www.esrf.fr/UsersAndScience/Experiments/XNP/ID13

(continued)

Table 4.1 (continued)

SR facility	Beamline	Energy range (keV)	Beam size	Techniques	Sample environment	Research areas	Web page
	ID21	2.1–10.0	180 nm–2 μm	mXRF, mXANES	In vacuum room temperature and cryogenic conditions (LN2) no in-situ capabilities	Medicine, cultural heritage, environmental sciences, earth and planetary sciences, materials and engineering, physics, life sciences, chemistry	https://www.esrf.fr/home/UsersAndScience/Experiments/XNP/beamline-snapshots/content/content-list/id21.html
	ID16A	17.0–33.6	30–400 nm	mXRF mXFCT	Room-temperature, in vacuum cryo-environment, in vacuum	Materials and engineering, medicine, environmental sciences, earth and planetary sciences, physics, life sciences	https://www.esrf.fr/UsersAndScience/Experiments/XNP/ID16A
	ID16B	6.0–33.0	50 nm–1 μm	mXRF mXFCT mXANES	He cryostat for XEOL high temperature furnace for in-situ nanotomography specific devices developed by users	Nanotechnology, earth and environmental sciences, bio-medical research	https://www.esrf.fr/home/UsersAndScience/Experiments/XNP/ID16B/overv.html

(continued)

Table 4.1 (continued)

SR facility	Beamline	Energy range (keV)	Beam size	Techniques	Sample environment	Research areas	Web page
LNSL, São Paulo, Brazil	XRF beamline	5.0–20.0	20 μm	mXRF mXFCT	Ambient nitrogen gas environment	Analytical chemistry, biomedicine, environmental geochemistry and materials science	https://lnls.cnpem.br/facilities/xrf-en/
MAX IV, Lund, Sweden	NanoMAX	8.0–15.0	30 nm	nXRF nXFCT	Vacuum, cryostation	Materials science, life science, earth science, nanoscience, physics, chemistry, biology	https://www.maxiv.lu.se/beamlines-accelerators/beamlines/nanomax/
NSRRC, Hsinchu, Taiwan	23A-X-ray nanoprobe (under construction)	4.0–15.0	40 nm	n-XRF, n-XAS	Vacuum	Materials science, physics, nanomaterials, semiconductor devices, energy materials, geoscience, chemistry, biomedical engineering	https://tpsbl.nsrrc.org.tw/bd_page.aspx?lang=en&port=23A&pid=1026

(continued)

Table 4.1 (continued)

SR facility	Beamline	Energy range (keV)	Beam size	Techniques	Sample environment	Research areas	Web page
Photon factory, Tsukuba, Japan	BL-4A Trace element analysis/X-ray microbeam station	6–18	5 μm (KB) 30 μm (polycapillary)	mXRF mXANES	Ambient	Biology, biomedicine, environment, geology, materials	http://pfwww.kek.jp/users_info/users_guide_e/station_spec_e/bl04/bl4a.html
	AR-NE7A: X-ray imaging and X-ray diffraction experiments at high pressure	Monochromatic mode: 10–60 white mode: 20–140	30 × 3 μm	mXRF	Ambient high pressure	Materials science Medical science X-ray optics	https://www2.kek.jp/imss/pf/eng/apparatus/bl/ne7a.html
SLRI, Nakhon Ratchasima, Thailand	BL8	1.25–10.0	10 μm	Full field mXRF	Ambient	Environmental science, life sciences, materials science	https://www.slri.or.th/en_web/bl8-x-ray.html [57]
	BL6b: Micro	4.0–10.0	50 μm	mXRF	Ambient	Environmental science, life sciences, materials science	[57]
SSRL, California, USA	2–3	4.9–23	2 μm	mXRF mXFCT mXAS	Ambient cryostat	Biology, materials, geology	https://www-ssrl.slac.stanford.edu/content/beam-lines/bl2-3

(continued)

Table 4.1 (continued)

SR facility	Beamline	Energy range (keV)	Beam size	Techniques	Sample environment	Research areas	Web page
SOLARIS, Krakow, Poland	POLYX	4.0–15.0	Polycapillary optics: 8–200 μm monocapillary: 2 μm	mXRF mXFCT mXAS	Air, helium, low vacuum (user setups)	Biomedical sciences, material science, physics, chemistry, earth science, environmental sciences, cultural heritage	https://synchrotron.uj.edu.pl/en_GB/linie-badawcze/polyx
SOLEIL, France	LUCIA	0.6–8.0	2.5 μm	mXRF, mXANES	Vacuum chamber He/N_2 cryostat	Earth and soil sciences, material sciences, chemistry, physics, life sciences	https://www.synchrotron-soleil.fr/en/beamlines/lucia
	PUMA	Micro beam endstation: 4–22 full field endstation 4–60	micro beam endstation: 5 μm Full field endstation 20 μm	mXRF, mXANES	Ambient	Archaeology, paleontology, paleo-environments, art history and conservation sciences, environmental sciences, earth sciences, biology, soft-matter and medicine	https://www.synchrotron-soleil.fr/en/beamlines/puma

(continued)

Table 4.1 (continued)

SR facility	Beamline	Energy range (keV)	Beam size	Techniques	Sample environment	Research areas	Web page
	NANOSCOPIUM	5.0–20.0	CX2: 100 nm CX3: 50 nm	nXRF, nXANES, nXFCT, ful field XRF imaging	Ambient	Earth science and geobiology, emvironmental sciences, biolohy-health, material science, cultural heritage	https://www.synchrotron-soleil.fr/en/beamlines/nanoscopium
SLS, Villigen, Switzerland	X05LA, environmental & materials sciences	3.0–23.0	1 μm	mXRF, mXANES	Ambient	Environmental science, material science	https://www.psi.ch/en/sls/microxas
SPring8, Saitama, Japan	BL37XU, trace element analysis	4.5–113	100 nm–1 μm	Full-field XRF imaging High energy nXRF and nXAS	Ambient	Material science, environmental science, geochemistry, biology, archaeology, forensic science	http://www.spring8.or.jp/wkg/BL37XU/instrument/lang-en/INS-0000000591/instrument_summary_view
	BL27SU, Soft X-ray photochemistry	B: 2.1–3.3	15 μm	mXRF	Low vacuum (1 Pa)	Material science	http://www.spring8.or.jp/wkg/BL27SU/instrument/lang-en/INS-0000001445

(continued)

Table 4.1 (continued)

SR facility	Beamline	Energy range (keV)	Beam size	Techniques	Sample environment	Research areas	Web page
SSRF, Shanghai, China	BL13W1 X-ray Imaging and Biomedical Applications Beamline	8.0–72.5	45 × 5 μm	XRFI, XFCT	Ambient	Biomedicine	http://e-ssrf.sari.ac.cn/beamlines/bl13w1/
	BL15U1 hard X-ray micro-focusing beamline	5.0–20.0	2 μm	mXRF, mXAS	Ambient	Life science material science earth science environmental science nanotechnology and industrial applications	http://e-ssrf.sari.ac.cn/beamlines/bl15u1/

4.6.3 Data Analysis

The software most frequently used for analyzing SRXFI data, PyMCA [55] and GeoPIXE [56] can handle most varying experimental conditions and types of samples. This adaptability in the software necessitates expert skills of the user. In ease-of-use and automation, it trails behind counterparts for lab-based systems. However, when a first set of analysis parameters has been established, further analyses can be automated. It should be noted, though, that for accurate quantification of several elements as, for example, in geological samples, with widely spanning concentration levels, a human examination is often needed. In the complex geological samples, peak-overlaps, broad peaks, and near-background peaks made assessment of the presence or absence of specific elements difficult. In addition, the overlap between emission peaks of several elements and near-background fluorescence may lead to an inaccurate representation of element distribution in fluorescence maps, complicating data interpretation [49].

4.7 Full-Field XRF Imaging

High performance position- and energy-sensitive X-ray detectors have boosted the development of advanced instruments based on the full field X-ray fluorescence imaging (FF-XRFI) method, providing a promising alternative to the conventional scanning approach for performing two-dimensional elemental imaging [58].

Compared to the SRXRFI scanning mode, where the sample is raster scanned and XRF spectra are recorded pixel by pixel, or in flying mode line by line, FF-XRFI works by using a broad X-ray beam to illuminate a large area (or even the full area) of a sample. The X-ray fluorescence induced by the primary radiation is detected on a position- and energy-sensitive detector, through a pinhole-collimator [58], an X-ray polycapillary [59], or Fresnel zoneplate [60], positioned between the sample and the detector. The selective detection of the characteristic X-rays hitting the detector allows to obtain the two-dimensional elemental distribution of the investigated samples (Fig. 4.3). Applications of FF-XRFI originate mainly from the cultural heritage scientific community, primarily relying on laboratory instruments of their own development as shown in Fig. 4.4 [58, 61].

A full-field X-ray fluorescence imaging (FF-XRFI) station was recently developed at the BL8 beamline of Synchrotron Light Research Institute (SLRI), Thailand. An incident synchrotron X-ray beam was monochromatized at 10 keV and has a moderate photon flux of 10^9 photons s^{-1} $(100\ mA)^{-1}$. The experimental beam size was 13×13 mm. An energy-dispersive pnCCD array detector equipped with polycapillary optics was employed, providing a full-frame image size of 12.3×12.3 mm comprising 256×256 detector pixels. In terms of XRF imaging quality, peak-to-background ratios in the range 7:1 (for K) to 250:1 (for Zn) were achieved and were comparable with those of the scanning XRF station at the BL6 beamline of SLRI. However, the FF-XRFI technique was more practical since a much larger image area could be recorded for an equal time [57].

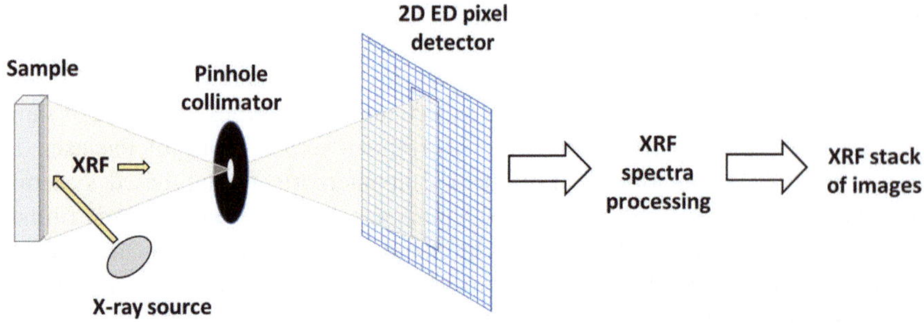

Fig. 4.3 A FF-XRFI setup. The sample is irradiated with X-rays from the source (e.g. synchrotron, X-ray tube). The X-ray fluorescence rays are collimated with a pinhole collimator and projected onto 2D energy dispersive pixel detector (e.g. 256 × 256 pixels). An XRF spectrum is recorded in each pixel and an XRF stack of images corresponding to particular X-ray fluorescence lines is obtained after spectra processing. Figure adopted and modified after [62]

Fig. 4.4 Element distribution maps of tin – lead glazed pottery obtained by a FF-XRFI laboratory setup. Figure adopted from Romano et al. [58] with the permission from ACS publications

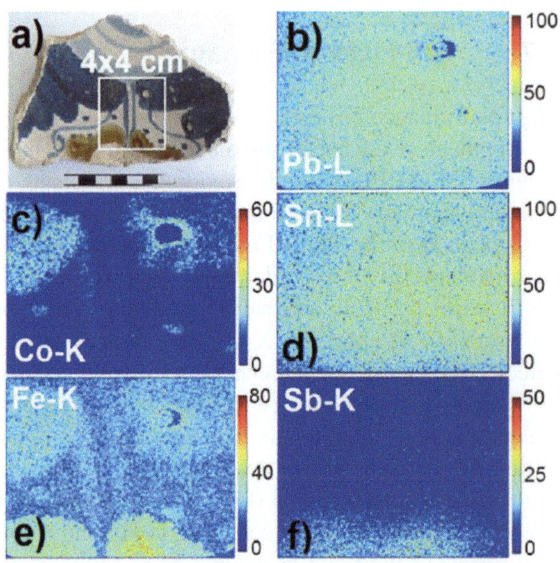

4.8 X-Ray Fluorescence Computed Tomography (XFCT)

X-ray fluorescence computed tomography (XFCT) is an imaging technique that enables multiplexed imaging without physically disturbing the samples. Besides 3D structural information via X-ray transmission microtomography, X-ray fluorescence detection is added to detect 3D element distribution utilizing filtered back-projection algorithms for image reconstruction, often with absorption corrections [63, 64].

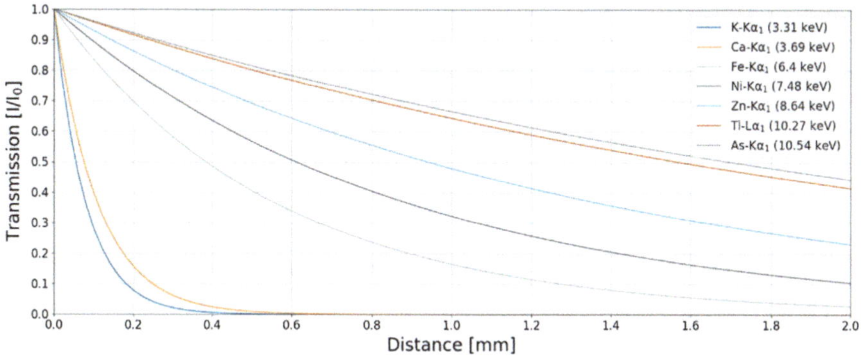

Fig. 4.5 Transmission of selected X ray emission lines through 2 mm of hydrated plant material (with an assumed composition of C7.3O33H59N0.7S0.15) with density 0.9 g cm^{-3}. This shows that lower energy characteristic (e.g. Kα1 or Lα1) fluorescence photons emitted from elements such as K or Ca, are absorbed within only 0.2 mm of plant material, whereas higher energy characteristic fluorescence photons emitted from elements such as Tl or as, can pass through more than 2 mm of plant material. Figure adopted from [66] under Creative Commons (CC BY) International

XFCT, however, faces limitations such as finite detector energy resolution, self-absorption effects, and long experiment durations. While XFCT excels at imaging high atomic number elements, like gold, gadolinium, and barium, with linear relationships between XRF intensity and element concentrations [65], it faces challenges with low-Z elements due to self-absorption effects (Fig. 4.5).

In plant sciences, XFCT is frequently used to image element distribution in plant organs, as revised by [66]. The advantage is that due to»virtual sectioning« no sample preparation is needed, but on the other hand, the measurements take time (depending on chosen lateral resolution and dwell time per pixel; 300 × 300 μm root cross-section with 1 μm lateral resolution and 50 ms dwell time per pixel would take approximately 1.25 h, not including dead time for motor movements).

4.8.1 Sample Preparation

The advantage of XFCT is that it needs little or no sample preparation [67]. In the case of biological samples, it is also possible to measure them in situ. However, care should be taken due to long measurements (XRF mode) and possible dehydration and shrinking of the samples, so cryofixation is recommended [68]. Flush-freezing of the samples can also be followed by controlled freeze-drying in case no cryo jet is available at the beamline [13].

4.8.2 Data Analysis

Recent advancements in X-ray fluorescence computed tomography (XFCT) have improved data analysis and reconstruction techniques. XRFtomo, a Python package, facilitates 3D data analysis and reconstruction of synchrotron-based XRF imaging [69]. An analytical simulation method for XFCT using GPU implementation has been developed, reducing computation time compared to Monte Carlo simulations [70]. Recent advancements have improved XFCT's capabilities for imaging lower-Z elements (e.g. Fe and Zn) through novel reconstruction methods and absorption correction algorithms [71].

Automated tomographic reconstruction is also possible in the IMOD software package, including the merging of tomograms taken around two orthogonal axes. Several developments enable the production of high-quality tomograms. When using fiducial markers for alignment, the markers to be tracked through the series are chosen automatically; if there is an excess of markers available, a well-distributed subset is selected that is most likely to track well [72].

4.9 X-Ray Absorption Near Edge (XANES) 2D and 3D Imaging

The possibility to scan X-ray energy in the vicinity of the absorption edge (50–100 eV) of a certain element provides a basis for X-ray Near-Edge Imaging (XANESI) and Computer Tomography (XANES-CT), which, besides localization of the element of interest in the sample, also provides 2D/3D information on its chemical speciation–oxidations state and ligand environment [73, 74]. Images are taken at different energies, yielding an image stack (Fig. 4.6). Since each image takes a few minutes, the number of images in the stack is a compromise between energy resolution and time consumed. In the nearest vicinity of the edge, the energy steps should be dense (at least 0.5 eV per step) in order not to miss characteristic pre-edge or post-edge spectral features.

The workflow of XANES imaging in XRF tomographic mode includes several steps, as shown in Fig. 4.7 (adapted according to [68]).

4.10 Conclusions

Synchrotron X-ray Fluorescence Imaging (SXRFI) is an advanced analytical technique that enables high-resolution, non-destructive imaging of element distribution in various materials. By utilizing the intense and focused monochromatic X-ray beams generated by synchrotron light sources, SXRFI provides detailed spatial distributions of elements at micro- and nanoscale, making it invaluable in scientific fields such as material science, environmental science, biology, medicine, and cultural heritage.

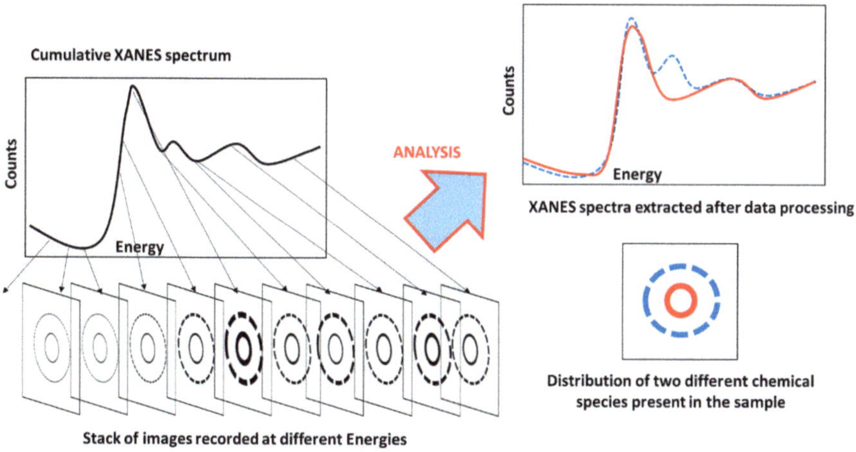

Fig. 4.6 A schematic presentation of XANES imaging. Images are recorded at different energies, yielding an image stack. The number of images in the stack is a compromise between XANES spectral energy resolution and measuring time. Two different chemical species are present in this imaginary sample, after the analysis yielding two different XANES spectra and distinct distribution patterns of each species depicted in blue and red

Recent advancements in synchrotron light technology and X-ray fluorescence detector systems have significantly enhanced SXRFI capabilities, including improved spatial resolution through nanofocused beams, increased sensitivity for detecting trace elements, shorter scanning time and the integration of multimodal imaging approaches.

The development of three-dimensional imaging techniques, such as XFCT, has expanded the scope of SXRFI, allowing researchers to reconstruct complex structural and biochemical information in three dimensions.

With decreased scanning time in situ and *in operando* studies have been made possible, enabling real-time observation of dynamic processes under controlled environmental conditions.

The application of machine learning and artificial intelligence in data processing is revolutionizing the analysis of SXRFI data, facilitating high-throughput studies and the extraction of quantitative information from complex datasets.

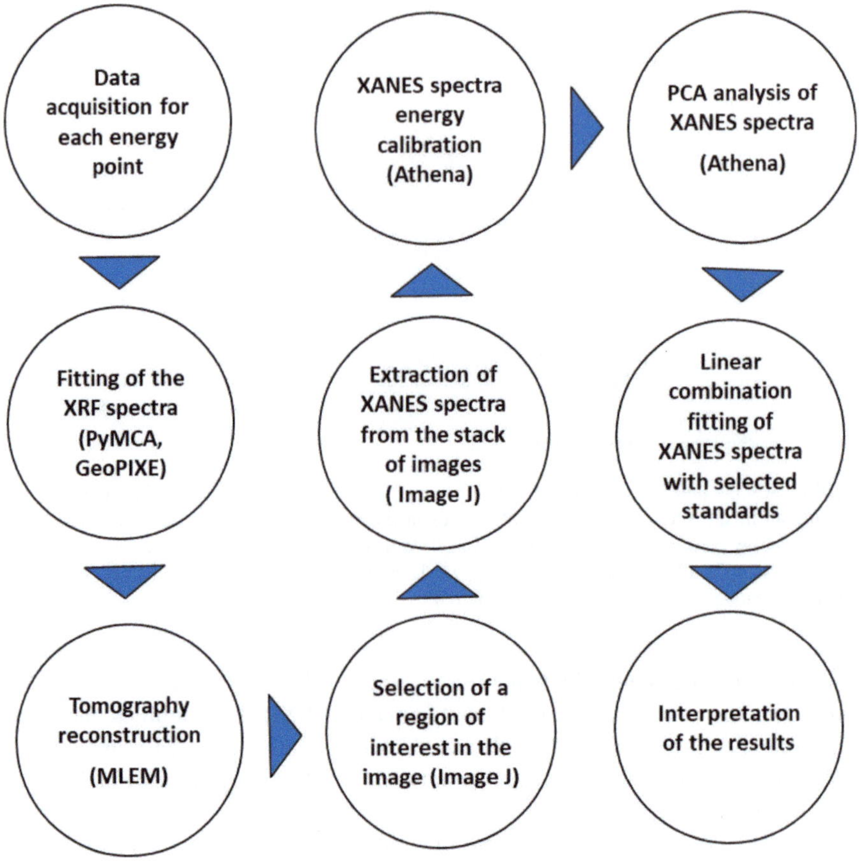

Fig. 4.7 Workflow diagram for the XANES tomography experiment and data analysis. Software used: PyMCA [55] or GeoPIXE [56], MLEM, ImageJ (https://fiji.sc/), Athena [75]

Acknowledgements The financial support by the Slovenian Research Agency (ARIS) through stable financing (P1-0212, Plant biology) is acknowledged.

References

1. Blewett, J.P.: Synchrotron radiation-early history. Int. Union Crystallogr. J. Synchrotron Radiat. J Synchrotron Rad **5**, 135–139 (1998)
2. Pollock, H.C.: The discovery of synchrotron radiation. Am. J. Phys. **51**, 278–280 (1983). https://doi.org/10.1119/1.13289
3. Mobilio, S., Boscherini, F., Meneghini, C.: Synchrotron radiation: basics, methods and applications. **799** (2015)
4. Balerna, A., Mobilio, S.: Introduction to synchrotron radiation. Synchrotron Radiat.: Basics, Methods Appl. 3–28 (2015). https://doi.org/10.1007/978-3-642-55315-8_1

5. Evans, J.: Brilliant opportunities across the spectrum. Phys. Chem. Chem. Phys. **8**, 3045–3058 (2006). https://doi.org/10.1039/B604369K
6. Carr, G.L., Hanfland, M., Williams, G.P.: Midinfrared beamline at the national synchrotron light source port U2B. Rev. Sci. Instrum. **66**, 1643–1645 (1995). https://doi.org/10.1063/1.1145870
7. Lattman, E.: A fourth generation synchrotron at SLAC? Proteins: Struct., Funct., Bioinf. (2000). https://doi.org/10.1002/(SICI)1097-0134(20000101)38:1
8. Van Grieken, R., Markowicz, A.: Handbook of X-Ray Spectrometry (2001)
9. Kump, P., Vogel-Mikuš, K.: Quantification of 2D elemental distribution maps of intermediate-thick biological sections by low energy synchrotron μ-X-ray fluorescence spectrometry. J. Instrum. **13**, 1–17 (2018). https://doi.org/10.1088/1748-0221/13/05/C05014
10. Nečemer, M., Kump, P., Vogel-Mikuš, K.: Use of X-ray fluorescence-based analytical techniques in phytoremediation (2011)
11. Kavčič, A., Mikuš, K., Debeljak, M., et al.: Localization, ligand environment, bioavailability and toxicity of mercury in Boletus spp. and Scutiger pes-caprae mushrooms. Ecotoxicol. Environ. Saf. **184** (2019). https://doi.org/10.1016/j.ecoenv.2019.109623
12. Pongrac, P., Arčon, I., Castillo-Michel, H., Vogel-Mikuš, K.: Mineral element composition in grain of Awned and Awnletted wheat (Triticum aestivum L.) cultivars: tissue-specific iron speciation and phytate and non-phytate ligand ratio. Plant. **9**:79 (2020). https://doi.org/10.3390/plants9010079
13. Vogel-Mikuš, K., Pongrac, P., Pelicon, P.: Micro-PIXE elemental mapping for ionome studies of crop plants. Int. J. PIXE **24**, 217–233 (2014). https://doi.org/10.1142/s0129083514400142
14. Koren, Š, Arčon, I., Kump, P., et al.: Influence of CdCl2 and CdSO4 supplementation on Cd distribution and ligand environment in leaves of the Cd hyperaccumulator Noccaea (Thlaspi) praecox. Plant. Soil. **370**, 125–148 (2013). https://doi.org/10.1007/s11104-013-1617-0
15. Laforce, B., Vermeulen, B., Garrevoet, J., et al.: Laboratory scale X-ray fluorescence tomography: instrument characterization and application in earth and environmental science. Anal. Chem. **88**, 3386–3391 (2016). https://doi.org/10.1021/ACS.ANALCHEM.6B00137
16. Masteghin, M.G., Gervais, T., Clowes, S.K., et al.: Benchmarking of X-ray fluorescence microscopy with ion beam implanted samples showing detection sensitivity of hundreds of atoms. Small. Methods. (2024). https://doi.org/10.1002/SMTD.202301610
17. Nečemer, M., Kump, P., Šket, P., et al.: A novel analytical technique suitable for the identification of plastics. Acta Chim. Slov. **60**, 701–705 (2013)
18. Janssens, K.H.: A mobile instrument for in situ scanning macro-XRF investigation of historical paintings. J. Anal. At. Spectrom. (2013)
19. Nečemer, M., Potočnik, D., Ogrinc, N.: Discrimination between Slovenian cow, goat and sheep milk and cheese according to geographical origin using a combination of elemental content and stable isotope data. J. Food Compos. Anal. **52**, 16–23 (2016). https://doi.org/10.1016/j.jfca.2016.07.002
20. Ocvirk, M., Nečemer, M., Košir, I.J.: The determination of the geographic origins of hops (Humulus lupulus L.) by multi-elemental fingerprinting. Food Chem. **277**, 32–37 (2019). https://doi.org/10.1016/j.foodchem.2018.10.070
21. Mahne Opatić, A., Nečemer, M., Kocman, D., Lojen, S.: Geographical origin characterization of slovenian organic garlic using stable isotope and elemental composition analyses. Acta. Chim. Slov. **64**, 1048–1055 (2017). https://doi.org/10.17344/acsi.2017.3476
22. Lilek, N., Kandolf Borovšak, A., Bertoncelj, J., et al.: Use of EDXRF elemental fingerprinting for discrimination of botanical and geographical origin of Slovenian bee pollen. X-Ray Spectrom. **51**, 186–197 (2022). https://doi.org/10.1002/xrs.3250
23. Kropf, U., Korošec, M., Bertoncelj, J., et al.: Determination of the geographical origin of Slovenian black locust, lime and chestnut honey. Food Chem. **121**, 839–846 (2010). https://doi.org/10.1016/j.foodchem.2009.12.094
24. Nečemer, M., Kump, P., Lazar, T.: Seal dies from the national museum of Slovenia: non-destructive analyses of medieval and early modern copper alloys. Archaeom. **62**, 965–973 (2020). https://doi.org/10.1111/arcm.12579

25. Hönicke, P., Kayser, Y., Beckhoff, B., et al.: Characterization of ultra-shallow aluminum implants in silicon by grazing incidence and grazing emission X-ray fluorescence spectroscopy. J. Anal. At. Spectrom. **27**, 1432–1438 (2012). https://doi.org/10.1039/C2JA10385K
26. Kayser, Y., Sá, J., Szlachetko, J.: Depth-resolved X-ray absorption spectroscopy by means of grazing emission X-ray fluorescence. Anal. Chem. **87**, 10815–10821 (2015). https://doi.org/10.1021/ACS.ANALCHEM.5B03346
27. Eckert, M.: Max von laue and the discovery of X-ray diffraction in 1912. Ann. Phys. **524**, A83–A85 (2012). https://doi.org/10.1002/ANDP.201200724
28. Staufer, T., Grüner, F.: Review of development and recent advances in biomedical X-ray fluorescence imaging. Int. J. Mol. Sci. **24**, 10990 (2023). https://doi.org/10.3390/IJMS241310990
29. Sparks, C.J., Raman, S., Ricci, E., et al.: X-ray fluorescence with synchrotron radiation. IEEE Trans. Nucl. Sci. **26**, 1368–1372 (1979). https://doi.org/10.1109/TNS.1979.4330388
30. Raimondi, P., Benabderrahmane, C., Berkvens, P., et al.: The extremely brilliant source storage ring of the European synchrotron radiation facility. Commun. Phys. **6** (2023). https://doi.org/10.1038/S42005-023-01195-Z
31. Raimondi, P.: ESRF-EBS: the extremely brilliant source project. Synchrotron. Radiat. News. **29**, 8–15 (2016). https://doi.org/10.1080/08940886.2016.1244462
32. Alianelli, L., Laundy, D., Alcock, S., et al.: Development of hard X-ray focusing optics at diamond light source. Synchrotron. Radiat. News. **29**, 3–9 (2016). https://doi.org/10.1080/08940886.2016.1198667
33. Nazaretski, E., Xu, W., Yan, H., et al.: Microscopy instrumentation and nanopositioning at NSLS-II: current status and future directions. Synchrotron. Radiat. News. **31**, 3–8 (2018). https://doi.org/10.1080/08940886.2018.1506233
34. Li, K., Ali, S., Wojcik, M., et al.: Tunable hard x-ray nanofocusing with Fresnel zone plates fabricated using deep etching. Optica. **7**, 410 (2020). https://doi.org/10.1364/OPTICA.387445
35. Jiang, H., Xie, J., He, Y., et al.: Multilayer Kirkpatrick-Baez focusing mirrors with phase compensation for sub-20 nm focusing at the hard X-ray nanoprobe beamline of SSRF. Opt. Express **32**, 13597 (2024). https://doi.org/10.1364/OE.514734
36. Takeuchi, A., Suzuki, Y.: Recent progress in synchrotron radiation 3D–4D nano-imaging based on X-ray full-field microscopy. Microscopy **69**, 259–279 (2020). https://doi.org/10.1093/JMICRO/DFAA022
37. Vanhoof, C., Bacon, J.R., Fittschen, U.E.A., Vincze, L.: Atomic spectrometry update: review of advances in X-ray fluorescence spectrometry and its special applications. J. Anal. At. Spectrom. **37**, 1761–1775 (2022). https://doi.org/10.1039/D2JA90035A
38. Mantouvalou, I., Lachmann, T., Singh, S.P.S.P., et al.: Advanced absorption correction for 3D elemental images applied to the analysis of pearl millet seeds obtained with a laboratory confocal micro X-ray fluorescence spectrometer. Anal. Chem. **89**, 5453–5460 (2017). https://doi.org/10.1021/acs.analchem.7b00373
39. Mejía-Ponce, L.V., Hernández-López, A.E., Reynoso-Cruces, S., et al.: Improvements to the X-ray spectrometer at the aerosol laboratory, Instituto de Física, UNAM. J. Nucl. Phys., Mater. Sci., Radiat. Appl. **6**, 57–60 (2018). https://doi.org/10.15415/JNP.2018.61009
40. Cirrincione, D., Antonelli, M., Aquilanti, G., et al.: A new collimated multichannel modular detection system based on Silicon drift detectors. Nucl. Instrum. Methods. Phys. Res. A. **1049** (2023). https://doi.org/10.1016/J.NIMA.2023.168118
41. Ryan, C., Siddons, D., Kirkham, R., et al.: The Maia detector journey: development, capabilities and applications. Microsc. Microanal. **24**, 720–721 (2018). https://doi.org/10.1017/S1431927618004099
42. Chen, W., Elliott, D., Giacomini, G., et al.: Improvement of SDD-Maia detector for X-ray fluorescence detection. Nucl. Sci. Symp. Med. Imaging Conf. (2017). https://doi.org/10.1109/NSSMIC.2017.8532602
43. Vanhoof, C., Bacon, J.R., Fittschen, U.E.A., Vincze, L.: Atomic spectrometry update–a review of advances in X-ray fluorescence spectrometry and its special applications. J. Anal. At. Spectrom. **36**, 1797–1812 (2021). https://doi.org/10.1039/D1JA90033A

44. Howard, D.L., Jonge, M.D.D., Afshar, N., et al.: The XFM beamline at the Australian synchrotron. J. Synchrotron Radiat. **27**, 1447–1458 (2020). https://doi.org/10.1107/S1600577520010152
45. Wang, S., Yang, Z., Ma, X., et al.: Emerging synchrotron radiation X-ray-sensitive probes for in situ bioimaging at the nanoscale. TrAC, Trends Anal. Chem. **170**, 117453 (2024). https://doi.org/10.1016/J.TRAC.2023.117453
46. Winarski, R.P., Holt, M.V., Rose, V., et al.: A hard X-ray nanoprobe beamline for nanoscale microscopy. J. Synchrotron Radiat. **19**, 1056–1060 (2012). https://doi.org/10.1107/S0909049512036783
47. Chen, S., Deng, J., Yuan, Y., et al.: The Bionanoprobe: hard X-ray fluorescence nanoprobe with cryogenic capabilities. J. Synchrotron Radiat. **21**, 66–75 (2014). https://doi.org/10.1107/S1600577513029676
48. Martínez-Criado, G., Villanova, J., Tucoulou, R., et al.: ID16B: a hard X-ray nanoprobe beamline at the ESRF for nano-analysis. J. Synchrotron Radiat. **23**, 344–352 (2016). https://doi.org/10.1107/S1600577515019839
49. Warlo, M., Bark, G., Wanhainen, C., et al.: Extreme-resolution synchrotron X-Ray fluorescence mapping of ore samples. Ore Geol. Rev. **140**, 104620 (2022). https://doi.org/10.1016/J.OREGEOREV.2021.104620
50. do Nascimento, C.W.A., Hesterberg, D., Tappero, R.: Imaging Zn and Ni distributions in leaves of different ages of the hyperaccumulator Noccaea caerulescens by synchrotron-based X-ray fluorescence. J. Hazard. Mater. **408**, 124813 (2021). https://doi.org/10.1016/J.JHAZMAT.2020.124813
51. Rumancev, C., Vöpel, T., Stuhr, S., et al.: Micro x-ray fluorescence analysis of trace element distribution in frozen hydrated HeLa cells at the P06 beamline at Petra III. Biointerphases. **16** (2021). https://doi.org/10.1116/6.0000593/124417
52. Budka, D., Mesjasz-Przybyłowicz, J., Przybyłowicz, W.J.: Micro-PIXE analysis: importance of biological sample preparation techniques. Radiat. Phys. Chem. **71**, 785–786 (2004). https://doi.org/10.1016/j.radphyschem.2004.04.091
53. van der Ent, A., Przybyłowicz, W.J., de Jonge, M.D., et al.: X-ray elemental mapping techniques for elucidating the ecophysiology of hyperaccumulator plants. New Phytol. **218**, 432–452 (2018)
54. Gianoncelli, A., Mikuš, K.V., Salomé, M., et al.: Difficulties and artefacts in cryo-fixation of ovarian tissues for X-ray fluorescence analyses. J. Anal. At. Spectrom. **38**, 1744–1750 (2023). https://doi.org/10.1039/D3JA00164D
55. Solé, V.A.A., Papillon, E., Cotte, M., et al.: A multiplatform code for the analysis of energy-dispersive X-ray fluorescence spectra. Spectrochim. Acta Part B At. Spectrosc. **62**, 63–68 (2007). https://doi.org/10.1016/j.sab.2006.12.002
56. Ryan, C.G.: Developments in Dynamic Analysis for quantitative PIXE true elemental imaging. Nucl. Instrum. Methods. Phys. Res. B. **181**, 170–179 (2001). https://doi.org/10.1016/S0168-583X(01)00374-3
57. Klysubun, W., Tarawarakarn, P., Thamsanong, N., et al.: A new full-field XRF imaging station at Synchrotron light research institute. J. Synchrotron Radiat. **30**, 655–661 (2023). https://doi.org/10.1107/S1600577523002722/RV5167SUP1.PDF
58. Romano, F.P., Caliri, C., Cosentino, L., et al.: Macro and micro full field x-ray fluorescence with an X-ray pinhole camera presenting high energy and high spatial resolution. Anal. Chem. **86**, 10892–10899 (2014). https://doi.org/10.1021/AC503263H
59. An, S., Krapohl, D., Norlin, B., Thungström, G.: Full-field X-ray fluorescence imaging with a straight polycapillary X-ray collimator. J. Instrum. **15**, P12033 (2020). https://doi.org/10.1088/1748-0221/15/12/P12033
60. Soltau, J., Soltau, J., Meyer, P., et al.: Full-field x-ray fluorescence imaging using a Fresnel zone plate coded aperture. Optica. **10**(1), 127–133 **10**:127–133 (2023). https://doi.org/10.1364/OPTICA.477809
61. Romano, F.P., Caliri, C., Cosentino, L., et al.: Micro X-ray fluorescence imaging in a tabletop full field-X-ray fluorescence instrument and in a full field-particle induced X-ray emission end

station. Anal. Chem. **88**, 9873–9880 (2016). https://doi.org/10.1021/ACS.ANALCHEM.6B02811/ASSET/IMAGES/LARGE/AC-2016-02811V_0011.JPEG
62. (2024) https://nucleus.iaea.org/sites/nuclear-instrumentation/Pages/Full-Field-XRF.aspx
63. Pereira, G.R., Rocha, H.S., Calza, C., et al.: Imaging system for XRF microtomography at LNLS-Brazil. IEEE Trans. Nucl. Sci. **56**, 1426–1429 (2009). https://doi.org/10.1109/TNS.2008.2010938
64. Pereira, G.R., Rocha, H.S., Calza, C., et al.: 3D elemental distribution images in biological samples by XRFμCT. X-Ray Spectrom. **40**, 260–264 (2011). https://doi.org/10.1002/XRS.1334
65. Kuang, Y., Pratx, G., Bazalova, M., et al.: First demonstration of multiplexed X-Ray fluorescence computed tomography (XFCT) imaging. IEEE Trans. Med. Imag. **32**, 262–267 (2013). https://doi.org/10.1109/TMI.2012.2223709
66. Spiers, K.M., Brueckner, D., Garrevoet, J., et al.: Synchrotron XFM tomography for elucidating metals and metalloids in hyperaccumulator plants. Metallomics. **14** (2022). https://doi.org/10.1093/MTOMCS/MFAC069
67. Pereira, G.R., Anjos, M.J., Rocha, H.S., et al.: Computed tomography and X-ray fluorescence CT of biological samples. Nucl. Instrum. Methods. Phys. Res. A. **580**, 951–954 (2007). https://doi.org/10.1016/J.NIMA.2007.06.113
68. Mijovilovich, A., Mishra, A., Brückner, D., et al.: MicroX-ray absorption near edge structure tomography reveals cell-specific changes of Zn ligands in leaves of turnip yellow mosaic virus infected plants. Spectrochim. Acta Part B At. Spectrosc. **157**, 53–62 (2019). https://doi.org/10.1016/J.SAB.2019.05.005
69. Marin, F., Chen, S., De Carlo, F., et al.: Processing X-ray fluorescence tomography data with XRFtomo. Microsc. Microanal. **26**, 1282–1284 (2020). https://doi.org/10.1017/S1431927620017572
70. Gao, B., Laforce, B., Dhaene, J., et al.: An analytical simulation method for X-ray fluorescence computed tomography. Radiat. Phys. Chem. **170**, 108622 (2020). https://doi.org/10.1016/J.RADPHYSCHEM.2019.108622
71. Gao, B., Laforce, B., Vincze, L., et al.: Quantitative reconstruction of polychromatic X-ray fluorescence computed tomography using transmission tomography. Anal. Chem. **93**, 2082–2089 (2021). https://doi.org/10.1021/ACS.ANALCHEM.0C03828
72. Mastronarde, D.N., Held, S.R.: Automated tilt series alignment and tomographic reconstruction in IMOD. J. Struct. Biol. **197**, 102–113 (2017). https://doi.org/10.1016/J.JSB.2016.07.011
73. Rau, C., Somogyi, A., Bytchkov, A., Simionovici, A.S.: XANES micro-imaging and tomography. **4503**, 249–255 (2002). https://doi.org/10.1117/12.452851
74. Price, S.W.T., Ignatyev, K., Geraki, K., et al.: Chemical imaging of single catalyst particles with scanning μ-XANES-CT and μ-XRF-CT. Phys. Chem. Chem. Phys. **17**, 521–529 (2014). https://doi.org/10.1039/C4CP04488F
75. Ravel, B., Newville, M.: ATHENA, ARTEMIS, HEPHAESTUS: data analysis for X-ray absorption spectroscopy using IFEFFIT. J. Synchrotron Radiat. **12**, 537–541 (2005). https://doi.org/10.1107/S0909049505012719

Recent Developments in Total Reflection X-Ray Fluorescence and Applications

5

Sangita Dhara and N. L. Mishra

Abstract

Total reflection X-ray fluorescence (TXRF) is a novel variant of energy dispersive X-ray fluorescence (EDXRF) with one significant geometric difference between them. Synchrotron Radiation is probably the most highlighted development in the field of TXRF which has opened up new analytical aspects of hyphenated techniques such as TXRF-X-ray absorption fine structures (TXRF-XAFS) which not only provides information of elemental concentration but also on speciation, local atomic and electronic structures of materials. Novel sample preparation and preconcentration approaches have also contributed significantly in increasing the analytical sensitivity and precision. These include dispersive liquid-liquid micro-extraction (DLLME), cloud point extraction (CPE), hollow fiber liquid phase microextraction (HF-LPME), etc. Further, use of membranes and nanomaterials for sample preconcentration have drastically improved the TXRF detection limits from picogram to femtogram level thereby, making it an excellent micro-analytical technique for analysis of micro-samples. Micro-TXRF is another upcoming variant of TXRF leading to elemental profiling at trace levels in the field of medical, biological, environmental, archeological, industrial and forensics.

S. Dhara (✉)
Fuel Chemistry Division, Bhabha Atomic Research Centre, Mumbai, India
e-mail: sdhara@barc.gov.in

Homi Bhabha National Institute, Anushakti Nagar, Mumbai, India

N. L. Mishra
Madan Mohan Malviya Technological University, Gorakhpur, Uttar Pradesh, India

V. K. Singh (ed.), *X-Ray Fluorescence Spectroscopy and Chemometrics*,
https://doi.org/10.1007/978-3-031-98375-7_5

5.1 Environment and Its Impact on Human Health

Our natural environment consists of earth atmosphere, land, water bodies, soil and different living things surrounding us. The quality of natural environment is very important for our good health, ease of living and survival e.g., the air we breathe in, the water we drink or the food we eat should be of good quality and their quality depends on environmental quality. In recent years, the quality of our environment has deteriorated drastically due to different human activities e.g., mining, chemical industries, power production, biological activities, hospital waste, use of fertilizers, fire crackers, transport vehicles, etc. These anthropogenic activities can pollute the environment with harmful microbes as well as hazardous chemicals. Further causes of environmental deterioration are due to the impact of human population increase and also due to improper protocol being followed in regulation of the activities to protect their impact on the environment. Improper disposal of waste into water, air and land has led to the degradation of aquatic systems due to accumulation of harmful chemicals in rivers, lakes, ocean and atmosphere. This has increased the spreading of air and water borne diseases, caused severe damage to crops and has disturbed the food chain.

As the human progress cannot be stopped, it is very difficult to completely stop these activities but certainly it is very much necessary to assess the impact of these anthropogenic and some natural activities on the environment and take the required remedial actions. These assessments can be made by evaluating the level of pollution by virus, bacteria, dust, chemical elements and their speciation in the environment.

5.2 Role of Trace Element in Environmental Quality Management

Trace elements play an important role in the assessment of the environmental quality because they can affect the human health and the quality of their life. Trace element as defined by IUPAC is any element that is having an average concentration less than 100 parts per million or 100 mg/Kg [1]. Elements including Cu, Zn, Fe, Mn, Mo, B, Co, Ni, As, Pb, Cr, Hg, Se, etc. have been studied extensively for decades to assess the environmental quality. Environmental pollution is the biggest concern due to rapid urbanization and industrialization. Various anthropogenic activities have lead to the generation of several hazardous substances into the air, soil and water. The increasing atmospheric greenhouse gas concentrations have lead to rising temperatures as well as alterations in global water cycle. The trace determination is important as it helps in assessing, managing and mitigating the impact of anthropogenic as well as natural activities on the ecosystem [2–4]. Knowing the nature and concentration of trace elements in environment helps to determine the level of pollution in the environment by different elements and also helps in the identification of the source of such pollution (natural or anthropogenic). The anthropogenic sources constitute chemical industries, power stations, vehicular traffic, municipal wastes, fire crackers, fertilizers, pesticides, etc., whereas natural sources include erosion of soil, weathering of rocks, dust storm, strong winds and sea winds.

In order to differentiate anthropogenic and natural sources of air pollutants, knowing the size of the particles (particulate matter or PM) containing these trace elements in air is also very much necessary. PM is categorized based on their aerodynamic equivalent diameter (AED), with PM_{10} referring to particles with a diameter ≤ 10 μm, $PM_{2.5} \leq 2.5$ μm, and $PM_{1.0}$ denoting ≤ 1.0 μm. Anthropogenic source particles are smaller in size whereas natural source particles are in comparatively bigger sizes. In addition, the knowledge of the oxidation state as well as chemical species of the elements, that is detected and quantified, is also necessary [5]. Figure 5.1 shows the size variation of different air particulate matters. In order to systematically evaluate the changes in the environmental quality with respect to time and location, a large number of samples are required to be collected from different locations at different time scales. This requires a large man power as well as huge sampling time. In addition, sometimes the amount of sample available to probe the atmospheric particles is very minute. Determination of trace elements in ocean helps in understanding the effects of climatic changes and the bio-geochemical cycles [6]. Study of ice cores or sediments provides historical data on environmental changes as well as anthropogenic impacts. They contain a wealth of information on the climatic changes and atmospheric aerosols [7]. Determination of trace elements like F, Cl, Pb, Hg, As, Cd, U, etc. in environmental samples, especially water samples, ensures the safety of drinking water and helps in regulation of these elements by various agencies e.g. World Health Organization (WHO), Environmental Protection Agency (EPA), etc. [8]. Trace elemental analysis of soil is vital for understanding soil contamination, its fertility, and suitability for agricultural purpose. The presence of micronutrients like B, Cl, Mn, Cu, Zn and Mo, in optimal quantities is critical for plant growth and crop yields. Monitoring these elements ensures optimal soil conditions for agriculture. Trace elemental analysis in flora and fauna helps to understand their accumulation through the food chains and its potential toxicity. Further an array of analytical techniques are required depending on the state of the sample (Solid, liquid, or gas), the elements to be detected and their concentration, its availability, the purpose of analysis.

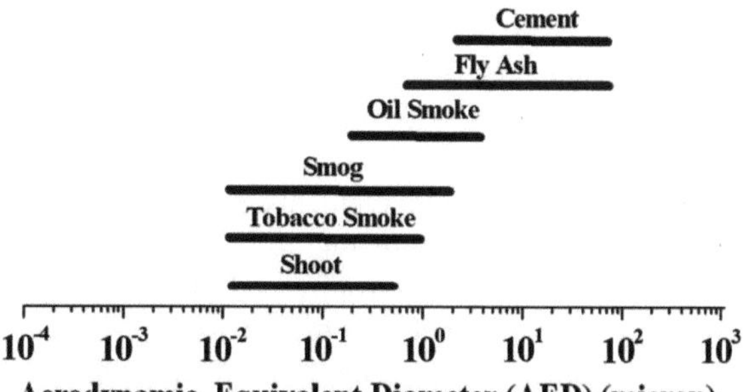

Fig. 5.1 The aerodynamic equivalent diameter (AED) of different air particulate matters

5.3 Ideal Trace Element Analytical Techniques

Trace elemental determinations are crucial in many areas due to their significant effect on the material properties. Though they are present in extremely small concentrations they can have a major impact on the physical, chemical, structural, biological properties of the material [3, 9, 10]. Determination of trace hazardous elements in environmental samples like water, soil, air, helps in assessing, controlling and reducing pollution in the environment [11–13]. In biological samples, such as body fluids, tissues, hair, etc. analysis of micro-nutrients and toxic elements are crucial for good human health, in order to monitor the deficiencies and toxicities in the body [14]. Quality control of the products, which again involves trace determinations, is an important step in any industrial process as well [15]. Checking the purity of the raw as well as the final industrial product involves determination of trace, minor and major elements, which are present in it.

As there are a vast number of environmental samples with different matrices, analytes and concentration it is very difficult for a single technique to satisfy all the above needs. A gamut of analytical techniques are available for trace determinations in environmental samples such as atomic absorption spectroscopy (AAS), inductively coupled plasma (ICP) spectroscopy with optical emission spectroscopy (OES), and mass spectrometry (MS), electroanalytical techniques, X-ray fluorescence (XRF), neutron activation analysis (NAA), particle induced X-ray emission (PIXE) [16]. The analytical techniques for trace and ultra-trace determination need the control of blank and verification of the analytical data. The error caused during the analysis can be mitigated by proper sampling, homogenization, storage, sample preparation and careful analysis. Each technique has their own unique features, strengths and limitations and it is crucial to select the appropriate method depending on the specific need. Atomic spectrometric techniques are quite sensitive however, their accuracy can be affected by the matrix of the sample, if proper care is not taken. X-ray and nuclear techniques provide the ease of non-destructive analysis but have very high limit of detections and matrix insensitivity. ICP based techniques have very high sensitivity but require cumbersome digestion of the sample. In Table 5.1, a comparison of some important analytical features of a few well established trace analytical techniques is tabulated. The contribution of different analytical techniques in terms of published articles in the field of environmental science for a period of 2006 to 2020 is shown in Fig. 5.2. It can be observed from the table that, XRF amongst all the other techniques has many advantageous features such as non-destructive analysis, minimum sample preparation, lesser spectral interference, versatility and very low maintenance and running cost for different sample matrices. However, the major disadvantages are lower sensitivity for light elements, sever matrix effects that can affect the accuracy and higher detection limits due to high scattering and background. With advancements in the field of electronics and optics, X-ray tubes and detectors with more efficiency are developed in order to expand the elemental range and improve the detection limits. Development of miniaturized and low power X-ray sources, novel X-ray optics devices, high transmission X-ray detector windows and improvement in signal detection systems are some of the recent advancements in XRF. Due to these developments, XRF is now comparable with other well established techniques for trace determinations. The various modes of operations in XRF includes: EDXRF, wavelength dispersive XRF (WDXRF), polarised EDXRF (P-EDXRF), micro-XRF (μ-XRF) and TXRF [17]. WDXRF employs diffraction by a dispersing crystal to separate the characteristic wavelengths emitted from the sample after its excitation by an X-ray source. A goniometer is used for maintaining the required angle between the diffraction device and the detector. Due to the high-resolving power of the crystals, characteristic lines with close energies can be detected without any interference with each other. As spectral interferences are avoided due to dispersion of emitted X-rays, the detectors with low intrinsic resolution such as proportional counters or scintillation detectors can be used. In EDXRF, only the X-ray source and detector are required. However, unlike WDXRF in EDXRF, the detectors with high intrinsic resolution such as semiconductors are employed. This detector with multichannel analyser (MCA) allows the simultaneous determination of the emitted X-rays with faster acquisition speed and display of data. In both WDXRF and EDXRF it has 2D geometry. However if the primary, secondary and the characteristic radiations from the sample are configured at mutually orthogonal angles it provides a 3D geometry and the main advantage of this geometry is that the scattered radiation cannot reach the detector because of polarization, so the sensitivity of polarised EDXRF (P-EDXRF) becomes better than that of 2D geometry. A specialized form of EDXRF is μ-XRF that provides elemental analysis with high spatial resolution down to micron scale. It is mainly used to evaluate the composition as well as distribution of the analytes within the sample. In μ-XRF a focused or collimated X-ray beam often achieved with a polycapillary optics or mirror, excites the sample. The focused beam is rastered across the sample surface and the data is collected at each discrete point to create a high resolution elemental distribution map of the analytes. TXRF, an advanced geometrical variant of XRF, is used widely in various fields, including environmental, archeological, material, biological and industrial. In TXRF the excitation under grazing incidence is emphasized and it has 90/0 degree geometry unlike 45/45 degree geometry in conventional EDXRF. This makes use of the fact that at very low glancing angles, the high background that would generally occur due to scatter from the sample support, is negligible. This low background makes detection limits in sub ppm or ppb level. The key features of TXRF analysis include simultaneous detection of elements from C to U, very low detection limits in the range of ppb level, requirement of very small sample size for analysis and negligible matrix effect [18]. The schematic representation of TXRF geometry is given in Fig. 5.3.

Table 5.1 Comparison of analytical features of techniques used for trace determinations

Features	XRF	ICP-MS	INAA	ICP-AES	ET-AES	Electro analytical
Sample						
Volume	2–5 μL/ 10–200 mg	2–5 mL	10–200 mg	2–5 mL	5 μL	2–5 mL
Analysis of solids	Digestion or suspension	Digestion	None	Digestion	Digestion or suspension	Digestion or suspension
Consumption	No	Yes	No	Yes	Yes	No
Detection						
Detection limits	1 ppm	<0.01 ppb	<1 ppb	<1 ppm	<0.1 ppb	ppb
Spectral interference	Very few	Severe	A Few	Moderate	None	None
Elemental limitations	Z < 10	H, C, N, O, F, P, S	Z < 9, Tl, Pb, Bi	Non-metals	Refractory elements	–
Isotope detection	No	Yes	No	No	No	No
Quantification						
Calibration	Several standards	Several standards	Pure element foils	Several standards	Standard for each element	–
Matrix effects	Severe	Severe	Negligible	Moderate	Moderate	Moderate
Measurement time	100–1000 s	~200 s	30 min-few days	~300 s	~300 s	~1000 s
Expenditure						
Capital cost	Low	very High	Very high	High	Low	Low
Maintenance	Seldom	Frequent	Seldom	Frequent	Seldom	Low
Running cost	Low	High	high	High	Low	Low

5.4 Suitability of TXRF for Trace Element Determinations in Environmental Samples

The two significant characteristic features of TXRF that are exploited for trace elemental analysis are high reflectivity on flat smooth sample support surfaces and the low penetration depth of the primary radiation. These two features drastically reduce the detection limits of TXRF to ng/mL level. The advantageous features of TXRF e.g. Low detection limits, simple quantification method, requirement of small sample mass and possibility of depth profiling, have made this technique highly efficient for trace and micro analysis [19–21]. TXRF has found its applications in a variety of fields including pharmaceutical, geological, biological, industrial, electronics, etc. Furthermore, its applications have been very significant in the field of environmental studies and pollution control. TXRF has found a wide

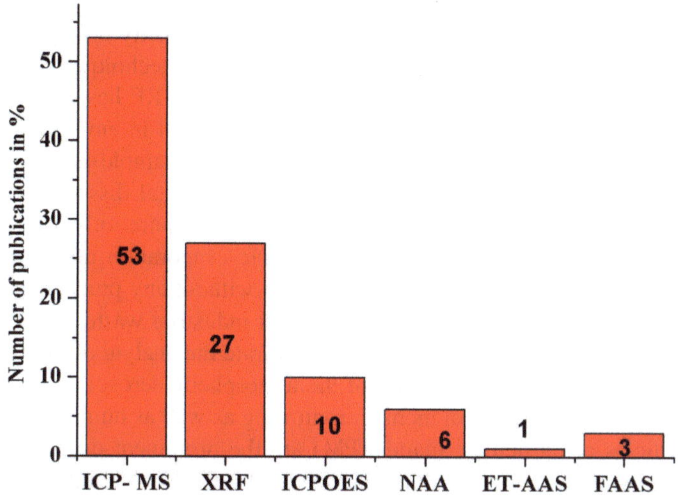

Fig. 5.2 Comparison of the contributions of different analytical techniques in terms of published articles in the field of environmental science for the period of 2006 to 2020

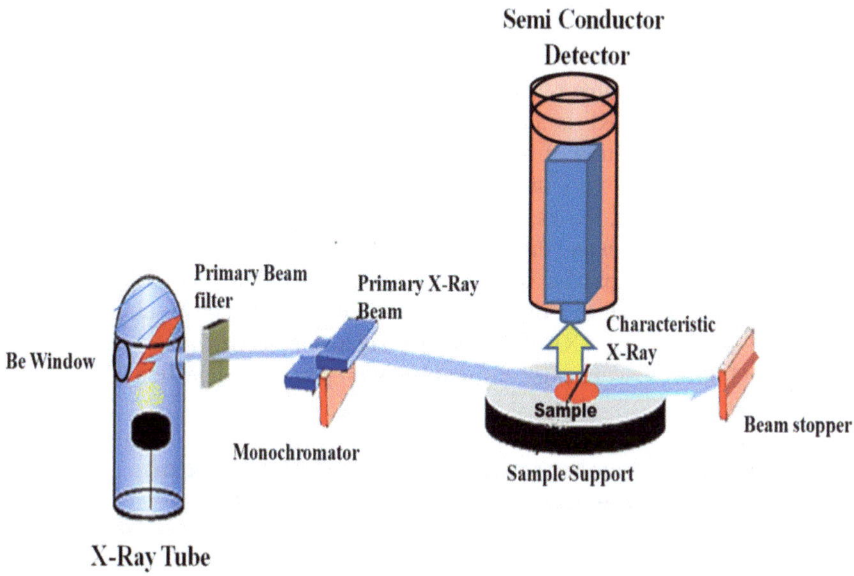

Fig. 5.3 Schematic representation of TXRF geometry

application for ultratrace analysis of environmental samples e.g. water, air borne particulate matter, fly ash, soils, and sediments. Water quality is of great concern to human and therefore, the need for suitable analytical technique for its assessment is growing continuously. The high sensitivity of TXRF has allowed it to be applied for the detection of trace elements in ng/mL level in environmental samples. Depending on the nature of the water sample the sample preparation steps may vary for its analysis by TXRF. Samples with low total dissolved salt (TDS) such as drinking water, rain water, ground water, etc. require only a few preparatory steps and can be pre-concentrated by simple evaporation after the addition of an internal standard or even analysed directly without any preconcentration. In samples with high TDS value especially sea and industrial waste water, chemical treatment is required to separate and pre-concentrate the analyte of interest. Determination of the chemical composition of the atmospheric aerosol particles, have a great significance on the environmental monitoring as well as human health. These aerosol particles or particulate matter (PM) are the prominent primary pollutants in the air and may contain heavy metals such as As, Cr, Pb, Ni and Cd, which can adversely affect human health [22, 23]. TXRF analysis requires smooth sample surface, therefore highly polished supports are used. For analysis of PM, the sample can be directly collected on these support and analyzed. Soil is another critical natural resource that plays a key role in determining the human health, and provides essential nutrients in the environment. Assessment of soil in terms of trace elemental analysis is required as it is associated with soil quality, crop health, water quality and finally, to human health as well as economic development. For analysis of soil samples, TXRF is an efficient and rapid analytical technique allowing simultaneous determination of many elements from Na to U in the periodic table. In addition, it requires very small sample quantities [24]. Biological indicators such as mosses, lichens, leaves, barks, and woods can also be used to evaluate the deposition, accumulation and distribution of heavy and toxic metals in the environment. Analysis of such passive samples has some benefits such as easy availability, low-cost, temporal resolution of environmental changes, and easy sampling procedures. The advantage of TXRF analysis for such samples is that these can either be measured directly, as powder suspension, slurry or after digestion.

5.5 Recent Developments in Sample Preparation Methodologies for Better Detection Limits

In recent years, with the developments in detector technology, optics and instrumentation, novel analytical methodologies, integration with automation for high-throughput analysis, incorporation of additional multivariate analysis methods such as fundamental parameters, partial least squares, principal components regression, multiple and step-wise regression and artificial intelligence (AI), TXRF has become highly suitable for performance of trace elemental analysis of various kinds of environmental samples. The continuous developmental work for creation of large facilities, such as synchrotron radiation (SR) and free electron laser (FEL), together with the significant improvements in X-ray optics have greatly expanded the application area of X-ray techniques for elemental analysis, spectroscopy as well as imaging studies. Several strategies have been followed for sample pre-treatment prior to TXRF analysis of environmental samples, e.g. acid digestion, microwave-assisted digestion, direct microdigestion onto the sample carrier, slurry sampling, suspensions, extraction, ultrasound-assisted extraction, microextraction, etc. Each approach displays its own advantages and disadvantages. Sample preparation in TXRF prior to analysis is necessary to achieve the following goals: remove matrix, thin film formation to ensure total reflection condition and minimize matrix effect, homogenous distribution of sample and internal standard, sample enrichment and avoid peak interference.

Sample carriers for TXRF analysis are generally circular disks of 30 mm diameter and 3 mm thickness. They are made of quartz glass, glassy carbon, sapphire, acrylic glass, Si wafer etc. In a recent advancement, Diamond-like carbon (DLC) sample supports presenting a smaller spot areas for analysis were used. A micropipette is generally used for drop casting of a liquid (μL) on TXRF sample supports. Q-tips and impactors are employed for direct deposition of solid particles on the supports. A deposition of a nanoliter or picoliter drop using an inkjet printer has emerged as an alternative to the traditional micropipettes. Recently, a new nanoliter droplet deposition unit was developed, which consists of an automatic pipette and a table equipped with three stepper motors and a vacuum pump. The three stepper motors allow the movement of the sample support in all the three directions [25].

5.5.1 Water Samples

Over several years, TXRF spectroscopy has become known as a versatile tool for elemental analysis across many application fields. The International Standard Organization (ISO) 20289 provides a method for determining the elemental concentration of impurities dissolved in water samples [26]. It exploits the advantages of TXRF spectroscopy to directly measure a dried residue on an X-ray reflector. Pure waters, such as rainwater, groundwater, tap and mineral water, can be analyzed almost directly. After acidification and addition of an internal standard, the sample is homogenized, evaporated and 10–50 μL is drop-casted on clean sample support. There are several studies and reviews reported for the determination of trace elements in water samples [27–33]. In ISO 20289, a chemical metrology for quantitative analysis, in order to obtain traceable results within a defined degree of accuracy and precision, is provided according to good practice and in compliance with ISO 17025. In addition, the standard presents a detailed description of sample preparation, contributions of the factors which affect the uncertainty, the general procedure to check instrumental calibration, determining limits of detection and quantification, and methods for spectra fitting.

Among the analysis of different water samples, a special mention is worth for the ice core and glacier samples which give paleoclimatic information on Earth's climate and environmental pollution at different timescales. Preliminary studies revealed that TXRF may be a very useful technique for the elemental analysis of ice cores [34]. A vertical timeline of past climate is stored and can be extracted from the ice-core layers on the basis of the elemental analysis in these layers. The first ever synchrotron radiation total X-ray fluorescence (SR-TXRF) and grazing incidence X-ray absorption spectroscopy (GI-XAS) experiments on mineral particles extracted from Alpine and Antarctic ice cores, dating back 160,000 yrs was carried out at the Stanford synchrotron radiation lightsource (SSRL) at the beamline 6–2 [35]. A dedicated high vacuum (10^{-7} Torr) experimental chamber was developed for this experiment. This chamber consisted of manipulator having six degrees of freedom with high precision for translation along X, Y and Z directions and angular accuracy. A silicon drift detector with thin AP 3.3 polymer entrance window was used which allows the collection of fluorescence signals down to energies as low as Boron Kα. The ocean sediment cores provide records dating back several million years whereas deep ice cores cover relatively shorter periods, with a high temporal resolution. The solid particulate matter in the ice core is a significant component as its compositional analysis can reveal temporal variations of the activity of dust sources and of their environmental conditions. These activities have consequences on global bio-geochemical cycles, and influence mineral alteration under different climatic conditions. The main challenge in such analysis is because the solid material available for analysis is in very small amounts. The particulate concentration in ice varies from 500 ng/g down to 20 ng/g. Extraction of this insignificant amount of insoluble fraction from ice core requires careful sample handling and clean protocols as describe by Cibin et al. [36]. However, the problems in frequent availability of beamline facilities and exorbitant cost for developing new facilities have caused hindrance for routine sample analysis. Therefore, development of an alternative lab based TXRF system was needed for such applications. The combination of powerful X-ray microfocus sources coupled to polycapillary optics (PolyCO) now makes it possible to achieve a focusing spot with diameters less than 50 μm. Moreover, it provides the intense radiation flux which is necessary to perform challenging elemental analysis in environmental samples. In the total external reflection X-ray fluorescence regime, the decreased detection limits and the lower noises are directly correlated to the lower measurable concentration. To apply this special technique in a simpler way, an incident parallel beam is required. This can be achieved by replacing the primary polycapillary full-lens with a semi-lens, which transforms the divergent source beam into a quasi-parallel one. This technique enables excellent results, sometime comparable with those obtained at SR facilities, even for samples with low elemental concentrations [37]. A prototype TXRF spectrometer using a polycapillary semi-lens was designed and assembled at XlabF. The optics were characterized according to the protocol developed and described by Hampai et al. [38]. The study presented a very peculiar capability of polycapillary optics to obtain a powerful quasi-parallel beam from a divergent source. This provided a great opportunity for elemental analysis using TXRF for a very low concentration samples. Using this, TXRF spectrometer, insoluble dust particles from an Antarctic ice core were analyzed successfully.

The spectrometer provided detection limits of nearly 10 ng for elements of interest such as Mg, Al, Si, S, Cl, K, Ca, Ti, Mn and Fe which is comparable to the detection limits obtained using a synchrotron radiation source.

As discussed earlier in this chapter, the uses of chemometric techniques are an alternate approach for quantitative analysis of multi-element samples having matrix effect and severe spectral interferences. There are different calibration methods proposed such as partial least squares (PLS), principal components regression (PCR), multiple linear regression and step-wise regression. In one such studies carried out for trace determination in water samples by TXRF in which the concentration range of elements varied from 0.05 to 2 mg/L, it was concluded that the PLS as regression method had the greatest advantages for such analysis [39]. While analysing water samples, the specific physicochemical properties such as matrix composition, viscosity, total dissolved salt (TSD), its crystallization behavior, etc. should also be taken into account[30]. Samples of low TDS (160–420 mg/L) e.g. drinking water, tap water, bottled water etc. an be analyzed without an chemical pretreatment and with high sensitivity, but in case of water samples with high salt contents (10−100 g/L) and suspended particles such as in river water, wastewater, sea water, etc. the preparation of the sample in the form of thin layer becomes difficult and it results in matrix effects and low sensitivity for trace element quantification. Sample deposition also plays a very critical role in the trace elemental analysis. In an interesting study, It was observed that pipetting 50μL of a sample showed poorer result than dropcasting 20 μL of the sample. While drying 50μL of a sample having high TDS, the condition of thin layer formation was not satisfied and the analysis was not performed under total reflection conditions anymore. Further, drying a large amount of sample with salt matrix, leads to inhomogeneous distribution of elements due to the accumulation of the salts in a ring shaped border that results in matrix effects. Recently, in one of the latest improvements in the reduction in sampled volume of liquid specimen from microliters to nanoliters and to picoliters has been achieved using the modified inkjet printers. The thermal inkjet picofluidic system (TIPS) is a prototype laboratory device which allows spotting of nominal volumes of 1, 5, 10, 35 and 250 picoliters. The size of the dried residue depends on the chemical properties and the surface roughness of the substrate. Therefore, samples were deposited on various sample carriers such as Si wafer, Acrylic glass, ultralene films and AP1 films and the residue size were studied. The study showed that the diameter of the dried residue varied as Si wafer > Acrylic glass > ultralene films > AP1 films. This observation was explained by the different chemical properties of these substrates. The Si wafer surface gets covered with a Si-oxide layer which is fairly hydrophilic and the aqueous sample spreads more on such surfaces. The acrylic glass is a polymer of methyl methacrylate which is less hydrophobic than the polypropylene films because of its ester groups. Hence, the chemical properties of the substrate have a strong influence on the wetting behavior of the aqueous standard. Auger analysis for elemental distribution showed micrometer level homogeneity in the sample residues [40, 41]. While using quarts sample support the surface can be made hydrophobic by covering the surface with 10–20 μL of silicone solution in isopropanol or by fluorocarbonpolymer coated sample supports. However, diamond like carbon (DLC) coated quartz sample supports are more hydrophobic. Using a DLC coated sample supports have the following advantages: (i) As DLC is more hydrophobic, it can reduce the size of the sample residue and enhance the fluorescent X-ray intensities, (ii) DLC films have high flatness and smoothness so high X-ray reflectivity is obtained and (iii) DLC films have high hardness and chemical-resistance, so they can be re-used by cleaning [42].

For samples with high salt content, dilution may also be useful to ensure the formation of a homogeneous thin layer on the reflector. Sample dilution with milli-Q water can help in the reduction of the water salinity as well as improve formation of a thin film [43]. Sample dilution with non-ionic detergent such as triton X−114, 100, etc. are useful for high mineralized water samples. Addition of a Triton X-100 solution to the sample inhibits crystallization process and a more uniform elemental distribution is obtained. A study showed that the dilution of a sample 100–1000 times with Triton X-100 resulted in a good agreement in the analytical results between the TXRF analysis and chemical analysis using ICPMS [44].

There are a number of preconcentration procedures reported for water samples such as solid phase extraction (SPE) including disc, filters, carbon nano tubes (CNT), multi- walled carbon nano tubes (MWCNT), graphene oxides, activated carbon, resins, metal organic framework (MOF), covalent organic framework (COF), and silica gel [45]. In a worth mentioning study, complexing the trace metals with sodium dibenzyldithiocarbamate (NaDBC) followed by automated solid phase extraction of the complexes was carried out for the determination of Mn, Fe, Ni, Cu, Zn, Se(IV), Pb and U in seawater. This method was further modified for As determination by arsenate reduction step performed using a L-cysteine solution followed by a complexation of As^{+3} with NaDBC [46]. In case of Dispersive Solid Phase Microextraction (DMSPE), a small amount of a solid sorbent (μg or mg range) is dispersed into the sample solution to extract the analytes of interest. The nature and properties of the solid sorbent are of prime importance in such studies. Often these sorbents are functionalized for a specific metal, metalloid or inorganic species [47]. After any microextraction procedure, use of a microanalytical technique is required for analysis as only a few microliters of liquid or ng of solids, are recovered to perform the measurement. Therefore, TXRF is a suitable technique for the analysis of such samples. Graphene oxides have excellent adsorptive properties and can be applied for selective separation and preconcentration of specific analyte after its modification with a functional group [48]. Magnetic solid phase extraction (MSPE) is another novel sample preparation methodology which uses magnetic nanoparticles to separate analytes from the sample solution. It is very simple and rapid process, uses less solvent than those used in traditional methods, it is eco-friendly and inexpensive. MSPE can be used with MOFs, CNTs, etc. The liquid phase solvent (LPSE) extraction method includes Dispersive Liquid-phase microextraction (DLPME), supported liquid membrane, hollow fibers, cloud point extraction, etc. In recent developments for preconcentration methodologies, which are sensitive, simpler, inexpensive, eco-friendly and faster, liquid-phase microextraction is an important development [49]. Cloud point

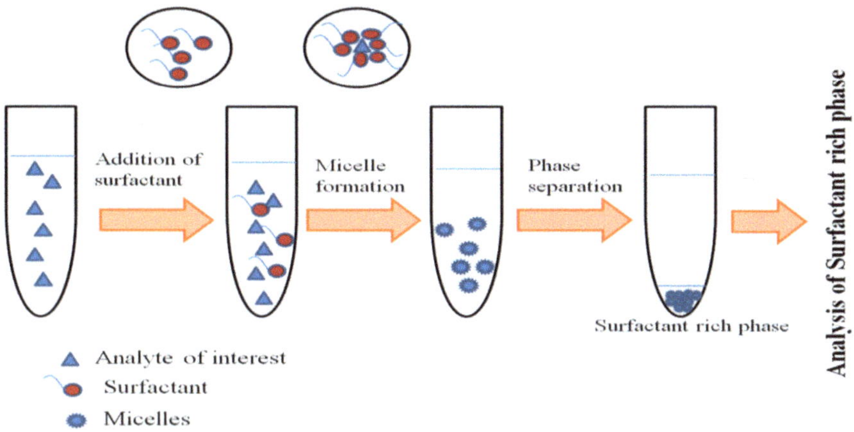

Fig. 5.4 Schematic representation of the cloud point extraction technique

extraction (CEP) also known as Micelle-mediated extraction (MME) is a surfactant based sample preparation methodology in which a hydrophobic species is formed by complexation with an organic ligand in a medium containing the surfactant and the metal complex which is extracted by incorporating into the micelle. Figure 5.4 shows schematic representation of the CEP technique. Another developmental work in TXRF based methodology for the determination of elements in natural water samples at ultra-trace concentration level, is the functionalisation of TXRF quartz sample supports with the extractant and directly immersing them into the sample solution for extraction of analyte on the support. The analytes get extracted and preconcentrated over the supports which can be used for measurement directly. The developed methodology is very fast, simple with less sample preparation steps involved. In Table 5.2 the detection limits obtained using different precocentration procedures for different elements analyzed using TXRF are tabulated.

5.5.2 Aerosols

Atmospheric aerosols have a great impact on the environment and therby on human health. These tiny particles are the primary cause of respiratory problems and their higher exposure can lead to severe cardiopulmonary diseases or even, lung cancer. The particles are small enough to penetrate deep inside our body and enter the bloodstream. Several studies over the past few years have shown that ultrafine particles are far more toxic than larger ones as they can reach the alveoli. PM are diverse and heterogeneous mixture of elements whose composition changes spatially as well as temporally. Knowledge of the elemental composition of aerosol particles is of interest not only because of its toxicity but also because it can be related to aerosol sources for origin. Due to this fact, regular monitoring of aerosol

Table 5.2 Detection limits obtained using different pre-concentration procedures for different elements determined using TXRF

S. No.	Pre-concentration procedure	DLs ng/mL	Elements analyzed
1	Reverse-phase chromatography	0.05–0.5	V, Mn, Fe, Ni, Cu, Zn, Pb, Cd, Mo, U [50, 51]
2	Solid phase extraction: Silica gel	0.005–0.02	Mn, Fe, Ni, Cu, Zn, Se(IV), Pb and U [52]
		0.010	As [46]
3	Solid phase microextraction: MWCNT	1–2	Cd, Pb [53]
		3	Cr (VI) [47]
4	Solid phase microextraction: graphene oxide	0.0035	Cr [48]
		0.40–0.014	U [54]
5	Magnetic solid phase extraction	0.29–26.1	Cu, Zn, Ge, As, Re, Au, Hg, Tl, Bi and Pb [55]
6	Liquid phase microextraction: hollow fiber	0.03	As (III) at pH 13; As (V) at pH 8.5 [49]
7	Liquid phase extraction	0.067	U [56]
8	Dispersive liquid phase microextraction	0.39	U [57]
9	Cloud point extraction	0.02	U [58]
		0.1	Th (IV) [59]
10	Functionalised TXRF support	0.013	U [60]
		0.05	As [61]

particles is required for environmental quality assessment. Elemental composition of PM can be determined by spectroscopic techniques such as AAS, ICP-MS, NAA, PIXE as well as XRF [62].

Particles from different sources have different composition as well as size. Both of these properties are dependent on each other. The trace elements originating from natural sources are generally found in coarse particles ($PM_{2.5}$–PM_{10}), whereas elements emitted from anthropogenic sources mostly prevail in fine particles ($PM_{2.5}$). The aerosol particles can be collected from the atmosphere by filtration, cascade impactors, cyclones or impingers on cellulose nitrate or polycarbonate or polytetrafluoroethylene (PTFE) filters. Cylones and impactors are the mechanical devices which are used to collect PM based on their aerodynamic diameter [63]. XRF has been identified as a promising surface analysis technique for the analysis of PM and can replace the laborious wet methods such as ICP-MS, ICP-AES [64]. The use of TXRF for the analysis of aerosol samples can be further beneficially used for such analysis as it has detection limts comparable with other conventional trace element analysis techniques and several other superior features. TXRF analysis of PM deposited on a filter paper was first reported in 1984 [65]. In this method the filter paper was digested with acid prior to analysis by TXRF. Digestion of the filters by microwave digestion or cold plasma ashing, is tedious, time consuming and has considerable risk of contamination. In another developmental work the filter samples were directly analyzed "as received" by TXRF without any chemical pretreatment to leach the elements from aerosol particles [22]. The filter was placed directly on the quartz supports and measured. The elements were quantified in percentage with respect to the total mass. Further, the filter paper can be removed completely by cold plasma asher prior to analysis [66]. In a new quantification methodology for the analysis of aerosol particles, the aerosol samples were directly collected on quartz, glassy carbon, Si wafer reflectors suitable for TXRF analysis [67, 68]. In this case, the reflectors act as impaction plates. The reflector surface had to be greased in order to reduce particle bounce-off by using a suspension of petroleum jelly in acetone, thus forming a thin sticky film. The reflectors are then ready for sampling and loaded into the impactor [69]. In a recent, the developmental work, a cascade impactor for size-fractionated analysis of aerosol metal content by TXRF was developed by Crazzolara et al. [70]. The major application of this newly developed impactor is to quantify low concentrations of heavy metals in atmospheric PM collected over periods of 1 h or less. The impactor was designed with the following aims: Firstly, to collect particles in individual size fractions (PM_{10}, $PM_{2.5}$ and $PM_{1.0}$) and quantify the metal content in them separately; secondly, to collect particles in circular areas with a diameter of 5 mm or less in the centre of the TXRF sample carrier for a full sample analysis, thereby increasing the overall sensitivity and lastly, to provide very low blank values and minimum cross-contamination. Use of synchrotron radiation (SR) as excitation source–in total reflection geometry further improves the detection limits [71]. Combination of SR-TXRF with TXRF-XANES essentially in the same experimental set up can be performed at the same beamline with the same experimental configuration to give elemental concentration as well as speciation. In a study of elements in atmospheric aerosols, the oxidation state of Fe was studied with TXRF-XANES using the same SR-TXRF set up [68, 71]. Figure 5.5 shows the various sampling techniques used for TXRF analysis of aerosol particulate matters.

5.5.3 Soil and Sediments

Soils and sediments are another source, which are available for the environmental monitoring of the past and the present. The change in elemental concentrations in the sediments can give valuable information about the change in environmental conditions. Sediments have a very complex matrix which requires a laborious sample preparation method to digest and extract the trace elements for their determination using conventional techniques. However, TXRF was applied for direct analysis of sediment samples even without dissolution. For direct TXRF analysis, the grain size and homogeneity of the samples play a crucial role. Further, deposition of the sample as a thin film is another prerequisite for TXRF analysis. The sample preparation procedure employed manual grinding of the sample using an

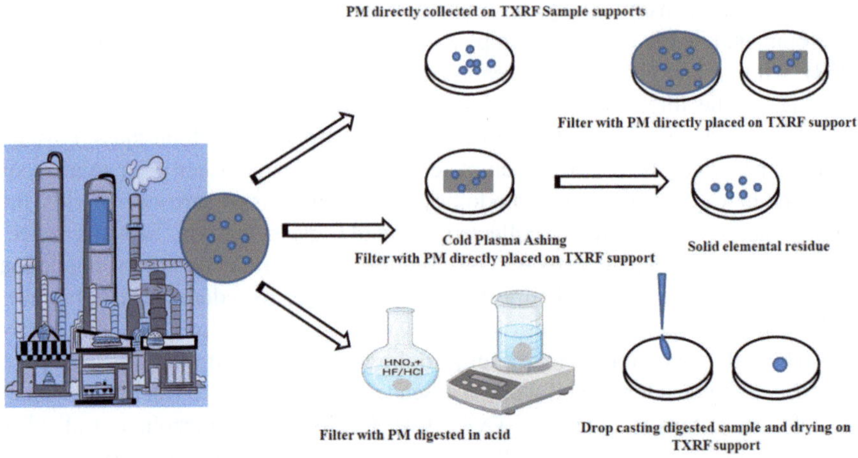

Fig. 5.5 The various sampling techniques used for TXRF analysis of aerosol particulate matters

agate mortar to reduce the grain size of the samples, followed by mixing it vigorously with Triton X solution to form a suspension. The solutions were thoroughly homogenized using shaker. Then, an internal standard was added for standardization and again this mixture was thoroughly homogenized using shaker. The various trace elements present in sediments were As, Sb, Cd, Cr, Pb, Hg, Ni, Se, Ag, V, Zn, Br, Ca, Mn, K, Rb, Th, Ti, Y, Sr, etc. It was observed that the detector active area is also a factor that has influence on the analytical parameters [72]. Further, a good agreement of results obtained by the TXRF analysis of suspended as well as digested soil samples showed that the preparation procedure does not significantly affect the results. Hence, using TXRF analytical technique both digested as well as direct samples can be analyzed in a good manner. The main advantages of TXRF is the possibility to perform direct sample analysis without digestion, thus reducing the sample preparation time and avoiding possibility of external contaminations [73]. With the development of small mobile bench top instruments, TXRF has emerged as a capable technique for on-site analysis of soils and sediments [24, 74]. This low power instrument has an air-cooled low power X-ray metal-ceramic tube and liquid nitrogen-free silicon drift detector (SSD) which allows mobile use for the measurements for on-site or in the field sample analysis.

Soil analysis can give the condition of soil and polluting sources including the concentration of the metals of toxicological interest. The potential effect of pollution of a sediment or soil can be determined in a more objective way from the identification of two types of soil fractions: one the mobile or bioavailable fractions, which are weakly linked to the sediment and therefore, easily released and the second one is the residual fraction, composed of metals representative of the geological structure of the sediments and is less available. Use of SR has many advantages as stated above and in one such study SR-TXRF has been used to study the soil contamination by heavy metals. The heavy metals were extracted from both the mobile and residual fractions and analyzed. The extraction from the mobile fraction was carried out using dilute HCl. Aliquots of sample and acid in 1: 3 ratio, respectively was agitated for 16 h at room temperature. Afterwards, the solution was filtered. This procedure releases the metals which are weakly attached to the sediment with little damage to the geological matrix. The metal extraction from the residual fraction is a bit tedious. The material remaining in the filter of the previous step is digested in a mixture of concentrated acids of HF + HNO_3 in a Teflon pump. After drying, it was re dissolved in dilute HCl. The extract was mixed with an internal standard and analyzed by TXRF. The soil was found to be contaminated with the elements V, Cr, Co, Cu, Zn and Pb, and majority of these were found to have appreciable bioavailability [75].

Clays and clay minerals are widespread natural materials whose characterization is needed in several scientific research areas such as geology, environmental sciences, soil science, archaeometry, material science, etc. Mineralogical and physical investigations along with the chemical analysis of clay samples, is crucial in their characterization. Though several methods are available for the chemical analysis of clays, TXRF is a fast and robust analytical method for the direct analysis of clays due to the reasons mentioned above. In a study, clay samples were analyzed directly after grinding the sample in a vibromilling system then suspending the fine particles obtained in 1% Triton X-100 solution, used as dispersant agent. After slurry preparation, internal standard solution was added to the suspension. The suspension was homogenized by a shaker and then placed in an ultrasonic bath. As discussed earlier the grain size of samples is very crucial and the particles having grain size lower than 10 μm is suitable for quantitative TXRF analysis [76]. In order to avoid the matrix effect, the thickness of the deposition should be lower than a critical thickness (t_{crit}) which is 50 nm for mineral powders [18]. Both the conditions were met in this study [76].

5.5.4 Biological Samples

Biomonitoring methods have also become an important part of pollution monitoring in the environment. Though trace elements present in water samples from different water bodies are an indicator of the environmental condition, such analysis cannot give information about the past pollution of water because the water contamination subsides once the pollution source is removed or stopped. However, the analysis of flesh of water organisms in water bodies can be very helpful for past information on environmental condition as the toxic elements consumed by water organisms, in case of water pollution, remain present in their body for longer duration. A detailed description of such studies is presented in the literature [27]. The advantage of biomonitors is that they are available widely. Due to their unique morphological and physiological features mosses are one of the useful bioindicators for a wide range of pollutants for the past 40 years. As they are resistant to many substances which are generally toxic to other plants, they tend to accumulate the contaminants and they exhibit large surface to volume ratio which promotes more intake of the atmospheric pollutants [77]. Trace elements present in Mosses have been used as an indicator and thereby to monitor the atmospheric pollution from several years. There is a direct correlation between the amount of trace elements in moss tissues and the atmosphere. The sample preparation procedure includes washing, freeze drying followed by grinding a weighed amount of the mosses collected from tress from a height greater than one meter. The ground samples were digested in HNO_3 and HCl mixture followed by digestion in H_2O_2 [77]. Figure 5.6 shows the comparison between the TXRF determined and reference elements concentration for the IAEA Standard reference material: IAEA-336 Lichen. In another work the mosses were presented as slurry samples for TXRF analysis wherein the complex digestion process was avoided. After air drying the mosses for five days, they were powdered and made into slurry by adding Millipore water and an internal standard. The slurries were homogenized using a vibrating vial shaker. The analytical results were compared with TXRF digested samples and a certified material. The agreement between slurry-TXRF and digestion-TXRF methods for element concentrations against the NIST 1575 pine needle standard reference material was comparable [79]. Many marine and fresh water species accumulate trace metals and metalloids from water and concentrate them several orders of magnitude greater than their level present in environment. Hence, these aquatic species serve as potentially useful organisms for monitoring the local environmental contamination. Microalgae species are key primary producers in aquatic ecosystems and accumulate many contaminants and are therefore an important biomonitor for pollution. While conventional analysis of macroalgae involves digestion of dried material in concentrated mineral acid and subsequent analysis, it is very time-consuming and labour-intensive. For TXRF analysis the samples were prepared as slurry samples from the freeze dried algae biomass for direct trace metal analysis. The concentrations of different metals including Mg, Zn, P, S, Cu, K, Ca, Mn and Fe were determined in the range of 6.8 ± 1.7–15,709 mg/kg [78].

Many studies have reported to use living plants as bio indicators that present qualitative and/or quantitative measurements in response to pollutants. The plants accumulate the pollutants within their tissues. Herbaceous species plants are good environmental pollution monitors as the elements present in these plants can give information about the heavy metal fraction accessible to biota. Such samples can be analyzed by TXRF in a simple manner. In one such study wild carrot plants have been collected from nearby areas of six sites located in two industrial corridors of Chicago city and their leaves were analyzed by TXRF. The leaves samples were cleaned, dried, microwave digested in acid and analyzed by TXRF [79]. The tree rings have been known to record the annual impressions of our abiotic environment, including pollution, and are commonly used bioindicators. For TXRF analysis of tree rings, these were submitted to an extraction procedure with concentrated nitric acid and hydrogen peroxide. The samples were successfully analysed using a XRF beamline at Brazilian Synchrotron Light Source Laboratory [80].

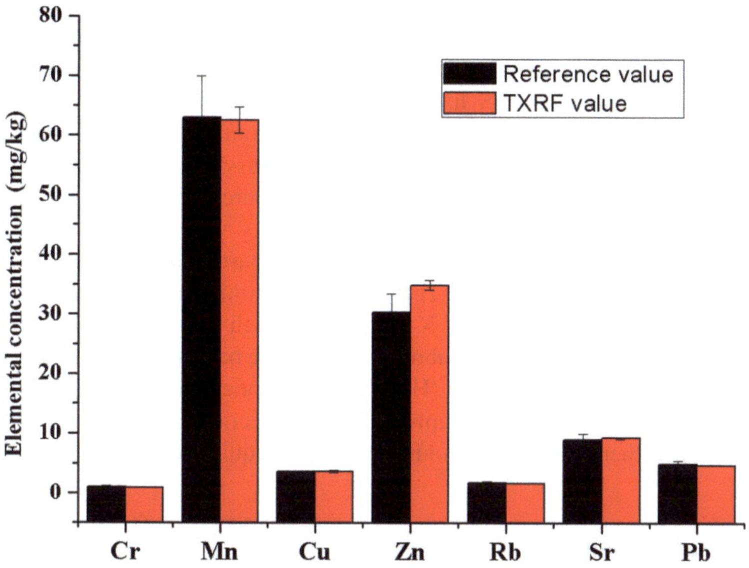

Fig. 5.6 A comparison between the TXRF determined and reference elements concentration for the IAEA: standard reference material: IAEA-336 Lichen

5.6 Role of Hyphenated TXRF Techniques for Elemental Analysis and Speciation

Micro-XRF in TXRF mode is also one of the recent trends in XRF analysis. Compared to normal XRF, in micro-XRF detailed information on elemental concentration as well as distributions are achieved. To achieve micro-XRF, X-ray focusing optics such as capillary optics, zone plates, and refractive lenses have been successfully used [38, 81]. With the development of third-generation synchrotron light sources, a significant improvement in the field of X-ray optics have been achieved. The high brilliance of SR sources has pushed the XRF detection limits many orders of magnitudes lower than that of lab based XRF. However, the number of available SR sources are relatively limited. Consequently, the frequent use of SR facilities would not be convenient option for routine analysis. As an alternative, a conventional source combined with a polycapillary semi-lens can provide a quasi-parallel beam intense enough for desktop TXRF analysis of low concentration samples [82]. Another study suggested that using a polycapillary X-ray half lens and a poly-CCC (Capillary Conic Collimator) lens, the XRF intensity was insufficient for performing TXRF analysis. This was due to the low primary X-ray intensity. When an alternative, the simplest device in X-ray optics for micro-XRF, which is a pinhole was used then a very high transmission efficiency was obtained although a pinhole did not have the focusing ability. Different pinholes with diameters of 1, 0.5 and 0.1 mm were designed to be attached to a silicon drift detector (SDD) [83].

X-ray absorption fine structure (XAFS) spectroscopy is a highly adapted method for speciation studies. It not only gives information on the electronic structure, but can also determine bond distances to neighboring atoms and the coordination. SR based XAFS have been used for characterization of aerosols [68, 71]. But, the application of SR is limited for routine analysis. Development of laboratory based XAFS will help in getting access to such studies more frequently. For such work laboratory based X-ray sources, novel optics materials with sufficient spectral resolution and high efficiency were required [84]. In one such study an existing spectrometer for laboratory-based XAFS experiments was further improved in various ways. This included optimization of the spectrometer components, such as a microfocus X-ray tube for high intensity, a large pixelated detector with photon counting capabilities for a large bandwidth and an optimized highly annealed pyrolytic graphite (HAPG)-based optics for XANES and EXAFS experiments, respectively [85]. Coupling of total reflection mode with XANES and EXAFS will further reduce the amount of sample requirement.

5.7 Conclusion

TXRF analysis has contributed in a very intense manner for environmental quality control for its sensitivity, accuracy, and versatility. It was been used widely for determining trace elements in different type of water, ice core, aerosol, sediments, soil, biological samples, etc. TXRF based environmental studies are of great value, but TXRF has to stand on firm footing and requires more intensive work in this area. There has been many advancements in the area of optics, instrumentation and quantification methods for TXRF analysis which have drastically improved the detection limits, extended the analytical range and improved the accuracy and precision of the analysis of environmental samples.

Acknowledgements The Authors express their sincere thanks to Dr. R. V. Pai, Head Fuel Chemistry Division, Prof. J. P. Saini, Hon'ble VC MMMUT Gorakhpur and Prof. P. P. Pande, Head, Department of Chemistry and Environmental Science, MMMUT Gorakhpur for their kind encouragement and support for this article.

References

1. McNaught A.D., Wilkinson, A.: IUPAC. Compendium of Chemical Terminology, 2nd edn (the "Gold Book"). Blackwell Scientific Publications, Oxford (1997). https://doi.org/10.1351/goldbook
2. Smith, A.C., Lacey, J.H.: Tracing anthropogenic climate and environmental change using stable isotopes. Quat. Sci. Rev. **346**, 109028 (2024)
3. Baranowska, I. (ed.): Handbook of Trace Analysis: Fundamentals and Applications. Springer (2016). https://doi.org/10.1007/978-3-319-19614-5
4. He, Z.L., Yang, X.E., Stoffella, P.J.: Trace elements in agroecosystems and impacts on the environment. J. Trace Elem. Med. Biol. **19**(2–3), 125–140 (2005)

5. Dongarrà, G., Manno, E., Varrica, D., Lombardo, M., Vultaggio, M.: Study on ambient concentrations of PM10, PM10–2.5, PM2.5 and gaseous pollutants. Trace elements and chemical speciation of atmospheric particulates. Atmos. Environ. **44** (39), 5244–5257 (2010)
6. Sohrin, Y., Bruland, K.W.: Global status of trace elements in the ocean. Trends in Anal. Chem. **3**(8), 1291–1307 (2011)
7. Röthlisberger, R., Bigler, M., Hutterli, M., Sommer, S., Stauffer, B., Junghans, H.G., Wagenbach, D.: Technique for continuous high-resolution analysis of trace substances in Firn and ice cores. Environ. Sci. Technol. **34**, 338–342 (2000)
8. Kumar, M., Singh, S., Mahajan, R.K.: Trace level determination of U, Zn, Cd, Pb and Cu in drinking water samples. Environ. Monit. Assess. **112**, 283–292 (2006). https://doi.org/10.1007/s10661-006-1069-6
9. Morrison, G.H., Risby, T.H.: Elemental trace analysis of biological materials. Crit. Rev. Anal. Chem. **8**(3), 287–320 (1979). https://doi.org/10.1080/10408347908542713
10. Frame, E.M.S.: Determination of trace impurities in materials. In: Briant, C.L. (ed.) Impurities in Engineering Materials, pp. 91–142. Marcel Dekker Inc., New York (1999)
11. Kanwar, V.S., Sharma, A., Srivastav, A.L., Rani, L.: Phytoremediation of toxic metals present in soil and water environment: a critical review. Environ. Sci. Pollut. Res. **27**, 44835–44860 (2020). https://doi.org/10.1007/s11356-020-10713-3
12. Singh, B.R., Steinnes, E.: Soil and Water Contamination by Heavy Metals, Soil Processes and Water Quality, 1st edn, pp. 233–263 (1994). https://doi.org/10.1201/9781003070184-6
13. Cheng, H.S., Pöschl, Y.U.: New multiphase chemical processes influencing atmospheric aerosols, air quality, and climate in the anthropocene. Acc. Chem. Res. **53**(10), 2034 (2020). https://doi.org/10.1021/acs.accounts.0c00246
14. Singh, V.K., Kawai, J., Tripathi, D.K.: X-Ray Fluorescence in Biological Sciences: Principles, Instrumentation, and Applications. John Wiley & Sons Ltd. (2022)
15. Uriano, G.A., Gravatt, C.C., Morrison, G.H.: The role of reference materials and reference methods in chemical analysis. Crit. Rev. Anal. Chem. **6**(4), 361–412 (1977). https://doi.org/10.1080/10408347708542696
16. Bulska, E., Ruszczyńska, A.: Analytical techniques for trace element determination. J. Phys. Sci. Rev. **2**(5), 20178002 (2017). https://doi.org/10.1515/psr-2017-8002
17. Marguí, E., Zawisza, B., Sitko, R.: Trace and ultratrace analysis of liquid samples by X-ray fluorescence spectrometry, TrAC trends anal. Chem. **53**, 73–83 (2014)
18. Klockenkämper, R., von Bohlen, A. (eds.): Total-Reflection X-Ray Fluorescence Analysis and Related Methods. John Wiley & Sons Inc., Hoboken (2015)
19. Dhara, S., Misra, N.L.: Elemental characterization of nuclear materials using total reflection X-ray fluorescence spectrometry, TrAC Trends Anal. Chem. **116**, 31–43 (2019)
20. Mishra, N.L., Dhara, S.: Total reflection X-ray fluorescence analysis of some biological, X-ray fluorescence in biological sciences: principles, instrumentation, and applications. In: Singh, V.K., Kawai, J., Tripathi, D.K., (eds.) pp 203–218 (2022), https://doi.org/10.1002/9781119645719.ch13
21. Stosnach, H.: Environmental trace-element analysis using a benchtop total reflection X-ray fluorescence spectrometer. Anal. Sci. **21**(7), 873–876 (2005)
22. Bontempi, E., Zacco, A., Benedetti, D., Borgese, L., Colombi, P., Stosnach, H., Finzi, G., Apostoli, P., Buttini, P., Depero, L.E.: Total reflection X-ray fluorescence for direct analysis of aerosol particle samples. Environ. Technol. **31**, 467–477 (2010)
23. Bilo, F., Borgese, L., Wambui, A., Assi, A., Zacco, A., Federici, S., Eichert, D.M., Tsuji, K., Lucchini, R.G., Placidi, D., Bontempi, E., Depero, L.E.: Comparison of multiple X-ray fluorescence techniques for elemental analysis of particulate matter collected on air filters. J. Aerosol Sci. **122**, 1–10 (2018)
24. Towett, E.K., Shepherd, K.D., Cadisch, G.: Quantification of total element concentrations in soils using total X-ray fluorescence spectroscopy (TXRF). Sci. Total. Environ. **463–464**, 374–388 (2013)

25. Wastl, A., Stadlbauer, F., Prost, J., Horntrich, C., Kregsamer, P., Wobrauschek, P., Streli, C.: Nanoliter deposition unit for pipetting droplets of small volumes for total reflection X-ray fluorescence applications. Spectrochim. Acta Part B **82**, 71–75 (2013)
26. Borgese, L., Bilo, F., Federici, S., Margui, E., Hase, T., Huang, Y., Beckhoff, B., Depero, L.E.: Summary of ISO standard 20289: total reflection X-ray fluorescence analysis of water. Surf. Interface Anal. **52**(3), 119–123 (2019). https://doi.org/10.1002/sia.6720
27. Mishra, N.L.: Advantages of total reflection X-ray fluorescence analysis over conventional trace analytical techniques in different areas of material characterization. In: Encyclopaedia of Analytical Chemistry. John Wiley & Sons (2024)
28. Dhara, S., Misra, N.L.: Application of total reflection X-ray fluorescence spectrometry for trace elemental analysis of rainwater. Pramana **76**(2), 361–366 (2011)
29. Wang, L.L., Yu, H.S., Li, L.N., Wei, X.J., Huang, Y.Y.: The development of TXRF method and its application on the study of trace elements in water at SSRF. Nucl. Instrum. Methods Phys. Res. B. **375**, 49–55 (2016)
30. Pashkova, G.V., Revenko, A.G.: A review of application of total reflection X-ray fluorescence spectrometry to water analysis. Appl. Spectrosc. Rev. **50**(6), 443–472 (2015). https://doi.org/10.1080/05704928.2015.1010205
31. Marguí, E., Tapias, J.C., Casas, A., Hidalgo, M., Queralt, I.: Analysis of inlet and outlet industrial wastewater effluents by means of benchtop total reflection X-ray fluorescence spectrometry. Chemosphere **80**(3), 263−270 (2010)
32. Sanyal, K., Dhara, S.: Recent advances in ultra-trace determination of uranium in natural water using total reflection X-ray fluorescence (TXRF) spectrometry. X Ray Spectrom. **53**(5), 326–339 (2024)
33. Garg, V.K., Tanta, A., Srivastav, A.L., Tiwari, M.K., Sharma, A., Kanwar, V.S.: Water quality assessment using synchrotron-based TXRF. WER **94**, e10759 (2022)
34. Misra, N.L., Tobler, L., Schwikowski, M.: Applicability of TXRF for Tarce Elemental Analysis in Ice Samples, PSI Scientific Report: Particles and Matter, p. 153 (2001) https://inis.iaea.org/collection/NCLCollectionStore/_Public/32/030/32030446.pdf#page=157
35. Cibin, G., Marcelli, A., Maggi, V., Sala, M., Marino, F., Delmonte, B., Albani, S., Pignotti, S.: First combined total reflection X-ray fluorescence and grazing incidence X-ray absorption spectroscopy characterization of aeolian dust archived in Antarctica and Alpine deep ice cores. Spectrochim. Acta Part B. **63**(12), 1503–1510 (2008)
36. Cibin, G., Marcelli, A., Maggi, V., Baccolo, G., Hampai, D., Robbins, P.E., Lied, A., Polese, C., D'Elia, A., Macis, S., Grilli, A., Raco, A.: Synchrotron radiation research and analysis of the particulate matter in deep ice cores: an overview of the technical challenges. Condens. Matter **4**(3), 61–71 (2019). https://doi.org/10.3390/condmat4030061
37. Cappuccio, G., Cibin, G., Dabagov, S.B., Filippo, A.D., Piovesan, G., Hampai, D., Maggi, V., Marcelli, A.: Challenging X-ray fluorescence applications for environmental studies at XLab frascati. Condens. Matter **3**(4), 33–42 (2018). https://doi.org/10.3390/condmat3040033
38. Hampai, D., Dabagov, S.B., Cappuccio, G., Cibin, G., Sessa, V.: X-ray micro-imaging by capillary optics. Spectrochim. Acta Part B **64**, 1180–1184 (2009)
39. Shulyumova, A., Maltsev, A., Umarova, N.: Multivariate calibration in TXRF analysis of water. X-Ray Spectrom. **47**(5), 396–404 (2018)
40. Fittschen, U.E.A., Havrilla, G.J.: Picoliter droplet deposition using a prototype picoliter pipette: control parameters and application in micro X-ray fluorescence. Anal. Chem. **82**, 297–306 (2010)
41. Fittschen, U.E.A., Bings, N.H., Hauschild, S., Förster, S., Kiera, A.F., Karavani, E., Frömsdorf, A., Thiele, J., Falkenberg, G.: Characteristics of picoliter droplet dried residues as standards for direct analysis techniques. Anal. Chem. **80**, 1967–1977 (2008)
42. Kunimura, S., Ohmoria, H.: A portable total reflection X-ray fluorescence spectrometer with a diamond-like carbon coated X-ray reflector. Analyst **137**, 312–314 (2012)
43. Yadav, V.B., Jha, S.K.: Status of trace and toxic elements pollution in creek ecosystem using TXRF method. J. Radioanal. Nucl. Chem. **295**(3), 1759–1762 (2013)

44. Pashkova, G.V., Revenko, A.G., Finkelshtein, A.L.: Study of factors affecting the results of natural water analyses by total reflection X-ray fluorescence. X-Ray Spectrom. **6**, 524–530 (2013)
45. Moradi, M., Yamini, Y., Kakehmam, J.: A review in the sample preparation of aqueous solutions combined with X-ray fluorescence detection. J. Iran. Chem. Soc. **12**, 831–838 (2015). https://doi.org/10.1007/s13738-014-0545-0
46. Staniszewski, B., Freimann, P.: A solid phase extraction procedure for the simultaneous determination of total inorganic arsenic and trace metals in seawater: sample preparation for total-reflection X-ray fluorescence. Spectrochim. Acta Part B **63**, 1333–1337 (2008)
47. Bahadir, Z., Bulut, V.N., Hidalgo, M., Soylak, M., Marguí, E.: Determination of trace amounts of hexavalent chromium in drinking waters by dispersive microsolid-phase extraction using modified multiwalled carbon nanotubes combined with total reflection X-ray fluorescence spectrometry. Spectrochim. Acta Part B **107**, 170–177 (2015)
48. Musielak, M., Serda, M., Gagor, A., Talik, E., Sitko, R.: Ultratrace determination and speciation of hexavalent chromium by EDXRF and TXRF using dispersive micro-solid phase extraction and tetraethylenepentamine graphene oxide. Spectrochim. Acta Part B **213**, 106863 (2024)
49. Majumder, S., Marguí, E., Roman-Ross, G., Chatterjee, D., HidalgoM,: Hollow fiber liquid phase microextraction combined with total reflection X-ray fluorescence spectrometry for the determination of trace level inorganic arsenic species in waters. Talanta **217**(1), 121005 (2020)
50. Prange, A., Boddeker, H., Kramer, K.: Determination of trace elements in river-water using total-reflection X-ray fluorescence. Spectrochim. Acta, B **48**(2): 207–215 (1993)
51. Schmidt, D., Gerwinski, W., Radke, I.: Trace metal determinations by total-reflection X-ray fluorescence analysis in the open Atlantic Ocean. Spectrochim. Acta, Part B **48**(2), 171–181 (1993)
52. Prange, A., Knöchel, A., Michaelis, W.: Multi-element determination of dissolved heavy metal traces in sea water by total-reflection X-ray fluorescence spectrometry. Anal. Chim. Acta **172**, 79–100 (1985)
53. Kocot, K., Zawisza, B., Marguí, E., Queralt, I., Hidalgo, M., Sitko, R.: Dispersive micro solid-phase extraction using multiwalled carbon nanotubes combined with portable total-reflection X-ray fluorescence spectrometry for the determination of trace amounts of Pb and Cd in water samples. J. Anal. At. Spectrom. **28**, 736–742 (2013)
54. Kocot, K., Pytlakowska, K., Talik, E., Krafft, C., Sitko, R.: Sensitive determination of uranium using β-cyclodextrin modified graphene oxide and X-ray fluorescence techniques: EDXRF and TXRF. Talanta **246**, 123501 (2022)
55. Romero, V., Costas-Mora, I., Lavillaa, I., Bendicho, C.: Insitu ultrasound-assisted synthesis of Fe3O4nanoparticles with simultaneous ion co-precipitation for multielemental analysis of natural waters by total reflection X-ray fluorescence spectrometry. J. Anal. At. Spectrom. **28**, 923–933 (2013)
56. Misra, N.L., Dhara, S., Singh Mudher, K.D.: Uranium determination in seawater by total reflection X-ray fluorescence spectrometry. Spectrochim. Acta **61B**(10–11), 1166–1169 (2006)
57. Shukla, V.K., Dhara, S., Mishra, N.L.: Total reflection X-ray fluorescence spectrometric determination of ultra-trace uranium in natural water samples using a dispersive liquid–liquid micro-extraction method. J. Anal. At. Spectrom. **35**, 1632–1640 (2020). https://doi.org/10.1039/D0JA00039F
58. Saha, A., Sanyal, K., Rawat, N., Deb, S.B., Saxena, M.K., Tomar, B.S.: Selective micellar extraction of ultratrace levels of uranium in aqueous samples by task specific ionic liquid followed by its detection employing total reflection X-ray fluorescence spectrometry. Anal. Chem. **89**(19), 10422–10430 (2017)
59. Saha, A., Sanyal, K., Kumari, K., Deb, S.B., Saxena, M.K.: Highly selective cloud point extraction of thorium(iv) ions in environmental samples for their ultra-trace level determination by total reflection X-ray fluorescence spectrometry, J. Anal. At. Spectrom. (2025). https://doi.org/10.1039/D4JA00348A

60. Sanyal, K., Dhara, S., Gumber, N., Pai, R.V.: A highly sensitive method for uranium quantification in water samples at ultra-trace level by total reflection X-ray fluorescence, after its direct pre-concentration on the surface of amidoxime functionalized quartz sample supports. Talanta **254**, 124129 (2023)
61. Sanyal, K., Chappa, S., Bahadur, J., Pandey, A.K., Mishra, N.L.: Arsenic quantification and speciation at trace levels in natural water samples by total reflection X-ray fluorescence after pre-concentration with N-methyl-d-glucamine functionalized quartz supports. J. Anal. At. Spectrom. **35**, 2770–2778 (2020)
62. Bilo, F., Cirelli, P., Borgese, L.: Elemental analysis of particulate matter by X-ray fluorescence methods: a green approach to air quality monitoring. TrAC Trends Anal. Chem. **170**, 117427 (2024)
63. Patel, P., Aggarwal, S.G.: On the techniques and standards of particulate matter sampling. J. Air Waste Manag. Assoc. **72**, 791–814 (2022). https://doi.org/10.1080/10962247.2022.2048129
64. Ogrizek, M., Kroflič, A., Šala, M.: Critical review on the development of analytical techniques for the elemental analysis of airborne particulate matter. Tren. Environ. Anal. Chem. **33**, e00155 (2022)
65. Ketelsen, P., Knöchel, A.: Multielement analysis of aerosol samples by X-ray fluorescence analysis with total reflecting sample holders. Fresen. Z. Anal. Chem. **317**, 333–342 (1984)
66. Wagner, A., Mages, M.: Total-Reflection X-ray fluorescence analysis of elements in size-fractionated particulate matter sampled on polycarbonate filters—composition and sources of aerosol particles in Göteborg. Sweden Spectrochim. Acta Part B **65**(6), 471–477 (2010)
67. Ebert, M., Dahmen, J., Hoffmann, P., Ortner, H.M.: Examination of clean room aerosol particle composition by total reflection X-ray analysis and electron probe microanalysis. Spectrochim. Acta Part B **52**(7), 967–975 (1997)
68. Fittschen, U.E.A., Meirer, F., Streli, C., Wobrauschek, P., Thiele, J., Falkenberg, G., Pepponi, G.: Characterization of atmospheric aerosols using Synchroton radiation total reflection X-ray fluorescence and Fe K-edge total reflection X-ray fluorescence-X-ray absorption near-edge structure. Spectrochim. Acta Part B **63**(12), 1489–1495 (2008)
69. Prost, J., Wobrauschek, P., Streli, C.: Quantitative total reflection X-ray fluorescence analysis of directly collected aerosol samples. X-Ray Spectrom. **46**(5), 454–460 (2017). https://doi.org/10.1002/xrs.2752
70. Crazzolara, C., Held, A.: Development of a cascade impactor optimized for size-fractionated analysis of aerosol metal content by total reflection X-ray fluorescence spectroscopy (TXRF). Atmos. Meas. Tech. **17**(7), 2183–2194 (2024)
71. Czömpöly, O., Börcsök, E., Groma, V., Pollastri, S., Osán, J.: Characterization of unique aerosol pollution episodes in urban areas using TXRF and TXRF-XANES. Atmos. Pollut. Res. **12**(11), 101214 (2021)
72. Bilo, F., Borgese, L., Cazzago, D., et al.: TXRF analysis of soils and sediments to assess environmental contamination. Environ. Sci. Pollut. Res. **21**, 13208–13214 (2014). https://doi.org/10.1007/s11356-013-2203-y
73. Stosnach, H.: Environmental trace-element analysis using a benchtop total reflection X-ray fluorescence spectrometer. Anal. Sci. **21**, 873–876 (2005)
74. Sirito de Vivesa, A.E., Brienza, S.M.B., Moreira, S., Zucchi, O.L.A.D., Barroso, R.C., Franco do Nascimento Filho, V.: Evaluation of the availability of heavy metals in lake sediments using SR-TXRF. Nucl. Inst. Methods Phys. Res. A. 579, 503−506 (2007)
75. García-Heras, M., Fernåndez-Ruiz, R., Tornero, J.D.: Analysis of archaeological ceramics by TXRF and contrasted with NAA. J. Archaeol. Sci. **24**, 1003–1014 (1997). https://doi.org/10.1006/jasc.1996.0178
76. Allegretta, I., Ciasca, B., Pizzigallo, M.D.R., Lattanzio, V.M.T., Terzano, R.: A fast method for the chemical analysis of clays by total-reflection x-ray fluorescence spectroscopy (TXRF). Appl. Clay Sci. **180**, 105201 (2019)
77. Zarazúa-Ortega, G., Poblano-Bata, J., Tejeda-Vega, S., Ávila-Pérez, P., Zepeda-Gómez, C., Ortiz-Oliveros, H., Macedo-Miranda, G.: Assessment of spatial variability of heavy metals in

metropolitan zone of Toluca Valley, Mexico, using the biomonitoring technique in mosses and TXRF analysis. Sci. World J. 426492 (2013). https://doi.org/10.1155/2013/426492
78. Natali, M., Zanella, A., Rankovic, A., et al.: Assessment of trace metal air pollution in Paris using slurry-TXRF analysis on cemetery mosses. Environ. Sci. Pollut. Res. **23**, 23496–23510 (2016). https://doi.org/10.1007/s11356-016-7445-z
79. Bungudu, J.I., Lorrie, M.M.: Determination and analysis of metals in freshwater microalgae (Chlorella vulgaris and Spirulina platensis) through total reflection X-ray fluorescence spectroscopy (TXRF). J. Appl. Chem. **8**(3), 32–38 (2021). https://doi.org/10.9734/ajacr/2021/v8i330194
80. Schmeling, M., Gende, M., Tovar, A.: Biomonitoring of urban industrial pollution using total reflection X-ray fluorescence. X-Ray Spectrom. (2024). https://doi.org/10.1002/xrs.3439
81. Nakano, K., Tanaka, K., Ding, X., Tsuji, K.: Development of a new total reflection X-ray fluorescence instrument using polycapillary X-ray lens. Spectrochim. Acta Part B **61**(10–11), 1105–1109 (2006)
82. Hampai, D., Dabagov, S.B., Polese, C., Liedl, A., Cappuccio, G.: Laboratory total reflection X-ray fluorescence analysis for low concentration samples. Spectrochim. Acta. Part B **101**, 114–117 (2014)
83. Tsuji, K., Kawamata, M., Nishida, Y., Nakano, K., Sasaki, K.: Micro total reflection x-ray fluorescence (μ-TXRF) analysis X-Ray. Spectrom. **35**(6), 375–378 (2006)
84. Jahrman, E.P., Holden W.M., Ditter, A.S., Mortensen, D.R., Seidler, G.T., Fister, T.T., Kozimor, S.A., Piper, L.F.J., Rana, J., Hyatt, N.C., Stennett, M.C.: An improved laboratory-based X-ray absorption fine structure and X-ray emission spectrometer for analytical applications in materials chemistry research. Rev. Sci. Instrum. **90**, 024106 (2019)
85. Schlesiger, C., Praetz, S., Gnewkow, R., Malzer, W., Kanngießer, B.: Recent progress in the performance of HAPG based laboratory EXAFS and XANES spectrometers. J. Anal. At. Spectrom. **35**, 2298–2304 (2020)

Data Analysis and Chemometrics in X-Ray Fluorescence and Applications

6

E. Marguí, E. de Almeida, E. Besalú, I. Queralt, Y. Bolaños, and F. L. Melquiades

Abstract

Proper characterization of some environmental, agronomic and industrial samples is still an analytical challenge. New pollutants and sample matrices have put pressure in analytical laboratories for the development of reliable methods that may be used in routine basis. In this context, it is interesting to mention the incorporation of new screening or quantitative low-cost analytical tools to solve analytical problems in the frame of the principles of green analytical chemistry. X-ray fluorescence (XRF) is a well-established analytical technique, which allows the analysis of solid samples with a minimum sample treatment, and recently, its application capability has been boosted up with chemometrics for quantification, classification, and identification of origin. It is interesting to remark that with the use of chemometric tools, the analytical performance of XRF may be also improved including the determination of trace elements by portable XRF systems or the determination of light elements. A myriad of studies utilizing chemometrics and XRF have been recently reported in the literature and in this chapter a comprehensive overview of them will be displayed. The

E. Marguí · E. Besalú
Department of Chemistry, University of Girona, Girona, Spain
e-mail: eva.margui@udg.edu

E. de Almeida
Laboratory of Nuclear Instrumentation (LIN), Center for Nuclear Energy in Agriculture (CENA), University of São Paulo (USP), Piracicaba, Brazil

I. Queralt
Institute of Environmental Assessment and Water Research, IDAEA-CSIC, Barcelona, Spain

Y. Bolaños
Cienfuegos Environmental Studies Center, Cienfuegos, Cuba

F. L. Melquiades (✉)
Department of Physics, State University of Londrina, Londrina, Brazil
e-mail: fmelquiades@uel.br

V. K. Singh (ed.), *X-Ray Fluorescence Spectroscopy and Chemometrics*,
https://doi.org/10.1007/978-3-031-98375-7_6

combination of XRF with chemometrics is undoubtedly a field of worth investigation which may open new interesting applications of XRF in many different fields in the near future.

6.1 X-Ray Fluorescence and Chemometrics

Analytical measurements are important in different fields in our daily lives including, for instance, from chemical pollutant monitoring in the environment, to quality control of raw and end industrial products. In all cases, the main goal of the analysis is reaching sufficient information with the appropriate quality for the end-users to make decisions and solve problems. Therefore, not always costly instrumental analytical techniques nor highly qualified trained analyst are required. Adequate analytical results can be obtained by methods with minimum sample treatment or by using cost-effective or portable instruments. In fact, in recent years, the conceptual advancement on green analytical chemistry (GAC) has moved in parallel with efforts to incorporate new screening or quantitative low-cost analytical tools to solve analytical problems. In this sense, techniques that afford either direct solid analysis or a minimum sample treatment must be highlighted. Spectroscopic methods which make use of electromagnetic radiation are adequate for non-destructive elemental analysis in solid samples measurements, particularly X-ray spectrometry.

Last decades, new technological developments in the XRF field, such as X-ray miniaturized sources, low-power micro-focus tubes, innovative X-ray optics devices and ultra-thin detector window for light elements detection have improved the XRF analytical performance of benchtop and portable XRF equipment. In addition, these advances afford operation simplicity at a low-cost design. Unfortunately, this kind of XRF systems present some weaknesses mostly related to the limited sensitivity for the determination of trace and ultratrace elements and matrix effects caused by elements present in the solid sample. In this sense, matrix correction models and adequate calibration approaches have to be developed to obtain meaningful results. In most cases, this is not an easy task and for this reason the use of XRF in some applications has been limited and questioned by many analytical chemists. In this context, additional research on the development of adequate quantification strategies and the use of chemometric tools can help to get more reliable XRF analytical results. XRF combined with chemometrics enhances a myriad of sample attributes evaluation, which is not conventionally investigated by EDXRF, such as major organic content determination [1], authenticity [2, 3] and classification (origin and type) of the sample [4–9]. In Fig. 6.1, a summary of main applications of XRF and chemometrics and fields of study is provided.

The first article dealing with the combined use of XRF and chemometrics date from 1972 and it was focussed on the classification of archaeological artifacts by applying pattern recognition to WDXRF data [10]. Until then, and above all in the

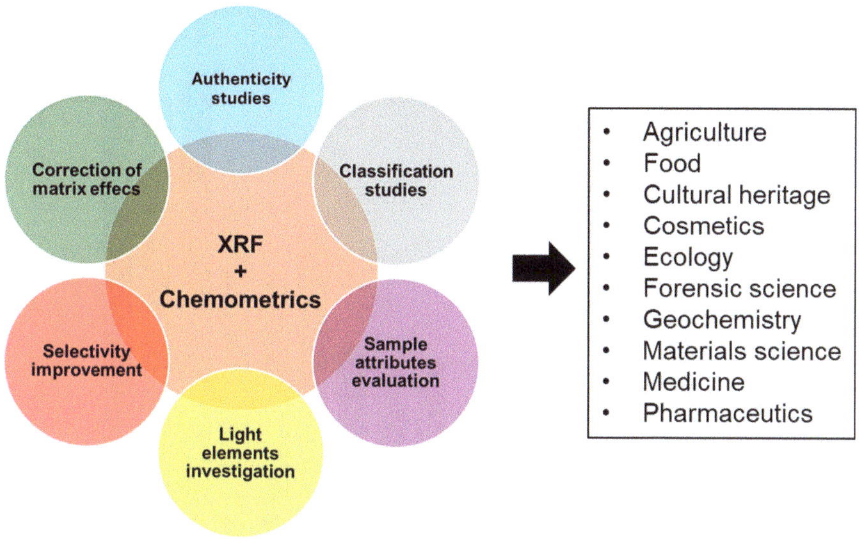

Fig. 6.1 Summary of main applications of XRF and chemometrics and fields of study

last fifteen years, research articles using X-ray fluorescence techniques and chemometric tools has increased significantly (roughly fourfold from 2009 to 2023), as shown in Fig. 6.2. This tendency may be attributed to the search for analytical procedures more in line with the GAC principles, the widespread use of advanced statistical methods with machine learning tools, high performance personal computer at low price as well as the possibility of new XRF applications, as cited above.

Additionally, besides the instrumental improvements that have been occurred in XRF last years [11], chemometrics may be considered a new edge-cutting methodology for XRF, which allow a better XRF analytical performance, particularly interesting when applied for the portable and low cost XRF devices. In this point of view, specifically the analysis and characterization of different environmental compartments, regular waste and e-waste, which are highly demanded in the environmental field, are promising application of allying XRF and chemometrics.

The objective of this chapter is to provide an overview of the basic principles and performance of most used linear multivariate statistical methods combined with XRF data for classification and quantification of complex samples. In the last section of the chapter some selected examples dealing with the combination of XRF and chemometrics have been also included as study cases. However, the applications can be extended to other similar samples and studies.

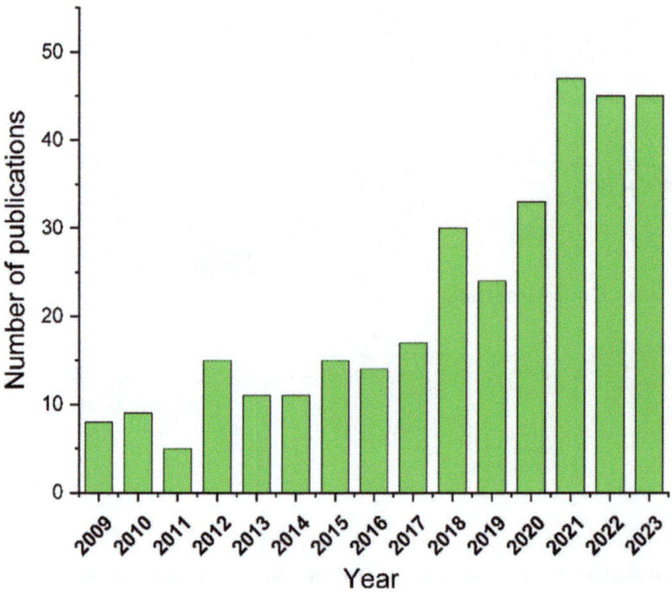

Fig. 6.2 Number of studies published in indexed journals applying "XRF or EDXRF or TXRF or WDXRF" and "LDA or SIMCA or KNN or PLS-DA or SVM or RF or MLR or PCR or PLS or ANN" acronyms, within "article title abstract and keywords" field on Scopus platform

6.2 Chemometrics

Chemometrics may be defined as the application of mathematical procedures and statistical concepts to evaluate data from complex chemical systems. In particular, it considers the application of matrix algebra calculations and multivariate statistical methods to convert raw data into information.

Multivariate statistical methods are used to analyze data which requires simultaneous determination of several species in a set of samples. They are excellent when the spectroscopic data has problems with the base line and/or interferences. The chemometric methods make use of special algorithms whose function is to find statistical correlations between data obtained in an experimental test, called independent variables (e.g. spectral data) and values of chemical or physical properties of samples, called dependent variables (e.g. concentration of an analyte or a target property of interest) [12, 13].

The main applications considering big data sets are exploratory analysis, classification models or quantification models with a myriad of machine learning tools. Figure 6.3 presents a flowchart with an approach rout to databases evaluation. The flowchart central branch indicates the steps to be followed to perform a complete study in a big data set, starting from the raw data up to the calibration model generation and its validation. These steps are data preparation and an exploratory analysis in the raw data considering preprocessing and variable selection application. Then, split the data into calibration and validation in order to apply the chemometric tools of interest and finishing with the evaluation of the prediction results. The figures of merit evaluation encompass a series of tests, for instance, criteria for outlier removal, linearity tests, quantification and detection limits determination, calculation of performance parameters. Finally, the external validation will confirm the usefulness of the statistical model. The "chemometric modeling algorithm" icon in Fig. 6.3 indicate the decision that must be considered depending on the aim at classifying the samples into classes or quantifying a parameter of interest. Examples of the most used chemometrics tools are listed in the flowchart.

When dealing with XRF data, the main signal from the samples are the spectrum dataset or the net peak intensities from the elements of interest which constitute the **X** matrix with independent variables. Depending on the objective, a **Y** matrix (usually a vector) with the results, classes or labels from the target variable must be available. Therefore, unsupervised methods are the ones in which just the **X** matrix is used in the calculation. Unlabeled data are evaluated by pattern recognition methods, but, if the classes are known, the information is used just to interpret the results. On the other hand, the supervised methods require an **X** matrix (signal) and a **Y** matrix (classes or quantitative results from the target variable) and classification models or regression models are generated.

When dealing with calibration and classification of complex systems the parsimony principle is always recommended. The parsimony principle is basic to all science and suggests choosing the simplest scientific explanation or method that fits the evidence or parameter of interest [14]. Applied to chemometrics, it means that as simple the method the better. Because of that, it is recommended to know your data, the correlations involved, the purposes of the data treatment and based on them begin with linear methods before trying higher orders ones. Moreover, it is suggested after the first modelling results, to apply a linearity test to verify if second order or non-linear models will be necessary.

The large majority of studies considering XRF present a linear behavior between the spectral dataset and the target of interest [15]. Then, the following sections will be focused on detailing linear methods for classification and quantification according to Fig. 6.3.

6.2.1 Data Preparation

Typical data representation in chemometrics consists of setting the experimental data in to a matrix $\mathbf{X} = \{X_{ij}\}$ where each line (i) contains different samples and each column (j) the variables. For instance, in the case of evaluating spectra, each line represents a spectrum and the columns are the photons count in each wavelength. However, if peak area or concentration values will be considered, each line represents a sample and each column the element quantified. Moreover, if there are missing data in the columns of the matrix, they must be filled following a criterion. There are some options such as the average of the column, the detection

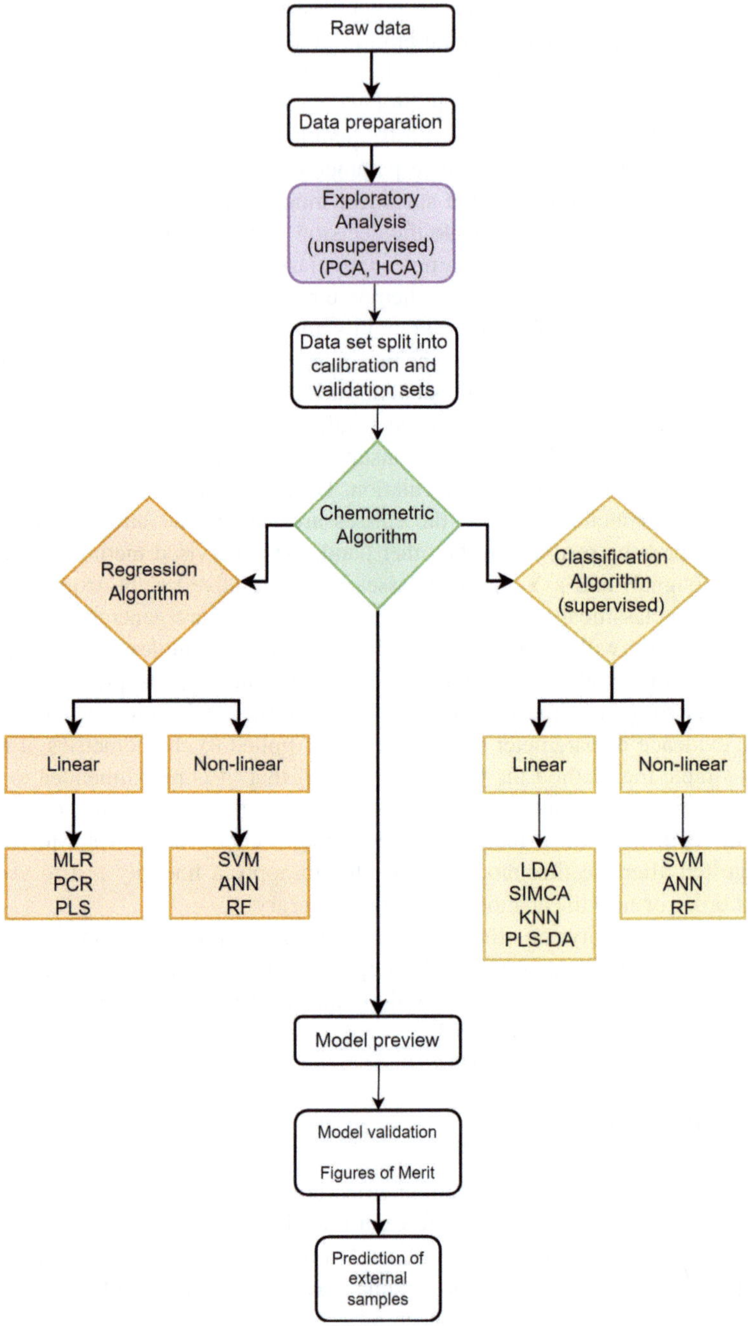

Fig. 6.3 Flowchart with the chemometrics analysis sequence

limit, a random number in the interval of the column, but avoid filling the missing data with zeros.

Once the raw matrix is organized, the machine learning methods usually require a preprocessing step. The basic operations are mean centering or autoscaling. In the case of spectral data some specific preprocessing may be applied like baseline correction, multiplicative scatter correction, derivation, smoothing. dos Santos et al. [16] tested 27 preprocessing combination in XRF spectra indicating that Pareto scaling and Poisson scaling have proved to be appropriate when few peaks have high contribution in the spectra. However, the literature brings out a huge variety of studies and the reader may consult the specialized method for the application of interest.

Another important task in chemometrics is splitting the samples into subsets for calibration and validation of the classification or calibration methods. In general, two-thirds of the samples are used in the calibration dataset. To stablish robust models, an intelligent choice of the data set must be conducted in such a way that representative samples complying the complete range of the samples are selected for the calibration set. There are some algorithms dedicated to data split and a critical study was conducted by Westad and Marini [17]. The most used is the Kennard-Stone algorithm. Special attention must be given to split the data in disbalanced datasets used for classification and the recommendation is to split the dataset by classes.

6.2.2 Exploratory Analysis

Exploratory analysis is pattern recognition methods applied to the **X** matrix. The methods are blind, i.e., the labels or classes are not used in the calculation, because of that they are considered unsupervised. The results are groupings or clusters identification in the dataset. The calculation may be conducted considering the distance between samples in a coordinate system or in the projection of the samples into a new axis direction.

Hierarchical cluster analysis (HCA) identifies similar homogeneous subgroups within the sample dataset considering the samples distance in the cartesian space of the variables [3, 9].

Principal component analysis (PCA) is a projection method which reduces the dimensionality of the dataset but identifying most of the data variance. It implies in rotating the axis of the coordinate system in order to find a new direction which contain the most of the data variability. The new axes are called the principal components (PCs). In algebraic terms, PCA is a decomposition matrix method. The **X** matrix is decomposed into scores (**S**) and loadings(**L**) matrixes, $\mathbf{X}_{nxm} = \mathbf{S}_{nxk}\mathbf{L}^{T}{}_{kxm} + \mathbf{E}_{nxm}$, being n the number of samples, m the number of descriptors and $k \leq m$ the number of PCs considered. The scores and loadings carry on information from the samples and the variables weight, respectively. The residuals matrix **E** contains the data percentage that was not captured by the number of PCs considered in the modelling. Particularly, PCA sheds light on the groupings in a dataset and their corresponding variable weight responsible for that grouping [18]. It has an interpretative power because both samples and variables are considered to extract the relevant information in detriment of dataset redundances. Understanding the PCA principle is fundamental because this method is the base for several others such as PLS and PLS-DA.

6.2.3 Linear Classification Methods–Qualitative Results

Classification methods are supervised standard recognition methods which involves predicting a discrete label or dummy variables for an input instance based on its responses from some measurement system. Typically, a binary discrete vector is required which contain the classes of interest from the corresponding samples (0 or 1; true or false, presence or absence of a contaminant, original or counterfeit samples). To reach the classification model a well-known calibration dataset must be available. It is strongly suggested that the number of samples in both classes be balanced to reach robust models. In the cases where data split into calibration and prediction sets are performed, the samples selection must be performed in each class individually and not in the complete dataset in order to avoid non balance, especially in the calibration set.

Classification models may be built by distance or by projection calculation. Linear discriminant analysis (LDA) and K-nearest neighbors (KNN) are examples of distance-based methods. LDA classify data points based in the distance to class centroids incorporating variance and co-variance information about the data. KNN operates under the assumption that similar points have small distances from its neighbors.

One of the most used methods based on projection is Partial Least Squares–discriminant analysis (PLS-DA) [19]. It is a variant of the PLS regression with the difference that the **Y** matrix is categorical, typically binary. The calculation consists in applying a PCA in **X** and **Y** matrixes and linear combinations are built to reach the calibration model [20]. Then, predicted values are calculated for **Y** and based in a Bayesian threshold the new samples are classified. The Bayesian threshold is selected after plotting the false positives and false negatives estimated distribution and evaluating where the curves cross each other [21].

6.2.4 Linear Regression Methods–Quantitative Results

Regression models may be linear or non-linear, based on classical or inverse methods, univariate or multivariate. In this topic will be detailed the multivariate linear methods considering inverse calibration procedure. The inverse calibration predicts the parameter of interest from the instrument data, in XRF, from the spectra data (responses). Besides that, inverse calibration methods consider that the highest source of error are in the parameter of interest instead of in the responses from the instrumentation, which nowadays are very precise [22]. Regression models are built from calibration procedures performing mathematical manipulation in the **X** matrix which contains the XRF measurement data from the samples and the **y** matrix, usually a vector with the response from the parameter of interest. In machine learning language **X** matrix are the *features* and **y** matrix the *targets.*

Multiple linear regression (MLR) is a well-known and very useful calibration model. It is excellent for well-behaved systems, i.e., systems with few interferents and low interaction among the analytes which do not have collinearity among the variables. To build the calibration model is considered, in matrixial terms, $\mathbf{y}_{nx1} = \mathbf{X}_{nxm}\ \boldsymbol{\beta}_{mx1} + \mathbf{E}_{nx1}$, illustrated in Fig. 6.4. The regression vector is determined by $\boldsymbol{\beta}_{mx1} = (\mathbf{X}^{T}\mathbf{X})^{-1}{}_{mxm}\ \mathbf{X}^{T}{}_{mxn}\ \mathbf{y}_{nx1}$. Therefore, an inverse matrix must be calculated, and the **X** matrix is required to be non-singular, because the MLR method do not eliminates data collinearity and all the information in the **X** matrix is used to predict **y**. Thus, the number of samples required in the calibration dataset to perform an MLR is 2^{k-1}, where k is the number of variables. MLR is not appropriate in cases where the number of variables is bigger than the number of samples.

In the case of performing a calibration model with XRF spectral data, which are typically acquired in multichannel analyzers with 1024 or 2048 channels, surely the number of variables will be higher than the number of samples. Moreover, XRF spectral data always present redundant information. For instance, the kα and kβ peaks have the same variability, the background region without peaks have almost the same intensity. These particularities bring out collinearity for the X matrix not permitting to calculate the inverse matrix. To circumvent this problem two options may be considered. Firstly, a previous variable selection could be performed with some variable selection method or based in the researcher experience and knowledge about the dataset. The second option, is to apply methods of dimensionality reduction in the dataset. As described in Sect. 6.2.2, PCA is one of the most used methods for dimensionality reduction and it is the fundament for principal component regression (PCR) and partial least squares Regression (PLS) methods. As

Parameter of interest (target) | XRF spectra data (features) | Regression vector | Residues vector

$$\begin{bmatrix} y_1 \\ y_2 \\ y_3 \\ \vdots \\ y_n \end{bmatrix} = \begin{bmatrix} 1 & x_1 & x_1^2 & \cdots & x_1^m \\ 1 & x_2 & x_2^2 & \cdots & x_2^m \\ 1 & x_3 & x_3^2 & \cdots & x_3^m \\ \vdots & \vdots & \vdots & \ddots & \vdots \\ 1 & x_n & x_n^2 & \cdots & x_n^m \end{bmatrix} \begin{bmatrix} \beta_0 \\ \beta_1 \\ \beta_2 \\ \vdots \\ \beta_m \end{bmatrix} + \begin{bmatrix} \varepsilon_1 \\ \varepsilon_2 \\ \varepsilon_3 \\ \vdots \\ \varepsilon_n \end{bmatrix},$$

Fig. 6.4 Matrixial representation of the inverse calibration method used in multiple linear regression, for instance, with XRF spectrum data

PCA generates PCs that carry out the data variability but filtering the collinearity, it reduces the dimensionality, e.g. from 2048 channels to few PCs (typically less than 10) turning viable the calculation of the inverse matrix and consequently the application of regression methods.

In PCR, the regression is performed in the scores matrix S from a PCA instead of in the original variables. The first step is to decompose the **X** matrix into scores and loading via PCA and then apply the MLR in the scores matrix, $\mathbf{y} = \mathbf{S}\,\boldsymbol{\beta} + \mathbf{E}$ in order to find the regression vector $\boldsymbol{\beta} = (\mathbf{S}^{T}\mathbf{S})^{-1}\ \mathbf{S}^{T}$ y. The PCR assumes that the parameter of interest in the y matrix is error free because just the X matrix is considered in the calculation.

In PLS method the **X** and **Y** matrixes are considered in the calculation and a practical aspect is that the errors, in the X matrix with the spectral data and the **Y** matrix with the reference results from the parameter of interest, are taken into account. Therefore, in PLS the PCA is applied in both **X** and **Y** matrix. Briefly, the PLS calculation [23] considers that the original variables are changed by a truncated subset of latent variables (LVs). As mentioned a PCA is operated in the features matrix: $\mathbf{X} = \mathbf{S}\ \mathbf{L}^{T} + \mathbf{E}_{X} = \Sigma\ \mathbf{s}_{A}\ \mathbf{l}_{A}{}^{T} + \mathbf{E}_{X}$ and in the target matrix: $\mathbf{Y} = \mathbf{U}\ \mathbf{Q}^{T} + \mathbf{Ey} = \Sigma\ \mathbf{u}_{A}\ \mathbf{q}_{A}{}^{T} + \mathbf{E}_{y}$ in which **S** and **U** are the scores in **X** and **Y**, and **L** and **Q** the loadings in **X** and **Y**, respectively and A the number of LVs. The residuals matrix **E** contains the non-modelled data comprised in the A LVs selected. After the PCA calculation a linear relation is stablished between the **X** and **Y** scores: $\boldsymbol{u}_A = \boldsymbol{\beta}_A \boldsymbol{s}_A$ being $\boldsymbol{\beta}_A$ is the regression vector obtained from $\boldsymbol{\beta}_A = (\mathbf{s}_A{}^{T}\ \mathbf{s}_A)^{-1}\ \mathbf{u}_A{}^{T}\ \mathbf{s}_A$.

The PLS iterative process may be simplified in the following steps: (i) standardize the data in **X** and **Y** using appropriate preprocessing methods; (ii) compute the weights for each variable calculating how much each variable represents to the parameter of interest; (iii) creates the LVs which are a linear combinations of the original variables weighted by the contribution of the parameter of interest (linear combinations on the scores and loadings of PCA from **X** and **Y**); (iv) perform a least square regression of the parameter of interest on the LVs; (v) iterate the process up to reach small differences in the weights. In summary, PLS is a powerful method for dealing with high dimensional data (e.g. spectral data) because it reduces the number of variables to a small set of uncorrelated components (LVs) and perform least square regression on these components.

6.2.5 Models Performance Evaluation and Figures of Merit

Evaluation metric for model performance including figures of merit is still under development for machine learning tools. However, for the most used methods International Union of Pure and Applied Chemistry (IUPAC) [24] and the American Society for Testing and Materials (ASTM) standard E1655–0017 [25] have established a series of metrics.

There are some performance parameters to evaluate multivariate models. Usually are considered the prediction errors, the determination coefficient of a dispersion plot and some ratio indicators such as ratio performance to deviation (RPD), the ratio of performance to interquartile distance (RPIQ) or the range error ratio (RER). The root mean square error (RMSE) is one of the most important parameters because it is related with the predicted values and the smaller the error, the better. The RMSE may be considered for the calibration set (RMSEC), the cross validation set (RMSECV) and for the validation or prediction set (RMSEP):

$$RMSEC = \sqrt{\frac{\sum_{i=1}^{nc}(y_i - \widehat{y_i})^2}{n_c}}$$

$$RMSECV = \sqrt{\frac{\sum_{i=1}^{nc}(y_i - \widehat{y_i})^2}{n_c - 1}}$$

$$RMSEP = \sqrt{\frac{\sum_{i=1}^{np}(y_i - \widehat{y_i})^2}{n_p - 1}}$$

where y_i represents the ith sample target parameter determined by conventional analyses, $\widehat{y_i}$ the target value predicted by the model, n_c is the number of calibration samples and n_p the number of prediction samples in the validation dataset.

The determination coefficient is usually reported from dispersion plots of measured versus predicted values. The closer to 1, the better. It is defined as:

$$R^2 = 1 - \frac{\sqrt{\sum_{i=1}^{n}(y_i - \widehat{y_i})^2}}{\sqrt{\sum_{i=1}^{n}(\overline{y}_i - y_i)^2}}$$

where $\overline{y}_i$ is the mean of the target parameter determined by conventional analyses.

The ratio indicators are used to stablish the model performance related to the conventional results. The one that better represent the spread of the data because it considers the interquartile range. RPIQ equation is given by [26]:

$$RPIQ = \frac{IQ}{RMSEP}$$

IQ is the difference between the third and first quartiles of y (IQ = Q3 − Q1), i.e. from the reference results from the parameter of interest. The following RPIQ intervals were adopted for its interpretation [27]: excellent models (RPIQ ≥ 2.5), very good models (2.5 > RPIQ ≥ 2.0), good model (2.0 > RPIQ ≥ 1.7), fair (1.7 > RPIQ ≥ 1.4), and very poor model (RPIQ < 1.4). The RPD is the ratio between the standard deviation in **y** and the RMSEP. It has the same interpretation as the R2 and brings out redundant information when both values are presented [28]. Finally, RER is equal to the range of the data values (the maximum value minus the minimum value) divided by the RMSEP. A RER value above 10 is roughly an indicator of a model with good predictive ability [29].

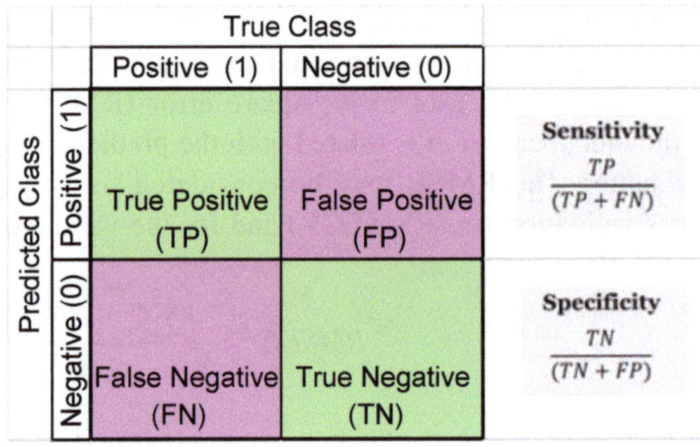

Fig. 6.5 Figures of merit in classification models: contingency tables and sensitivity and specificity equations

6.2.5.1 Figures of Merit in Classification Models

In the case of classification models the figures of merit are: the contingency table also named of confusion matrix, the specificity, sensibility and accuracy (Fig. 6.5). Sensibility is the model capability to classify correctly a sample as positive since it is true positive. Specificity is the model capability to identify correctly negative samples given they are true negative. Accuracy is a statistical parameter that solely measure the performance of a classification model. Accuracy is the ratio between the number of samples correctly classified, independently of the class, by the total number of samples [30].

6.2.5.2 Figures of Merit in Regression Models

The figures of merit evaluated in regression problems are: accuracy, precision, linearity, systematic errors (bias), sensitivity (SEN), the inverse of analytical sensitivity (γ^{-1}), limit of detection (LOD) and limit of quantification (LOQ) [18].

The accuracy expresses the concordance level between the predicted or measured value and the reference value. The RMSEC and RMSEP values are the most used and the last is the most robust because it considers the prediction samples. A complementary parameter is the relative error of predicted values in percentage, RE (%) given by

$$RE(\%) = 100\sqrt{\frac{\sum_{i=1}^{n_p}(y_i - \hat{y}_i)^2}{\sum_{i=1}^{n_p} y_i^2}}$$

The precision expresses the concordance level of a series of independent measurements in the same sample under prescribed conditions. It may be calculated in three levels: repeatability, intermediary precision and reproducibility [31]. At least the repeatability ought to be evaluated and coefficient of variation is applied

$$CV(\%) = 100\frac{SD}{\overline{y}}$$

where SD is the standard deviation among the replicates.

The linearity of the data is correlated with the random distribution of the residues and may be evaluated in a quantitative way by normality, homoscedasticity and independency testes over the residuals [32]. The Durbin-Watson (DW) test [33] is a tool that attest the independency of residuals and consequently the data linearity. DW test evaluates the lack of tendency in the residues considering the autocorrelations in an augmented partial residues plot [34]. The test examines the null hypothesis that there is no correlation between the successive residuals, versus the alternative hypothesis that the correlation exists. Probability values greater than 0.05 associated to the DW statistics indicates absence of non-linearities.

$$DW = \frac{\sum_{i=1}^{n_c-1}(e_{i+1} - e_i)^2}{\sum_{i=1}^{n_c-1} e_i^2}$$

where e_i is the ith residual and n is the number of calibration samples.

Bias are all the error components which are not randomly distributed. The mean bias for the prediction dataset is calculated. The bias affects the accuracy and the confidence limits interval determination. The occurrence of systematic errors (bias) according to E1655 ASTM [25], may be investigated by a t-test according to the following equations:

$$bias = \frac{\sum_{i=1}^{n_p}(y_i - \hat{y}_i)}{n_p}$$

$$SDV = \sqrt{\frac{\sum_{i=1}^{n_p}[(y_i - \hat{y}_i) - bias]^2}{n_p - 1}}$$

$$t = \frac{|bias|\sqrt{n_p}}{SDV}$$

if the t-calculated is greater than the t-critic at $n_p - 1$ degrees of freedom with a 95% significance level, there is an indication that the systematic errors present in the model are significant.

The basis for the sensitivity, the detection and quantification limits calculation in multivariate regression consider the NAS (net analyte signal) [35, 36]. The NAS is defined as the fraction of the analytical signal exclusively linked to the parameter of interest and orthogonal to the interferents space. It is determined from the norm of the regression vector in PCR and PLS analysis.

The sensitivity corresponds to the signal fraction which is responsible by the increase in one concentration unit from the parameter of interest. In linear multivariate models in which a regression vector is calculated, the sensitivity is determined by [22]:

$$SEN = \frac{1}{\|\boldsymbol{b}\|}$$

where $\|\boldsymbol{b}\|$ represents the $\boldsymbol{b}$ Euclidean norm.

The inverse of analytical sensitivity indicates the method resolution by providing an estimation of the minimum difference which is discernible by the analytical method considering just the random experimental noise as source of error.

The Detection limit (DL) is the lowest concentration of the property of interest which may be detected, but not necessarily quantified. The Quantification limit (QL) represents the lowest concentration of the property of interest which is quantified. For linear regression models considering the spectra data as the **X** matrix, LOD and LOQ are determined using the norm of the regression vector **b** and, an experimental noise from the background region the XRF spectrum:

$$DL = 3.3\|\boldsymbol{b}\|\delta_X$$
$$QL = 10\|\boldsymbol{b}\|\delta_X$$

The first application of the NAS concept was performed in NIR data [33, 37]. Some applications with EDXRF data were conducted with samples from soil and plants studies [1, 38, 39].

6.3 Cases of Study

6.3.1 Classification and Authentication of Food Samples

In addition to safety and nutritional purposes, information on elemental content in food samples is useful to confirm their authenticity, to find out their geographical origin but also to detect food fraud.

So far, several research attempts have been carried out to find reliable methods that can correctly differentiate product provenance. Most of these studies are based on the combined use of elemental composition measurements with chemometric tools [40].

In this sense it is interesting to mention the combined used of exploratory, classification and discrimination approaches such as principal component analysis (PCA), cluster analysis partial least square discriminant analysis (PLS-DA) and linear discriminant analysis (LDA) with XRF spectrometry. A high number of contributions have been published in the last years about, for example, the application for the classification of wines [41], teas [42] and coffee samples [43] according to their geographical origin and type.

Usually, in these studies, elemental concentrations determined by the XRF method are used as variables to carry out the chemometric analysis. However, unlike other atomic spectroscopic techniques used to determine the elemental composition of coffee samples (i.e., ICP-MS, ICP-OES, FAAS), EDXRF provides a continuous spectrum as an output, which could be used as a sample fingerprint. This approach has been widely used in molecular analysis such as near infrared spectroscopy (NIRS) but has been little explored in combination with EDXRF spectrometry in the food field. Only, in very few contributions the potential of using XRF spectra in combination with multivariate analysis have been explored

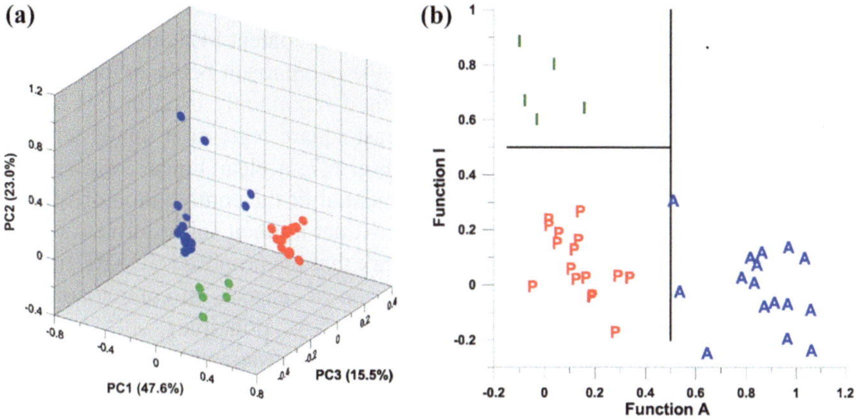

Fig. 6.6 Classification of coffee samples (I: Instant coffee (green), P: Pure coffee (red), A: Coffee with additives (blue)) using different chemometric approaches and EDXRF analysis: **a** PCA analysis using elemental concentrations (K, Ca, Mn, Fe, Cu, Zn, Rb and Sr) determined by EDXRF method and **b** leave-one-out procedure using PLS-DA from the raw EDXRF spectrographic data (adapted from Ref. [43])

including for instance the use of PLS-DA and EDXRF spectra to verify the geographical origin of varieties of Spanish honeys with a protected denomination of origin [6], beans seeds [44] and tomatoes [45].

In a recent contribution, a comparison of the discrimination potential of several EDXRF + chemometric approaches has been carried out for the specific case of classification of different coffee types (pure, instant and with additives) [43]. As it is shown in Fig. 6.6, results from the different chemometric approaches explored demonstrated that the application of PLS-DA on the raw EDXRF spectra can be a promising approach. In this case, fitting the EDXRF spectra and the determination of the elemental concentrations are not necessary, and the discriminating potential is similar.

6.3.2 Soil Attributes Identification

Traditionally, soil chemical analysis requires various steps from the sample preparation to the instrumental analysis, which includes sample drying, grinding, sieving, extraction or digestion, and finally, the measurement mostly by a spectroscopy technique. These methods have been evolved for decades to better predict soil quality attributes. Although current methods for soil analysis properly answer soil quality status, those methods are laborious, time consuming and costly. In this context, XRF has been recently applied for soil quality attributes. In fact, XRF commonly is utilized for total nutrient content determination, in which some nutrient unambiguously detected by their corresponding characteristic X-ray fluorescence signals. However, due to advances in chemometrics tools, implicitly information in the soil X-ray spectrum has been obtained and correlated to soil attributes such as pH [46], the bioavalibility nutrient content [47] and organic matter [39]. For instance, in Fig. 6.7 it is shown the comparison between total carbon content measured by the Walkley Black method (commonly used in soil laboratories) and values obtained using a partial least square regression model with the whole EDXRF spectra. As it is shown in Fig. 6.7, a clear linear correlation is found between both data sets showing the potential of the combination of spectral EDXRF data with multivariate calibration to extract implicit soil information from the spectra. Moreover, it is interesting to remark that the proposed EDXRF method entails less sample preparation and consumption of reagents in comparison with the reference method and it is more line with the GAC principles.

In another study, Kaniu et al. [48] estimated the bioavailable macronutrient (C, N, Na, Mg and P) fraction in soils by utilizing chemometrics and XRF. Specifically, PCA, PLS and ANNs were evaluated. It is highlighted that this strategy affords a direct solid analysis and expeditious semi-quantitative soil quality evaluation. In addition, this method allowed the bioavailable fraction of low atomic number element evaluation at trace concentration. Similarly, Kaniu [49] utilizing PLS and ANNs evaluated macro (NO^{3-}, $SO_4{}^{2-}$, $H_2PO_4{}^{-}$) and also micronutrient (Fe, Cu and Zn). Tavares [50] evaluate the Compton normalization, and the multivariate tools (a) multiple linear regression and (b) partial least square regression to correct the matrix effect in the XRF analysis and for better predict the soil fertility attribute, in which the multivariate strategies improved the model for exchangeable K and Ca, clay, and cation exchange capacity predictions compared to univariate approach.

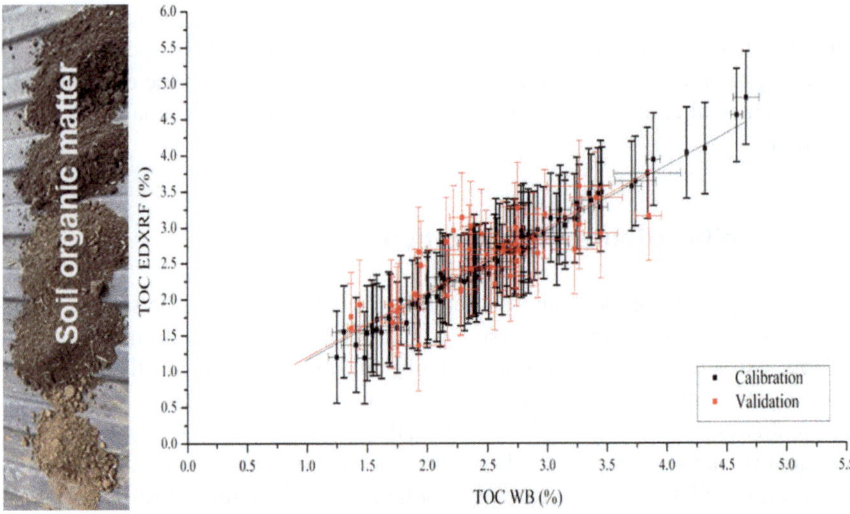

Fig. 6.7 Illustration of total organic carbon (TOC) concentration measured by the Walkley Black (WB) method versus predicted values obtained using a partial least square regression model with the whole EDXRF spectra (adapted from Ref. [39])

6.3.3 E-waste Analysis

Electronic waste (e-waste) worldwide production attained the staggering figure of 62 million of tonnes per year (2022), and it is estimates to reach 82 million of tonnes by 2030. Along with these dramatic figures, the amount of adequately collected or recycled e-waste is only ca. one quarter of this amount [51].

E-waste contains various hazardous chemical content, including organic and inorganic compounds, being them harmful to human being and environment [52]. Particularly, the potentially toxic elements present in e-waste, such as Al, Cr, Ni Cu, Zn, Cd, Pb, Hg, are of concern. Within a myriad of analytical techniques, XRF stands out among them due to direct solid analysis and minimum sample preparation features. Although, the XRF technique has been utilized for e-waste analysis for total elemental concentration [53] and also together with other analytical techniques for plastic identification from e-waste [54], recent studies bring the use of chemometrics and XRF to characterize e-waste [55, 56]. As example, Daniel and co-workers [56] utilized laser ablation-inductively coupled plasma-mass spectrometry (LA-ICP-MS), laser-induced breakdown spectroscopy (LIBS), and micro-X-ray fluorescence spectroscopy (micro-XRF) techniques to investigate elemental composition of printed circuit boards from mobile phones. In a first study, these techniques were combined with PCA to investigate the chemical composition on the surface and depth profiling of PCB samples. After, PLS regression was used to data fusion of analytical information from LIBS and micro-XRF analysis with the aim of Cu quantification in PCB samples. This approach was tested with two PCBs represented in 40 subsamples as a proof of concept. As it is shown in Fig. 6.8, examining the trueness obtained with the proposed fusion approach, promising results were obtained for Cu analysis with recovery values in the range of 80–120%.

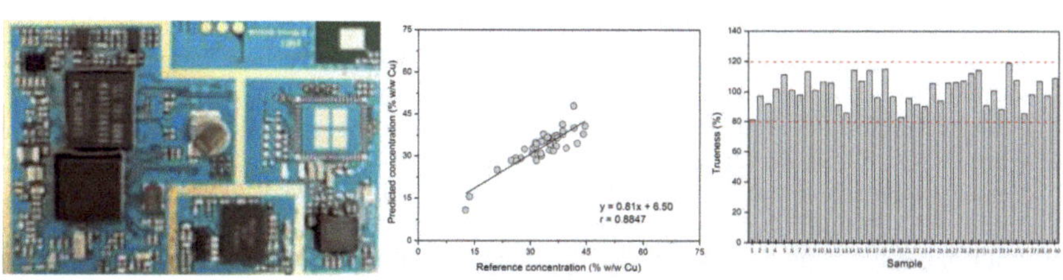

Fig. 6.8 Correlation graph (Reference versus predicted Cu concentrations (in % w/w)) and trueness evaluation using micro-XRF + LIBS data fusion. Figure adapted from Andrade et al. [56] with the permission from Elsevier

6.4 Final Considerations

The combination of XRF data with chemometric expands the analytical application of the technique, insert it in the GAC and turns easier and faster the determination of attributes, light elements and trace elements making use of machine learning tools. Moreover, when dealing with huge datasets in which the parameter of interest is known, machine learning tools are overlapping convention calibration and turning faster and cheaper industrial processes.

In the age of Artificial Intelligence and cloud data storage and operation, a tendency is the migration of process that were performed locally, in a specific laboratory or spectrometer, to automatic process based on dataset in the cloud and data interchange. However, this step must be made with criteria and with total knowledge of the machine learning tools, its potentialities and limitations to reach confidence results. Nevertheless, the future of XRF analysis combined with chemometrics is very promising with improved solutions in different areas and application.

References

1. Melquiades, F.L., Bortoleto, G.G., Marchiori, L.F.S., Bueno, M.I.M.S.: Direct determination of sugar cane quality parameters by X-ray spectrometry and multivariate analysis. J. Agric. Food. Chem. **60**(43), 10755–10761 (2012). https://doi.org/10.1021/jf302471b
2. Melendez-Perez, J.J., et al.: Forensic application of X-ray fluorescence spectroscopy for the discrimination of authentic and counterfeit revenue stamps. Appl. Spectrosc. **70**(11), 1910–1915 (2016). https://doi.org/10.1177/0003702816645352
3. Panebianco, S., et al.: Feasibility study of tomato fruit characterization by fast XRF analysis for quality assessment and food traceability. Food. Chem. **383**, 132364 (2022). https://doi.org/10.1016/j.foodchem.2022.132364
4. Fiamegos, Y., Dumitrascu, C., Papoci, S., de la Calle, M.B.: Authentication of PDO paprika powder (Pimentón de la Vera) by multivariate analysis of the elemental fingerprint determined by ED-XRF. A feasibility study. Food. Control. **120**, 107496 (2021). https://doi.org/10.1016/j.foodcont.2020.107496
5. Armetta, F., et al.: The silver collection of San Gennaro treasure (Neaples): a multivariate statistic approach applied to X-ray fluorescence data. Spectrochim. Acta. Part. B At Spectrosc. **180**, 106171 (2021). https://doi.org/10.1016/j.sab.2021.106171
6. Ghidotti, M., Fiamegos, Y., Dumitrascu, C., de la Calle, M.B.: Use of elemental profiles to verify geographical origin and botanical variety of Spanish honeys with a protected denomination of origin. Food. Chem. **342**, 128350 (2021). https://doi.org/10.1016/j.foodchem.2020.128350
7. Kogou, S., Lee, L., Shahtahmassebi, G., Liang, H.: A new approach to the interpretation of <scp>XRF</scp> spectral imaging data using neural networks. X-Ray Spectrom. **50**(4), 310–319 (2021). https://doi.org/10.1002/xrs.3188
8. Chen, T., Qi, X., Si, Z., Cheng, Q., Chen, H.: An energy dispersive X-ray fluorescence spectrometry approach for the identification of geographical origin of wheat flour. Int. J. Food Eng. **17**(11), 877–884 (2021). https://doi.org/10.1515/ijfe-2020-0265
9. Williams, H.F.L., Liu, K.: Testing XRF discrimination of marine and terrestrial flood deposits in Southeastern Texas coastal marshes. J. Coast. Res. **37**(6) (2021). https://doi.org/10.2112/JCOASTRES-D-21-00046.1
10. Kowalski, F.H., Schatzki, B.R., Stross, T.F.: Classification of archaeological artifacts by applying pattern recognition to trace element data. Anal. Chem. **44**(13), 2176–2180 (1972)

11. Marguí, E., Queralt, I., de Almeida, E.: X-ray fluorescence spectrometry for environmental analysis: basic principles, instrumentation, applications and recent trends. Chemosphere **303**, 135006 (2022). https://doi.org/10.1016/j.chemosphere.2022.135006
12. Ferreira, M.M.C., Faria, C.G., Paes, E.T.: Oceanographic characterization of northern São Paulo coast: a chemometric study. Chemom. Intell. Lab. Syst. **47**(2), 289–297 (1999). https://doi.org/10.1016/S0169-7439(98)00216-0
13. Ferreira, M.M.C., Antunes, A.M., Melgo, M.S., Volpe, P.L.O.: Quimiometria I: calibração multivariada, um tutorial. Quim Nova. **22**(5), 724–731 (1999). https://doi.org/10.1590/S0100-40421999000500016
14. Gori, M., Betti, A., Melacci, S.: Learning principles. In: Machine Learning, pp. 53–111. Elsevier (2024). https://doi.org/10.1016/B978-0-32-389859-1.00009-X.
15. Panchuk, V., Yaroshenko, I., Legin, A., Semenov, V., Kirsanov, D.: Application of chemometric methods to XRF-data–a tutorial review. Anal. Chim. Acta. **1040**, 19–32 (2018). https://doi.org/10.1016/j.aca.2018.05.023
16. dos Santos, F.R., de Oliveira, J.F., Bona, E., Barbosa, G.M.C., Melquiades, F.L.: Evaluation of pre-processing and variable selection on energy dispersive X-ray fluorescence spectral data with partial least square regression: a case of study for soil organic carbon prediction. Spectrochim. Acta. Part. B. At. Spectrosc. **175**, 106016 (2021). https://doi.org/10.1016/j.sab.2020.106016.
17. Westad, F., Marini, F.: Validation of chemometric models–a tutorial. Anal. Chim. Acta. **893**, 14–24 (2015). https://doi.org/10.1016/j.aca.2015.06.056
18. Valderrama, P., Braga, J.W.B., Poppi, R.J.: Variable selection, outlier detection, and figures of merit estimation in a partial least-squares regression multivariate calibration model. A case study for the determination of quality parameters in the alcohol industry by near-infrared spectroscopy. J. Agric. Food. Chem. **55**(21), 8331–8338 (2007). https://doi.org/10.1021/jf071538s
19. Alves, F.C.G.B.S., Valderrama, P.: Ultraviolet spectroscopy and supervised pattern recognition methods for authentication of transgenic and non-transgenic soybean oils. Anal. Methods. **7**(22), 9702–9706 (2015). https://doi.org/10.1039/C5AY02440D
20. Brereton, R.G., Lloyd, G.R.: Partial least squares discriminant analysis: taking the magic away. J. Chemom. **28**(4), 213–225 (2014). https://doi.org/10.1002/cem.2609
21. Bishop, C.M.: Pattern Recognition and Machine Learning, 1st edn. Springer, Chennai (2013)
22. Brereton, R.G.: Introduction to multivariate calibration in analytical chemistry. Analyst. **125**(11), 2125–2154 (2000). https://doi.org/10.1039/b003805
23. Geladi, P., Kowalski, B.R.: Partial least-squares regression: a tutorial. Anal. Chim. Acta. 185, 1–17 (1986). https://doi.org/10.1016/0003-2670(86)80028-9
24. Currie, L.A.: Nomenclature in evaluation of analytical methods including detection and quantification capabilities1Adapted from the international union of pure and applied chemistry (IUPAC) document "Nomenclature in evaluation of analytical methods including detection." Anal. Chim. Acta. **391**(2), 105–126 (1999). https://doi.org/10.1016/S0003-2670(99)00104-X
25. Standard Practices for Infrared Multivariate Quantitative Analysis E1655–05 ASTM. Conshohocken (2012)
26. Bellon-Maurel, V., Fernandez-Ahumada, E., Palagos, B., Roger, J.-M., McBratney, A.: Critical review of chemometric indicators commonly used for assessing the quality of the prediction of soil attributes by NIR spectroscopy. TrAC Trends Anal. Chem. **29**(9), 1073–1081 (2010). https://doi.org/10.1016/j.trac.2010.05.006
27. Nawar, S., Mouazen, A.M.: Predictive performance of mobile vis-near infrared spectroscopy for key soil properties at different geographical scales by using spiking and data mining techniques. Catena (Amst). **151**, 118–129 (2017). https://doi.org/10.1016/j.catena.2016.12.014
28. Minasny, A., Budiman, McBrateney.: Why you don't need to use RPD. Pedometron (33) (2013)
29. Teixeira dos Santos, C.A., et al.: Raman spectroscopy for wine analyses: a comparison with near and mid infrared spectroscopy. Talanta. **186**, 306–314 (2018). https://doi.org/10.1016/j.talanta.2018.04.075

30. López, M.I., Callao, M.P., Ruisánchez, I.: A tutorial on the validation of qualitative methods: from the univariate to the multivariate approach. Anal. Chim. Acta. **891**, 62–72 (2015). https://doi.org/10.1016/j.aca.2015.06.032
31. Pryseley, A., Mintiens, K., Knapen, K., Van der Stede, Y., Molenberghs, G.: Estimating precision, repeatability, and reproducibility from Gaussian and non-Gaussian data: a mixed models approach. J. Appl. Stat. **37**(10), 1729–1747 (2010). https://doi.org/10.1080/02664760903150706
32. de Souza, S.V.C., Junqueira, R.G.: A procedure to assess linearity by ordinary least squares method. Anal. Chim. Acta. **552**(1–2), 25–35 (2005). https://doi.org/10.1016/j.aca.2005.07.043
33. Centner, V., de Noord, O.E., Massart, D.L.: Detection of nonlinearity in multivariate calibration. Anal. Chim. Acta. **376**(2), 153–168 (1998). https://doi.org/10.1016/S0003-2670(98)00543-1
34. Olivieri, A.C.: Practical guidelines for reporting results in single-and multi-component analytical calibration: a tutorial. Anal. Chim. Acta. **86**, 10–22 (2015). https://doi.org/10.1016/j.aca.2015.01.017
35. Lorber, A., Faber, K., Kowalski, B.R.: Net analyte signal calculation in multivariate calibration. Anal. Chem. **69**(8), 1620–1626 (1997). https://doi.org/10.1021/ac960862b
36. Bro, R., Andersen, C.M.: Theory of net analyte signal vectors in inverse regression. J. Chemom. **17**(12), 646–652 (2003). https://doi.org/10.1002/cem.832
37. Braga, J.W.B., Poppi, R.J.: Figures of merit for the determination of the polymorphic purity of carbamazepine by infrared spectroscopy and multivariate calibration. J. Pharm. Sci. **93**(8), 2124–2134 (2004). https://doi.org/10.1002/jps.20109
38. Manhas Verbi Pereira, F., Rodrigues Pereira-Filho, E., Maretti Silveira Bueno, M.I.: Development of a methodology for calcium, iron, potassium, magnesium, manganese, and zinc quantification in teas using X-ray spectroscopy and multivariate calibration. J Agric Food Chem. **54**(16), 5723–5730 (2006). https://doi.org/10.1021/jf0603782
39. Morona, F, dos Santos, F.R., Brinatti, A.M., Melquiades, F.L.: Quick analysis of organic matter in soil by energy-dispersive X-ray fluorescence and multivariate analysis. Appl. Radiat. Isot. **130**, 13–20 (2017). https://doi.org/10.1016/j.apradiso.2017.09.008
40. Aleixandre-Tudo, J.L., Castello-Cogollos, L., Aleixandre, J.L., Aleixandre-Benavent, R.: Chemometrics in food science and technology: a bibliometric study. Chemom. Intell. Lab. Syst. **222**, 104514 (2022). https://doi.org/10.1016/j.chemolab.2022.104514
41. Vitali Čepo, D., Karoglan, M., Borgese, L, Depero, L.E., Marguí, E., Jablan, J.: Application of benchtop total-reflection X-ray fluorescence spectrometry and chemometrics in classification of origin and type of Croatian wines. Food. Chem. X. **13**, 100209 (2022). https://doi.org/10.1016/j.fochx.2022.100209.
42. Rajapaksha, D., et al.: XRF to support food traceability studies: classification of Sri Lankan tea based on their region of origin. X-Ray Spectrom. **46**(4), 220–224 (2017). https://doi.org/10.1002/xrs.2748
43. Weinberger, M., et al.: Critical evaluation of energy dispersive X-ray fluorescence spectrometry for multielemental analysis of coffee samples: sample preparation, quantification and chemometric approaches. Spectrochim. Acta. Part. B. At. Spectrosc. **215**, 106898 (2024). https://doi.org/10.1016/j.sab.2024.106898
44. Allegretta, I., et al.: TXRF spectral information enhanced by multivariate analysis: a new strategy for food fingerprint. Food Chem. **401**, 134124 (2023). https://doi.org/10.1016/j.foodchem.2022.134124
45. Galvan, D., et al.: Energy-dispersive X-ray fluorescence combined with chemometric tools applied to tomato and sweet pepper classification. Food. Control. 143, 109326 (2023). https://doi.org/10.1016/j.foodcont.2022.109326
46. dos Santos Teixeira, A.F., et al.: Tropical soil pH and sorption complex prediction via portable X-ray fluorescence spectrometry. Geoderma. **361**, 114132 (2020). https://doi.org/10.1016/j.geoderma.2019.114132

47. Tavares, T.R., et al.: Estimating plant-available nutrients with XRF sensors: towards a versatile analysis tool for soil condition assessment. Geoderma. **439**, 116701 (2023). https://doi.org/10.1016/j.geoderma.2023.116701
48. Kaniu, M.I., Angeyo, K.H., Mwala, A.K., Mangala, M.J.: Direct rapid analysis of trace bioavailable soil macronutrients by chemometrics-assisted energy dispersive X-ray fluorescence and scattering spectrometry. Anal. Chim. Acta. **729**, 21–25 (2012). https://doi.org/10.1016/j.aca.2012.04.007
49. Kaniu, M.I., Angeyo, K.H., Mwala, A.K., Mwangi, F.K.: Energy dispersive X-ray fluorescence and scattering assessment of soil quality via partial least squares and artificial neural networks analytical modeling approaches. Talanta. **98**, 236–240 (2012). https://doi.org/10.1016/j.talanta.2012.06.081
50. Tavares, T.R., et al.: Assessing soil key fertility attributes using a portable X-ray fluorescence: a simple method to overcome matrix effect. Agron. **10**(6), 787 (2020). https://doi.org/10.3390/agronomy10060787
51. [Online]. https://unitar.org/about/news-stories/press/global-e-waste-monitor-2024-electronic-waste-rising-five-times-faster-documented-e-waste-recycling
52. Parvez, S.M., et al.: Health consequences of exposure to e-waste: an updated systematic review. Lancet. Planet. Health. **5**(12), e905–e920 (2021). https://doi.org/10.1016/S2542-5196(21)00263-1
53. Lancaster, S.T., et al.: Evaluation of X-ray fluorescence for analysing critical elements in three electronic waste matrices: a comprehensive comparison of analytical techniques. Waste Manag. **190**, 496–505 (2024). https://doi.org/10.1016/j.wasman.2024.10.015
54. Puype, F., Ackerman, L.K., Samsonek, J.: Evaluation of direct analysis in real time–high resolution mass spectrometry (DART-HRMS) for WEEE specific substance determination in polymeric samples. Chemosphere **232**, 481–488 (2019). https://doi.org/10.1016/j.chemosphere.2019.05.166
55. Ferreira, D.S., Pereira, F.M., Olivieri, V.A.C., Pereira-Filho, E.R.: Electronic waste analysis using laser-induced breakdown spectroscopy (LIBS) and X-ray fluorescence (XRF): Critical evaluation of data fusion for the determination of Al, Cu and Fe. Anal. Chim. Acta. **1303**, 342522 (2024). https://doi.org/10.1016/j.aca.2024.342522
56. Andrade, D.F., de Almeida, E., de Carvalho, H.W.P., Pereira-Filho, E.R., Amarasiriwardena, D.: Chemical inspection and elemental analysis of electronic waste using data fusion-application of complementary spectroanalytical techniques. Talanta. **225**, 122025 (2021). https://doi.org/10.1016/j.talanta.2020.122025.5

Data Analysis and X-Ray Fluorescence in Environmental Toxicology

7

Sofia Pessanha, Pedro Catalão Moura, and Sofia Barbosa

Abstract

Over the past four decades, X-ray fluorescence based analytical techniques have been deeply employed for a vast range of applications that extend from academic to industrial settings. Among these applications, the field of environmental toxicology has relied in XRF as a trustworthy and sensitive tool for purposes such as geological studies of soil and sediments, vegetable analyses, air and water pollutants characterization and waste management. Adding to its non-destructive nature and lack of sample preparation, large quantities of data are produced during XRF analyses. In this way, suitable methodologies for data processing and interpretation are necessary and newer approaches are under constant demand. This chapter aims to review the current tools and future trends used to statistically process data previously collected through XRF devices in the field of environmental toxicology. The main advantages and disadvantages of each technique are equally addressed. Finally, newer mathematical models whose preliminary results denote an auspicious future, namely, recurrent, deep and convolution neural networks-based algorithms, are described and properly framed in the environmental toxicology field.

S. Pessanha (✉) · P. C. Moura
Laboratory for Instrumentation, Biomedical Engineering and Radiation Physics, (LIBPhys-LA-REAL), NOVA School of Science and Technology, NOVA University of Lisbon, 2829-516 Caparica, Portugal
e-mail: sofia.pessanha@fct.unl.pt

S. Barbosa
GeoBioTec (UNL) - GeoBiosciences, GeoTechnologies and GeoEngineering, Department of Earth Sciences, NOVA School of Science and Technology, NOVA University of Lisbon, 2829-516 Caparica, Portugal

V. K. Singh (ed.), *X-Ray Fluorescence Spectroscopy and Chemometrics*,
https://doi.org/10.1007/978-3-031-98375-7_7

7.1 Introduction

X-ray fluorescence (XRF)-based analytical techniques have been playing a pivotal role in both scientific and civil communities. Their key characteristics, which include non-destructive analyses, rapid results, and analytical flexibility, combined with sustainable advantages offered by straightforward procedures of sample preparation and analysis, make these techniques particularly valuable for a vast range of applications. Focusing on the field of environmental applications, XRF-based techniques have been employed in geological studies of soils and sediments [1], vegetal analyses [2], air and water pollutants characterization [3], and even overall waste products exploratory studies [4]. In spite of the type of environmental analysis, the utilization of XRF techniques produces large amounts of data that need appropriate mining techniques. As a result, the need for accurate and efficient data processing methodologies and newer approaches remains a pressing demand within the scientific community. This chapter, therefore, aims to explore the current procedures and future trends of XRF data processing tools by addressing the most relevant studies in key fields of application and the tools employed by researchers. Emphasis will be given to the advantages and evolution of these methods with a focus on how they meet the growing demands for more sophisticated data analysis.

Any analysis performed with an XRF-based technique produces large amounts of data that need to be processed. The majority of the procedures for data processing rely on statistical tools of different natures. These tools are numerous, and the selection of the most suitable one depends on the aim of the study and the type of data under processing. By framing the statistical methods available for data processing, three major group of tools arise: methods of univariate analysis (UVA), methods of bivariate analysis (BVA), and methods of multivariate analysis (MVA) [5]. In the particular case of benchtop micro-XRF analysis [6], there are two main types of data to be considered in this context that are generated and capable of being analyzed and interpreted statistically: (1) elemental semiquantitative quantities present in the samples derived from spectral interpretation and analysis, and (2) data obtained from image processing of 2D map surveys [7, 8]. In both cases, univariate, bivariate, and multivariate statistical tools can potentially be applied, besides specific image processing tool to be utilized in case (2) [9]. Univariate analysis refers to tools used to describe or characterize a set of observations of a single variable or attribute within a given population. Among the main tools of univariate analysis, one can mention the measures of central tendency, like the mean, median, and mode, the measures of dispersion, like the range, the variance, and the standard deviation, the descriptive methods, like the geometric mean, and harmonic mean, and even the inferential methods, like the one-sample t-test, and the one-sample sign test. The results of this kind of analysis are often displayed through a numeral value with a specific significate or significance [10].

Bivariate analysis, on the other hand, groups all the tools used to statistically process, compare, and identify the correlation between the attributes of two distinct variables or two distinct populations of one variable. Regression- and correlation-based tools are the most common methodologies in the bivariate analysis, namely, linear regression, least squares regression, Pearson correlation, Spearman correlation, Chi-squared test, and many others. The results are often visualized through graphs or plots like a scatterplot, a linear graph, and other models that can explain bivariate relations [11].

Finally, the multivariate analysis involves the observation and processing of a significant number of multiple variables or attributes from either a single population or multiple populations. It aims to identify potential relationships, patterns, or co-dependencies among these variables. This group is often considered the largest one since it compiles the biggest number of methods, namely, variance and covariance analysis, principal component analysis (PCA), correspondence analysis (CA), redundancy analysis (RA), linear discriminant analysis (LDA), clustering systems, and even artificial intelligence-based methods like artificial neural networks, and deep learning methods. Results from multivariate analysis can be displayed through all types of visualization tools, ranging from simple tables to more informative multidimensional graphics [12].

Despite all the available tools to statistically process a dataset, the most suitable method depends largely on the type of the data and the purpose of the study. Due to its complexity and to the vast range of features explorable in a set of XRF data, researchers have been employing different types of processing tools each exhibiting strengths and limitations depending on the attributes under investigation. Additionally, newer and more robust methods are constantly being tested and even developed, aiming to take fully advantage of XRF results.

7.2 Univariate Analysis

Univariate statistical methods are employed to characterize or explore a set of observations in regard to a single variable or attribute of that set. These methods can be categorized into four main groups: measures of central tendency, measures of dispersion, descriptive methods, and inferential methods. The addressed tools are summarized in Table 7.1.

Table 7.1 Statistical tools under the scope of univariate analysis

Univariate analysis	
Measures of central tendency	Mean
	Median
Measures of dispersion	Range
	Standard deviation
	Variance
Measures of symmetry	Skewness
	Kurtosis
Inferential methods	One-sample T-test

7.2.1 Measures of Central Tendency and Dispersion

Measurements of central tendency, like the mean and the median, are tools that provide a characteristic central value for a probability distribution, while the measures of dispersion (range, standard deviation and variance), provide information on how dispersed the data included in a population is.

The mean of a population corresponds to a numeric quantity that indicates the center of the data, i.e., is an intermediary value in regard to the quantities of both minimum and maximum numbers of the dataset. It is calculated by summing all the values in a dataset and then dividing them by the total number of observations. The median, in its turn, corresponds to the value that separates a data set or population in two equal halves. It differs from the mean since it is not influenced by how small or how large are the minimum or the maximum values of the population. These central tendency tools are largely used in XRF-based works, namely, to characterize the variability of elements like potassium or calcium among distinct samples [13], to assess the average content of each element in specific populations [14], and as a way of expressing the limits of detection and quantification for specific elements [15].

Despite their wide use, these tools are not robust measures when it comes to outliers in the datasets. Prudence is advised especially in the case of one or more outliers, and asymmetric distributions. Additionally, they often fail in properly describing a data set as two completely distinct populations can have the exact same mean value. This way, dispersion tools like range, standard deviation or variance allow to better explore the populations since they consider the variability among the values.

The range is the difference between the smallest and the largest value of a dataset. Its main advantage is the easiness of the calculation, nonetheless, a range is often poorly informative if provided alone. To overcome this disadvantage, the range is commonly accompanied by the standard deviation or the variance. From all the dispersion methods, is one of the most used, being a measure of how spread the data is also providing information on the skewness of the population. As a dispersive tool, the variance of a population provides information on how spread the observations are from the mean value. In terms of advantage, variance is very robust to algebraic manipulation, however, its units are often different from the units of the observations, making it less practical in some scenarios.

All these tools are widely employed during the processing of XRF data, namely, to represent the concentration intervals specific elements in datasets [16], or to investigate the variability of the results achieved through central tendency measures [17]. Messager et al. [18], in their work on the suitability of mosses for gauging metal pollution throughout decades of exposure, used some of the mentioned tools, namely, the mean and standard deviation to determine the coefficient of variation. This parameter allowed the authors to describe the overall variability of the measurements across moss mats on a given tree [18]. Finally, the mean and standard deviation were also used by Gallardo et al. [19] to assess bromine accumulation in two agricultural soils fumigated with a methyl bromide pesticide and

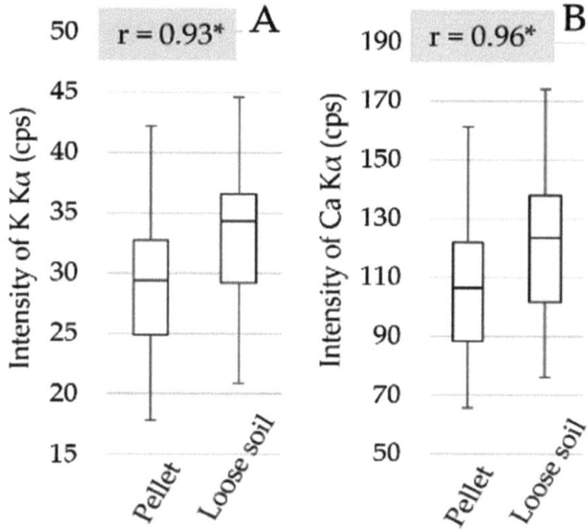

Fig. 7.1 Boxplot chart to illustrate the mean values, the range, the standard deviation and even the Pearson correlation (addressed in due time) for the intensity levels of Kα emission lines for K and Ca present in samples of soil. Figure adapted from Tavares et al. [13] under Creative Commons Attribution (CC BY) License

irrigated with regenerated waste water. These tools allowed the authors to verify if the used solid suspension TXRF method was unbiased by comparing the achieved mean and standard deviation values with the certified values [19].

Both central tendency and dispersion tools are often used together and mostly illustrated in boxplot graphs like the one developed by Sun et al. [20] to plot the range of detection for elements like Si, Al, Fe, Ca, K, Ti, Mn, Zr, S, Sr, Rb, V, Zn, Cr, Ni, Cu, Y, Pb, Nb, and Th. This procedure was used by the authors to verify that their concentrations in samples of soils, exceed the detection limits of the device used during the analyses [20]. Boxplot graphs were also used by Tavares et al. [13] to illustrate the mean values, the range, the standard deviation and even the Pearson correlation (addressed in Sect. 7.3.2.1) for the intensity levels of Kα emission lines for K and Ca present in samples of soil [13] (Fig. 7.1). Another common tool used to represent dispersion results in the ternary diagram. Rowe et al. [21], during their work on the assessment of the percentual amounts of specific elements in geological samples of mudrocks, a type of siliciclastic sedimentary rocks, employed several central tendency and dispersion tool to characterize the quantity of certain elements in the samples. Additionally, they also explored chemostratigraphic shifts in sequences of mudrocks based on their mineral composition, namely, relative variation in CaO, Al_2O_3 and SiO_2 assumed to reflect bulk mineralogical changes in calcite, clay and quartz. These results that were plotted in a ternary graph (Fig. 7.2) [21].

7.2.2 Measures of Symmetry

The assessment of how symmetrically a population or a distribution behaves is often important and even necessary to the application of other statistical tool that

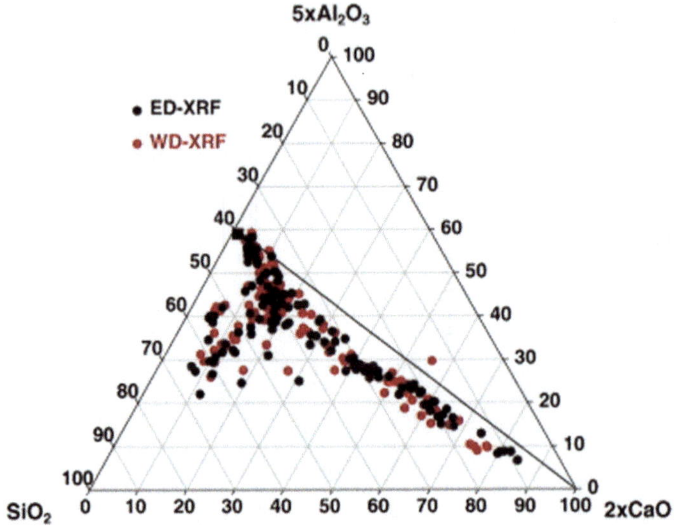

Fig. 7.2 Ternary plot of the relative variations in CaO, Al_2O_3 and SiO_2 known for reflecting bulk mineralogical changes in calcite, clay and quartz. Figure adapted from Rowe et al. [21] with the permission from Elsevier

require, for example, a normal distribution. Skewness and Kurtosis are the most used measures of symmetry.

7.2.2.1 Skewness and Kurtosis

Skewness and Kurtosis are the most common measures of symmetry. Skewness allows the assessment of symmetry of a dataset or distribution relatively to the centre value, i.e., it evaluates how unsymmetric the population is relatively to its mean. Kurtosis, on the other hand, measures how light- or heavy-tailed relatively to a normal distribution, the dataset is.

The work of Idris [22] is an example of the usefulness of both kurtosis and skewness for datasets produced with XRF analyses. In his work, the author aimed to test between-bottle homogeneity of certified reference materials for later utilization in the full assessment of macro-, micro- and trace elements in environmental samples. To do so, the author employed a wavelength dispersive X-ray fluorescence spectrometer and tested bottles of different origins for the same certified reference material. Then, based on the detected quantities of the present elements, the authors determined the average concentration per element and used skewness and kurtosis to evaluate the symmetry of this distribution. The skewness exhibited values <1, indicating symmetrical distribution of elemental content. The kurtosis, in its turn, exhibited values around one and two, indicating light- to heavy-tailed relative to the normal distribution [22].

Tepanosyan et al. [23] used both measures of central tendency and measures of dispersion to evaluate the content of Pb, Mo and Ti in Yerevan urban soils. Their analysis aimed to infer the historical impacts of human activity on the soil composition. The results showed that Ti levels corresponded to the natural geological makeup of the region, while Mo concentrations reflected sources of industrial pollution. In contrast, Pb distribution was aligned with urbanized areas, highlighting the influence of human development. The determination of the datasets skewness was also very useful to the authors since they could eliminate from their datasets, elements exhibiting highly skewed distribution [23].

7.2.3 Inferential Methods

Inferential methods are often considered to be transition tools among the UVA and the BVA. In fact, this class of statistical methods uses a small sample to infer conclusions on a larger population. These tools, often based on multivariate tests, differ from the descriptive methods since they allow to draw some conclusions on the characteristics of a population considering a portion of it. They are equally used to test hypotheses and produce statistical propositions.

7.2.3.1 One-Sample T-Test

The t-test is a statistical hypothesis test and is often used to assess the statistically significance between two populations, for example. In its one-sample variation, it is classified as a univariate tool and is used to verify whether the mean of a data set exhibits a value specified in a null hypothesis. Additionally, this test is only valid for datasets that follow a normal distribution, in this way, it must be preceded by a normality assessment test.

Some XRF-based data has been explored with the one-sample T-test. Pedrão et al. [24], employed the one-sample T-test to compare the suitability of XRF to identify and quantify major elements like Fe, Ti and Ca in comparison with ICP-OES. To do so, the authors analyzed samples of marine sediments with an XRF spectrometer and applied the one-sample T-test to the concentration levels achieved with both analytical techniques for the mentioned elements (p value > 0.05). This statistical tool proved that both techniques exhibit similar results meaning that they are fully capable of identifying and quantifying the targeted elements. In parallel, the authors also determined several linear regressions to translate the relationship between the amount of Ti and Ca ($R^2 = 0.875$), and Fe and Ca ($R^2 = 0.755$). In this way, they used the Bland–Altman (B-A plot), a tool widely used to compare the results achieved through two different methods, to comparatively verify the accuracy of the results achieved by both linear regressions and T-test (Fig. 7.3). This plot revealed all the values used in the mentioned tools were within the 95% confidence interval, verifying the correctness of the results [24].

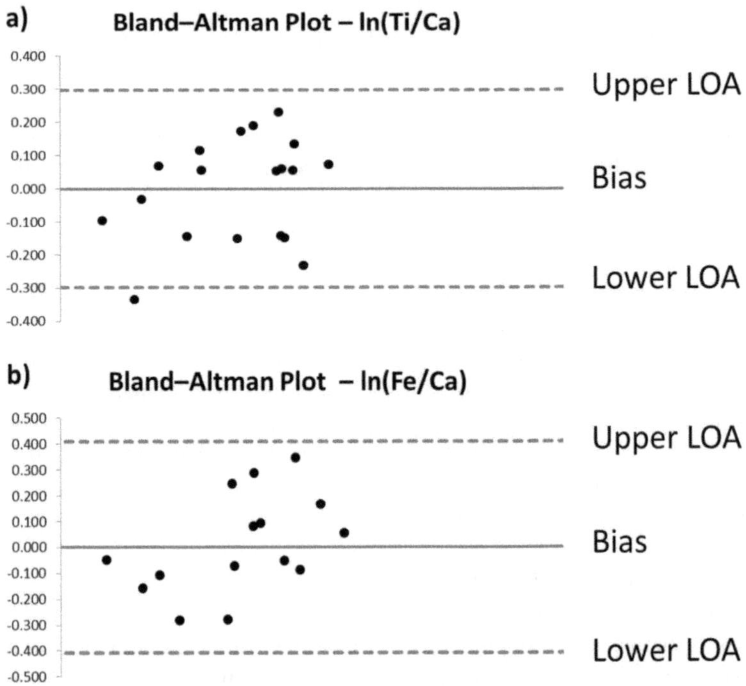

Fig. 7.3 Bland—Altman plot for the **a** ln (Ti/Ca) and **b** ln (Fe/Ca) data used by Pedrão et al. [24] as a way of comparatively assessing the results achieved with both the linear regression and the one-sample T-test tools [24]. Figure adapted from Pedrão et al. [24] under Creative Commons Attribution (CC BY) License

7.3 Bivariate Analysis

BVA exhibits evident advantages when compared with UVA since it allows to explore a specific population in regard to two characteristic parameters or variables, and even to assess eventual patterns or relationships among those variables. The majority of bivariate methods is grouped in two main classes, the regression methods and the correlation methods. The addressed tools are summarized in Table 7.2.

Table 7.2 Statistical tools addressed under the scope of bivariate analysis

Bivariate analysis	
Regression methods	Linear regression
	Partial least squares regression
Correlation methods	Pearson correlation
	Spearman correlation

7.3.1 Regression Methods

Regression methods are employed to predict relationships among at least two variables. These variables often include a dependent one and one or more independent parameters. Among all the regression methods, one can list linear, logistic polynomial, stepwise, or lasso regressions. Linear regression is the simplest method and, consequently, the most used due to its vast applicability.

7.3.1.1 Linear Regression

Linear regression estimates the linear relationship among a dependent and an independent variable by plotting a linear curve whose statistical model, often described by the slope and the y-intercept, which describes that association.

Hutton et al. [25], aimed to enhance the capabilities of XRF spectrometry by four orders of magnitude as a way of tracing heavy metals in solutions previously prepared. They successfully applied this statistical tool to estimate the linear relationship between the concentration of the targeted elements and the respective peak height in the XRF spectrum [25]. Kikongi et al. [26] also used linear regression to achieve a relationship among two specific variables, the measured concentrations (%w/w) and the predicted concentrations (%w/w). The authors aimed at exploring several proceedings of curve-fitting in order to improve the quantification of light elements, in this way, they plotted linear regressions in order to identify the relationship among the measured concentration and the concentration values obtained with a previous approach [26]. Interestingly, linear regression-based data processing has been so widely employed that Schauer et al. [27] even developed and disclosed a dedicated software, to which they called the Munich Procedure, and respective documentation whose main goal is to process data specifically collected with an XRF-base device. This tool, if employed worldwide, could enable a discussion among different applications and between independent groups, a positive occurrence in the academic field [27].

7.3.1.2 Partial Least Squares Regression

Partial least squares (PLS) regression allows to identify relationships between two distinct variables, specifically, the predicted variables and the observable variables. In terms of operating procedure, PLS finds a linear regression model that relates both variables by projecting them, both the predicted and the observable variables, into a new space, i.e., this tool tries to find the multidimensional direction in the space of the first variable that explains the maximum variance direction in the space of the second variable.

PLS also to be very useful in the research work developed by Kirsanov et al. [28]. Here, the authors aimed at solving the issues often faced during simultaneous quantitative analysis of lanthanides, namely, Ce, Pr, Nd, Sm, Eu and Gd in previously prepared mixed solutions. The authors claimed that the regular calibration procedures failed to provide reasonable precision, while using the PLS tool allowed them to circumvent spectral inferiorities and yielded adequate calibration models for both energy dispersive XRF and total reflection XRF [28]. Finally,

Javadi et al. [29] used a PLS regression-based statistical model to predict attributes of geological samples previously analyzed in an XRF device. The authors claim that this approach enabled them to characterize several soil attributes with coefficient of determination above 0.95, denoting the suitability of the PLS for XRF data processing [29].

Figure 7.1 presents the box-plot charts Tavares et al. [13] aimed to predict exchangeable nutrients in samples of soils with a portable x-ray fluorescence spectrometer. In their work, the application of the arithmetic mean proved to be very useful to characterize the variability of elements like potassium or calcium among distinct soils [13]. In a more recent work, the same group also explored geological samples of agricultural nature with XRF-based technologies and employed the mean to achieve a mean spectrum for each analyzed sample, as well as, to assess the average content of each element in the samples. As a way of developing a model capable of accurately predicting key soil fertility attributes, the authors based their modeling procedure on the PLS tool. This approach allowed them to successfully overcome the matrix effects and, in overall, to achieve predictions regarding the element's presence with R^2 values ranging between 0.87 and 0.96 [14].

7.3.2 Correlation Methods

A correlation method is used to explore any kind of relationship between two random variables. Additionally, these methods have a predictive nature which can be exploited in practical scenarios. Four main correlation methods can be used, the Pearson correlation, Kendal rank correlation, Spearman correlation or even Point-Biserial correlation. In the field of XRF, the Pearson correlation has rendered interesting results.

7.3.2.1 Pearson Correlation

Also known as Pearson correlation coefficient, this correlation tool allows to explore a relationship among two variables on a linear manner. To do so, it considers the ratios among the covariance of both parameters as well as the product of the respective standard deviations. In simple terms, Pearson correlation is a linear normalization of the data's covariance.

In order to explore the direct correlation between metal concentrations in plant-based antidiabetic drugs, Zamir et al. [30] applied the Pearson correlation to data previously collected with an XRF-based device. Metals like Mn, Cu, and Zn were considered for the study and this correlation tool allowed the authors to detect a high correlation between Zn and Cu [30]. Tóth et al. [31], in their turn, employed the Pearson correlation to explore data collected from soil samples using XRF and assess the soil's clay content and salinity. Rubidium was found to be the best classifier enabling to verify an inverse relationship between the rubidium concentration and the soil clay content and salinity [31]. Samples of soils and, specifically, assess elevated concentrations of potentially toxic elements in their composition, was also

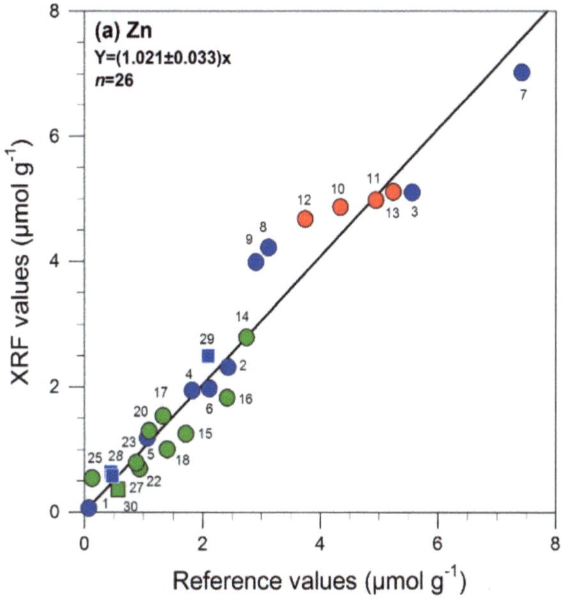

Fig. 7.4 Calibration curve between reference values and XRF values for the concentration of Zn in soil samples and respective Pearson correlation coefficient. Figure adapted from Mejía-Piña et al. [32] with the permission from Elsevier

the target samples of the work developed by Barsby et al. [17]. The authors used a combined methodology of measures of central tendency and dispersion for the use of concentrations determined by XRF and geostatistical analysis to investigate the spatial variability of pseudo-total and bio-accessible concentrations of As, Cd, Co, Cr. Cu, Ni, Pb, U, V and Zn in Northern Ireland soils. Then, the Pearson correlation coefficient allowed them to explore eventual correlations between independent elements and assess how the presence of some of them can influence the presence of other [17]. Finally, Mejía-Piña et al. [32] aimed their work at proving that handheld XRF devices can be precise, accurate and reliable enough to analyze at least 27 elements present in samples of sediments. Their work benchmarked the performance of handheld XRF devices against conventional laboratory-based XRF spectrometers and achieved successful results also due to the utilization of the Pearson correlation coefficient that allowed them to explore relationships between reference values and XRF-based values for all the target elements. One of the graphs plotted by the authors is included in Fig. 7.4 for illustrative purposes [32].

7.3.2.2 Spearman Correlation

The Spearman correlation has equally been largely used due to its robustness and capacity of finding correlations between variables beyond linearity. In fact, this correlation tool is a nonparametric measure of the relationship, i.e., the statistical dependence, between two variables and differs from the Pearson Correlation since it can be used for scenarios of nonlinear correlations.

In the field of XRF, the Spearman correlation has been gaining relevance. Declercq et al. [33], for example, used this tool to assess the relationship between the concentration levels of the elements determined using a XRF device to analyze soil samples, and known certified concentrations theoretically predicted. The Spearman correlation allowed the authors to establish the mention relationship and prove the suitability of the developed protocol for purposes of elemental quantification [33]. Intending to analyze a vast amount of reference materials of geological and biological origin, Fedeli et al. [34], used an XRF spectrometer and, to the collected data, applied several statistical tools, being one of them the calculation of the median values. The authors noted that, for some samples, XRF largely overestimated the certified values of concentration. To fix this issue, the authors based their approach on the strong statistically significant correlations between the tested elements and determined the median values that were later divided by the certified values to correct the concentration overestimation. The Spearman coefficient allowed to assess the correlation between the concentrations of the target elements in the reference materials and in the soil samples measured with XRF [34].

7.4 Multivariate Analysis

Most datasets are often complex and require the analysis of multiple variables to be adequately processed. In this way, MVA includes the most used tools, namely, factorial analysis, unsupervised classification, parametric and non-parametric tests, supervised and unsupervised "machine learning" tests, and "deep learning" tests. The addressed tools are summarized in Table 7.3.

Table 7.3 Summary of the statistical tools addressed under the scope of multivariate analysis

Multivariate analysis		
Factorial analysis	Simple factorial analysis	
	Factorial analysis of mixed data	
	Multiple correspondence analysis	
	Principal component analysis	
Unsupervised classification	K-Means	
	Hierarchical classification	
Parametric and non-parametric tests	Variance	ANOVA
		MANOVA
		One-Way ANOVA
Machine learning methods	Supervised	
	Unsupervised	
	Deep learning	

7.4.1 Factorial Analysis

Factorial analysis (FA) is one of the many MVA tools which is often used to assess the variability among the observations of a data set by using multiple variables correlated in terms of factors. Factors are underlying variables that explain patterns in the data. They correspond to latent variables, i.e., unobserved variables that help to account for the relationships between observed variables in a dataset. Factor analysis groups together correlated variables, attributing their covariance to one or more underlying factors, making it easier to interpret large datasets by reducing the complexity associated to multiple original variables. This category is often split into simple factorial analysis (SFA) and factorial analysis of mix data (FAMD).

7.4.1.1 Simple Factorial Analysis

In simple factorial analysis, the behavior of observed variables is modeled to explore and predict the behavior of latent variables. This approach is useful in cases where a large number of observed variables reflect the behavior of a smaller set of underlying variables, such as in scenarios where systematic interdependence exists between observed and unobserved parameters.

Figueroa et al. [35], explored rock samples collected from an ore exploration site and analyzed them with an XRF device in order to classify them and distinguish among them based on their content in specified elements. In this study, the elements of Ca, Cu, Fe, K, Mn, Pb, Rb, Sr, Ti and Zn were considered as variables to characterize the distinct samples, task that was successfully achieved due to the application of factorial analysis, as claimed by the authors [35]. Factorial analysis and, in specific, principal component analysis, proved to be very useful in the work of Aida et al. [36]. The authors aimed at surpassing the challenges often found during the production of elemental maps with XRF, in this way, they applied factorial analysis to the XRF spectra fully distinguish among the considered samples and improve the overall quality of the maps [36]. Aiming to distinguish among distinct samples of soils, Kaniu et al. [37] applied the factorial analysis tool to the spectra previously collected with an XRF device. The authors claimed that the statistical procedure enabled them to fully characterize the different samples in terms of elemental content [37].

7.4.1.2 Factorial Analysis of Mixed Data

Factorial analysis of mixed data (FAMD) is a specific approach of factorial analysis whose applicability is focused on datasets in which a group of observations can be described in both qualitative and quantitative parameters.

A relevant example of the utility of FAMD in the processing of XRF data is the work of Drobíková et al. [38]. Here, the authors aimed at characterizing the fine-grained waste materials produced by blast furnaces during operation periods. The authors prepared and analyzed the power samples with an XRF device, and the produced data were processed through FAMD. This statistical tool allowed them to prove the existence of a direct correlation among the concentrations of specific elements, namely, iron, and the expected content of the samples [38].

7.4.1.3 Multiple Correspondence Analysis

A third factorial analysis tool is the multiple correspondence analysis (MCA). This technique allows the processing of nominal categorical data, and its major advantage is the capacity to represent underlying structure in the population.

Multiple correspondence analysis was used by Rosin et al. [39] in their work focused on spatializing soil elemental concentration. The authors employed this analytical technique as a way of crosschecking the cluster maps generated with the XRF elemental results against the soil samples themselves. This approach allowed the author to evaluate the alignment of the spatial information obtained during the analyses and the pedological characteristics of the studied area [39].

7.4.1.4 Principal Component Analysis

In spite of having gained relevance as an unsupervised machine learning technique, principal component analysis (PCA) is a tool of factorial analysis, seeking to identify patterns by summarizing the variance in a dataset so, it aims to reduce the dimensionality of data and uncover underlying structures. In this way, it is a factorial method because it identifies important components or structures within data and aims at data reduction and structure discovery Additionally, it is a non-parametric method and does not assume an underlying model of the data. It also differs from traditional factor analysis once instead of using the latent variables (factors) it seeks to maximize the total variance explained by new, uncorrelated variables which are the principal components, transforming the data to align with directions of maximum variance. Principal components are, then, linear combinations of the original variables that capture as much variance as possible.

In the field of XRF, PCA has established itself as a very relevant and useful tool for data processing. Aida et al. [36], for example, relied on this tool to process their XRF-collected data. The authors noted that their elemental maps produced by micro XRF were not clear due to high background intensity, in this way, they applied the PCA to recreate new imagens solely based on the principal components with large eigenvalues improving, in this way, the contrast and spatial resolution of the maps [36]. Figure 7.5a, b explain the principal component scores of the corresponding spectra shown in the first (t1) or second (t2) score vector, respectively. These plots showed that group 1 and 2 consisted of spectra including an Fe peak and other peaks, respectively.

Another example is the work of Xu et al. [40]. The authors used PCA to reduce the dimension of data previously collected from a total of 15 potentially toxic elements present in soils. With just three variables an explained variance of over 70% was achieved so, these were used as new dataset for the following statistical analysis, namely, for the application of K-means clustering [40]. Soils was also the type of sample explored with a wavelength-dispersive X-ray fluorescence spectrometer by Custo et al. [41]. The authors aimed at determining the elemental composition of different soils and used this information to differentiate among them through principal component analysis. Interestingly, this statistical tool allowed the authors to cluster the soils in terms of closeness to the elements most preponderant in their

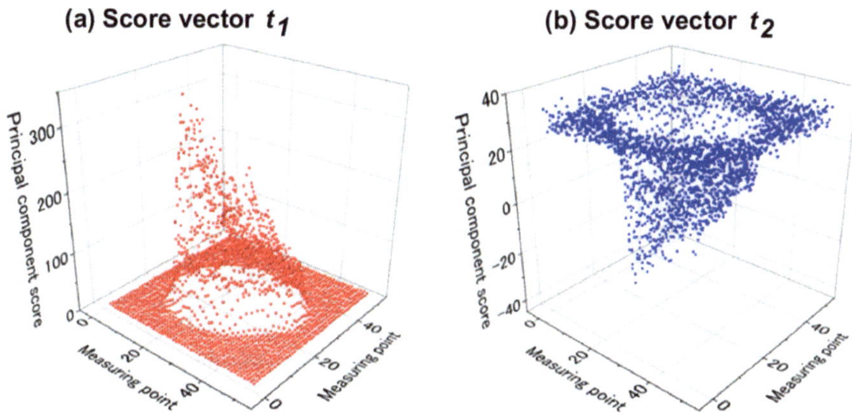

Fig. 7.5 Principal component scores of corresponding spectra explained by score vector t_1 (**a**) and t_2 (**b**). Figure adapted from Aida et al. [36] with the permission from Elsevier

composition. A total explained variance of 70.2% was achieved by the authors [41].

7.4.2 Unsupervised Classification

Unsupervised classification tools are increasingly implemented due to their significant contribution to the development of artificial intelligence algorithms. These methods enable the extraction of patterns from unlabeled data, i.e., they classify or cluster unlabeled data based on underlying hidden features of the dataset. Two of the main unsupervised classification tools are K-means and hierarchical classification.

7.4.2.1 K-means

K-means is an unsupervised classification method primarily used for clustering observations of a given dataset. In specific, it partitions the dataset into clusters with each observation assigned to the cluster whose centroid (mean) is nearest-it can classify unknown data into existing clusters, enabling to explore unlabeled observations.

In the field of x-ray fluorescence spectrometry, several studies have benefited from the utilization of k-means tool. Niu et al. [42], used this statistical tool as a way of solving the issues with overlapping spectral peaks during the analysis of heavy metals with similar energies. To do so, the authors gathered data collected from soil samples that exhibited the described issue and used the k-means procedure to produce new parameters that were later used or further statistical processing [42]. Xu et al. [40] applied k-means clustering to investigate the sources and controlling factors of 15 potentially toxic elements (PTEs) in 6862 topsoil samples. Their study aimed to enhance understanding of the toxic effects of these elements on organisms and assess methodologies for improving environmental management practices [40].

7.4.2.2 Hierarchical Classification

Hierarchical classification allows to classify in one fixed structure a subject and often unlabeled dataset. The observations are, then, assigned to a specific class in the available finite classes considering the intrinsic characteristics of each observation.

This statistical tool has been largely used to process XRF data. Mbaye et al. [43], for example, used hierarchical classification as a way of studying similarities among the concentrations of the elements detected with an XRF spectrometer in samples of dry herbal mass. The main goal was to detect eventually toxic elements in tea leaves, namely K, Mg, Cu, Zn, As, Sr, Fe, Co, Ca, P, S, Pb, Rb, and Mn, and by applying the hierarchical clustering, the authors could successfully relate all the elements [43]. Renda et al. [44], in an independent study, used hierarchical classification for a specific purpose, group in clusters samples collected from glass paintings. To do so, the authors employed an XRF spectrometer to analyze the samples and the collected data was treated with several statistical procedures, one of them being principal component analysis. Then, in order to group in clusters, the new components achieved during the dimension reduction procedure were hierarchical classified, leading their grouping in accordance with their similarities [44]. Finally, hierarchical classification was applied by Grant et al. [45] in order to cluster the elements detected during their analyses, as depicted in Fig. 7.6. In their study, the authors aimed at exploring and identifying a total of 36 elements using x-ray fluorescence spectrometry in 25 distinct samples of rice. The authors claim that this statistical tool proved to be a rapid method for the analysis of the dataset, correctly clustering rice samples [45].

7.4.3 Parametric and Non-parametric Tests

A parametric statistical test allows to make assumptions about the parameters of a population distribution considering a sample of that same population. They are applied to datasets that exhibit a normal distribution. Non-parametric tests, in their turn, allow to explore a population whose nature is not necessarily of normal behavior. Normality and Variance tests are the most relevant examples of parametric and non-parametric methods.

7.4.3.1 Variance

The variance corresponds to the squared deviation from the mean of a given variable. Its determination allows to fully assess how dispersive the observations of the dataset are and, consequently, properly address the population behavior. Several tests have been used in the XRF field to determine and explore the variance of a given dataset, as addressed below.

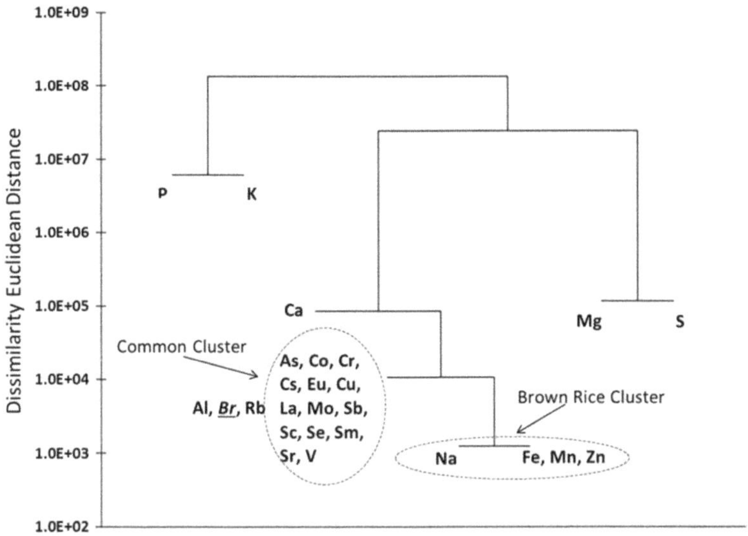

Fig. 7.6 Cluster analysis of elements in brown rice. Figure adapted from Grant et al. [45] with the permission from Springer

The Analysis Of Variance (ANOVA) procedures, correspond to a group of statistical models whose application is helpful to explore differences among means. ANOVA settles its working principle on the law of total variance in which the variance of a given variable is considered to be composed of variations coming from different sources—it tests how similar or even equals are the means of several populations. In the work developed by Ravansari et al. [46], the authors aimed at investigating the impact of organic matter present in soil samples during analyses to trace metals with XRF spectrometry. In this way, the authors applied the ANOVA methodology on the carbon-spiked samples to successfully determine the fraction of pXRF measurement variance attributable to geochemical variability, sample heterogeneity, and analytical variability [46].

Quantifying variation of metals in urban soils is also crucial for efficient environmental and public health management. Meng et al. investigated three metals, Al, Ca and Pb, to gauge the different levels of geogenic and anthropogenic controls in urban topsoil of Greater London Authority area. These three metal elements clearly demonstrated the different spatial distribution affected by human activities using ANOVA [47]. Also aiming to explore soils with XRF, Walser et al. [48] identify and quantify the elemental composition of samples collected from four urban locations. The utilization of ANOVA allowed the authors to identify significant differences in terms of concentration across all soils because of increasing soil moisture or grain size heterogeneity. Significance levels of around 95% were achieved during the utilization of ANOVA procedure [48]. Soil samples were also explored in the work of Chen et al. [49]. The authors aimed at developing a net methodology to predict the clay content of soils based on data collected with visible near-infrared spectroscopy, X-ray diffraction and portable XRF. To do so, they base their analysis on the quantitative determinations obtained with the devices and on several statistical tools. The utilization of the ANOVA technique allowed them to test for significant differences between different device combinations and evaluation which configuration provided the most accurate results [49].

Multivariate analysis of variance (MANOVA) also consists of comparing how similar are the means of distinct populations. Results of geological origin are among the most explored with MANOVA procedure. Silva et al. [50], in their study whose aim was to characterize geological samples of organic-rich soils from humid subtropical climate locations, employed MANOVA to the data collected with a portable XRF device. The variance values achieved through the test helped the authors to check the differences between the geological and pedological units for each soil [50].

The Kruskal–Wallis test is often called the one-way ANOVA test. It is used to assess whether unknow samples come from the same distribution, in this way, it does not assume the normality of the data. In practical terms, a significant result indicates that one or more samples stochastically dominate one other sample but does not define where that dominance occurs. Declercq et al. [33], for example, employed this statistical tool as a way of assessing the variability of the distributions previously achieved considering the soil elemental (Pb, Cd, As, Zn, Cu, Ni, Fe, Mn, Cr, Ca, K, S, P, Al, and Mg) concentrations [33].

7.4.4 Machine Learning Methods

For the past decade, and due to the increased attention to artificial intelligence-based methodologies, several statistical tools have gained additional attention due to is capability of prediction and of learning. These procedures are commonly grouped in three main categories: supervised machine learning tools, unsupervised machine learning tool and deep learning tools, as summarized in Table 7.4.

7.4.4.1 Supervised Statistical Methods

Supervised statistical methods based their working principle on exploring and finding relationships between variables whose nature can be of dependent or independent. Two main groups are defined in the category of supervised methods: regression and classification tools.

Regression

Regression tools aim to identify or estimate relationships between dependent and independent variables as a way of classifying the behaviour and, eventually, imposing some kind of predictive model. In accordance with the number and nature of the considered variables, regression can be categorized in simple, multiple, or logistic.

Linear regression or simple regression identifies a linear relationship among a dependent variable and an independent variable. This relationship is often translated

Table 7.4 Classification of the most relevant machine learning tools

Machine learning tools			
Supervised	Regression	Simple regression	Linear
			Non-Linear
		Multiple regression	Linear
			Non-Linear
		Logistic regression	
	Classification	Support vector machine	
		k-Nearest neighbors	
		Gaussian Naïves Bayes	
Unsupervised	Clustering	K-Means	
		K-Medoids	
		Mean shift	
	Dimension reduction	Principal component analysis	
		Singular value decomposition	
		Independent component analysis	
Deep learning	Sum product network		
	Recurrent neural network		
	Deep neural network		
	Convolutional neural network		

through the slope and y-interception of a curve. The adjustment of linear regression to XRF-generated data has been largely used and has been very helpful to the field, nonetheless, it is a bivariate analysis tool and not a multivariate analysis one.

Similarly to the simple regression, the multiple regression tools allow to explore relationships among a dependent variable and two or more independent variables. Its employment often involves defining a mathematical equation which translates the mentioned relationship with predictive capacities, i.e., with the capacity of outputting the value correspondent to the input value or values. A relevant example of the utility of multiple regression in the field of XRF is the work of Tavares et al. [51]. During the analysis of soil samples in regard to their content in nutrients, the authors faced some challenges related to the differences in the matrix of the soils from different collection sites. To overcome such challenges, the authors developed a few predictive models based on several statistical tools, being the multiple regression one of them. This approach allowed to mitigate the matrix effects and determine the content of the soils in terms of Ca and K with an high predictive performance of $R^2 \geq 0.84$ [51]. A final example of the usefulness of multiple regression in the field of XRF analysis is the work of Silva et al. [52]. Here, the authors aimed at characterizing the elemental properties of soils on a predictive manner and, to do so, they based their data processing on the multiple linear regression. This tool allowed the authors to remove the least relevant variables among all the available ones and, consequently, improve the predictive capacity of the developed model [52].

Table 7.5 Confusion matrix metrics from Bayesian approximation model, adapted from McManus et al. [53]

Metric	
Overall accuracy	0.63
Kappa	0.26
ROC	0.52
Sensitivity	0.21
Specificity	0.9

Being a regression model, logistic regression also allows to exploit relationships among dependent and independent variables. This method models the log-odds of the scenario under analysis by considering linear combinations of the independent variables involved. The outcome are often logarithmic or exponential-logarithmic curves. McManus et al. [53] explored the logistic regression tool as a way to treat the data collected from different mining locations and characterize the mining life cycle [53]. Table 7.5 presents the confusion matrix obtained after the application of this model. In this table, important metrics, such as, accuracy (correctly classified), Kappa (agreement between observed classifications and predicted classifications), Sensitivity (proportion of true positives), Specificity (proportion of true negatives) and Receiver Operating Characteristics, ROC, (trade-off between Sensitivity and Specificity). Xu et al. [54] introduced a logistic regression model to improve the sorting ability of an XRF-based sensor for the detection of copper in mining soils. Additionally, the authors used the regression to explore the relationship between the grades of Cu measured with ICP and XRF This model outperformed a Support Vector Machine model (see Sect. 4.4.1.2) [54].

Classification

Classification tools allow to find functions that divide or categorize the observations of the dataset into classes or clusters based on specific parameters of the population. One of the most used classification tools is the decision tree model, also known as hierarchical classification, allowing to classify a subject and often unlabeled dataset in one fixed structure. The observations are, then, assigned to a specific class in the available finite classes considering the intrinsic characteristics of each observation.

Williams et al. [55] used hierarchal cluster analysis (HCA) to determine the effectiveness of XRF in distinguishing between wash over in coastal lakes and marsh sediments. Prior to clustering, elemental data was normalized to unit range (0–1) to prevent elements with large numeric concentration values from dominating in distance-based clustering functions [55].

Support vector machines (SVM), another very used classification method, provide a statistical model capable of grouping the observations of a population into several clusters by linearly or nonlinearly dividing those observations. In the linear SVM, the observations and, consequently, the clusters can be divided by a straight line, fact that cannot be verified in the nonlinear SVM. The lines responsible for driving the clusters are often described through mathematical equations with a predictive nature. The work of Chakraborty et al. [56] is a relevant example of the usefulness of SVM

Table 7.6 Classification summary of land use/land change type using a validation set of n = 106 and 4 different classification models. Adapted from Chakraborty et al. [56]

Model	Agriculture	Converted	Forest	Misclassification rate
Random Forest				
Agriculture	36	3	4	
Converted	2	26	0	
Forest	0	1	34	0.09
Linear SVM				
Agriculture	34	2	7	
Converted	1	27	0	
Forest	1	1	33	0.11
Non-linear SVM				
Agriculture	36	3	4	
Converted	1	27	0	
Forest	0	0	35	0.07
Regression tree				
Agriculture	33	4	6	
Converted	7	21	0	
Forest	1	1	33	0.17

to process XRF data. Here, the authors aimed at exploring the suitability of four distinct machine learning techniques for the distinction of elemental data collected from soil samples through XRF. Authors compared algorithms previously developed considering random forest, regression trees, and linear and non-linear SVM, and all the tested tools exhibited similar and very good capabilities in differentiating the samples [56] (Table 7.6).

Xu et al. [54] also used an SVM model to improve the sorting ability of an XRF-based sensor for the detection of copper in mining soils. The results achieved with this learning algorithm were compared with a parallel analysis performed with a logistics regression-based algorithm. Interestingly, authors claimed that the second procedure outperforms the SVM-based approach [54].

Another classification tool of interest is called k-nearest neighbours. This non-parametric supervised classification tool that allows to classify the observations existent in a dataset having per base a plurality vote of its neighbours. In this way, each value is assigned to the class or cluster most common among its nearest neighbours. Muras et al. [57], having as goal the development of a machine learning algorithm capable of generating maps of chemical elements detected with a XRF spectrometer, used several statistical tools for the development of that algorithm. Among all the methods used, k-nearest neighbours exhibited the highest R^2 coefficient [57].

Similarly to the previous methods, Naïve Bayes tool is also a classifier whose working principle is based on considering that the features used during the classification process of the observations are conditionally independent. Yetis et al. [58]

aimed their work at classify eventual pathologies caused by the exposure to fluoride in groundwater and, to do so, the authors developed a machine learning algorithm based on Naïve Bayes classifiers. Interestingly, the authors claimed that the developed procedure, when applied to previously collected data, could improve the discriminant power of the variables measured with the XRF spectrometer [58].

7.4.4.2 Unsupervised Statistical Methods

Unsupervised methods are tools that enable to classify and divide in clusters the observations of a specific population. The main difference between these methods and the ones described in Sect. 4.4.1 is related to the labeling of the samples. The supervised tools require labeled samples in order to classify them but, unsupervised methods, in their turn, can classify distinct and unlabeled observations. These methos are often divided into clustering and dimension reduction tools. The clustering procedures include K-means, K-medoids and mean shifts, all of them already addressed in other subsections. In terms of dimension reduction, the most used are Principal Component Analysis (PCA) and Independent Component Analysis (ICA).

Dimension Reduction

Techniques of dimension reduction are useful to transform data that is represented in a high-dimensional space into a new form of that same data represented in a low-dimensional space. An elevated number of dimensions can end up in very complex statistical tools to extract meaningful results from the population, in this sway, if a dimension reduction is applied to the original data, and if some properties of that data are preserved during the process, then, the new form of data is considerably easier to process. Principal component analysis (PCA) and independent component analysis (ICA) are the main examples of dimension reduction techniques.

When applied to a high-dimensional dataset, and despite also being a factorial tool as previously mentioned, PCA linearly transforms the observations onto a new system of coordinates lower in dimensions. These dimensions, often called principal components, represent the largest variation in the data. This clustering and dimension reduction technique has been widely used to treat XRF data. Li et al. [59], for example, employed PCA as a way of extracting features from data analysis with distinct analytical techniques. To do so, the authors considered soil samples and, in specific, their elemental composition as a way of comparing the accuracy of the results achieved for all the devices. The authors claim that PCA exhibit the best extraction performance [59]. With a similar approach, Allegretta et al. [60] used the XRF spectra collected from samples of meteorites to plot a PCA graph capable of clustering the distinct samples. Interestingly, the authors could successfully group all the samples in three clusters, namely, the meteorite samples with elevated concentrations of iron, the meteorites with composition similar to earth stone, and the samples wrongly classified as meteorites, as depicted in Fig. 7.7 [60].

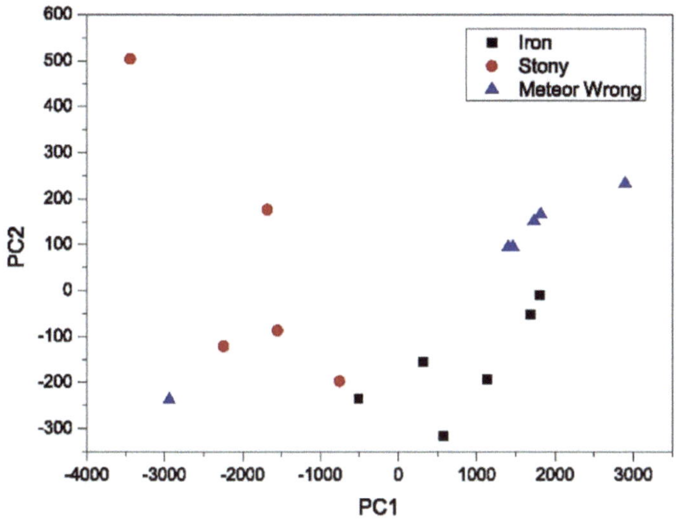

Fig. 7.7 Plot of PCA scores achieved for three types of geological samples analyzed with XRF. Figure adapter from Allegretta et al. [60] with the permission from Elsevier

ICA, in its turn, is a statistical method capable of separating a multivariate dataset into additive components, i.e., the method decomposes the dataset into statistically independent components that, if summed back into a single set, the original population is obtained. An example of the usefulness of this statistical tool is the work of Monakhova et al. [61]. Here, the authors aimed at demonstrating the suitability of ICA to process data previously collected with a XRF spectrometer from samples used in automotive industry and known for their hazardous impact on the environment. To do so, the authors gathered spectra from distinct samples and used them for the independent component analysis. Interestingly, the authors claim that ICA reliably separated groups of objects and allowed to increase the percentage of correct predictions for new samples not included into the model. Additionally, the author verified that ICA provided results comparable with specialized discriminant analysis methods like the linear discriminant analysis [61].

7.4.4.3 Future Trends: Deep Learning Methods

Many statistical tools have been tested and employed to evaluate data originated from XRF spectrometry measurements in environmental toxicology applications. This way, there is a constant demand for innovative procedures that allow to explore even further the elemental determinations obtained with XRF. Newer techniques, namely, recurrent neural network (RNN), deep neural network (DNN) and convolutional neural network (CNN) have been explored in regard to their potential for application to XRF-based data.

Recurrent Neural Network

A recurrent neural network is a model capable of processing and converting a sequential data input into a specific sequential data output. This algorithm bases its learning process in the recurrent unit which corresponds to a hidden form of memory that is updated at every new input. In this way, this methodology allows to create a model capable of sequential predictions.

Fang et al. [62] explored a total of 15 types of minerals and mining locations for fine mineral sorting (over 3000 spectra). Given the vast variety of ores, this study aims to minimize early-stage manual data labelling and improve classification accuracy by deriving subclass information about the spectral properties. This involved identifying potential subcategories within the same ore type that may differ due to their geographical origins and depositional environments. Consequently, the original Invariant Information Clustering (IIC) framework has been adapted by replacing the CNN model by an RNN model, suitable for handling sequential spectral data. Figure 7.8 depicts the structure of the RNN-overhead-xgboost structure, that exhibited a sorting accuracy of 99.21% and a prediction time of just 0.0831 s [62].

Deep Neural Network

A deep neural network (DNN) is a learning methodology that bases its learning process not in just one layer of data to produce the output layer, but in more than three, input layers of data. In these models, every layer of nodes is responsible for training a specific group of features based on the previous layer's output. This

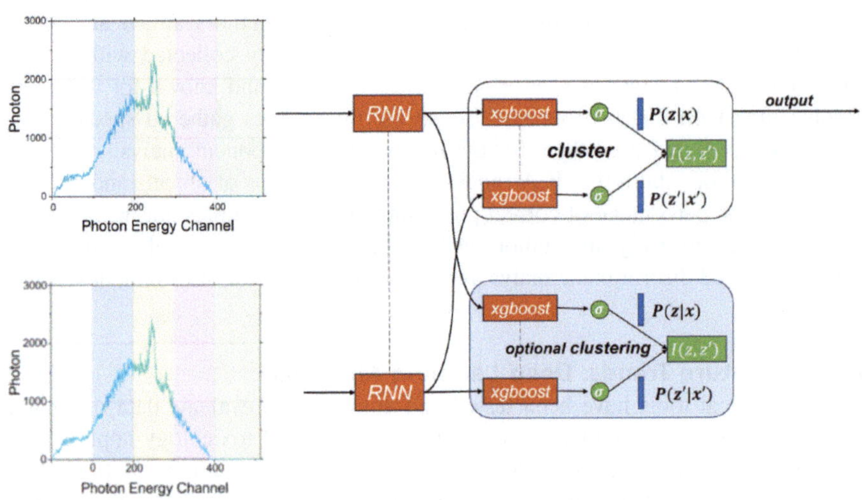

Fig. 7.8 Structure of the RNN-overhead-xgboost structure. The input data, a 512-dimensional sequence or spectra is augmented with random noise to generate transformed spectra producing a 128-dimensional feature vector. Figure adapted from Fang et al. [62] with the permission from Royal Society of Chemistry

approach allows that, the deepest the nodes are, they can recognize more complex features from the data.

Yang et al. [63] developed a new procedure for quantitative analysis of soil samples using XRF. This method combines pre-processing and concentration prediction. The pre-processing method was based on iterative adaptive window empirical wavelet transform and Gaussian convolution (IAWEWT-GC) for identification and removal of spectral background. In order to accurately quantify PTEs in the pre-processed spectra, a hierarchical deep neural network (HDNN), depicted in Fig. 7.9, was designed enabling quantitative regression of multiple elements simultaneously. Due to the success of the results reported by the authors, this approach has an evident potential to be widely implemented and used in combination with XRF in environmental toxicology [63].

Convolutional Neural Network

Convolutional Neural Networks are particularly relevant to the field of XRF since they are optimized models in terms of image processing, recognition, classification and segmentation. In fact, this methodology bases its learning principle in filter optimization whose shared-weight architecture provide the outputs considering the input layers. As mentioned, it's very used in scenarios of higher-layer features like image and video processing due to its capability of minimizing the work required to process complex datasets.

An example of how useful CNN can be in the near future for the processing of XRF data is the work of Zhang et al. [64]. Aiming to quantify the concentration of

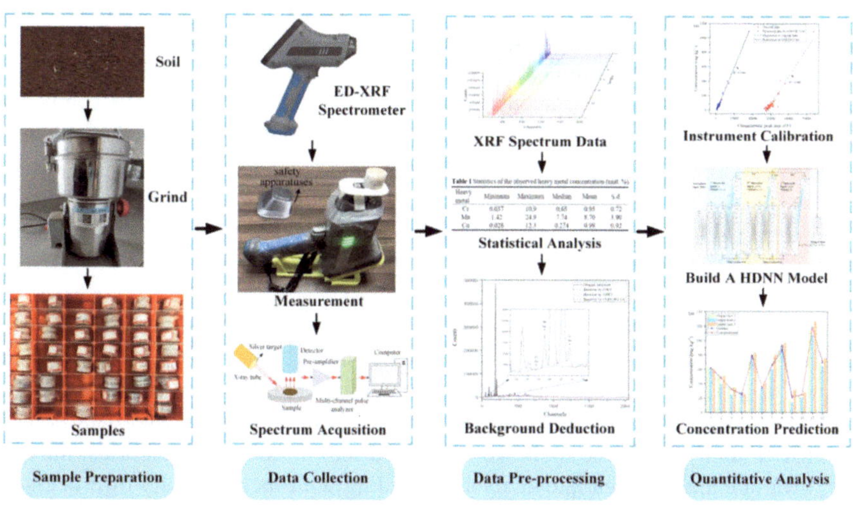

Fig. 7.9 Flowchart of the experimental arrangement to detect soil samples by energy dispersive X-ray fluorescence and using the IAWEWT-GC background deduction method and HDNN concentration prediction method for quantitative analysis. Figure adapted from Yang et al. [63] with the permission from Royal Society of Chemistry

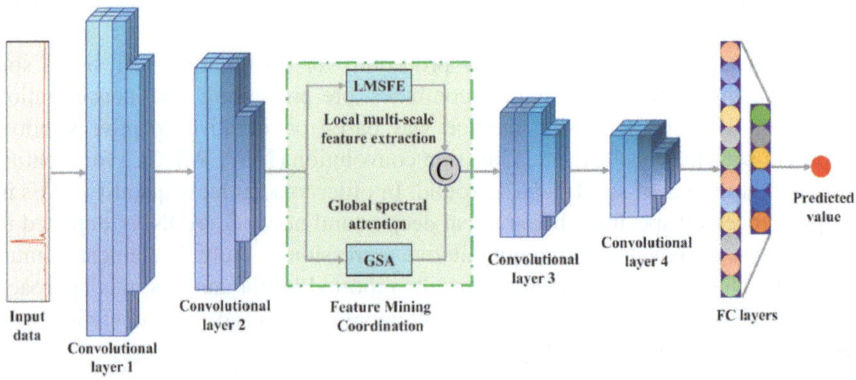

Fig. 7.10 Architecture of the spectral prediction network for heavy metals concentration in soil. Figure adapted from Fang et al. [64] with the permission from Royal Society of Chemistry

heavy metals in geological matrixes, the authors analysed soil samples with a XRF spectrometer and focused their methodology on five elements, Ti, Mn, Cu, Zn and Pb. Figure 7.10 shows the overall architecture of the model. XRF one-dimensional data are taken as input and the initial extraction of features and control of spatial dimensions are first executed by convolutional layer 1 and convolutional layer 2. Second, after the initial feature extraction, the spectral feature vectors are fed into the Feature Mining Coordination (FMC) module to ensure the deep feature mining and feature compensation. Third, the enhanced spectral features are sent to convolutional layer 3 and convolutional layer 4 to achieve further dimension expansion and spectral feature vector compression. Finally, the accurate prediction of soil heavy metals concentration is realized through two fully connected layers, where a dropout layer is added prevent overfitting. The success of the algorithm was confirmed by comparing its performance with other already certified procedures [64].

7.5 Conclusions

With the increasing complexity of data generated by the continuous advancement of XRF-based technologies, the demand for newer and more robust statistical techniques to process such data has become essential. This chapter aimed to review the statistical methodologies employed for processing environmental data derived from XRF analysis. These tools were grouped into three main categories: univariate, bivariate and multivariate analysis according with their capacity of processing one, two or more than two variables simultaneously.

Under the univariate analysis, a total of four categories were considered, namely, measures of central tendency, measures of dispersion, measures of symmetry and inferential methods. All these tools are used to statistically process a single variable and allow to explore average values, explore errors of dispersion, the symmetry of a dataset dispersion and a few other parameters. For the bivariate analysis, regression and correlation methods were considered. Most of these tools allow to compare, correlate or find relationships between two specific variables. They have been largely used to process XRF-originated data, specifically, for the construction of calibration curves. Multivariate analysis formed a central component of this work, given its increasing importance in XRF data processing. These techniques allow to explore relationships between more than two variables. Tools such as factorial analysis, unsupervised classification tools, parametric and non-parametric tests were thoroughly addressed. Furthermore, machine learning methods, supervised and unsupervised were also carefully described and discussed regarding their potential to process XRF data.

This way, this chapter provides a source of information for the scientific community dedicated to environmental toxicology field and, overall, to all the researchers focused on statistically processing complex and large XRF datasets.

References

1. Bilo, F., Borgese, L., Cazzago, D., Zacco, A., Bontempi, E., Guarneri, R., Bernardello, M., Attuati, S., Lazo, P., Depero, L.E.: TXRF analysis of soils and sediments to assess environmental contamination. Environ. Sci. Pollut. Res. **21**, 13208–13214 (2013)
2. McGladdery, C., Weindorf, D.C., Chakraborty, S., Li, B., Paulette, L., Podar, D., Pearson, D., Kusi, N., Duda, B.: Elemental assessment of vegetation via portable X-ray fluorescence (PXRF) spectrometry. J. Environ. Manage. **210**, 210–225 (2018)
3. Marguí, E., Kregsamer, P., Hidalgo, M., Tapias, J., Queralt, I., Streli, C.: Analytical approaches for Hg determination in wastewater samples by means of total reflection X-ray fluorescence spectrometry. Talanta **82**, 821–827 (2010)
4. Schlummer, M., Vogelsang, J., Fiedler, D., Gruber, L., Wolz, G.: Rapid identification of polystyrene foam wastes containing hexabromocyclododecane or its alternative polymeric brominated flame retardant by X-ray fluorescence spectroscopy. Waste Manag. Res. J. Sustain. Circular Econ. **33**, 662–670 (2015)
5. Denis, D.J.: Applied Univariate, Bivariate, and Multivariate Statistics Using Python: A Beginner's Guide to Advanced Data Analysis, Hoboken, New Jersey. Wiley, USA (2021)
6. Barbosa, S., Dias, A., Pacheco, M., Pessanha, S., Almeida, J.A.: Investigating metals and metalloids in soil at micrometric scale using μ-XRF spectroscopy—a case study. Eng **4**, 136–150 (2023)
7. Barbosa, S., Dias, A., Durão, D., Grilo, J., Baptista, G., Cagiza, J., Pessanha, S., Simão, J., Almeida, J.: Exploring high-resolution chemical distribution maps of incompatible and scarce metals in a nepheline syenite from the massif of "Serra de Monchique" (Portugal, Iberian Peninsula). Minerals **12**, 1178 (2022)
8. Barbosa, S., Dias, A., Ferraz, A., Amaro, S., Brito, M.G., Almeida, J.A., Pessanha, S.: The dual paradigm of mining waste: "from ecotoxicological sources to potential polymetallic resources"—an example from Iberian pyrite belt (Portugal) †. Materials Proceedings **5**, 23 (2021)
9. Flude, S., Haschje, M., Storey, M.: Application of benchtop micro-XRF to geological materials. Mineral. Mag. **81**, 923–948 (2017)
10. Cleff, T.: Univariate data analysis. In: Exploratory Data Analysis in Business and Economics. Springer, Cham (2014)
11. Bertani, A., Paola, G., Russo, E., Tuzzolino, F.: How to describe bivariate data. J. Thorac. Dis. **10**, 1133–1137 (2018)

12. Backhaus, K., Erichson, B., Gonsler, S., Weiber, R., Weiber, T.: Multivariate Analysis: An Application-Oriented Introduction, New York. Springer, USA (2021)
13. Tavares, T.R., Nunes, L.C., Alves, E.E.N., Almeida, E., Maldaner, L.F., Krug, F.J., Carvalho, H.W.P., Molin, J.P.: Simplifying sample preparation for soil fertility analysis by X-ray fluorescence spectrometry. Sensors **19**, 5066 (2019)
14. Tavares, T.R., Mouazen, A.M., Alves, E.E.N., Santos, F.R., Melquiades, F.L., Carvalho, H.W.P., Molin, J.P.: Assessing soil key fertility attributes using a portable X-ray fluorescence: a simple method to overcome matrix effect. Agronomy **10**, 787 (2020)
15. Turner, A., Solman, K.R.: Analysis of the elemental composition of marine litter by field-portable-XRF. Talanta **159**, 262–271 (2016)
16. Luo, L., Chu, B., Li, Y., Xu, T., Wang, X., Yuan, J., Sun, J., Liu, Y., Bo, Y., Zhan, X., Wang, S., Tang, L.: Determination of Pb, As, Cd and trace elements in polluted soils near a lead–zinc mine using polarized X-ray fluorescence spectrometry and the characteristics of the elemental distribution in the area. X-Ray Spectrom. **41**, 133–143 (2012)
17. Barsby, A., McKinley, J.M., Ofterdinger, U., Young, M., Cave, M.R., Wragg, J.: Bioaccessibility of trace elements in soils in Northern Ireland. Sci. Total Environ. **433**, 398–417 (2012)
18. Messager, M.L., Davies, I.P., Levin, P.S.: Development and validation of in-situ and laboratory X-ray fluorescence (XRF) spectroscopy methods for moss biomonitoring of metal pollution. MethodsX **8**, 101319 (2021)
19. Gallardo, H., Queralt, I., Tapias, J., Candela, L., Margui, E.: Bromine and bromide content in soils: analytical approach from total reflection X-ray fluorescence spectrometry. Chemosphere **156**, 294–301 (2016)
20. Sun, F., Bakr, N., Dang, T., Pham, V., Weindorf, D.C., Jiang, Z., Li, H., Wang, Q.B.: Enhanced soil profile visualization using portable X-ray fluorescence (PXRF) spectrometry. Geoderma **358**, 113997 (2020)
21. Rowe, H., Hughes, N., Robinson, K.: The quantification and application of handheld energy-dispersive x-ray fluorescence (ED-XRF) in mudrock chemostratigraphy and geochemistry. Chem. Geol. **324–325**, 122–131 (2012)
22. Idris, A.M.: Between-bottle homogeneity test of new certified reference materials employing wavelength dispersive X-ray fluorescence spectrometry. BMC Chem. **13**, 23 (2019)
23. Tepanosyan, G., Sahakyan, L., Zhang, C., Saghatelyan, A.: The application of Local Moran's I to identify spatial clusters and hot spots of Pb, Mo and Ti in urban soils of Yerevan. Appl. Geochem. **104**, 116–123 (2019)
24. Pedrão, G.A., Costa, K.B., Toledo, F.A.L., Tomazella, M.O., Jovane, L.: Semi-quantitative analysis of major elements and minerals: clues from a late Pleistocene core from campos Basin. Appl. Sci. **11**, 6206 (2021)
25. Hutton, L.A., O'Neil, G.D., Read, T.L., Ayres, Z.J., Newton, M.E., Macpherson, J.V.: Electrochemical X-ray fluorescence spectroscopy for trace heavy metal analysis: enhancing x-ray fluorescence detection capabilities by four orders of magnitude. Anal. Chem. **86**, 4566–4572 (2014)
26. Kikongi, P., Salvas, J., Gosselin, R.: Curve-fitting regression: improving light element quantification with XRF. X-Ray Spectrom. **46**, 347–355 (2017)
27. Schauer, M., Siegmund, F., Helfert, M., Drake, B.L.: The Munich Procedure—standardising linear regression documentation in p-XRF research. Software Impacts **21**, 100660 (2024)
28. Kirsanov, D., Panchuk, V., Goydenko, A., Khaydukova, M., Semenov, V., Legin, A.: Improving precision of X-ray fluorescence analysis of lanthanide mixtures using partial least squares regression. Spectrochim. Acta, Part B **113**, 126–131 (2015)
29. Javadi, S.H., Mouazen, A.M.: Data fusion of XRF and Vis-NIR using outer product analysis, granger-Ramanathan, and least squares for prediction of key soil attributes. Sensors **13**, 2023 (2021)
30. Zamir, R., Islam, N., Faruque, A.: Comparison of toxic metal concentrations in antidiabetic herbal preparations (ADHPs) available in Bangladesh using AAS and XRF analytical tools. Scientific World J. **2019**, 715484 (2019)

31. Tóth, T., Kovács, A., Rékási, M.: XRF-measured rubidium concentration is the best predictor variable for estimating the soil clay content and salinity of semi-humid soils in two catenas. Geoderma **342**, 106–108 (2019)
32. Mejía-Piña, K.G., Huearta-Diaz, M.A., González-Yajimovich, O.: Calibration of handheld X-ray fluorescence (XRF) equipment for optimum determination of elemental concentrations in sediment samples. Talanta **161**, 359–367 (2016)
33. Declercq, Y., Delbecque, N., Grave, J., Smedt, P., Finke, P., Mouazen, A.M., Nawar, S., Vandenberghe, D., Meirvenne, M., Verdoodt, A.: A Comprehensive study of three different portable XRF scanners to assess the soil geochemistry of an extensive sample dataset. Remote Sensing **11**, 2490 (2019)
34. Fedeli, R., Lella, L.A., Loppi, S.: Suitability of XRF for routine analysis of multi-elemental composition: a multi-standard verification. Methods and Protocols **7**, 53 (2024)
35. Figueroa-Cisterna, J., Bagur-González, M.G., Morales-Ruano, S., Carrillo-Rosúa, J., Marín-Peinado, F.: The use of a combined portable X ray fluorescence and multivariate statistical methods to assess a validated macroscopic rock samples classification in an ore exploration survey. Talanta **85**, 2307–2315 (2011)
36. Aida, S., Matsuno, T., Hasegawa, T., Tsuji, K.: Application of principal component analysis for improvement of X-ray fluorescence images obtained by polycapillary-based micro-XRF technique. Nucl. Inst. Methods Phys. Res. B **402**, 267–273 (2017)
37. Kaniu, M.I., Angeyo, K.H.: Challenges in rapid soil quality assessment and opportunities presented by multivariate chemometric energy dispersive X-ray fluorescence and scattering spectroscopy. Geoderma **241–242**, 32–40 (2015)
38. Drobíková, K., Vallová, S., Motyka, O., Kutláková, K.M., Plachá, D., Seidlerová, J.: Effects of binder choice in converter and blast furnace sludge briquette preparation: environmental and practical implications. Waste Manage. **79**, 30–37 (2018)
39. Rosin, N.A., Demattê, J.A.M., Carvalho, H.W.P., Rodriguez-Albarracín, H.S., Rosas, J.T.F., Novais, J.J., Dalmolin, R.S.D., Alves, M.R., Facioni, R., Tziolas, N., Mallah, S., Mello, D.C., Francelino, M.R.: Spatializing soil elemental concentration as measured by X-ray fluorescence analysis using remote sensing data. CATENA **240**, 107988 (2024)
40. Xu, H., Croot, P., Zhang, C.: Discovering hidden spatial patterns and their associations with controlling factors for potentially toxic elements in topsoil using hot spot analysis and K-means clustering analysis. Environ. Int. **151**, 106456 (2021)
41. Custo, G., Boeykens, S., Cicerone, D., Vázquez, C.: Combining XRF analysis and chemometric tools for apreliminary classification of argentine soils. X-Ray Spectrom. **31**, 132–135 (2002)
42. Niu, C., Li, F., Ren, S., Lu, A.: Application of the k-means algorithm and the Gaussian mixture model (GMM)—african vulture optimization algorithm (AVOA) for the resolution of overlapping energy-dispersive X-ray fluorescence (EDXRF) peaks. Anal. Lett. **57**, 2629–2643 (2024)
43. Mbaye, M., Traore, A., Ndao, A.S., Wague, A.: Multivariate statistical techniques to determine essential and toxic elements in biological samples by x-ray fluorescence. Instrum Sci. Technol. **43**, 369–378 (2015)
44. Renda, V., Nardo, V.M., Anastasio, G., Caponetti, E., Vasi, C.S., Saladino, M.L., Armetta, F., Trusso, S., Ponterio, R.C.: A multivariate statistical approach of X-ray fluorescence characterization of a large collection of reverse glass paintings. Spectrochimica Acta Part B **159**, 105655 (2019)
45. Grant, C.N., Dennis, H.T., Antoine, J.M.R., Hoo-Fung, L.A., Lalor, G.C.: Agglomerative hierarchical clustering analysis of twenty-six rice samples analysed by instrumental neutron activation analysis and other techniques. J. Radioanal. Nucl. Chem. **297**, 233–239 (2013)
46. Ravansari, R., Lemke, L.D.: Portable X-ray fluorescence trace metal measurement in organic rich soils: pXRF response as a function of organic matter fraction. Geoderma **319**, 175–184 (2018)

47. Meng, Y., Cave, M., Zhang, C.: Identifying geogenic and anthropogenic controls on different spatial distribution patterns of aluminium, calcium and lead in urban topsoil of Greater London Authority area. Chemosphere **238**, 124541 (2020)
48. Walser, S.L., Sirkovich, E.C., Richardson, J.B., McStay, A.E., Perdrial, N.: Moisture, organic matter, and large particle correction foraccurate Pb portable X-ray fluorescence assessment inurban soils. X-Ray Spectrom. **52**, 72–82 (2022)
49. Chen, Y., Gao, S., Jones, E.J., Singh, B.: Prediction of soil clay content and cation exchange capacity using visible near-infrared spectroscopy, portable X-ray fluorescence, and X-ray diffraction techniques. Environ. Sci. Technol. **55**, 4629–4637 (2021)
50. Silva, M.P.N., Guedes, C.C.F., Melo, V.F., Mascarenhas, R.O., Salvador, F.A.S.: Evaluating geostatistical methods along with semi-destructive analysis for forensic provenancing organic-rich soils in humid subtropical climate. Forensic Sci. Int. **341**, 111508 (2022)
51. Tavares, T.R., Almeida, E., Junior, C.R.P., Guerrero, A., Fiorio, P.R., Carvalho, H.W.P.: Analysis of total soil nutrient content with X-ray fluorescence spectroscopy (XRF): assessing different predictive modeling strategies and auxiliary variables. AgriEngineering **5**, 680–697 (2023)
52. Silva, S.H.G., Teixeira, A.F.S., Menezes, M.D., Guilherme, L.R.G., Moreira, F.M.S., Curi, N.: Multiple linear regression and random forest to predict and map soil properties using data from portable X-ray fluorescence spectrometer (pXRF). Ciência e Agrotecnologia **41**, 648–664 (2017)
53. McManus, S., Rahman, A., Coombes, J., Horta, A.: Comparison of interpretation uncertainty in spatial domains using portable x-ray fluorescence and ICP data. Appl. Comput. Geosci. **12**, 100067 (2021)
54. Xu, Y., Klein, B., Li, G., Gopaluni, B.: Evaluation of logistic regression and support vector machine approaches for XRF based particle sorting for a copper ore. Miner. Eng. **192**, 108003 (2023)
55. Williams, H.F.L., Beaubouef, C.E., Liu, K., Culligan, N., Riedlinger, L.: Testing XRF identification of marine washover sediment beds in a Coastal Lake in Southeastern Texas, USA. Mar. Geol. **443**, 106705 (2022)
56. Chakraborty, S., Li, B., Weindorf, D.C., Deb, S., Acree, A., De, P., Panda, P.: Use of portable X-ray fluorescence spectrometry for classifying soils from different land use land cover systems in India. Geoderma **338**, 5–13 (2019)
57. Miras, J.R., Gacto, M.J., Blanc, M.R., Arroyo, G., López, L., Torres, J.C., Martín, D.: Machine learning regression algorithms for generating chemical element maps from X-ray fluorescence data of paintings. Chemom. Intell. Lab. Syst. **248**, 105116 (2024)
58. Yetis, A.D., Yesilnacar, M.I., Atas, M.: A machine learning approach to dental fluorosis classification. Arab. J. Geosci. **14**, 95 (2021)
59. Li, F., Xu, L., You, T., Lu, A.: Measurement of potentially toxic elements in the soil through NIR, MIR, and XRF spectral data fusion. Comput. Electron. Agric. **187**, 106257 (2021)
60. Allegretta, I., Marangoni, B., Manzari, P., Porfido, C., Terzano, R., Pascale, O., Senesi, G.S.: Macro-classification of meteorites by portable energy dispersive X-ray fluorescence spectroscopy (pED-XRF), principal component analysis (PCA) and machine learning algorithms. Talanta **212**, 120785 (2020)
61. Monakhova, Y.B., Tsikin, A.M., Mushtakova, S.P.: Independent components analysis as an alternative to principal component analysis and discriminant analysis algorithms in the processing of spectrometric data. J. Anal. Chem. **70**, 1055–1061 (2015)
62. Fang, Z., Song, S., Wang, H., Yan, H., Lu, M., Chen, S., Li, S., Liang, W.: Mineral classification with X-ray absorption spectroscopy: a deep learning-based approach. Miner. Eng. **217**, 108964 (2024)
63. Yang, W., Li, F., Lyu, S., Zhang, Q., Zhao, Y.: Quantitative analysis of heavy metals in soil via hierarchical deep neural networks with X-ray fluorescence spectroscopy. J. Anal. At. Spectrom. **38**, 1830 (2023)
64. Zhang, Q., Li, F., Yang, W.: A deep spectral prediction network to quantitatively determine heavy metal elements in soil by X-ray fluorescence. J. Anal. At. Spectrom. **39**, 478 (2024)

Benchtop Micro X-Ray Fluorescence (μ-XRF) for Plant Analysis

8

Katarina Vogel-Mikuš

Abstract

X-ray fluorescence (XRF) has recently become a powerful tool for the non-destructive elemental analysis of plant materials, providing critical insights into the composition and distribution of major and trace elements. This chapter explores the application of XRF analytical techniques in plant biology studies, particularly emphasizing the use of benchtop μ-XRF systems. Benchtop μ-XRF systems with lateral resolution down to 5 μm offer accessible and efficient elemental analyses that are suitable for routine laboratory studies and elemental screening of plant organs and tissues in environmental, agricultural, and biological studies. On the other hand, synchrotron μ-XRF (SXRFI) with monochromatic excitation provides significantly higher spatial resolution (down to nanometric scale) and sensitivity (down to sub μg g^{-1} level), enabling detailed mapping of elemental distributions at the cellular and subcellular levels. This capability is essential for studying the mechanisms of nutrient uptake, metal transport, and the localization of potentially toxic elements in plants. The chapter discusses recent advancements in the application of XRF in plant biology with an emphasis on the element distribution in plant organs by benchtop μ-XRF systems. In addition, it addresses the challenges in sample preparation, quantification methods, and the analysis and interpretation of complex data sets.

K. Vogel-Mikuš (✉)
Biotechnical Faculty, Department of Biology, University of Ljubljana, Ljubljana, Slovenia
e-mail: Katarina.VogelMikus@bf.uni-lj.si

Jozef Stefan Institute, Ljubljana, Slovenia

V. K. Singh (ed.), *X-Ray Fluorescence Spectroscopy and Chemometrics*,
https://doi.org/10.1007/978-3-031-98375-7_8

8.1 Introduction

X-ray Fluorescence (XRF) has recently become a powerful tool for the non-destructive elemental analysis of plant materials, providing critical insights into the composition and distribution of major and trace elements [1, 2]. This chapter explores the application of the XRF analytical techniques in plant biology studies, particularly emphasizing the use of benchtop μ-XRF systems.

Benchtop μ-XRF systems with lateral resolution down to 5 μm offer accessible and efficient elemental analyses that are suitable for routine laboratory studies and elemental screening of plant organs and tissues in environmental, agricultural, and biological studies. On the other hand, synchrotron μ-XRF (SXRFI) with monochromatic excitation provides significantly higher spatial resolution (down to nanometric scale) and sensitivity (down to sub μg g^{-1} level), enabling detailed mapping of elemental distributions at the cellular and subcellular levels. This capability is essential for studying the mechanisms of nutrient uptake, metal transport, and the localization of potentially toxic elements in plants.

The chapter discusses recent advancements in the application of XRF in plant biology with an emphasis on the element distribution in plant organs by benchtop μ-XRF systems. In addition, it addresses the challenges in sample preparation, quantification methods, and the analysis and interpretation of complex data sets.

8.2 Plant Mineral Metabolism

Plant growth and development are largely determined by nutrient availability; therefore, to ensure better productivity of crop plants, it becomes essential to understand the dynamics of nutrient uptake, transport, accumulation, compartmentation, and their biological interactions [3].

In higher plants, 16 elements (N, K, Ca, Mg, P, S, B, Cu, Fe, Mn, Mo, Zn, and Cl) besides C, O, and H are considered essential, while the requirement for Si, Na, Ni, Co, and Se is restricted to a limited number of plant species [4]. These elements are thus considered beneficial for plant growth and development. Progress in research defining the essentiality or benefits of particular elements for plants was closely related to the development of analytical methods and purification of chemicals [5]. Plants can also accumulate non-essential elements because Cd, Pb and As, for example, can mimic essential elements and are inadvertently transported over the membrane via transporters for essential mineral nutrients. These elements are considered potentially toxic for plant metabolism since, in higher concentrations, they may lethally interfere with various processes in the cells [6].

In order to maintain homeostasis with optimal cell functioning and integrity and/or to avoid toxicity, proper allocation of elements at organ, tissue, cellular, and subcellular levels is necessary, with proper element/metal speciation and ligand environment presenting a key to bioavailability and homeostasis in organisms. Studies of element localization, as well as speciation and ligand environment, are therefore crucial to reveal the mechanisms of element homeostasis, transport, and

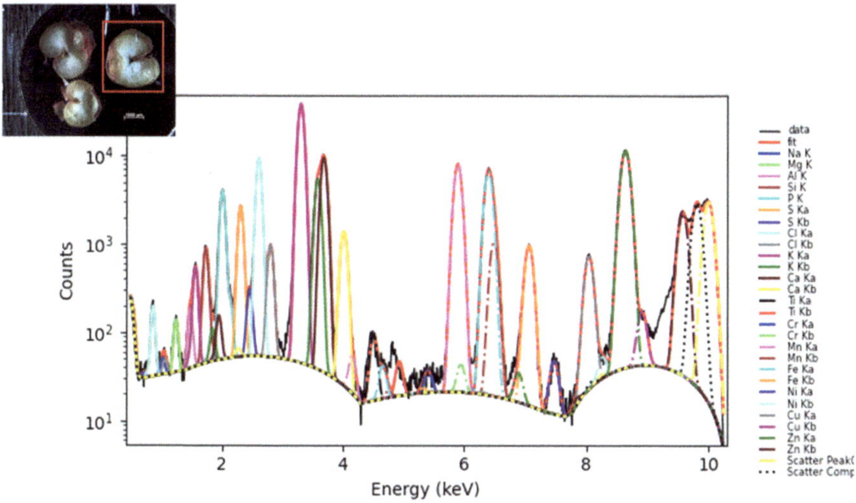

Fig. 8.1 Cumulative XRF spectrum recorded on a wheat grain cross-section (upper left image, in the red square) at XRF beamline of Elettra Synchrotron Trieste. The excitation energy was 10.0 keV, the beam size was 100 μm, the dwell time was 5 s pix^{-1}, map 28 × 59 px. The X-rays were detected by Peltier-cooled SDD (Nano GmbH, XFlash 5030, energy resolution 131 eV at 5.9 keV)

also tolerance and toxicity [7–9]. Moreover, studies of localization and speciation of trace elements, for example, in the grains of staple food crops, are also of high applicative value, allowing the determination of element concentrations and ligand environment in particular tissues and different milling fractions without potential contamination due to mechanical mixing of tissues [10, 11]. Disturbances in element allocation in plant organs may be indicative of plant diseases [12] or mineral malnutrition [13] or may indicate processes associated with plant cell death [14].

8.3 X-Ray Fluorescence

X-ray fluorescence is induced by the excitation (ionization) of atoms in the tightly bound inner K (for the elements $20 < Z < 50$) and L ($Z > 50$) atomic shells with energies that must exceed the binding energies of the K and L electrons, through the process called the photoelectric effect. Characteristic K, L and M X-ray photons are typical of particular elements, which provides the basis for distinguishing between elements in X-ray spectrometry [15]. The quality of the spectra and, consequentially, the limits of detection (LOD) for particular elements depends on the excitation source—monochromatic (e.g. radioisotope sources, synchrotron source) vs. polychromatic (e.g. X-ray tube) (Figs. 8.1 and 8.2) and detector systems. Silicon drift diode (SDD) detectors can nowadays achieve excellent energy resolution

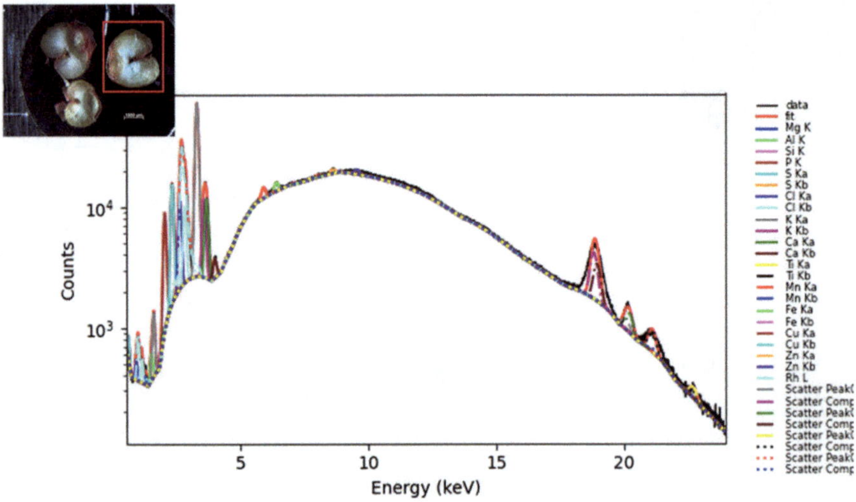

Fig. 8.2 Cumulative XRF spectrum recorded on a wheat grain cross-section (upper left image) with benchtop µ-XRF instrument (XGT-9000, Horiba) equipped with Rh tube, 30 kV, 1000 mA and Peltier cooled SDD (specifications not disclosed by Horiba). The beam was focused with a polycapillary lens to 15 µm. The map of 128 × 117 pixels was recorded on a grain with 500 ms dwell time per pixel. Polychromatic excitation yields a considerable background in the middle z-region, also due to the scattering of X-rays on wheat grain starchy parts. This significantly increases LOD for middle z-elements in starchy material compared to synchrotron µ-XRF

(down to 125 eV at 5.9 keV), significantly contributing to a better deconvolution of overlapping spectral lines [16].

8.4 Elemental Profiling by XRF

For bulk elemental analyses of plant material and other environmental samples such as for example geological materials, Energy Dispersive X-ray Fluorescence Spectrometry (EDXRF) based on radioisotope or X-ray tube excitation and Total-Reflection X-ray Spectrometry based on monochromatic X-ray tube excitation are commonly used, with the second being a method of choice when trace elements are to be analyzed. Because the plant material needs to be digested before the TXRF measurement, during which the organic matrix is eliminated, the background is much lower in the TXRF compared to EDXRF, making TXRF more sensitive. [17, 18]. Figure 8.3 compares detection limits (LOD) for various EDXRF and TXRF systems when measuring plant material.

In plant biology, bulk elemental EDXRF is especially convenient for fast element profiling of plant vegetative or reproductive structures. Element profiling combined with clustering (Fig. 8.4) or multivariate statistical analysis is, for example, useful for detecting plant varieties with enhanced essential trace element concentrations for biofortification purposes [19], to determine the geographical

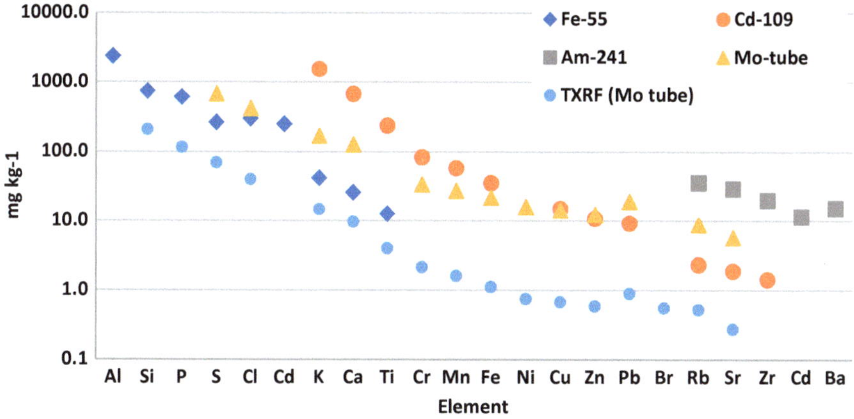

Fig. 8.3 Limits of detection (LOD) in mg kg^{-1} of different EDXRF (radioisotope excitation with Fe-55, Cd-109 and Am-241, Mo-X-ray tube excitation) and TXRF systems (Courtesy: Jozef Stefan Institute)

origin of plants and food products [20] or for correlating plant elementomics with genomics [21, 22] and physiological parameters [23] in the studies of plant responses to the environment. In addition, EDXRF is very useful for rapidly detecting potentially toxic elements like Pb, Cd, As, and others [18].

Compared to ICP-MS, EDXRF analysis provides faster and cheaper service of major and trace elements [1, 18], while it is not sensitive enough to detect ultra-trace elements (> 1 ppm). Additionally, the simultaneous multi-elemental profiling enables the detection of the elements that are not expected to be present in the sample. Sample preparation is relatively simple, with no need for technically demanding chemical digestion procedures [17].

In recent years, hand-held XRF spectrometers (Fig. 8.5) have emerged as valuable tools for real-time, non-destructive elemental analysis of plant material. These devices offer comparable performance to benchtop XRF systems for measuring essential (Si, P, S, Cl, K, Ca, Mn, Fe, Cu and Zn) [19, 25] as well as potentially toxic (Ni, Cr, Pb, As, Hg) [23, 26] elements in biological materials. Hand-held XRF systems have been successfully applied to analyze silicon content in crops [27], track real-time uptake of Zn and Mn in soybean plants [28] and analyze herbarium plant material for potential metal hyperaccumulation [29]. The technique has also proven effective for measuring calcium and potassium concentrations in intact apples and pears, revealing spatial variability within individual fruits [30]. Hand-held XRF provides a cost-effective, high-throughput alternative to traditional destructive methods, allowing for in situ analysis and time-resolved studies of living plant tissues. Due to the rapid development of excitation sources and detection systems, this technology shows promise for enhancing our understanding of plant nutrition, elemental distribution, and temporal patterns in nutrient uptake across various plant species and organs.

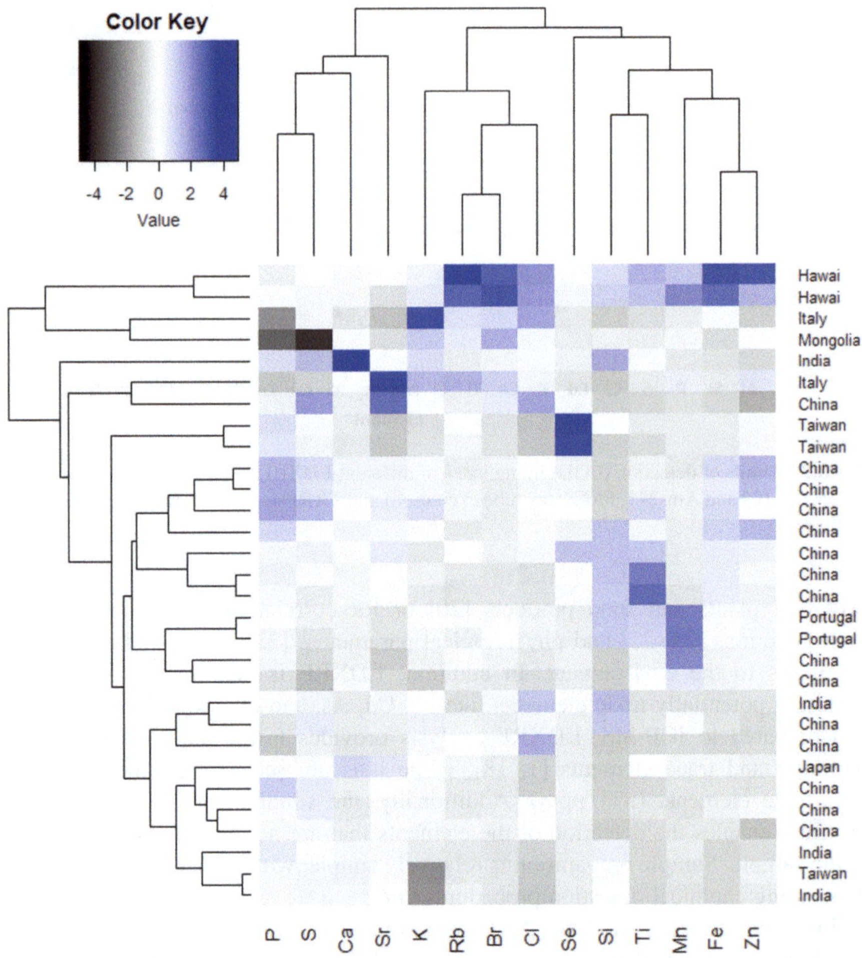

Fig. 8.4 Elemental profiling of *Spirulina* algae used as food supplements obtained from different countries to determine the batches with higher mineral nutrient concentrations. The two-way clustering based on Euclidian distances with a heat chart was produced in »R« on z-normalized data. The samples from Hawaii form a distinct cluster due to higher Rb, Br, Cl, Ti, Mn, Fe and Zn concentrations. Two samples from Taiwan are rich in Se, and the samples from Portugal are rich in Mn. Data were taken from [24] under Creative Common (CC BY) International

8.5 Imaging of Element Distribution by μ-XRF

8.5.1 Imaging with Benchtop μ-XRF Instruments

Inventions in X-ray polycapillary optics have allowed the development of a range of commercially available benchtop micro-XRF (μ-XRF) instruments that can produce X-ray spot sizes down to 5 μm on the sample, allowing major and

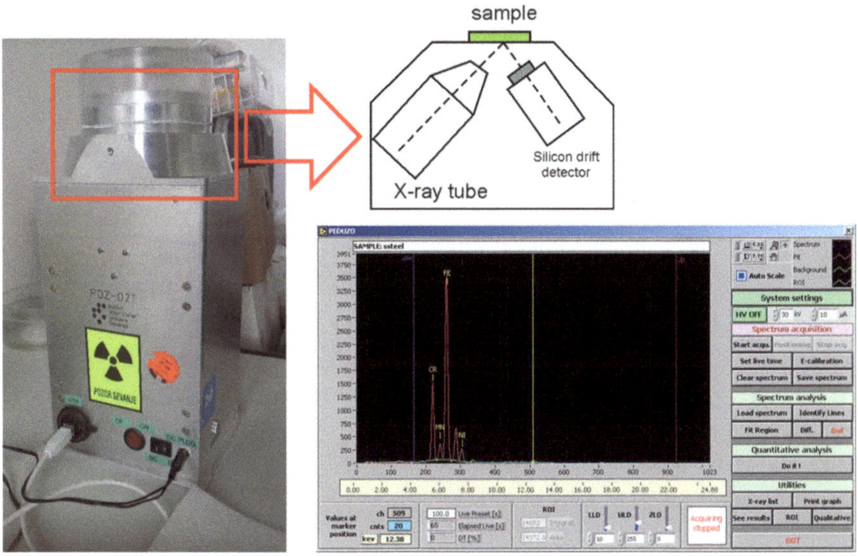

Fig. 8.5 A custom-made hand-held spectrometer PDZ-O2T equipped with Rh tube (Moxtek Inc., USA) and SDD detector (Amptek, USA) with field programable gate array multichannel analyzer. Spectra acquisition and quantitative analysis is performed in Labwiev. Quantification is based on fundamental parameters [31]. Courtesy: Jozef Stefan Institute

trace-element analysis in various materials with minimal preparation steps [32]. In Table 8.1 basic characteristics of the most frequently used μ-XRF systems in plant biology studies are presented.

Despite being less sensitive compared to SXRFI (Figs. 8.1 and 8.2) and laser ablation inductively coupled mass spectrometry (LA-ICPMS), the advantage of benchtop μ-XRF systems is the accessibility and non-destructive analysis. The major drawback of using synchrotron radiation-based μ-XRF imaging (SXRFI) in plant science is the limited beamtime provided by synchrotron facilities due to highly competitive calls for proposals at SXRFI beamlines. Within these calls, the proposals from the plant science community usually compete with those from other research fields, such as material science, medicine and cultural heritage, yielding limited beamtime only for highly ambitious state-of-the-art projects. Besides, the development of SXRFI instrumentation in the recent few years has headed in the direction of nanofocusing, where it is, in most cases, impossible to scan larger objects (a few cm). The scan size at nano beamlines usually ranges around 1 × 1 mm, suitable for tissue, cellular and sub-cellular levels. Only a few beamlines in the world (e.g. X27A of the National Synchrotron Light Source, XFM of ANSTO and P06, DESY) are able to provide high-resolution scans of large objects such as plant leaves [33, 34].

Table 8.1 Commercially available μ-XRF instruments and their basic characteristics

Producer	Bruker	iXRF systems	Horiba	EDAX
Model	Tornado M4	Atlas	XGT-9000	Orbis
Tube	50 kV, 30 W	50 kV, 50 W, 1 mA, Rh	50 kV, 50 W, 1 mA, Rh	50 kV, 50 W, 1 mA, Rh
Spot size	20 μm, polycapillary optics	5 μm, polycapillary optics	15 μm polycapillary optics	30 μm polycapillary optics
Detector	XFlash ® silicon drift detector 30 mm^2, Energy resolution: <145 eV at at 5.9 keV	4 Silicon Drift Detectors (SDD) (25 mm^2 to 70 mm^2 area), ≤130–145 eV resolution at 5.9 keV	SDD (data not disclosed by the company)	SDD 30–50 mm^2 140 eV resolution at 5.9 keV
Geometry	Top-down beam (Perpendicular)	Top-down beam (Perpendicular)	Top-down beam (Perpendicular)	Top-down beam (Perpendicular)
Transmission detector	No	No	Yes (NaI)	No
Sample view	Two simultaneous live images from above with different magnifications for sample overview and precise positioning, lateral fisheye camera for the sample chamber overview	Three Sample Positioning and Analysis Cameras	Two Sample Positioning and Analysis Cameras	Dual CCD video: 10× and 75× with 3× digital zoom
Sample environment	Air, Vacuum, Helium for Solids, liquids, and powders	Air, Vacuum, Helium for Solids, liquids, and powders	Air, Vacuum, Helium for Solids, liquids, and powders	Air, Vacuum,
Map area	190 × 160 mm^2	220(X) × 200(Y) mm	100 mm × 100 mm	720 mm × 270 mm
Dimensions	815 (W) × 680 (D) x (H) 580 mm, 130 kg*	951 (L) × 556 (W) × 564 (H) mm	680 (W) × 860 (D) × 760 (H) mm	not disclosed
Reference	https://www.bruker.com/en/products-and-solutions/elemental-analyzers/micro-xrf-spectrometers/m4-tornado.html	https://www.ixrfsystems.com/atlas_m/	https://www.horiba.com/esp/scientific/products/detail/action/show/Product/xgt-9000-1909/	https://www.edax.com/-/media/ametekedax/files/micro_xrf/product_bulletins/orbis-micro-xrf-elemental-analyzer.pdf?dmc=1&la=en

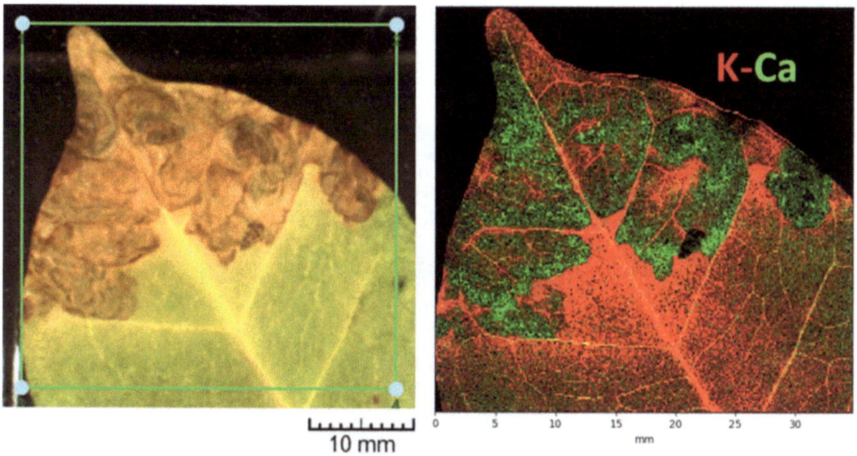

Fig. 8.6 Imaging of element distribution of a coffee (*Coffea arabica*) leaf infected by fungal pathogens. Depletion of K and accumulation of Ca are seen in dead cells. μ-XRF imaging was performed with XGT-9000 (Horiba) equipped with Rh tube (50 kV, 1000 mA, 15 μm beam spot size), 256 × 243 pixels, acquisition time 50 ms pixel^{-1}, step size 136 μm

In plant biology studies, benchtop μ-XRF systems are thus especially convenient for analyzing element distribution in various plant species for various purposes [28, 29, 35–37].

Here, two case studies are presented (Figs. 8.6 and 8.7). In both cases, the plant material was flush frozen and freeze-dried prior to analysis [38]. μ-XRF imaging was performed with XGT-9000 (Horiba, Japan) instrument equipped with Rh tube (50 kV, 1 mA, 50 W) (https://www.horiba.com/int/scientific/products/detail/action/show/Product/xgt-9000-1909/).

8.5.2 Full-Field Imaging

As already mentioned in the Chap. 4, high performances position- and energy-sensitive X-ray detectors have boosted the development of advanced instruments based on the Full Field X-Ray Fluorescence Imaging (FF-XRFI) method, providing a promising alternative to the conventional scanning approach for performing two-dimensional elemental imaging [40]. Compared to classical μ-XRF set-ups the main advantage of full field imaging is decreased acquisition time, making it especially convenient to image larger objects. To date, the majority of the studies using FF-XRFI deal with cultural heritage, while in biological studies, only a few cases are reported [41].

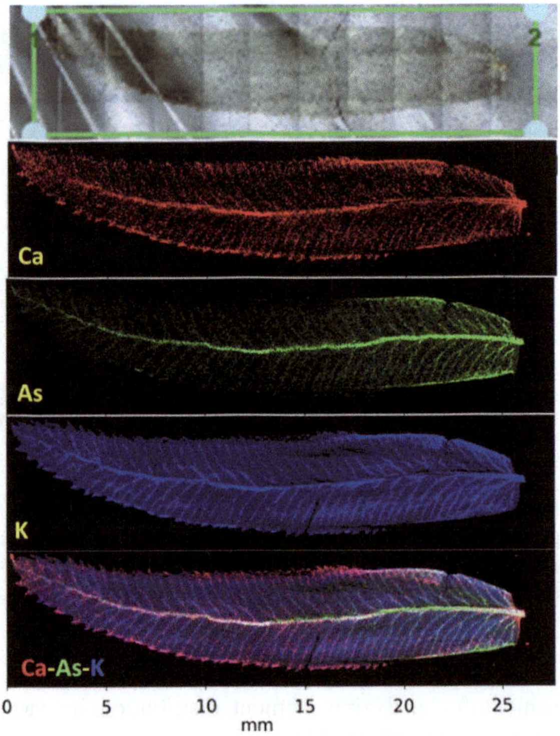

Fig. 8.7 Ca, As and K distribution in a leaf of a fern *Pteris vittata*, an As hyperaccumulator. μ-XRF imaging was performed with XGT-9000 (Horiba) equipped with Rh tube (50 kV, 1000 mA, 15 μm beam spot size), 512 × 124 pixels, acquisition time 100 ms $pixel^{-1}$, step size 54 μm (sample provided by Johannes van Elteren). As accumulates primarily in the main and lateral veins, as already shown by [39]

8.5.3 Confocal μ-XRF

Confocal micro X-ray fluorescence (CMXRF) spectroscopy is a non-destructive, element-specific technique that is used to analyze the in-depth elemental composition of a sample. It is based on the confocal overlap of two polycapillary lens foci (excitation and detection channels) (Fig. 8.8), creating a depth-sensitive probing volume. Three-dimensional resolved element distribution images can be obtained by measuring the fluorescence intensity as a function of the three-dimensional position. The confocal design of the instrument allows for depth-resolved analysis, in most cases, with a lateral resolution of a few tenth micrometres. By moving the z-axis of the sample stage, the depth profile of the sample can be probed. In scanning mode, when moving the sample stage in x and y positions, image stacks can be obtained and reconstructed in a 3D space. This makes CMXRF spectroscopy a valuable tool for a wide range of applications, especially when samples with an inhomogeneous distribution of elements and a relatively light matrix have to be analyzed, which are typical examples in materials science, geology, and biology [42].

While probing the in-dept profile of a non-homogeneous sample, incoming (excitation) and out-going (fluorescence) X-rays are absorbed in the upper layers

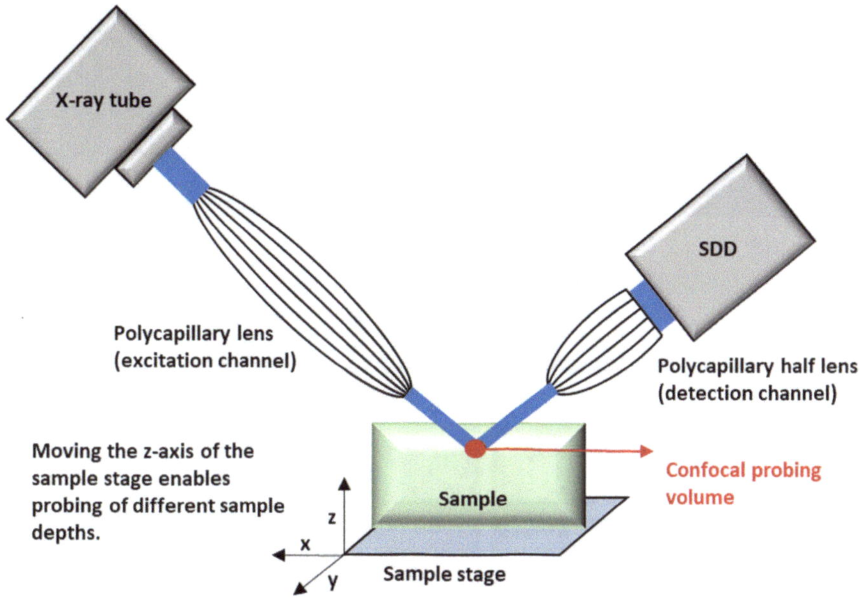

Fig. 8.8 Confocal micro-X-ray fluorescence (CMXRF) is based on the confocal overlap of two polycapillary lens foci, creating a depth-sensitive probing volume. Adapted according to [42]

of the sample, distorting the obtained in-depth images. For this purpose, absorption correction has to be applied [43] as in the presented case study on high- and low-iron pearl millet seeds (Fig. 8.9), where an algorithm for absorption correction was developed [43].

8.6 Sample Preparation for μ-XRF

Sample preparation is one of the crucial steps in any XRF analysis. In imaging XRF mode, besides elemental composition, the key goal is to preserve element distribution as much as possible, similar to the sample in in vivo state. It is clear that for micro and nano-scaled 2D XRF imaging, plant samples prepared by cryofixation followed by measuring in a frozen-hydrated state yield the best results. For the instruments that do not possess cryo-stage, however, measuring freeze-dried specimens presents a good alternative. However, care should be taken that freeze-drying is performed slowly, otherwise the protoplasts may shrink and stick to the cell walls, altering the structure of the sample [38].

The literature on the application of μ-XRF systems in imaging of element distribution in plant organs in the scale of few 10 μm, reports the use of plant material in various states—from fresh, freeze-dried to herbarized (air-dried). Each approach has some advantages and some disadvantages. For the analysis of live plant organs, one has to consider that they can dehydrate during the longer runs and that they

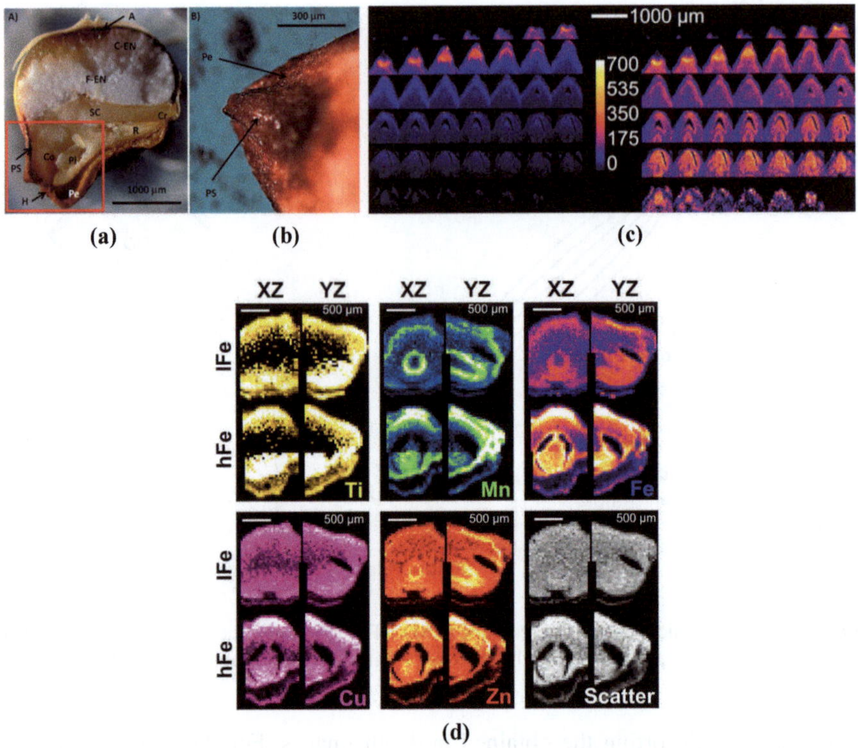

Fig. 8.9 Confocal imaging of element distribution in high (hFe) and low (lFe) iron pearl-millet seed. **a** A cross-section of pearl millet seed. Endosperm: C-EN-corneous endosperm, F-EN-floury endosperm; Embryo: Sc-scutellum. R-radicule, Pl-plumule, Co-coleoptile, Cr-coleorhiza. H-hillum, PS-pigment strand, A-aleurone, Pe-pericarp. **b** top view of the pigment strand in an intact seed submitted to the confocal analysis. **c** Montage of the Fe Kα intensity of the virtual slices through the germ region of the lFe seed measured from the top. left: measured data, right: corrected data; the top virtual slice is in the top left corner, the bottom slice in the bottom, right. **d** XZ and YZ slices through the high Fe and low Fe seed for comparison of the Ti, Mn, Fe, Cu, Zn Kα fluorescence and the scattered intensity. Figure adapted from [43] with permission from ACS Publications

might be exposed to radiation damage (especially with synchrotron light), which can affect structural and biochemical composition [44] A special measuring chamber used for the µ-XRF and chlorophyll fluorescence kinetic measurements of living samples was developed by [36]. Chlorophyll fluorescence, being a very stress-sensitive plant parameter, was, however, not significantly altered after the µ-XRF imaging using Tornado M4, Bruker instrument with a 15 µm beam spot size [36].

On the other hand, leaves soaked in the 95% alcohol solution were imaged in comparison with fresh ones, yielding a quite similar distribution patterns of Ca, K and Mn. In addition, it is apparent that for both K and Ca the drying of the leaves resulted in lower absorption of the low-energy fluorescence X-rays and scattering due to water. This results in better 'definition' in fine scale patterns of the distribution of these elements. This is especially visible in the Ca maps where Ca-crystalline deposits lining the fine vasculature of the leaves are more visible in the dried ethanol-immersed leaf [29].

It is possible to conclude that element distribution at the organ and tissue level is not that much affected by leaf drying (comparing fresh and freeze-dried leaves) and that the elimination of water may increase the sensitivity of the measurement due to decreased scattering of X-rays on water molecules, which is especially important with polychromatic excitation. In addition, dry tissues are less prone to radiation damage since there is no liquid media to carry around free radicals that are formed during X-ray irradiation. Additional experiments are, however needed with plant roots.

8.7 Quantification Procedures

Many benchtop μ-XRF instruments come with already implemented quantitative analysis using fundamental parameter-based 'standardless' quantification algorithms [32]. Quantitative analysis of internal reference materials and international standards shows that such instruments can provide highly reproducible data, especially with alloys, but for biological and geological samples, standardless quantification is not accurate.

To quantify μ-XRF images obtained on plant organs (e.g. leaves, roots) a chemical approach using matrix match standard reference materials can be used. In a study by [36], foil standards that reproduce the leaf matrix were prepared by dissolving polyvinyl alcohol (PVA, Ultimaker, Geldermalsen, The Netherlands) in double distilled water (20% w/v). PVA was chosen as a matrix because, among available polymers, it resembles the organic matrix quite well with its formula [CH2CH(OH)]n. Besides, it is very well water soluble and, thus, homogeneously miscible with the metal solutions until dryness. Metal salts were added to this solution (0.5 to 5 mmol L^{-1}). The mixtures of PVA and metal salts were then pipetted into a 3D-printed grid of 1×1 cm^2 squares and left to dry in a clean room. Foil layers were stuck to reproduce 200 and 400 μm thickness, and the metal content of the foil was validated with ICP-MS. In the leaves, however, the veins are thicker than the mesophyll parts, so the thickness of each tissue separately should be taken into account in true quantitative imaging.

In SXRFI, quantification procedures are not widely available at the μ-XRF beamlines and usually, the elemental distribution images are expressed only in counts s^{-1} normalized to incoming beam intensity (I0). Element concentrations correlate with atomic emission line intensities; however, true quantitative images must consider the local sample thickness and the local residual matrix. For this purpose two fundamental-parameter (FP) based models for quantification of 2D elemental distribution maps of intermediate-thick biological samples by synchrotron low energy μ-X-ray fluorescence spectrometry (SR-μ-XRF) were developed and applied to the elemental analysis in experiments with monochromatic focused photon beam excitation at two low energy X-ray fluorescence beamlines—TwinMic, Elettra Sincrotrone Trieste, Italy, and ID21, ESRF, Grenoble, France [45]. The models assume intermediate-thick biological samples composed of measured elements, the sources of the measurable spectral lines, and by the residual matrix (C, O, N, H), which affects the measured intensities through absorption. The necessary condition that enables quantification by this approach is the measurement of the absorption signal with a CCD camera or photodiode. This is only possible in low-energy SR regimes since, at higher energies, biological samples in the form of 20–50 μm cuttings become transparent for transmitting X-rays. In the first model, a fixed residual matrix of the sample (e.g. cellulose) is assumed, while in the second model, the residual matrix is obtained by the iteration refinement of elemental concentrations and an adjusted residual matrix, similarly as described in [31] (Fig. 8.10). The absorption of the incident focused beam in the biological sample at each scanned pixel position, determined from the output of a photodiode or a CCD camera, is applied as a control in the iteration procedure of quantification. A similar approach could also be utilized in quantification procedures with benchtop μ-XRF instruments, especially those equipped with transmission detectors.

8.8 Conclusions

XRF is now a powerful tool for the non-destructive elemental analysis of plant materials, providing critical insights into the composition and spatial distribution of major and trace elements.

Benchtop and hand-held EDXRF systems offer accessible and efficient non-destructive elemental analysis of plant bulk macro and micro elemental analyses in environmental, agricultural, and biological studies. For trace elements, TXRF can be used due to better sensitivity.

To better understand element uptake, transport and localization, XRF imaging techniques are employed. Synchrotron μ-XRF with monochromatic excitation provides spatial resolution down to nanometric scale and sensitivity down to sub-ppm levels, enabling detailed mapping of elemental distributions at the cellular and subcellular levels. The disadvantage of SXRFI is very limited beamtime accessibility to the plant biology community and limited scan size, so it is difficult to analyze elemental distribution at the organ level. For this purpose, benchtop μ-XRF instruments with beam spot sizes down to 5 μm can be employed.

In benchtop μ-XRF analysis, plant samples can be prepared in different ways, from measuring live or dried material. Freeze-drying has some advantages since it reduces the X-ray scattering on water molecules and increases sensitivity to detect trace elements.

Quantification of elemental distribution can be performed with matrix match standards or based on fundamental parameters. With the FP approach, the thickness of the sample and residual matrix should also be measured/calculated.

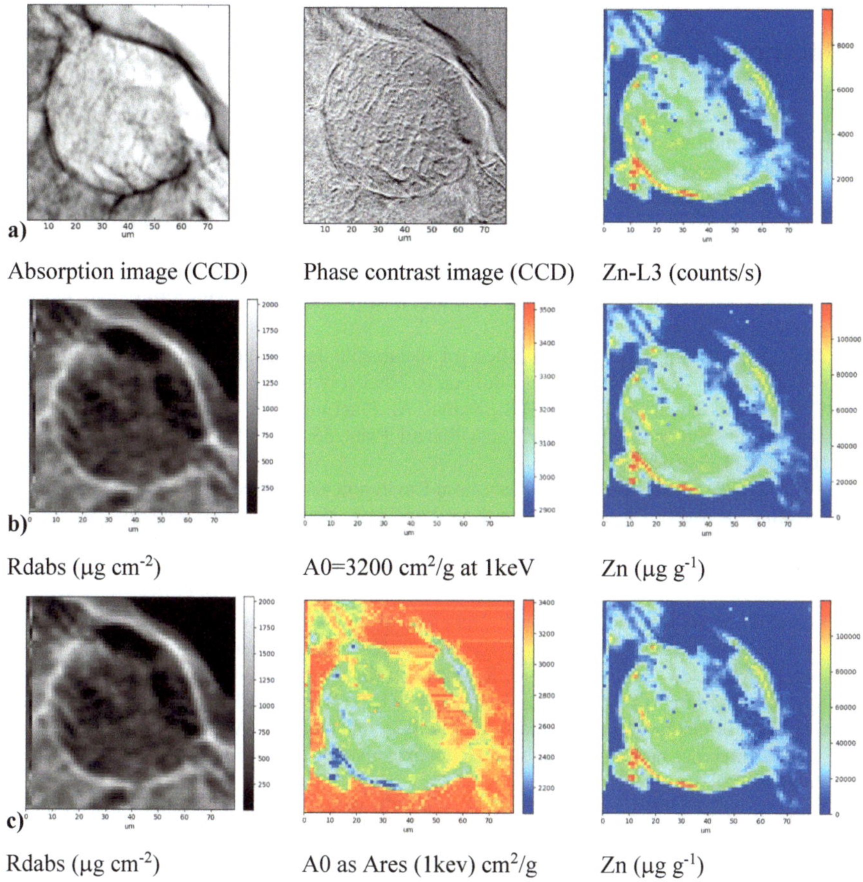

Fig. 8.10 Quantitative Zn distribution in an epidermal cell of the leaf of a Zn hyperaccumulator *N. praecox*. The sample was prepared by flush freezing in liquid propane cooled with liquid nitrogen followed by freeze-drying. The measurements were performed at TwinMic beamline, Elettra at 2.1 keV. Dwell time was 13 s $pixel^{-1}$, lateral resolution 1.2 μm. **a** parameters obtained directly from the measurements: absorption and phase contrast obtained by CCD camera and Zn-L3 line intensity (counts s^{-1}). **b** elemental map quantified according to a calculation model with a fixed residual matrix (value $A0 = 3200\ cm^2\ g^{-1}$ at 1 keV), **c** elemental map quantified according to a calculation model with a residual matrix obtained by the iteration refinement of elemental concentrations and the pixel-to-pixel adjusted residual matrix. The differences in pixel-to-pixel absolute Zn concentrations are a consequence of pixel-to-pixel matrix correction. More details on quantification procedures can be found from [45]

Acknowledgements The financial support by the Slovenian Research Agency (ARIS) through stable financing (P1-0212, Plant Biology) is acknowledged. TwinMic and XRF beamlines at Synchrotron Elettra, Trieste, are acknowledged for the provision of the beamtime (20200328, 20110043).

References

1. Guild, G.E., Stangoulis, J.C.R.: EDXRF for screening micronutrients in lentil and sorghum biofortification breeding programs. Plant Soil **463**, 461–469 (2021). https://doi.org/10.1007/S11104-021-04922-Z
2. Singh, V.K., Sharma, N., Singh, V.K.: Application of X-ray fluorescence spectrometry in plant science: solutions, threats, and opportunities. X-Ray Spectrom. **51**, 304–327 (2022). https://doi.org/10.1002/XRS.3260
3. Kumar, S., Kumar, S., Mohapatra, T.: Interaction between macro- and micro-nutrients in plants. Front. Plant Sci. **12**, 665583 (2021). https://doi.org/10.3389/FPLS.2021.665583/BIBTEX
4. White, P.J., Brown, P.H.: Plant nutrition for sustainable development and global health. Ann. Bot. **105**, 1073–1080 (2010). https://doi.org/10.1093/aob/mcq085
5. Singh, U.M., Sareen, P., Sengar, R.S., Kumar, A.: Plant ionomics: a newer approach to study mineral transport and its regulation. Acta Physiol. Plant. **35**, 2641–2653 (2013). https://doi.org/10.1007/s11738-013-1316-8
6. Adriano, D.C.: Trace Elements in Terrestrial Environments. Biogeochemistry, Bioavailability, and Risks of Metals. Springer-Verlag, New York, N.Y, New York, NY (2001)
7. Vogel-Mikuš, K., Arčon, I., Kodre, A.: Complexation of cadmium in seeds and vegetative tissues of the cadmium hyperaccumulator Thlaspi praecox as studied by X-ray absorption spectroscopy. Plant Soil **331**, 439–451 (2010). https://doi.org/10.1007/s11104-009-0264-y
8. Kodre, A., Vogel-Mikuš, K., Arčon, I., Debeljak, M.: EXAFS study of mercury ligand environment in mycorrhizal plant roots. In: The 16th International Conference on X-ray Absorption Fine Structure, 23–28 August, 2015, Karlsruhe : program, p. 92 (2015)
9. Kavčič, A., Murn, T., Budič, B., Vogel-Mikuš, K.: Effects of selenium on the uptake and speciation of mercury in plants and further in the food chains. In: Abstract book, 18th International Conference on Heavy Metals in the Environment, pp. 449–450. Universitet Gent, Ghent, Belgium (2016)
10. Vogel-Mikuš, K., Pelicon, P., Vavpetič, P., et al.: Elemental analysis of edible grains by micro-PIXE: common buckwheat case study. Nucl. Instrum. Methods Phys. Res. B **267**, 2884–2889 (2009). https://doi.org/10.1016/j.nimb.2009.06.104
11. Pongrac, P., Vogel-Mikuš, K., Potisek, M. et al.: Mineral and trace element composition and importance for nutritional value of buckwheat grain, groats, and sprouts. In: Molecular Breeding and Nutritional Aspects of Buckwheat. Elsevier Inc., pp. 261–269 (2016)
12. Verbi Pereira, F.M., Bastos Pereira Milori, D.M.: Investigation of the stages of citrus greening disease using micro synchrotron radiation X-ray fluorescence in association with chemometric tools. J. Anal. At. Spectrom. **25**, 351–355 (2010). https://doi.org/10.1039/B920980H
13. El-Jendoubi, H., Vázquez, S., Calatayud, Á. et al.: The effects of foliar fertilization with iron sulfate in chlorotic leaves are limited to the treated area. A study with peach trees (Prunus persica L. Batsch) grown in the field and sugar beet (Beta vulgaris L.) grown in hydroponics. Front Plant Sci 5 (2014). https://doi.org/10.3389/fpls.2014.00002
14. Vogel-Mikuš, K., Pongrac, P.: Imaging of potassium and calcium distribution in plant tissues and cells to monitor stress response and programmed cell death. Methods Mol. Biol. **2447**, 233–246 (2022). https://doi.org/10.1007/978-1-0716-2079-3_19
15. Van Grieken, R., Markowicz, A.: Handbook of X-Ray Spectrometry (2001)
16. Utica, G., Fabbrica, E., Carminati, M., et al.: ARDESIA-16: a 16-channel SDD-based spectrometer for energy dispersive X-ray fluorescence spectroscopy. J. Instrum. **16**, P07057 (2021). https://doi.org/10.1088/1748-0221/16/07/P07057
17. Nečemer, M., Kump, P., Vogel-Mikuš, K.: Energy dispersive X-Ray fluorescence analysis of biological materials. X-Ray Fluorescence in Biological Sciences: Principles, Instrumentation, and Applications, 309–325 (2022).https://doi.org/10.1002/9781119645719.CH21

18. Nečemer, M., Kump, P., Ščančar, J., et al.: Application of X-ray fluorescence analytical techniques in phytoremediation and plant biology studies. Spectrochim. Acta Part B At. Spectrosc. **63**, 1240–1247 (2008). https://doi.org/10.1016/j.sab.2008.07.006
19. Pongrac, P., Arčon, I., Castillo-Michel, H., Vogel-Mikuš, K.: Mineral element composition in grain of awned and awnletted wheat (Triticum aestivum l.) cultivars: Tissue-specific iron speciation and phytate and non-phytate ligand ratio. Plants 9 (2020). https://doi.org/10.3390/plants9010079
20. Nečemer, M., Potočnik, D., Ogrinc, N.: Discrimination between Slovenian cow, goat and sheep milk and cheese according to geographical origin using a combination of elemental content and stable isotope data. J. Food Compos. Anal. **52**, 16–23 (2016). https://doi.org/10.1016/j.jfca.2016.07.002
21. Ricachenevsky, F.K., Punshon, T., Lee, S., et al.: Elemental profiling of rice FOX lines leads to characterization of a new Zn plasma membrane transporter, OsZIP7. Front. Plant Sci. **9**, 366183 (2018). https://doi.org/10.3389/FPLS.2018.00865/BIBTEX
22. Huang, X.Y., Salt, D.E.E.: Plant Ionomics: from elemental profiling to environmental adaptation. Mol. Plant **9**, 787–797 (2016). https://doi.org/10.1016/J.MOLP.2016.05.003
23. Fourati, E., Vogel-Mikuš, K., Bettaieb, T., et al.: Physiological response and mineral elements accumulation pattern in Sesuvium portulacastrum L. subjected in vitro to nickel. Chemosphere **219**, 463–471 (2019). https://doi.org/10.1016/j.chemosphere.2018.12.081
24. Rutar, J.M., Hudobivnik, M.J., Nečemer, M., et al.: Nutritional quality and safety of the spirulina dietary supplements sold on the slovenian market. Foods **11**,(2022). https://doi.org/10.3390/foods11060849
25. Bueno Guerra, M.B., De Almeida, E., Carvalho, G.G.A., et al.: Comparison of analytical performance of benchtop and handheld energy dispersive X-ray fluorescence systems for the direct analysis of plant materials. J. Anal. At. Spectrom. **29**, 1667–1674 (2014). https://doi.org/10.1039/C4JA00083H
26. Roberts, A.A., Guimarães, D., Tehrani, M.W., et al.: A field-based evaluation of portable XRF to screen for toxic metals in seafood products. X-Ray Spectrom. (2023). https://doi.org/10.1002/XRS.3407
27. Howell, D., Tan, D., Ulitzka, S., Quinnell, R.: Comparative analysis and evaluation of handheld x-ray fluorescence spectrometry to measure silicon in crop plants. Proc West Mark Ed Assoc Conf **87**,(2020). https://doi.org/10.3390/PROCEEDINGS2019036087
28. Montanha, G.S., Rodrigues, E.S., Marques, J.P.R., et al.: X-ray fluorescence spectroscopy (XRF) applied to plant science: challenges towards in vivo analysis of plants. Metallomics **12**, 183–192 (2020). https://doi.org/10.1039/C9MT00237E
29. van der Ent, A., Casey, L.W., Purwadi, I., Erskine, P.D.: Laboratory μ-X-ray fluorescence elemental mapping of herbarium specimens for hyperaccumulator studies. Plant Soil **493**, 663–671 (2023). https://doi.org/10.1007/S11104-023-06201-5/FIGURES/5
30. Kalcsits, L.A.: Non-destructive measurement of calcium and potassium in apple and pear using handheld X-ray fluorescence. Front. Plant Sci. **7**,(2016). https://doi.org/10.3389/FPLS.2016.00442/PDF
31. Kump, P.: A somewhat different fundamental parameters approach to EDXRF quantification of solid environmental samples. X-Ray Spectrom. **53**, 16–26 (2024). https://doi.org/10.1002/XRS.3352
32. Flude, S., Haschke, M., Storey, M.: Application of benchtop micro-XRF to geological materials. Mineral. Mag. **81**, 923–948 (2017). https://doi.org/10.1180/MINMAG.2016.080.150
33. Van Der Ent, A., Kopittke, P.M., Paterson, D.J., et al.: Distribution of aluminium in hydrated leaves of tea (Camellia sinensis) using synchrotron- and laboratory-based X-ray fluorescence microscopy. Metallomics **12**, 1062–1069 (2020). https://doi.org/10.1039/C9MT00300B
34. Punshon, T., Tappero, R., Ricachenevsky, F.K., et al.: Contrasting calcium localization and speciation in leaves of the Medicago truncatula mutant cod5 analyzed via synchrotron X–ray techniques. Plant J. **76**, 627–633 (2013). https://doi.org/10.1111/TPJ.12322

35. Fittschen, U.E.A., Kunz, H.H., Höhner, R., et al.: A new micro X-ray fluorescence spectrometer for in vivo elemental analysis in plants. X-Ray Spectrom. **46**, 374–381 (2017). https://doi.org/10.1002/XRS.2783
36. Mijovilovich, A., Morina, F., Bokhari, S.N., et al.: Analysis of trace metal distribution in plants with lab-based microscopic X-ray fluorescence imaging. Plant Methods **16**, 1–21 (2020). https://doi.org/10.1186/S13007-020-00621-5/FIGURES/9
37. Rodrigues, E.S., Gomes, M.H.F., Duran, N.M., et al.: Laboratory microprobe X-ray fluorescence in plant science: Emerging applications and case studies. Front. Plant Sci. **871**, 376011 (2018). https://doi.org/10.3389/FPLS.2018.01588/BIBTEX
38. Vogel-Mikuš, K., Pongrac, P., Pelicon, P.: Micro-PIXE elemental mapping for ionome studies of crop plants. Int. J. PIXE **24**, 217–233 (2014). https://doi.org/10.1142/s0129083514400142
39. Ryan, C.G., Kirkham, R., de Jonge, M.D., et al.: The Maia detector and event mode. Synchrotron Radiat. News **31**, 21–27 (2018). https://doi.org/10.1080/08940886.2018.1528430/ASSET//CMS/ASSET/4FBDEF3B-BF34-4563-AD16-F3CC974DFF79/08940886.2018.1528430.FP.PNG
40. Romano, F.P., Caliri, C., Cosentino, L., et al.: Macro and micro full field x-ray fluorescence with an X-ray pinhole camera presenting high energy and high spatial resolution. Anal. Chem. **86**, 10892–10899 (2014). https://doi.org/10.1021/AC503263H
41. Klysubun, W., Tarawarakarn, P., Thamsanong, N., et al.: A new full-field XRF imaging station at Synchrotron Light Research Institute. J. Synchrotron Radiat. **30**, 655–661 (2023). https://doi.org/10.1107/S1600577523002722/RV5167SUP1.PDF
42. Heimler, K., Gottschalk, C., Vogt, C.: Confocal micro X-ray fluorescence analysis for the non-destructive investigation of structured and inhomogeneous samples. Anal. Bioanal. Chem. **415**, 5083–5100 (2023). https://doi.org/10.1007/S00216-023-04829-X/FIGURES/5
43. Mantouvalou, I., Lachmann, T., Singh, S.P., et al.: Advanced absorption correction for 3D elemental images applied to the analysis of pearl millet seeds obtained with a laboratory confocal micro x-ray fluorescence spectrometer. Anal. Chem. **89**, 5453–5460 (2017). https://doi.org/10.1021/acs.analchem.7b00373
44. Montanha, G.S., Marques, J.P.R., Santos, E., et al.: Physiological responses of plants to in vivo X-ray damage from X-ray fluorescence measurements: insights from anatomical, elemental, histochemical, and ultrastructural analyses. Metallomics **15**,(2023). https://doi.org/10.1093/MTOMCS/MFAD034
45. Kump, P., Vogel-Mikuš, K.: Quantification of 2D elemental distribution maps of intermediate-thick biological sections by low energy synchrotron μ-X-ray fluorescence spectrometry. J. Instrum. **13**,(2018). https://doi.org/10.1088/1748-0221/13/05/C05014

X-Ray Fluorescence Analysis of Vegetation Tissues via Chemometric Tools

9

John Tsado Mathew, Jonathan Hussaini, Abel Inobeme, Yakubu Azeh, Musah Monday, Elijah Yanda Shaba, Etsuyankpa Muhammad Bini, Tanko Ezekiel, Jibrin Mohammed Ndejiko, Charles Oluwaseun Adetunji, Muhammad Aishetu Ibrahim, Musa Safiyanu Tanko, and Amos Mamman

Abstract

X-ray fluorescence (XRF) is a powerful, non-destructive analytical technique widely used in various fields including environmental science for the elemental analysis. In this chapter, we explore the application of XRF for analysing vegetation tissues. Vegetation analysis is crucial for understanding nutrient dynamics, pollution levels, and overall plant health, making accurate and efficient analytical methods essential. The chemometric tools, such as principal component analysis and partial least squares regression, were to enhance the interpretation of the complex data generated by X-ray fluorescence. These multivariate statistical methods allow for the extraction of meaningful patterns and relationships from the large datasets typically produced in XRF analysis.

J. T. Mathew (✉) · J. Hussaini · Y. Azeh · M. Monday · M. A. Ibrahim · A. Mamman
Department of Chemistry, Ibrahim Badamasi Babangida University Lapai, Lapai, Niger, Nigeria
e-mail: johntsadom@gmail.com; jmathew@ibbu.edu.ng

A. Inobeme
Department of Chemistry, Edo State University Uzairue, Auchi, Edo, Nigeria

E. Y. Shaba · T. Ezekiel
Department of Chemistry, Federal University of Lafia, Lafia, Nasarawa, Nigeria

E. M. Bini · M. S. Tanko
Department of Chemistry, Federal University of Technology Minna, Minna, Nigeria

J. M. Ndejiko
Department of Microbiology, Ibrahim Badamasi Babangida University Lapai, Lapai, Niger, Nigeria

C. O. Adetunji
Applied Microbiology, Biotechnology and Nanotechnology Laboratory, Department of Microbiology, Edo State University Uzairue, Auchi, Edo, Nigeria

V. K. Singh (ed.), *X-Ray Fluorescence Spectroscopy and Chemometrics*,
https://doi.org/10.1007/978-3-031-98375-7_9

By integrating XRF with chemometric techniques, we can better differentiate between various plant species, identify stress indicators, and monitor environmental changes reflected in vegetation tissues. This method holds significant potential for applications in environmental monitoring, agricultural management, and ecological research, where understanding the elemental makeup of plants is crucial. Future studies will focus on refining these techniques to improve sensitivity and expand the range of detectable elements, ultimately contributing to more comprehensive environmental assessments.

9.1 Introduction

The analysis of vegetation tissues is essential in environmental science, agriculture, and ecological research, as it provides valuable insights into plant health, soil quality, and environmental pollution. Traditionally, elemental analysis of plant tissues has relied on techniques such as atomic absorption spectroscopy (AAS), inductively coupled plasma mass spectrometry (ICP-MS), and neutron activation analysis (NAA). While these methods are highly accurate, they often require extensive sample preparation, are time-consuming, and can be expensive. In recent years, X-ray fluorescence (XRF) has gained prominence as a powerful alternative due to its ability to perform non-destructive, rapid, and cost-effective elemental analysis. XRF's efficiency and ease of use make it an attractive option for researchers and practitioners seeking to analyze the elemental composition of vegetation tissues, thereby offering a practical solution for large-scale and routine applications [1, 2].

XRF is a spectroscopic technique that operates by exciting atoms within a sample using high-energy X-rays, which then emit characteristic secondary (fluorescent) X-rays. Each element emits X-rays with unique energy levels, enabling both qualitative and quantitative analysis of the sample's elemental composition. XRF offers several advantages over traditional methods, such as AAS and ICP-MS. These advantages include minimal sample preparation, the ability to simultaneously analyze a wide range of elements, and the capability to analyze both solid and liquid samples. These features make XRF particularly well-suited for the analysis of complex biological matrices like vegetation tissues. By providing rapid, non-destructive, and comprehensive elemental analysis, XRF has become an invaluable tool in environmental science, agriculture, and ecological research, offering a practical and efficient alternative to more traditional, labor-intensive methods [3].

The application of XRF in vegetation analysis has garnered significant attention for its ability to detect and quantify essential nutrients, trace elements, and pollutants within plant tissues. This capability is particularly relevant in agricultural studies, where the elemental composition of plants can provide critical insights into soil health, nutrient availability, and crop yield. By analyzing the elemental content of vegetation, researchers can assess the effectiveness of fertilization practices, identify nutrient deficiencies, and optimize agricultural productivity. Moreover, XRF has become a valuable tool in environmental monitoring, allowing scientists to assess the impact of industrial pollution on vegetation. It offers a reliable method for tracking the accumulation of heavy metals and other contaminants in plants, which is crucial for understanding the environmental impact of industrial activities and implementing appropriate remediation strategies. The non-destructive and rapid nature of XRF makes it an ideal choice for both agricultural and environmental applications, providing a practical and efficient means of monitoring and managing plant and soil health [4].

Despite its numerous advantages, XRF analysis faces challenges in data interpretation, particularly when applied to complex biological samples like vegetation tissues. Issues such as overlapping spectral peaks, matrix effects, and the necessity for precise calibration can complicate the accurate quantitative analysis of elements. These challenges can lead to potential inaccuracies in elemental concentration measurements and limit the effectiveness of XRF in certain applications. To overcome these obstacles, the integration of chemometric tools with XRF data analysis has become increasingly essential. Chemometrics involves using statistical and mathematical methods to extract meaningful information from complex data sets. Techniques such as principal component analysis (PCA), cluster analysis, and partial least squares regression (PLSR) are commonly employed to enhance the accuracy and reliability of XRF data interpretation. By applying these chemometric techniques, researchers can more effectively deconvolute overlapping peaks, correct for matrix effects, and improve the precision of element quantification, thereby maximizing the utility of XRF in vegetation tissue analysis. This integration represents a significant advancement in the field, enabling more accurate and comprehensive studies of plant health, soil quality, and environmental contamination [5].

The combination of XRF and chemometrics offers a powerful approach for analyzing vegetation tissues, allowing researchers to gain more accurate and comprehensive insights into the elemental composition of plants. Chemometric techniques like PCA can identify patterns and correlations within XRF data, making it possible to distinguish between different plant species, growth conditions, or environmental influences. This capability is particularly valuable in ecological and agricultural studies where subtle differences in elemental composition can have significant implications. Cluster analysis, another chemometric tool, can group similar samples based on their elemental profiles, providing insights into the similarities and differences among various plant populations or treatment conditions. Additionally, partial least squares regression (PLSR) can be used to build predictive models that link elemental composition to specific biological or environmental factors, such as nutrient availability or pollutant exposure. By integrating these chemometric techniques with XRF, researchers can enhance the interpretation of complex data sets, leading to more precise and meaningful conclusions in vegetation analysis [6].

Moreover, the use of chemometric tools with XRF enables the simultaneous analysis of multiple elements, offering a holistic view of elemental interactions within plant tissues. This approach is particularly valuable in studies focused on trace elements and pollutants, where detecting subtle changes in elemental composition is crucial for understanding environmental and ecological processes. The ability to analyze multiple elements together allows researchers to explore interactions and correlations that might be overlooked when elements are studied individually. This comprehensive perspective is essential for gaining insights into nutrient dynamics, pollutant accumulation, and the overall health of ecosystems. Additionally, the integration of chemometrics with XRF facilitates the efficient handling and interpretation of large data sets. This capability is particularly beneficial in extensive field studies or long-term monitoring programs, where vast amounts of data are generated. Chemometric tools streamline the processing of complex data, making it easier for researchers to identify trends, patterns, and relationships within the data, ultimately leading to more informed and accurate conclusions in vegetation analysis [7].

One of the key applications of XRF combined with chemometrics in vegetation analysis is the assessment of heavy metal contamination in plants. Toxic metals like lead (Pb), cadmium (Cd), and arsenic (As) can be detrimental to plant health and pose significant risks to human health when they enter the food chain. XRF, coupled with chemometric analysis, provides a powerful tool for monitoring the uptake and accumulation of these metals in plants. This approach offers valuable insights for environmental risk assessment and pollution management by identifying contamination hotspots and tracking the movement of heavy metals through ecosystems. Furthermore, this combined methodology can be applied to study the effects of soil amendments, fertilizers, and other agricultural practices on the elemental composition of crops. By analyzing how these practices influence the presence of essential nutrients and potentially harmful elements, researchers can contribute to the development of more sustainable and efficient farming practices. This holistic understanding aids in optimizing crop management strategies while minimizing environmental impact and ensuring food safety [8].

In addition, the application of XRF for the analysis of vegetation tissues offers numerous advantages, including rapid, non-destructive, and cost-effective elemental analysis. However, the complexity of biological samples necessitates the use of advanced data interpretation techniques. The integration of chemometric tools with XRF analysis represents a significant advancement in this field, enabling more accurate and comprehensive insights into the elemental composition of vegetation. This combined approach holds great promise for a wide range of applications, from agricultural research and environmental monitoring to ecological studies and beyond. As technology continues to advance, the role of XRF and chemometrics in vegetation analysis is likely to expand, offering new opportunities for research and practical applications in the coming years [9]. The schematic representation of the chapter is presented (Fig. 9.1).

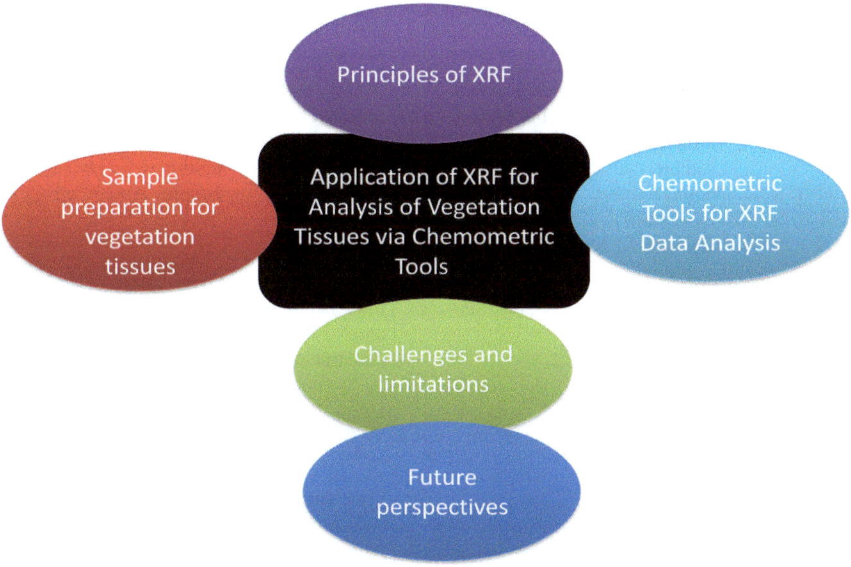

Fig. 9.1 Schematic representation of chapter

9.2 Principles of X-Ray Fluorescence (XRF)

XRF is an analytical technique used to determine the elemental composition of materials. It involves directing high-energy X-rays at a sample, causing atoms within the sample to become excited and eject inner-shell electrons. This excitation results in the emission of secondary, or fluorescent, X-rays that are characteristic of specific elements. The energy and intensity of these emitted X-rays are measured by a detector, allowing for both qualitative and quantitative analysis of the sample's elemental makeup. XRF is widely used due to its non-destructive nature, minimal sample preparation, and ability to analyze multiple elements simultaneously. However, challenges such as overlapping peaks, matrix effects, and the need for accurate calibration must be addressed for precise results. This Table 9.1 summarizes the core principles of XRF, highlighting the process from excitation to data interpretation, along with the benefits and challenges associated with each stage [10].

9.3 Sample Preparation for Vegetation Tissues

The preparation of vegetation tissues for XRF analysis is a vital process that directly impacts the accuracy and reliability of the results. Proper sample preparation is essential to accurately represent the elemental composition of the tissue and minimize potential interferences during analysis. The process generally involves

Table 9.1 The principles of X-ray fluorescence

Principle	Description	Key Points	Advantages	Challenges	References
X-ray excitation	High-energy X-rays are directed at a sample, causing atoms to eject inner-shell electrons	Interaction between X-rays and sample atoms	Non-destructive; minimal sample preparation	Potential radiation damage in sensitive samples	[11]
Fluorescence emission	Atoms release energy as secondary (fluorescent) X-rays when outer electrons fill inner-shell vacancies	Element-specific X-ray emissions	Elemental identification and quantification	Overlapping peaks can complicate spectrum interpretation	[12, 13]
Detection of emitted X-rays	The emitted X-rays are detected by an energy-dispersive detector, which measures the energy and intensity of the fluorescent X-rays	Energy dispersive vs. wavelength dispersive	Simultaneous detection of multiple elements	Detector resolution and efficiency limitations	[14]
Qualitative and quantitative analysis	The detected X-rays are analyzed to determine the elemental composition and concentration in the sample	Calibration and data processing	Accurate elemental analysis across a wide range of concentrations	Requires calibration standards and correction for matrix effects	[15]
Data interpretation and reporting	Results are interpreted using software, often integrating chemometric tools for enhanced analysis	Use of software and chemometrics	Detailed elemental maps and statistical data interpretation	Complex data may require advanced statistical techniques	[16]

several key stages: collection, drying, grinding, and homogenization. Each of these steps plays a critical role in ensuring that the sample is representative and uniform, which is necessary for effective XRF analysis. By carefully executing these stages, researchers can obtain precise and reproducible data, making sample preparation a foundational element in successful XRF studies of vegetation tissues [17].

The first step in sample preparation is the collection of vegetation tissues. It is essential to collect samples in a manner that minimizes contamination and ensures representativeness. Samples should be taken from different parts of the plant, such as leaves, stems, or roots, depending on the study's focus. The choice of tissue type is important because elemental concentrations can vary significantly between different parts of the plant. To preserve the integrity of the samples, they should be stored in clean, contamination-free containers and, if necessary, kept cool or frozen to prevent degradation. Proper labeling and documentation during collection are also crucial to ensure traceability and accurate analysis [18, 19].

Once the samples have been collected, the next step is to remove moisture from the tissues. Vegetation samples contain a high percentage of water, which can interfere with XRF analysis by affecting the accuracy of the measured elemental concentrations. Drying the samples is, therefore, an essential step. This can be achieved by air-drying, oven-drying, or freeze-drying, depending on the specific requirements of the study. Oven-drying at temperatures between 60 °C and 80 °C is commonly used, as it effectively removes moisture while minimizing the risk of altering the elemental composition of the tissues. Freeze-drying, although more time-consuming and costly, is preferred when it is essential to preserve the chemical integrity of the sample, especially for heat-sensitive elements [20].

After drying, the vegetation tissues need to be ground into a fine powder. This step is critical for achieving homogeneity, which is necessary for accurate XRF analysis. The particle size of the sample can significantly influence the results, as larger particles may cause uneven distribution of elements and result in poor analytical precision. Grinding can be done using a mortar and pestle or mechanical grinders, depending on the hardness of the sample and the desired fineness. Ideally, the resulting powder should have a uniform particle size of less than 100 μm to ensure that the sample is homogeneous and that the X-rays interact uniformly with the material during analysis [20].

Homogenization is a critical process in sample preparation that ensures the even distribution of ground material, so that any sub-sample selected for analysis accurately represents the entire sample. This step is essential because even small variations in the elemental composition within the sample can result in significant analytical errors. To achieve homogenization, thorough mixing is required, which can be done using mechanical mixers or through manual shaking. Once the sample is homogenized, a sub-sample is carefully taken for XRF analysis. It is important to execute this sub-sampling step with precision to avoid introducing any bias that could affect the results. To further enhance the reliability and consistency of the analysis, multiple sub-samples may be taken and analysed [21].

The quality of sample preparation directly impacts the accuracy, precision, and reproducibility of XRF results. Poor sample preparation can lead to a variety of issues, including matrix effects, where the sample matrix (the material surrounding the element of interest) affects the X-ray emission, leading to inaccurate readings. Additionally, if the sample is not homogenous or contains large particles, it can result in spotty or uneven readings, which can compromise the reliability of the data. Furthermore, improper drying can lead to retained moisture, which can dilute the concentration of elements and affect the sensitivity of the analysis. On the other hand, over-drying or excessive grinding can lead to the loss of volatile elements, thereby skewing the results. Therefore, careful attention to each step of the sample preparation process is essential to ensure that the XRF analysis yields accurate and meaningful data [17, 22].

In general, sample preparation is a vital component of XRF analysis when working with vegetation tissues. Each step, from collection to homogenization, must be carefully executed to ensure that the sample accurately represents the elemental composition of the plant material. By minimizing contamination, ensuring uniformity, and carefully controlling moisture and particle size, researchers can obtain reliable and reproducible results from XRF analysis. As such, rigorous sample preparation is foundational to the success of XRF studies in environmental monitoring, agricultural research, and ecological assessments [17].

9.4 XRF Applications in Vegetation Analysis

Table 9.2 provides a concise overview of how XRF is used in vegetation analysis, highlighting the applications, key elements, benefits, and challenges associated with each use case. XRF is widely used in vegetation analysis to assess the elemental composition of plant tissues. One key application is nutrient analysis, where XRF determines the concentration of essential elements like nitrogen, phosphorus, and potassium, aiding in agricultural management. XRF is also employed to monitor trace elements such as Cu and Zn, which are crucial for plant health. Additionally, XRF plays a vital role in assessing heavy metal contamination, identifying toxic elements like lead and cadmium in polluted environments. Environmental impact studies utilize XRF to analyze changes in plant elemental composition due to environmental factors, providing insights into ecosystem health. Lastly, XRF can differentiate plant species based on their unique elemental profiles, supporting biodiversity studies and species identification. These applications demonstrate XRF's versatility and importance in vegetation research [7].

9.5 Chemometric Tools for XRF Data Analysis

XRF spectroscopy is a powerful technique for the elemental analysis of complex samples such as vegetation tissues. However, the raw data generated by XRF instruments can be intricate and challenging to interpret due to the complexity of biological matrices, overlapping spectral peaks, and the presence of multiple elements. Chemometric tools have emerged as essential methods for enhancing the analysis of XRF data, allowing researchers to extract meaningful information from complex datasets, improve the accuracy of quantitative analyses, and identify patterns that would be difficult to discern through traditional methods. These tools

Table 9.2 Various applications of XRF in vegetation analysis

Application	Description	Key elements analyzed	Benefits	Challenges	References
Nutrient analysis in plants	Assessing the concentration of essential nutrients within plant tissues	N, P, K, Ca, Mg	Supports agricultural productivity and crop management	Detection limits for lighter elements can be challenging	[23]
Trace element monitoring	Identifying and quantifying trace elements present in vegetation	Cu, Zn, Mn, Fe	Important for studying plant physiology and soil health	Requires precise calibration and matrix corrections	[24, 25]
Heavy metal contamination assessment	Evaluating the accumulation of toxic heavy metals in plants due to pollution	Pb, Cd, As, Hg	Essential for environmental monitoring and pollution control	Overlapping peaks and matrix effects can affect accuracy	[26]
Environmental Impact Studies	Analyzing the effects of environmental factors on plant elemental composition	Various elements, depending on the study area	Useful for tracking changes in ecosystems over time	Variability in environmental conditions can complicate analysis	[27]
Plant species differentiation	Distinguishing between different plant species based on their elemental profiles	Multiple elements, depending on species	Aids in biodiversity studies and taxonomy	Requires robust chemometric analysis for pattern recognition	[28]

integrate statistical, mathematical, and computational techniques to optimize the interpretation of XRF spectra, providing more robust and reliable results [12].

9.5.1 Principal Component Analysis (PCA)

PCA is one of the most widely used chemometric tools in XRF data analysis. PCA is a multivariate statistical technique that reduces the dimensionality of complex datasets while preserving the variance within the data. By transforming the original correlated variables into a smaller set of uncorrelated variables called principal components, PCA allows researchers to identify patterns, trends, and outliers in the data. In the context of XRF analysis, PCA can be used to differentiate between plant species, identify environmental stressors, or detect the presence of pollutants. For example, by analyzing the elemental composition of plant tissues from different regions, PCA can reveal distinct patterns that correlate with specific environmental conditions or pollution levels, facilitating the interpretation of complex environmental data [29, 30].

9.5.2 Cluster Analysis

Cluster analysis is another important chemometric tool used in the analysis of XRF data. It groups samples into clusters based on their similarities in elemental composition, allowing researchers to classify vegetation samples into meaningful categories. This technique is particularly useful in environmental monitoring, where it can help identify areas with similar pollution levels or determine the impact of different environmental factors on plant health. For example, cluster analysis can be used to group plant samples from various locations based on their heavy metal content, identifying regions that are heavily contaminated and those that are relatively clean. By grouping similar samples together, cluster analysis simplifies the interpretation of large datasets and aids in making informed decisions regarding environmental management and pollution control [31].

9.5.3 Partial Least Squares Regression (PLSR)

PLSR is a powerful chemometric tool for quantitative analysis in XRF studies. PLSR is particularly useful when dealing with datasets that have a large number of variables, as it can handle multicollinearity and small sample sizes more effectively than traditional regression methods. In XRF data analysis, PLSR is used to build predictive models that correlate the spectral data with the concentration of specific elements in the sample. This is particularly valuable in studies where direct measurement of certain elements is challenging due to overlapping peaks or matrix effects. By applying PLSR, researchers can create robust calibration models that improve the accuracy of quantitative XRF analysis, even in complex biological matrices like plant tissues. This enhances the ability to predict the concentration of key nutrients, trace elements, or contaminants in vegetation samples [32, 33].

9.5.4 Pattern Recognition and Classification

Pattern recognition and classification techniques are critical chemometric tools for the analysis of XRF data, especially when dealing with large and complex datasets. These techniques involve identifying patterns in the data that correspond to specific classes or categories, such as different plant species, growth conditions, or levels of contamination. Techniques such as discriminant analysis, support vector machines (SVM), and neural networks are commonly used for pattern recognition in XRF studies. For instance, discriminant analysis can classify plant samples based on their elemental profiles, helping to distinguish between species or identify plants subjected to different environmental conditions. Similarly, SVM and neural networks can be trained to recognize patterns in XRF spectra, enabling the automated classification of samples and the detection of subtle differences in elemental composition that might be overlooked using traditional methods [34].

9.5.5 Advantages and Challenges of Chemometric Tools

The integration of chemometric tools into XRF data analysis offers numerous advantages, including the ability to handle large and complex datasets, improve the accuracy of quantitative analyses, and identify subtle patterns that are not immediately apparent. By using chemometric techniques, researchers can gain deeper insights into the elemental composition of vegetation tissues, leading to more informed decisions in fields such as agriculture, environmental monitoring, and ecological research. However, there are also challenges associated with the use of these tools. Chemometric analysis requires a solid understanding of statistical and mathematical principles, as well as expertise in the use of specialized software. Additionally, the interpretation of results can be complex, and there is a risk of overfitting models if the analysis is not conducted carefully. Despite these challenges, the benefits of chemometric tools in enhancing the interpretation of XRF data make them indispensable in modern analytical chemistry [35].

In addition, chemometric tools play a crucial role in the analysis of XRF data, enabling researchers to extract meaningful information from complex datasets and improve the accuracy and reliability of their findings. Techniques such as PCA, cluster analysis, PLSR, and pattern recognition provide powerful methods for exploring and interpreting the elemental composition of vegetation tissues. While the application of these tools requires specialized knowledge and careful implementation, their ability to enhance XRF data analysis makes them invaluable in a wide range of scientific and environmental studies. As XRF technology and chemometric methods continue to evolve, their combined use will likely lead to even greater advancements in the field of vegetation analysis [34].

9.6 Vegetation in Industrially Contaminated Areas

In industrially contaminated areas, XRF is used to analyze vegetation for detecting and quantifying pollutants. Key elements studied include heavy metals such as lead (Pb), cadmium (Cd), arsenic (As), and mercury (Hg), which are indicative of environmental contamination levels. XRF helps assess how plants accumulate these contaminants, providing insights into bioaccumulation and the potential risks to human health through the food chain. It is also valuable for monitoring environmental impacts over time and evaluating the effectiveness of remediation strategies. While XRF offers significant benefits, such as non-destructive analysis and simultaneous detection of multiple elements, challenges include managing overlapping

Table 9.3 The analysis of vegetation in industrially contaminated areas using XRF

Aspect	Description	Key Elements Analyzed	Benefits	Challenges	References
Identification of contaminants	Detecting and quantifying heavy metals and pollutants in plant tissues	Pb, Cd, As, Hg	Provides insight into contamination levels and sources	Overlapping peaks can complicate detection	[4]
Assessment of bioaccumulation	Evaluating how plants accumulate and store contaminants from the environment	Various heavy metals and trace elements	Helps understand the impact of pollution on plant health	Variability in plant uptake and contamination levels	[37]
Monitoring environmental impact	Tracking changes in elemental composition due to industrial activities	Multiple elements, depending on industrial processes	Assesses the long-term effects of pollution on vegetation	Requires long-term monitoring for accurate results	[38, 39]
Risk assessment for human health	Analyzing the potential transfer of contaminants from plants to humans	Heavy metals and toxic elements	Essential for evaluating food chain contamination risks	Complexity of matrix effects and bioavailability issues	[30]
Remediation and management	Evaluating the effectiveness of remediation strategies and pollution management	Elements targeted by remediation efforts	Aids in developing and accessing pollution mitigation strategies	Effectiveness of remediation can vary depending on methods used	[40]

peaks and variability in plant uptake. Accurate assessment requires careful interpretation and long-term monitoring to fully understand the contamination's impact on both vegetation and human health. This table provides an overview of how XRF is utilized to study vegetation in areas affected by industrial contamination, highlighting key aspects, elements analyzed, benefits, and challenges associated with each aspect (Table 9.3) [36].

9.7 Monitoring of Agricultural Fields

Monitoring agricultural fields is crucial for optimizing crop production, managing soil health, and ensuring sustainable farming practices. Effective monitoring involves assessing various factors such as soil nutrients, plant health, and environmental conditions. Advanced analytical techniques, including XRF, play a pivotal role in this process by providing detailed insights into the elemental composition of both soil and plant tissues. This information is invaluable for making informed decisions regarding fertilization, pest control, and overall crop management [41, 42].

9.7.1 Soil Nutrient Analysis

Soil nutrient analysis is a fundamental aspect of monitoring agricultural fields. Adequate nutrient levels are essential for healthy crop growth and optimal yield. Traditional soil testing methods, such as chemical extraction and colorimetric assays, can provide information about soil fertility, but they often involve labor-intensive procedures and may not capture the full spectrum of elements present. XRF offers a rapid and non-destructive alternative for soil nutrient analysis. By analyzing soil samples, XRF can detect and quantify essential elements such as phosphorus (P), potassium (K), calcium (Ca), and magnesium (Mg). This information helps farmers tailor their fertilization practices, ensuring that crops receive the right balance of nutrients. Additionally, XRF can identify trace elements and potential contaminants, aiding in the assessment of soil quality and safety [43].

9.7.2 Plant Health Assessment

Monitoring plant health is critical for maximizing crop yield and minimizing losses due to disease or nutrient deficiencies. XRF can be used to analyze plant tissues to determine the concentration of essential nutrients and trace elements. For example, deficiencies in nutrients such as nitrogen (N), phosphorus (P), or potassium (K) can lead to poor plant growth and reduced yields. By regularly analyzing plant samples, farmers can detect nutrient imbalances early and take corrective actions. XRF also helps identify the presence of toxic elements or pollutants that may affect plant health. This is particularly important in areas with potential contamination from industrial activities or excessive use of chemicals. Regular plant health monitoring using XRF can guide adjustments in fertilization practices and improve overall crop management strategies [44].

9.7.3 Environmental Impact and Pollution Monitoring

Agricultural fields can be impacted by environmental factors such as pollution and contamination from nearby industrial activities. Monitoring the levels of heavy metals and other pollutants in soil and plants is essential for assessing the impact of these factors on crop production and food safety. XRF is an effective tool for detecting and quantifying contaminants like lead (Pb), cadmium (Cd), and arsenic (As) in both soil and plant tissues. By integrating XRF analysis into routine monitoring practices, farmers can identify areas of contamination and take appropriate measures to mitigate its effects. This includes adjusting agricultural practices, such as changing fertilization strategies or implementing soil remediation techniques. Environmental monitoring using XRF helps ensure that crops are safe for consumption and that farming practices are environmentally sustainable [45].

9.7.4 Precision Agriculture and Yield Optimization

Precision agriculture involves the use of technology and data-driven approaches to optimize crop production and resource use. XRF can be integrated into precision agriculture systems to provide detailed information about soil and plant composition across different areas of a field. This data allows for the creation of detailed nutrient maps, which can guide variable-rate application of fertilizers and other inputs. By applying nutrients precisely where they are needed, farmers can enhance crop growth and yield while minimizing waste and environmental impact. XRF also supports the development of precision irrigation strategies by providing insights into soil moisture and nutrient availability. The integration of XRF with other precision agriculture tools, such as GPS and remote sensing, enables a more targeted and efficient approach to farming [46].

9.7.5 Challenges and Future Directions

While XRF offers significant advantages for monitoring agricultural fields, there are challenges to consider. The accuracy of XRF analysis can be affected by factors such as sample preparation, matrix effects, and calibration. Proper sample handling and calibration are essential for reliable results. Additionally, XRF technology requires specialized equipment and expertise, which may not be readily available in all agricultural settings. Despite these challenges, advancements in XRF technology and its integration with other analytical and data management tools are expected to enhance its application in agriculture. Future developments may include portable XRF devices for on-site analysis, improved data processing algorithms, and increased automation. As technology continues to evolve, XRF will play an increasingly important role in supporting sustainable and efficient agricultural practices [47, 48].

In general, the monitoring of agricultural fields using X-ray Fluorescence (XRF) provides valuable insights into soil and plant composition, supporting optimized crop management and sustainable farming practices. By analyzing soil nutrients, assessing plant health, monitoring environmental impact, and integrating with precision agriculture systems, XRF helps farmers make informed decisions and enhance crop production. Addressing the challenges associated with XRF technology and embracing future advancements will further improve its effectiveness and accessibility, ultimately contributing to more sustainable and productive agricultural systems [49].

9.8 Challenges and Limitations

The application of X-ray Fluorescence (XRF) combined with chemometric tools offers powerful capabilities for analyzing the elemental composition of vegetation tissues. However, several challenges and limitations must be addressed to ensure accurate and reliable results [7].

One of the primary challenges in XRF analysis of vegetation tissues is the presence of matrix effects. Vegetation tissues are complex biological matrices with varying compositions, including water content, organic matter, and a diverse range of elements. These factors can influence the XRF signal, leading to inaccuracies in the quantification of elements. Matrix effects can cause issues such as absorption or enhancement of X-rays by other elements in the sample, which complicates the interpretation of spectra. Chemometric tools can partially mitigate these effects by applying mathematical corrections and calibrations, but completely eliminating matrix effects remains difficult, especially when dealing with heterogeneous samples [7].

XRF spectra often contain overlapping peaks, particularly when analyzing samples with multiple elements present at similar energy levels. This overlap can make it challenging to accurately distinguish and quantify individual elements, especially in complex samples like vegetation tissues. For example, elements such as potassium (K) and calcium (Ca) have closely spaced peaks that can interfere with each other. While chemometric techniques like Principal Component Analysis (PCA) or Partial Least Squares Regression (PLSR) can help deconvolute overlapping signals, the effectiveness of these methods depends on the quality of the calibration models and the complexity of the sample. In cases where peaks are heavily overlapped, even advanced chemometric tools may struggle to provide accurate results [50].

Accurate XRF analysis relies heavily on proper calibration and standardization. Developing calibration curves requires the use of reference materials that closely match the matrix of the vegetation tissues being analyzed. However, obtaining such reference materials can be challenging, particularly for unique or complex samples. Additionally, calibration must account for variations in sample thickness, particle size, and other physical properties that can affect XRF measurements. The use of chemometric tools can improve calibration by allowing for the inclusion of multiple variables in the calibration model, but creating robust and reliable models requires significant expertise and may not always be feasible for all types of vegetation tissues [51].

XRF is generally less sensitive to elements present in low concentrations, particularly lighter elements such as nitrogen (N), oxygen (O), and sulfur (S). These elements are often critical for understanding plant health and nutrient status, yet their low fluorescence yields make them difficult to detect with XRF. Chemometric tools can enhance the detection of low-concentration elements by improving signal processing and data interpretation, but there are inherent limitations in the sensitivity of XRF technology itself. This limitation may require the use of complementary analytical techniques, such as Inductively Coupled Plasma Mass Spectrometry (ICP-MS), to obtain a complete elemental profile of vegetation tissues [17, 52].

While chemometric tools significantly enhance the capability of XRF analysis, they also introduce complexity into the data analysis process. Applying chemometric methods such as PCA, cluster analysis, or PLSR requires a deep understanding of statistical and mathematical principles. Additionally, interpreting the results from chemometric analyses can be challenging, particularly when dealing with large and complex datasets. There is also a risk of overfitting models, where the chemometric tool becomes too tailored to the training data, leading to inaccurate predictions for new samples. Properly applying chemometric tools requires not only technical expertise but also careful consideration of the specific research question and the characteristics of the samples being analysed [53].

In addition, while the application of X-ray Fluorescence for the analysis of vegetation tissues via chemometric tools offers many advantages, it also presents several challenges and limitations. Matrix effects, overlapping spectral peaks, calibration issues, sensitivity to low concentration elements, and the complexity of chemometric analysis are all factors that must be carefully managed to obtain accurate and reliable results. Addressing these challenges requires a combination of technical expertise, advanced analytical techniques, and careful experimental design. As XRF technology and chemometric methods continue to evolve, it is likely that some of these limitations will be mitigated, further enhancing the utility of XRF for vegetation tissue analysis [54, 55].

9.9 Future Perspectives

The integration of X-ray Fluorescence (XRF) with chemometric tools has already revolutionized the analysis of vegetation tissues, offering a robust method for elemental analysis in environmental and agricultural research. However, the future holds even more potential as technological advancements and new methodologies emerge. These future perspectives highlight the directions in which this field is likely to evolve, promising enhanced capabilities, greater accessibility, and broader applications [56].

One of the most exciting future prospects is the continuous improvement in XRF technology itself. Innovations in detector sensitivity, resolution, and portability are likely to enhance the accuracy and efficiency of XRF analysis. For instance, developments in silicon drift detectors (SDDs) could allow for better detection of lighter elements, which are currently challenging to measure with traditional XRF systems. Additionally, the advent of more portable and handheld XRF devices will make it easier to conduct in-situ analyses in the field, reducing the need for extensive sample preparation and transport. These advancements will expand the application of XRF in real-time monitoring of vegetation health, nutrient status, and environmental contamination [57].

The future of XRF analysis in vegetation studies will likely involve its integration with other complementary analytical techniques. Combining XRF with methods such as Inductively Coupled Plasma Mass Spectrometry (ICP-MS), Raman spectroscopy, or Near-Infrared Spectroscopy (NIRS) could provide a more comprehensive understanding of the elemental and molecular composition of plant tissues. Such multi-modal approaches could offer deeper insights into plant physiology, stress responses, and interactions with the environment. The use of machine learning algorithms to fuse data from multiple sources could further enhance the accuracy and interpretability of results, leading to more precise and holistic assessments of vegetation health and contamination levels [58].

As chemometric tools continue to evolve, they are expected to become even more powerful in handling complex datasets generated by XRF. Future developments in chemometrics could include more sophisticated algorithms for data analysis, capable of better handling non-linear relationships, complex interactions, and large-scale datasets. Machine learning and artificial intelligence (AI) are likely to play a significant role in these advancements, providing automated and adaptive methods for pattern recognition, classification, and prediction [59]. These tools could significantly reduce the time and expertise required to interpret XRF data, making advanced analytical techniques more accessible to researchers in various fields. Furthermore, the development of user-friendly software platforms that integrate XRF analysis with advanced chemometrics will likely broaden the adoption of these methods [13, 22].

The future application of XRF and chemometrics in vegetation analysis will likely see a significant expansion in precision agriculture and environmental monitoring. As agricultural practices increasingly rely on data-driven decision-making, XRF's ability to provide rapid, non-destructive, and detailed elemental analysis will become a cornerstone of precision farming. By integrating XRF data with geospatial technologies, farmers will be able to create highly detailed nutrient maps, enabling precise application of fertilizers and other inputs, thereby reducing waste and environmental impact [60]. In environmental monitoring, XRF could be used for continuous monitoring of vegetation in sensitive areas, providing early warning signs of pollution or ecological changes. The ability to monitor plant health and contamination levels in real-time could revolutionize how we manage agricultural fields and natural ecosystems, leading to more sustainable practices [54].

For the full potential of XRF and chemometrics to be realized, there will need to be broader adoption and standardization of these techniques across various fields. This will involve the development of standardized protocols for sample preparation, calibration, and data analysis to ensure consistency and comparability of results. Collaborative efforts between academia, industry, and regulatory bodies will be essential in establishing these standards. Furthermore, as XRF technology becomes more affordable and user-friendly, it is expected that its use will expand beyond research institutions to include routine applications in agriculture, environmental management, and even industry. Educational initiatives aimed at training the next generation of scientists and practitioners in the use of XRF and chemometric tools will also be critical in driving this broader adoption [38].

In addition, the future of X-ray Fluorescence for the analysis of vegetation tissues via chemometric tools is bright, with numerous advancements on the horizon. Improvements in technology, integration with other analytical methods, enhanced chemometric techniques, and broader adoption will all contribute to making XRF an even more powerful tool for environmental and agricultural research [61]. As these developments unfold, XRF and chemometrics are poised to play an increasingly important role in advancing our understanding of plant health, environmental contamination, and sustainable agricultural practices [54].

9.10 Conclusion

The application of X-ray Fluorescence (XRF) for the analysis of vegetation tissues, enhanced by chemometric tools, represents a significant advancement in environmental and agricultural research. XRF provides a rapid, non-destructive method for determining the elemental composition of plant tissues, which is crucial for understanding nutrient dynamics, pollution impacts, and overall plant health. Chemometric tools further enhance the utility of XRF by enabling the interpretation of complex data sets, addressing challenges such as matrix effects and overlapping spectral peaks. While there are challenges associated with XRF analysis, including the need for proper calibration and the limitations in detecting low-concentration elements, ongoing technological advancements and the integration of machine learning are expected to overcome these hurdles. The future holds promising developments in XRF technology, including greater portability, improved sensitivity, and broader adoption in precision agriculture and environmental monitoring. In addition, the synergy between XRF and chemometric tools offers a powerful approach for the detailed and accurate analysis of vegetation tissues. As this field continues to evolve, it will play an increasingly important role in supporting sustainable agricultural practices, environmental protection, and the broader understanding of plant-environment interactions.

References

1. Genchi, G., Carocci, A., Lauria, G., Sinicropi, M.S., Catalano, A.: Nickel: human health and environmental toxicology. Int. J. Environ. Res. Public Health **17**(3), 679 (2020). https://doi.org/10.3390/ijerph17030679
2. Adetunji, C.O., Inobeme, A., Singh, K.R.B., Bodunrinde, R.E., Mathew, J.T., Olaniyan, O., Ogundolie, F.A., Singh, J., Nayak, V., Singh, R.P.: Genomic analysis of heavy metal-resistant genes in wastewater treatment plants. In book: Microbial Community Studies in Industrial Wastewater Treatment. Taylor & Francis eBooks, pp. 83–97 (2023), https://doi.org/10.1201/9781003354147-8
3. Planeta, K., Kubala-Kukus, A., Drozdz, A., Matusiak, K., Setkowicz, Z., Chwiej, J.: The assessment of the usability of selected instrumental techniques for the elemental analysis of biomedical samples. Sci. Rep. **11**(1), 3704 (2021). https://doi.org/10.1038/s41598-021-82179-3
4. He, S., Niu, Y., Xing, L., Liang, Z., Song, X., Ding, M., Huang, W.: Research progress of the detection and analysis methods of heavy metals in plants. Front. Plant Sci. **15**, 1310328 (2024). https://doi.org/10.3389/fpls.2024.1310328
5. Yatsuk, O., Ferretti, M., Gorghinian, A., Fiocco, G., Malagodi, M., Agostino, A., Gulmini, M.: Data from multiple portable XRF units and their significance for ancient glass studies. Molecules (Basel, Switzerland) **27**(18), 6068 (2022). https://doi.org/10.3390/molecules27186068
6. Panebianco, S., Pellegriti, M.G., Finocchiaro, C., Musumarra, A., Barone, G., Caggiani, M.C., Cirvilleri, G., Lanzafame, G., Pulvirenti, A., Scordino, A., Mazzoleni, P.: XRF analysis searching for fingerprint elemental profile in south-eastern Sicily tomatoes. Sci. Rep. **13**(1), 13739 (2023). https://doi.org/10.1038/s41598-023-40124-6
7. Mathew, J.T., Inobeme, A., Adekoya, M.A., Musah, M., Azeh, Y., Otori, A.A., Shaba, E.Y., Mamman, A.: Processing techniques, test methods, and regulatory standards of smart textiles. In book: Smart Textiles from Natural Resources, Elsevier Inc., 1, pp. 237–272 (2024). https://doi.org/10.1016/B978-0-443-15471-3.00016-9
8. Montanha, G.S., Rodrigues, E.S., Marques, J.P.R., de Almeida, E., Dos Reis, A.R., Pereira de Carvalho, H.W.: X-ray fluorescence spectroscopy (XRF) applied to plant science: challenges towards in vivo analysis of plants. Metallomics: Integrated Biometal Sci. **12**(2), 183–192 (2020). https://doi.org/10.1039/c9mt00237e
9. Tian, X., Xie, Q., Fan, M., Chai, G., Li, G.: Identification of heavy metal pollutants and their sources in farmland: an integrated approach of risk assessment and X-ray fluorescence spectrometry. Sci. Rep. **12**(1), 12196 (2022). https://doi.org/10.1038/s41598-022-16177-4
10. Kharbach, M., Alaoui Mansouri, M., Taabouz, M., Yu, H.: Current application of advancing spectroscopy techniques in food analysis: data handling with chemometric approaches. Foods (Basel, Switzerland) **12**(14), 2753 (2023). https://doi.org/10.3390/foods12142753
11. Utazi, E.B., Oyewole, O.A., Yakubu, J.G., Tsado, P.Y., Omoregie, I.P., Adetunji, C.O., Mathew, J.T., Igiku, V., Eniola, K.I.T., Yerima, M.B.: Toxicological effects of nanobiofertilizer on water body, water quality, lower plants, zooplanktons, and beneficial microorganisms. in book: handbook of agricultural biotechnology. Wiley **5**, 201–232 (2024). https://doi.org/10.1002/9781394211548.ch8
12. Sauer, K., Zizak, I., Forien, J.B., Rack, A., Scoppola, E., Zaslansky, P.: Primary radiation damage in bone evolves via collagen destruction by photoelectrons and secondary emission self-absorption. Nat. Commun. **13**(1), 7829 (2022). https://doi.org/10.1038/s41467-022-34247-z
13. Pushie, M.J., Pickering, I.J., Korbas, M., Hackett, M.J., George, G.N.: Elemental and chemically specific X-ray fluorescence imaging of biological systems. Chem. Rev. **114**(17), 8499–8541 (2014). https://doi.org/10.1021/cr4007297
14. Adetunji, C.O., Olaniyan, O T, Inobeme, A., Adeyomoye, O., Mathew, J.T., Akinbo, O.: Recent advances in the characterization and application of graphene in the food industry.

In book: Sensing and Artificial Intelligence Solutions for Food Manufacturing. CRC Press, pp. 153–164 (2023). https://doi.org/10.1201/9781003207955-11
15. Khan, M.S.I., Oh, S.W., Kim, Y.J.: Power of scanning electron microscopy and energy dispersive X-ray analysis in rapid microbial detection and identification at the single cell level. Sci. Rep. **10**(1), 2368 (2020). https://doi.org/10.1038/s41598-020-59448-8
16. Singh, P., Datta, M., Ramana, G.V., Gupta, S.K., Malik, T.: Qualitative comparison of elemental concentration in soils and other geomaterials using FP-XRF. PLoS ONE **17**(5), e0268268 (2022). https://doi.org/10.1371/journal.pone.0268268
17. Ngulube, P.: Improving the quality of reporting findings using computer data analysis applications in educational research in context. Heliyon **9**(9), e19683 (2023). https://doi.org/10.1016/j.heliyon.2023.e19683
18. Pushie, M.J., Sylvain, N.J., Hou, H., Hackett, M.J., Kelly, M.E., Webb, S.M.: X-ray fluorescence microscopy methods for biological tissues. Metallomics: Integrated Biometal Sci. **14**(6), mfac032 (2022). https://doi.org/10.1093/mtomcs/mfac032
19. Wong, P.B., Wiley, E.O., Johnson, W.E., Ryder, O.A., O'Brien, S.J., Haussler, D., Koepfli, K.P., Houck, M.L., Perelman, P., Mastromonaco, G., Bentley, A.C., Venkatesh, B., Zhang, Y.P., Murphy, R.W., G10KCOS: Tissue sampling methods and standards for vertebrate genomics. GigaScience **1**(1), 8 (2012).https://doi.org/10.1186/2047-217X-1-8
20. Shaba, E.Y., Tijani, J.O., Jacob, J.O., Suleiman, M.A.T., Mathew, J.T.: Preparation, characterization, adsorptive and antimicrobial properties of $Fe_3O_4@SiO_2@ZnO$ nanocomposite. Colloids Surf. A **686**, 1–17 (2024). https://doi.org/10.1016/j.colsurfa.2024.133190
21. Li, S., Shen, J., Bishop, T.F.A., Viscarra Rossel, R.A.: Assessment of the effect of soil sample preparation, water content and excitation time on proximal X-ray fluorescence sensing. Sensors (Basel, Switzerland) **22**(12), 4572 (2022). https://doi.org/10.3390/s22124572
22. Inobeme, A., Shahnawaz, M., Adetunji, C.O., Mathew, J.T., Adetuyi, B.O., Popoola, O.A., Olaitan, F.Y., Akinbo, O., Kolawole, O.M., Oyewole, O.A., Eniola, K.I.T., Yerima, B.M.: Microplastic as a Pathogenic Vector to Cause Diseases in Marine Biota. In book: Microplastic Pollution. Springer, Singapore, 1, pp. 321–349 (2024). https://doi.org/10.1007/978-981-99-8357-5_19
23. Zhang, K., Banerjee, K.: A review: sample preparation and chromatographic technologies for detection of aflatoxins in foods. Toxins **12**(9), 539 (2020). https://doi.org/10.3390/toxins12090539
24. Mathew, J.T., Musah, M., Azeh, Y., Muhammed, M.: Removal of some toxic metals from pharmaceutical wastewater using geopolymer/Fe_3O_4/ZnOnanocomposite: Isotherm, Kinetics and Thermodynamic Studies. Confluence Univ. J. Sci. Technol. **1**(1), 51–94 (2024). https://doi.org/10.5455/CUJOSTECH.240706
25. Furey, G.N., Tilman, D.: Plant biodiversity and the regeneration of soil fertility. Proc. Natl. Acad. Sci. U.S.A. **118**(49), e2111321118 (2021). https://doi.org/10.1073/pnas.2111321118
26. Dastych, M., Jr., Šenkyřík, M., Dastych, M., Novák, F., Wohl, P., Maňák, J., Kohout, P.: Trace element status (Zinc, Copper, Selenium, Iron, Manganese) in patients with long-term home parenteral nutrition. Ann. Nutr. Metab. **69**(2), 120–124 (2016). https://doi.org/10.1159/000450763
27. Inobeme, A., Adetunji, C.O., Maliki, M., Onyeachu, B.I., Kelani, T., Eziukwu, C.A., Olori, Edu., Mathew, J.T., Bamigboye, M.O.: Strategies to synthesize, advantages, and disadvantages of pharmaceutical nanoparticles. In: book: Nanotechnology for Drug Delivery and Pharmaceuticals. Academic Press is an imprint of Elsevier, pp. 371–402 (2023), https://doi.org/10.1016/B978-0-323-95325-2.00006-7
28. Das, S., Sultana, K.W., Ndhlala, A.R., Mondal, M., Chandra, I.: Heavy metal pollution in the environment and its impact on health: exploring green technology for remediation. Environ. Health Insights **17**, 11786302231201260 (2023). https://doi.org/10.1177/11786302231201259
29. Han, W., Chen, L., Su, X., Liu, D., Jin, T., Shi, S., Li, T., Liu, G.: Effects of soil physicochemical properties on plant species diversity along an elevation gradient over alpine grassland on the qinghai-tibetan plateau. China. Frontiers Plant Sci. **13**, 822268 (2022). https://doi.org/10.3389/fpls.2022.822268

30. Oh, M., Park, H.S., Um, S., Yang, T.J., Kim, S.H.: A comparative phytochemical study of nine Lauraceae species by using chemometric data analysis. PLoS ONE **17**(9), e0273616 (2022). https://doi.org/10.1371/journal.pone.0273616
31. Guccione, P., Lopresti, M., Milanesio, M., Caliandro, R.: Multivariate analysis applications in X-ray diffraction. Crystals **11**, 12 (2021). https://doi.org/10.3390/cryst11010012
32. Inobeme, A., Adetunji, C.O., Mathew, J.T., Ajai, A.I., n Inobeme, J., Oyedolapo, B:. Bioremediation of Chlorinated Compounds. Springer Nature Switzerland AG 2023 M. P. Shah (ed.), Modern Approaches in Waste Bioremediation, pp. 101–115 (2023). https://doi.org/10.1007/978-3-031-24086-7_5
33. Du, R., Luo, L., Hudson, L.G., Nozadi, S., Lewis, J.: An adjusted partial least squares regression framework to utilize additional exposure information in environmental mixture data analysis. J. Appl. Stat. **50**(8), 1790–1811 (2022). https://doi.org/10.1080/02664763.2022.2043254
34. Adetunji, C.O., Akinbo, O. Mathew, J.T., Egbuna, C., Inobeme, A., Olotu, T.M., Ajayi, O.O., Dauda, W., Ghazanfar, S., Ogundolie, F.A., Adetunji, J.B., Adetuyi, B.O., Ajenifujah-Solebo, S.O., Aborode, A.T.: Nanotechnology: history, trends and modern applications. In book: Handbook of Agricultural Biotechnology, John Wiley & Sons, pp. 1-17 (2024). https://doi.org/10.1002/9781119836513.ch1
35. Festa, G., Scatigno, C., Armetta, F., Saladino, M.L., Ciaramitaro, V., Nardo, V.M., Ponterio, R.C.: Chemometric tools to point out benchmarks and chromophores in pigments through spectroscopic data analyses. Molecules (Basel, Switzerland) **27**(1), 163 (2021). https://doi.org/10.3390/molecules27010163
36. Mendoza, L.C., Nolos, R.C., Villaflores, O.B., Apostol, E.M.D., Senoro, D.B.: Detection of heavy metals, their distribution in Tilapia spp., and health risks assessment. Toxics **11**(3), 286 (2023). https://doi.org/10.3390/toxics11030286
37. Asiminicesei, D.M., Fertu, D.I., Gavrilescu, M.: Impact of heavy metal pollution in the environment on the metabolic profile of medicinal plants and their therapeutic potential. Plants (Basel, Switzerland) **13**(6), 913 (2024). https://doi.org/10.3390/plants13060913
38. Manisalidis, I., Stavropoulou, E., Stavropoulos, A., Bezirtzoglou, E.: Environmental and health impacts of air pollution: a review. Front. Public Health **8**, 14 (2020). https://doi.org/10.3389/fpubh.2020.00014
39. Adetunji, C.O., Akinbo, O., Mathew, J.T., Egbuna, C., Inobeme, A., Olotu, T.M., Ajayi, O.O., Dauda, W., Ghazanfar, S., Ogundolie, F.A., Adetunji, J.B., Adetuyi, B.O., Ajenifujah-Solebo, Shakirat, O., Aborode, A.T.: Relevance of Nanomaterials Derived From Medicinal Plants for Marine and Terrestrial Environments: Recent Advances. In book: Handbook of Agricultural Biotechnology. Wile, pp. 1,311–326 (2024). https://doi.org/10.1002/9781119836483.ch12
40. Bala, S., Garg, D., Thirumalesh, B.V., Sharma, M., Sridhar, K., Inbaraj, B.S., Tripathi, M.: Recent strategies for bioremediation of emerging pollutants: a review for a green and sustainable environment. Toxics **10**(8), 484 (2022). https://doi.org/10.3390/toxics10080484
41. Hou, D., Bolan, N.S., Tsang, D.C.W., Kirkham, M.B., O'Connor, D.: Sustainable soil use and management: an interdisciplinary and systematic approach. Sci. Total Environ. **729**, 138961 (2020). https://doi.org/10.1016/j.scitotenv.2020.138961
42. Adetunji, C.O., Mathew, J.T., Inobeme, A., Taiwo, O., Olaniyan, O., Shakira, G., Wadazani, D.: Application of biosensors for detection and monitoring of water quality. In: Book: Organic and Inorganic Materials Based Sensors, John Wiley & Sons, Inc., pp. 1065–1077 (2024), https://doi.org/10.1002/9783527834266.ch46
43. Tavares, T.R., Nunes, L.C., Alves, E.E.N., Almeida, E., Maldaner, L.F., Krug, F.J., Carvalho, H.W.P., Molin, J.P.: Simplifying sample preparation for soil fertility analysis by X-ray fluorescence spectrometry. Sensors (Basel, Switzerland) **19**(23), 5066 (2019). https://doi.org/10.3390/s19235066
44. Imakumbili, M.L.E., Semu, E., Semoka, J.M.R., Abass, A., Mkamilo, G.: Plant tissue analysis as a tool for predicting fertiliser needs for low cyanogenic glucoside levels in cassava roots: an assessment of its possible use. PLoS ONE **15**(2), e0228641 (2020). https://doi.org/10.1371/journal.pone.0228641

45. Mathew, J.T., Adetunji, C.O., Inobeme, A., Musah, M., Shaba, E.Y., Azeh, Y., Abulude, F.O., Mamman, A.: Organic Compounds in Atmospheric Aerosols: Atmospheric Aerosols Properties, pp. 43–59. Sources and Detection. Nova Science Publishers, Inc (2022)
46. Alengebawy, A., Abdelkhalek, S.T., Qureshi, S.R., Wang, M.Q.: Heavy metals and pesticides toxicity in agricultural soil and plants: ecological risks and human health implications. Toxics **9**(3), 42 (2021). https://doi.org/10.3390/toxics9030042
47. Loria, N., Lal, R., Chandra, R.: Handheld in situ methods for soil organic carbon assessment. Sustainability **16**, 5592 (2024). https://doi.org/10.3390/su16135592
48. Acquah, G.E., Hernandez-Allica, J., Thomas, C.L., Dunham, S.J., Towett, E.K., Drake, L.B., Shepherd, K.D., McGrath, S.P., Haefele, S.M.: Portable X-ray fluorescence (pXRF) calibration for analysis of nutrient concentrations and trace element contaminants in fertilisers. PLoS ONE **17**(1), e0262460 (2022). https://doi.org/10.1371/journal.pone.0262460
49. Terzano, R., Denecke, M.A., Falkenberg, G., Miller, B., Paterson, D., Janssens, K.: Recent advances in analysis of trace elements in environmental samples by X-ray based techniques (IUPAC Technical Report). Pure and applied chemistry. Chimie pure et appliquee **91**(6), 1029–1063 (2019). https://doi.org/10.1515/pac-2018-0605
50. Pino, L.K., Searle, B.C., Yang, H.Y., Hoofnagle, A.N., Noble, W.S., MacCoss, M.J.: Matrix-matched calibration curves for assessing analytical figures of merit in quantitative proteomics. J. Proteome Res. **19**(3), 1147–1153 (2020). https://doi.org/10.1021/acs.jproteome.9b00666
51. Feng, X., Zhang, H., Yu, P.: X-ray fluorescence application in food, feed, and agricultural science: a critical review. Crit. Rev. Food Sci. Nutr. **61**(14), 2340–2350 (2021). https://doi.org/10.1080/10408398.2020.1776677
52. Inobeme, A., Mathew, J.T., AJai, I.A., Adetunji, C.O., Mann, A., Jacob, O.J., Shaba, E.Y., Okonkwo, O.S., Edema, G., Efosa, J., Osarenren, E.B., Evbuomwan, L., Tedunjaiye, O.H.: MXenes for catalysis and electrocatalysis. In: Book: MXenes: Next Generation 2D Materials: Fundamentals and Applications. John Wiley & Sons, p. 231 (2024)
53. Moktan, H., Jayarathna, S., Cho, S.H.: Sensitivity enhancement of an experimental benchtop x-ray fluorescence imaging system by deploying a single crystal cadmium telluride detector system optimized for high flux x-ray operations. Nuclear instruments & methods in physics research. Section A, Accelerators, spectrometers, detectors and associated equipment, 1025, 166198 (2022). https://doi.org/10.1016/j.nima.2021.166198
54. Mathew, J.T., Adetunji, C.O., Inobeme, A., Musah, M., Azeh, Y., Otori, A.A., Shaba, E.Y., Mamman, A., Tanko, E.: Removal of Heavy Metals Using Bio-remedial Techniques. Springer Nature Switzerland AG 2023 M. P. Shah (ed.), Modern Approaches in Waste Bioremediation, pp. 117–130 (2023). https://doi.org/10.1007/978-3-031-24086-7_6.
55. González-Domínguez, R., Sayago, A., Fernández-Recamales, Á.: An overview on the application of chemometrics tools in food authenticity and traceability. Foods (Basel, Switzerland) **11**(23), 3940 (2022). https://doi.org/10.3390/foods11233940
56. Inobeme, A., Ajai, A.I., Adetunji, C.O., Maliki, M., Mathew, J.T., Mann, A., Osarenre, J.E., Inobeme, J., Adekoya, M., Eziukwu, C.: Oxidation and advanced oxidation processes in pharmaceutical wastewater treatment. In: Book: Development in Wastewater Treatment Research and Processes: Innovative Trends in Removal of Refractory Pollutants from Pharmaceutical Wastewater, Elsevier Inc., pp. 157–169 (2024). https://doi.org/10.1016/B978-0-323-99278-7.00008-0
57. Radočaj, D., Šiljeg, A., Marinović, R., Jurišić, M.: State of major vegetation indices in precision agriculture studies indexed in web of science: a review. Agriculture **13**, 707 (2023). https://doi.org/10.3390/agriculture13030707
58. Pilakouta, M., Trapali, M., Kallithrakas-Kontos, N.: XRF calibration with low-cost samples and implementation for quantification of inorganic elements in lipsticks. MethodsX **12**, 102704 (2024). https://doi.org/10.1016/j.mex.2024.102704
59. Ungureanu, E.L., Mocanu, A.L., Stroe, C.A., Duță, D.E., Mustățea, G.: Assessing health risks associated with heavy metals in food: a bibliometric analysis. Foods (Basel, Switzerland) **12**(21), 3974 (2023). https://doi.org/10.3390/foods12213974

60. Abimbola, O.F., Adetunji, C.O., Dare, B.J., Mathew, J.T., Inobeme, A., Ghazanfar, S., Olaniyan, O.T, Ajiboye, M.D., Dauda, W.: Packaging of animal-based food products: the need for intelligent freshness sensors. In: Book: Sensing and Artificial Intelligence Solutions for Food Manufacturing. CRC Press, pp. 89–102 (2023). https://doi.org/10.1201/9781003207955-6
61. Mathew, J.T., Musah, M., Azeh, Y., Muhammed, M.: Development of Fe_3O_4 nanoparticles for the removal of some toxic metals from pharmaceutical wastewater. Caliphate J. Sci. Technol. (CaJoST) **6**(1), 26–34 (2024). https://doi.org/10.4314/cajost.v6i1.4

X-Ray Fluorescence in Analyzing Agricultural Contamination via Chemometric Tools

10

John Tsado Mathew, Abel Inobeme, Yakubu Azeh, Musah Monday, Elijah Yanda Shaba, Tanko Ezekiel, Etsuyankpa Muhammad Bini, Musa Safiyano Tanko, Jibrin Mohammed Ndejiko, Jonathan Hussaini, Charles Oluwaseun Adetunji, Muhammad Aishetu Ibrahim, and Amos Mamman

Abstract

The increasing concern over agricultural contamination, driven by the rapid expansion of industrial activities and the widespread use of agrochemicals, has necessitated the development of reliable and efficient analytical techniques for monitoring and assessing contaminants in agricultural environments. X-Ray fluorescence (XRF) spectroscopy has emerged as a valuable tool in this context, offering a non-destructive, rapid, and cost-effective method for the analysis of a wide range of elements in various matrices, including soil, plant tissues, and agricultural products. This review explores the application of XRF in agriculture, with a particular focus on its use for measuring contamination and the integration of chemometric tools to enhance the accuracy and interpretability of the results. Chemometric tools have been increasingly applied in conjunction

J. T. Mathew (✉) · Y. Azeh · M. Monday · J. Hussaini · M. A. Ibrahim · A. Mamman
Department of Chemistry, Ibrahim Badamasi Babangida University, Lapai, Nigeria
e-mail: jmathew@ibbu.edu.ng

A. Inobeme
Department of Chemistry, Edo State University Uzairue, Auchi, Edo State, Nigeria

E. Y. Shaba · T. Ezekiel
Department of Chemistry, Federal University of Technology Minna, Minna, Nigeria

E. M. Bini · M. S. Tanko
Department of Chemistry, Federal University of Lafia, Lafia, Nasarawa State, Nigeria

J. M. Ndejiko
Department of Microbiology, Ibrahim Badamasi Babangida University, Lapai, Nigeria

C. O. Adetunji
Applied Microbiology, Biotechnology and Nanotechnology Laboratory, Department of Microbiology, Edo State University Uzairue, Auchi, Edo State, Nigeria

V. K. Singh (ed.), *X-Ray Fluorescence Spectroscopy and Chemometrics*,
https://doi.org/10.1007/978-3-031-98375-7_10

with XRF analysis. Chemometrics involves the application of mathematical and statistical methods to optimize experimental design, enhance data interpretation, and improve the accuracy of quantitative analysis. The integration of chemometric methods with XRF has proven to be particularly effective in agricultural contamination studies. By employing these tools, researchers can develop robust calibration models that account for the variability inherent in agricultural samples, thereby improving the reliability of contaminant quantification.

10.1 Introduction

Agriculture is the backbone of human civilization, providing the primary means of food production, raw materials, and economic stability for billions of people worldwide. However, modern agricultural practices, while significantly boosting productivity, have also led to various environmental challenges, particularly the contamination of soils, water sources, and crops. Contaminants such as heavy metals, pesticide residues, and other pollutants can accumulate in the environment, posing serious risks to human health, ecosystem balance, and food security. These concerns have driven the need for reliable, efficient, and cost-effective methods to monitor and assess contamination in agricultural settings [1, 2].

Among the various analytical techniques available, X-Ray fluorescence (XRF) spectroscopy has emerged as a powerful tool for the elemental analysis of agricultural samples. XRF is a non-destructive technique that provides rapid, multi-elemental analysis with minimal sample preparation. It operates by exposing a sample to high-energy X-Rays, which cause the atoms in the sample to emit secondary (or fluorescent) X-Rays. These emitted X-Rays are characteristic of the elements present in the sample, allowing for qualitative and quantitative analysis of its elemental composition [3].

The application of XRF in agriculture is particularly relevant for detecting and quantifying heavy metals such as lead (Pb), cadmium (Cd), arsenic (As), and mercury (Hg), which are common contaminants in soils and crops. These metals can originate from various sources, including industrial emissions, the use of contaminated water for irrigation, and the application of certain fertilizers and pesticides. Once these metals enter the food chain, they can have detrimental effects on human health, causing conditions such as neurological disorders, kidney damage, and cancer. Therefore, accurate monitoring of these contaminants is crucial for ensuring food safety and environmental protection [4].

In addition to heavy metals, XRF can also be used to assess the nutrient content of soils and crops, providing valuable information for optimizing fertilizer use and improving crop yields. For instance, elements such as potassium (K), phosphorus (P), and calcium (Ca) are essential for plant growth, and their concentrations in soils and plant tissues can be effectively measured using XRF. This capability makes XRF an important tool not only for contamination analysis but also for general agricultural management and sustainability. Despite its advantages, the application of XRF in agriculture is not without challenges. Agricultural samples, such as soils and plant tissues, are often heterogeneous in nature, with varying matrix compositions, moisture content, and particle sizes. These variations can affect the accuracy and precision of XRF measurements, leading to potential errors in the quantification of contaminants and nutrients. To overcome these challenges, the integration of chemometric tools with XRF analysis has gained prominence [5].

Chemometrics involves the application of statistical and mathematical techniques to optimize experimental design, enhance data interpretation, and improve the accuracy of analytical results. In the context of XRF, chemometric methods such as Principal Component Analysis (PCA), Partial Least Squares (PLS) regression, and multivariate calibration are used to correct for matrix effects, normalize data, and develop robust calibration models. These techniques allow researchers to account for the variability inherent in agricultural samples, resulting in more reliable and accurate measurements. The integration of chemometrics with XRF has opened new possibilities for agricultural contamination studies. For example, the combination of XRF and chemometrics has been successfully used to detect and quantify trace levels of heavy metals in soils and crops, even in the presence of complex matrices. Moreover, chemometric-enhanced XRF analysis has been applied to assess the nutrient status of soils and crops, providing insights into soil fertility and crop health [6, 7].

10.2 Overview of Analytical Techniques in Agriculture

The advancement of agricultural science relies heavily on the ability to accurately analyze the composition of soils, plants, and agricultural products. As agricultural practices evolve to meet the growing demands of a burgeoning global population, so too must the methods used to ensure the safety, quality, and sustainability of agricultural production. A wide range of analytical techniques have been developed and refined over the years to address these needs, each with its own set of advantages and limitations. This overview highlights some of the most commonly used analytical techniques in agriculture, emphasizing their applications, benefits, and challenges [8].

10.2.1 Spectroscopic Techniques

Spectroscopic methods are widely used in agricultural analysis due to their ability to provide detailed information about the chemical composition of samples. Among the most commonly used spectroscopic techniques are [9].

10.2.1.1 X-Ray Fluorescence (XRF) Spectroscopy

XRF is a non-destructive analytical technique that measures the elemental composition of a sample by detecting the characteristic secondary X-Rays emitted by elements when exposed to high-energy X-Rays. XRF is particularly useful for detecting heavy metals in soils and plants, analyzing nutrient content, and monitoring contamination. Its rapid analysis time, minimal sample preparation, and ability to analyze multiple elements simultaneously make it a popular choice in agricultural studies [10, 11].

10.2.1.2 Atomic Absorption Spectroscopy (AAS)

AAS is a technique used to measure the concentration of specific elements in a sample by detecting the absorption of light by free atoms. It is highly sensitive and is often used for the analysis of heavy metals in agricultural samples. However, AAS typically analyzes one element at a time, making it less efficient than other multi-elemental techniques like XRF and ICP-OES [3].

10.2.1.3 Near-Infrared Spectroscopy (NIRS)

NIRS is a non-destructive technique that measures the absorption of near-infrared light by molecular bonds in organic compounds. It is widely used in agriculture to assess the chemical composition of soils, crops, and animal feed. NIRS is particularly valued for its speed and ability to analyze large sample sets with minimal preparation [12].

10.2.1.4 Chromatographic Techniques

Chromatography is essential for separating and analyzing complex mixtures of compounds, particularly in the context of pesticide residue analysis and the detection of organic contaminants [13].

10.2.1.5 Gas Chromatography (GC)

GC is used to separate volatile compounds in a sample, which are then detected by various detectors, such as a mass spectrometer (GC–MS) or a flame ionization detector (GC-FID). GC is highly effective for analyzing pesticide residues, volatile organic compounds, and fatty acids in agricultural products [14, 15].

10.2.1.6 High-Performance Liquid Chromatography (HPLC)

HPLC is similar to GC but is used for the separation of non-volatile and thermally labile compounds. It is widely used in agriculture for the analysis of plant secondary metabolites, mycotoxins, and other organic contaminants. When coupled with detectors like mass spectrometry (HPLC–MS), it provides high sensitivity and specificity [16].

10.2.1.7 Mass Spectrometry

Mass spectrometry (MS) is a highly sensitive technique that measures the mass-to-charge ratio of ions to identify and quantify compounds in a sample:

- ***Inductively Coupled Plasma Mass Spectrometry (ICP-MS)***

ICP-MS combines the capabilities of ICP with mass spectrometry, offering unparalleled sensitivity for detecting trace elements and isotopes in agricultural samples. It is particularly useful for analyzing heavy metals and rare earth elements in soils and plants [17].

- ***Liquid Chromatography-Mass Spectrometry (LC–MS)***

LC–MS is a powerful tool for identifying and quantifying complex organic molecules, such as pesticides, hormones, and plant metabolites. It is widely used in residue analysis and the study of plant biochemistry [18].

10.2.1.8 Electrochemical Techniques

Electrochemical methods, including ion-selective electrodes and voltammetry, are used to measure the concentration of specific ions or molecules in agricultural samples. These techniques are particularly useful for soil and water analysis, where they can measure parameters such as pH, nutrient levels, and salinity [19].

10.2.1.9 Microbiological and Molecular Techniques

In addition to chemical analysis, microbiological and molecular techniques play a crucial role in agricultural research:

- ***Polymerase Chain Reaction (PCR)***

PCR is a molecular technique used to amplify and detect specific DNA sequences, allowing for the identification of pathogens, genetically modified organisms (GMOs), and microbial communities in agricultural systems [20].

The selection of an analytical technique in agriculture depends on the specific objectives of the study, the nature of the samples, and the required sensitivity and accuracy. While spectroscopic and chromatographic methods are widely used for their versatility and reliability, other techniques, including electrochemical and molecular methods, offer valuable insights into specific aspects of agricultural systems. As agriculture continues to evolve, so too will the analytical techniques used to support sustainable and productive agricultural practices (Fig. 10.1) [3].

10.3 Role of Chemometrics in Agricultural Analysis

Agriculture is a complex system where multiple variables, such as soil composition, climate conditions, crop genetics, and management practices, interact to influence productivity, quality, and sustainability. Traditional analytical methods, while effective at measuring individual parameters, often struggle to handle the complexity and variability inherent in agricultural systems. This is where chemometrics plays a crucial role. Chemometrics, the application of mathematical and

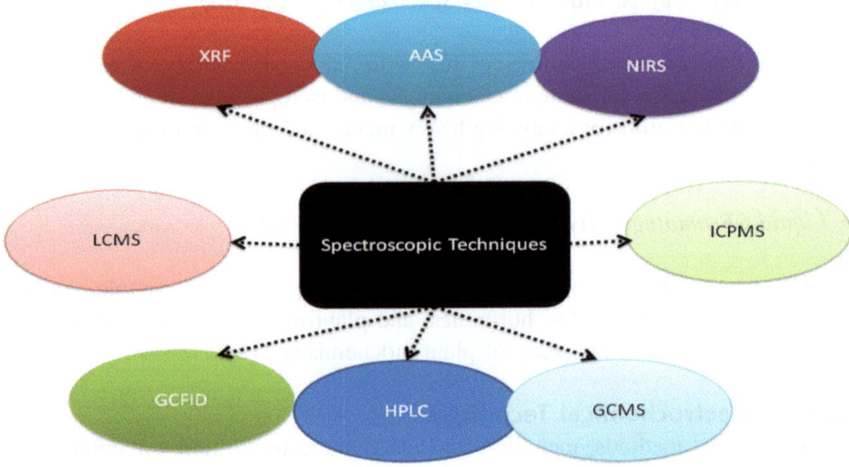

Fig. 10.1 Spectroscopic techniques

statistical methods to design experiments, optimize processes, and analyze chemical data, has become an indispensable tool in agricultural analysis. By integrating chemometric techniques with conventional analytical methods, researchers and practitioners can extract meaningful information from complex datasets, leading to better decision-making and improved agricultural outcomes [21].

- ***Enhancing Data Interpretation***

Agricultural data is often multidimensional, involving multiple variables that need to be analyzed simultaneously. Chemometrics provides the tools to handle this complexity, enabling the extraction of relevant information from large datasets. For example, Principal Component Analysis (PCA) is a widely used chemometric technique that reduces the dimensionality of data while retaining the most important variance. In agriculture, PCA can be applied to datasets containing information on soil properties, crop yields, and environmental factors to identify patterns and relationships that may not be apparent through univariate analysis. By highlighting the key variables that contribute to variability in the data, PCA helps in identifying the factors that most significantly impact agricultural outcomes, such as soil fertility, crop health, or pest prevalence [22, 23].

- ***Optimizing Experimental Design***

Designing experiments that efficiently and effectively test hypotheses is a critical aspect of agricultural research. Chemometrics offers tools such as Design of Experiments (DOE) and Response Surface Methodology (RSM) to optimize experimental design. DOE allows researchers to systematically vary multiple factors simultaneously, rather than testing one factor at a time, leading to a more comprehensive understanding of how different variables interact. In agricultural studies, this might involve varying levels of fertilizers, irrigation, and planting density to determine the optimal conditions for crop growth. RSM further refines this approach by modeling the relationships between the factors and the response variables, allowing for the identification of optimal conditions with fewer experimental runs. These methods save time and resources while providing more robust and reliable results [24].

- ***Improving Quantitative Analysis***

Quantitative analysis in agriculture often involves the measurement of nutrient concentrations, contaminant levels, or other chemical parameters in soils, plants, and agricultural products. The variability in sample matrices, such as moisture content or organic matter, can complicate these measurements. Chemometric techniques, particularly multivariate calibration methods like Partial Least Squares (PLS) regression, are used to account for these matrix effects and improve the accuracy of quantitative analysis. PLS regression can model the relationship between the measured spectral data and the concentration of target analytes, even in the presence of interfering substances. This approach is widely used in techniques like Near-Infrared Spectroscopy (NIRS) and X-Ray Fluorescence (XRF) spectroscopy, where chemometrics enables more precise quantification of elements and compounds in heterogeneous agricultural samples [25].

- ***Enhancing Contamination Detection***

In agricultural settings, detecting contaminants such as heavy metals, pesticides, and mycotoxins is essential for ensuring food safety and environmental protection. Chemometrics enhances the sensitivity and specificity of contamination detection by improving the interpretation of complex analytical signals. For instance, in XRF or ICP-OES analysis, chemometric tools can be used to deconvolute overlapping spectral lines, correct for background noise, and compensate for matrix effects, leading to more accurate identification and quantification of contaminants. Moreover, chemometrics can be used to develop predictive models that relate environmental or management practices to contamination levels, enabling proactive risk management in agricultural production [26].

- ***Advancing Genomic and Metabolomic Studies***

The integration of chemometrics with genomic and metabolomic data is opening new frontiers in agricultural research. By analyzing the relationships between genetic markers, metabolite profiles, and phenotypic traits, chemometrics helps to identify biomarkers associated with desirable traits such as disease resistance, drought tolerance, or enhanced nutritional content. This information can be used in breeding programs to develop new crop varieties with improved characteristics, contributing to food security and resilience in the face of climate change [27, 28].

10.4 Principles of XRF Spectroscopy

XRF spectroscopy is a powerful analytical technique widely used for elemental analysis in various fields, including agriculture, environmental science, geology, and material science. It is a non-destructive method that provides qualitative and quantitative information about the elemental composition of a sample. The underlying principles of XRF spectroscopy are based on the interaction between X-Rays and the atoms within a sample, leading to the emission of characteristic fluorescent X-Rays. Understanding these principles is essential to appreciate how XRF works and its applications [29].

- ***Detection and Measurement***

The XRF instrument detects the emitted fluorescent X-Rays and measures their energy and intensity. The energy of the X-Rays corresponds to the specific elements present in the sample, while the intensity (or count rate) of the X-Rays provides quantitative information about the concentration of those elements [30].

- ***Wavelength Dispersive XRF (WDXRF)***

In WDXRF, the emitted X-Rays are dispersed by a crystal before they reach the detector, which measures their wavelengths. WDXRF generally offers higher resolution and greater precision than EDXRF but is more complex and slower.

- ***Sample Preparation and Matrix Effects***

Sample preparation plays a critical role in XRF spectroscopy because the accuracy of the results can be influenced by factors such as particle size, surface roughness, and sample homogeneity. Proper sample preparation ensures that the X-Ray beam interacts consistently with the sample, leading to more reliable measurements [31, 32].

- ***Matrix Effects***

It refers to the influence of other elements and the overall composition of the sample, can also affect the accuracy of XRF analysis. These effects can cause either enhancement or absorption of the emitted X-Rays, leading to inaccurate quantification. To minimize matrix effects, calibration curves using standard reference materials with known compositions are often employed. Additionally, modern

XRF instruments incorporate software with algorithms to correct for these effects [33].

- ***Quantitative Analysis***

Quantitative XRF analysis involves measuring the intensity of the characteristic X-Rays and relating it to the concentration of the element in the sample. This relationship is usually linear, allowing for the creation of calibration curves that can be used to determine unknown concentrations in a sample [34].

- ***Calibration***

Accurate calibration is crucial for quantifying elements in a sample. Calibration involves measuring the X-Ray intensities of standard samples with known concentrations and constructing a calibration curve that correlates intensity with concentration [35].

- ***Detection Limits***

The detection limit of XRF refers to the lowest concentration of an element that can be reliably detected. Detection limits vary depending on the element, the sample matrix, and the type of XRF instrument used [10, 36].

- ***Interference***

Spectral overlaps, where the X-Ray lines of different elements are close to each other, can cause interference. Advanced software and correction algorithms help mitigate these interferences [37].

10.5 Advantages and Limitations of XRF in Agriculture

Table 10.1 provides a clear comparison of the advantages and limitations of XRF, highlighting its impact on agricultural practices.

XRF spectroscopy is an analytical technique that has gained widespread use in agriculture due to its ability to quickly and accurately measure the elemental composition of various samples. However, like any analytical method, it comes with both advantages and limitations that must be considered in agricultural applications. One of the most significant advantages of XRF is its non-destructive nature. Samples remain intact after analysis, allowing for further testing or archiving. This is particularly useful in agriculture where preserving samples for future reference or additional testing is important [38].

XRF can simultaneously detect and quantify a wide range of elements, from heavy metals to essential nutrients. This capability makes it an efficient tool for comprehensive soil and plant analysis, enabling quick assessment of nutrient status

Table 10.1 Advantages and limitations of XRF in agriculture

Aspects	Advantage	Limitations	Impact in agriculture	Refs.
Non-destructive analysis	XRF is a non-destructive technique, allowing samples to be analyzed without being altered or consumed	Cannot provide information on chemical bonding or molecular structure	Enables repeated testing and preservation of samples, which is valuable for archival purposes and quality control	[43, 44]
Multi-elemental analysis	Capable of detecting multiple elements simultaneously, making it efficient for analyzing complex agricultural samples like soil and plant tissues	Less sensitive to light elements such as carbon, nitrogen, and oxygen, which are critical in organic compounds	Facilitates comprehensive nutrient and contaminant profiling in soils, plants, and fertilizers	[45]
Minimal sample preparation	Requires little to no sample preparation, reducing time and labor costs associated with the analysis	Sample homogeneity and surface texture can affect the accuracy of the results, requiring careful calibration	Increases throughput in routine agricultural testing and rapid decision-making	[46]
Field Portability	Portable XRF devices are available, allowing for on-site analysis and immediate results in the field	Portable XRF devices may have lower sensitivity and resolution compared to laboratory-based systems, limiting their use for trace element analysis	Supports precision agriculture by providing real-time data for site-specific management practices	[47, 48]

and potential contaminants. XRF requires little to no sample preparation, which reduces the time and labor involved in the analysis process. This is advantageous in agricultural settings where large numbers of samples need to be processed quickly, such as in soil testing laboratories. Portable XRF devices are available, allowing for on-site analysis in agricultural fields. This portability facilitates real-time decision-making, supporting practices like precision agriculture where timely information on soil and crop conditions is crucial [39].

XRF is less sensitive to light elements such as carbon, nitrogen, and oxygen, which are critical in organic materials and soil composition. This limitation can restrict the ability to fully analyze all components of agricultural samples, particularly those related to organic matter and nutrients essential for plant growth. The accuracy of XRF results can be influenced by the homogeneity and surface texture of the sample. Inhomogeneous samples or those with uneven surfaces may produce inaccurate readings, necessitating careful calibration and, in some cases, additional sample preparation [40, 41].

While XRF is effective at detecting a broad range of elements, it may struggle with elements present in very low concentrations, particularly with portable devices. This limitation can be a challenge when precise quantification of trace elements is required. XRF provides information on elemental composition but not on the chemical forms or compounds in which those elements exist. This is a significant limitation in agricultural research, where understanding the bioavailability of nutrients or the speciation of contaminants is important [42].

10.6 Chemometric Tools in XRF Analysis

XRF spectroscopy is widely recognized for its ability to determine the elemental composition of various materials, including agricultural samples such as soil, plants, and fertilizers. However, the complexity of agricultural matrices and the need for precise and reliable data have driven the integration of chemometric tools with XRF analysis. Chemometrics, the application of statistical and mathematical methods to chemical data, enhances the interpretability and accuracy of XRF results, particularly when dealing with complex, multi-elemental datasets. This integration is crucial for making informed decisions in agricultural research and practice. Before performing chemometric analysis, it is essential to preprocess the XRF data to ensure accuracy and consistency. Raw XRF data can be influenced by various factors such as sample heterogeneity, instrumental noise, and matrix effects. Chemometric tools are employed to preprocess the data through techniques like normalization, baseline correction, and smoothing [49].

Principal component analysis (PCA) is a fundamental chemometric tool used in XRF analysis to reduce the dimensionality of complex datasets while retaining most of the variance. In agricultural studies, PCA is often used to identify patterns and relationships among different elements in a sample. For example, in soil analysis, PCA can help to differentiate between soils from different locations based on their elemental composition, or to identify the key elements that contribute to soil fertility. PCA works by transforming the original variables (i.e., the elemental concentrations) into a new set of uncorrelated variables called principal components. These components capture the maximum variance in the data, allowing researchers to visualize and interpret complex relationships in a simpler, lower-dimensional space. This visualization can reveal clusters, trends, or outliers that are not immediately apparent in the raw data [50].

Partial least squares regression (PLSR) is another widely used chemometric tool in XRF analysis, particularly for quantitative analysis. PLSR is a multivariate regression technique that models the relationship between a set of independent variables (XRF spectral data) and dependent variables (elemental concentrations). Unlike traditional regression methods, PLSR can handle collinear and noisy data, making it well-suited for complex agricultural matrices where multiple elements may be present at varying concentrations. In agricultural applications, PLSR is often used to develop calibration models that predict the concentration of specific elements in a sample based on its XRF spectrum. These models are particularly valuable when analyzing heterogeneous samples, such as soils or plant tissues, where matrix effects can complicate direct interpretation of the spectra. PLSR can also be used to correct for matrix effects, improving the accuracy and robustness of the XRF analysis [51].

Cluster analysis is a chemometric technique used to group similar samples based on their elemental composition. In XRF analysis, cluster analysis can help to identify natural groupings within a dataset, such as different types of soils, crops, or fertilizers. This technique is particularly useful in agricultural studies where samples from different regions, treatments, or conditions need to be compared [52].

By grouping samples with similar elemental profiles, cluster analysis can provide insights into the underlying factors that influence the composition, such as soil type, environmental conditions, or agricultural practices. This information can be used to tailor management practices to specific conditions, improving crop yield and soil health. Discriminant analysis is another chemometric tool used in XRF analysis to classify samples into predefined categories. In agricultural research, discriminant analysis can be used to distinguish between different soil types, crop varieties, or contamination levels based on their XRF spectra. This technique is particularly useful when the goal is to develop predictive models that can classify new samples based on their elemental composition [53].

For example, discriminant analysis can be used to develop a model that differentiates between soils with different nutrient levels or to classify crops based on their mineral content. By applying this model to new samples, researchers can quickly and accurately determine the category to which the sample belongs, aiding in decision-making processes such as soil amendment or crop selection [54, 55].

Multivariate curve resolution (MCR) is a chemometric technique used to resolve overlapping peaks in XRF spectra, which is common when analyzing complex agricultural samples. MCR works by decomposing the spectral data into a set of pure component spectra and their corresponding concentrations, allowing for the accurate identification and quantification of elements even in the presence of spectral interference. In agricultural XRF analysis, MCR can be used to separate the contributions of different elements in a sample where the spectral lines overlap, such as in soils with high levels of iron and manganese. By resolving these overlaps, MCR improves the accuracy of the analysis and ensures that all relevant elements are correctly identified and quan [56].

10.7 Application of XRF in Measuring Contamination in Agriculture

Table 10.2 summarizes the key aspects of using XRF for measuring contamination in agriculture, highlighting both the advantages and limitations associated with this technique.

XRF spectroscopy is a powerful analytical technique widely used in agriculture to measure contamination by detecting the elemental composition of soil, water, and crops. Its application is particularly valuable in monitoring heavy metals, pesticide residues, nutrient imbalances, and soil salinity, which are critical factors in agricultural productivity and environmental health [57].

One of the primary uses of XRF in agriculture is the detection of heavy metals such as lead, cadmium, and arsenic. These toxic elements can accumulate in soil and crops, posing significant health risks to both humans and animals. XRF enables rapid, on-site analysis of these contaminants with minimal sample preparation, making it a convenient tool for regular monitoring. XRF is also employed to assess pesticide residues by detecting inorganic elements associated with certain pesticides. While XRF cannot directly identify organic pesticide compounds, it can indirectly measure contamination by identifying elemental markers, helping

Table 10.2 Application of XRF in measuring contamination in agriculture

Contaminant type	XRF application	Advantages	Limitations	Refs.
Heavy metals	Detection of heavy metals like lead, cadmium, and arsenic in soil, water, and crops	Rapid, on-site analysis; minimal sample preparation; high sensitivity for heavy elements	Limited detection of elements at very low concentrations; potential interference from matrix effects	[45]
Pesticides residues	Indirect measurement by detecting elemental markers in pesticide residues on crops and soils	Non-destructive analysis; can detect multiple elements simultaneously	XRF cannot directly identify organic compounds; limited to detecting inorganic elements associated with pesticides	[59]
Nutrient imbalance	Measurement of essential and toxic nutrient levels in soil and plant tissues (e.g., zinc, copper)	Simultaneous multi-element analysis; supports precision agriculture by providing rapid nutrient profiling	Less sensitive to light elements like nitrogen and carbon; may require calibration for accurate results	[60]
Soil salinity	Assessment of soil salinity through detection of elements like sodium, chlorine, and magnesium	Quick field assessment; aids in managing soil salinity and preventing crop damage	May struggle with low concentrations; interference from overlapping peaks in complex matrices	[61]

farmers manage pesticide application and reduce environmental impact. In addition to contaminants, XRF is used to measure nutrient levels in soil and plant tissues, including essential elements like zinc and copper. This capability supports precision agriculture, where accurate nutrient profiling is crucial for optimizing fertilizer use and improving crop yield. However, XRF is less sensitive to lighter elements such as nitrogen and carbon, which are also important for plant growth [58].

Finally, XRF can assess soil salinity by detecting elements like sodium and chlorine. High soil salinity can severely impact crop growth, and XRF provides a quick way to monitor salinity levels, allowing for timely intervention. Despite its advantages, XRF has limitations, such as reduced sensitivity to light elements and potential interference from complex sample matrices. However, it remains a valuable tool in managing agricultural contamination, contributing to safer and more sustainable farming practices [57].

10.8 Integration of XRF with Chemometrics for Contamination Detection

The integration of XRF spectroscopy with chemometric tools represents a significant advancement in the field of contamination detection in agriculture. XRF spectroscopy is well-regarded for its ability to rapidly and non-destructively determine the elemental composition of agricultural samples, including soil, water, and plant tissues. However, the complexity of these samples and the need for precise, reliable data necessitate advanced data analysis techniques. Chemometrics, the application of statistical and mathematical methods to chemical data, enhances the interpretability and accuracy of XRF results, particularly when dealing with complex matrices and subtle variations in contamination levels. This integration is vital for making informed decisions in agricultural management and environmental protection [61].

Normalization techniques adjust the XRF data to a common scale, which is essential when comparing samples with different elemental concentrations. This step is crucial in contamination detection, where the concentration of contaminants can vary widely across different samples or locations. XRF spectra can contain background noise that obscures the true elemental peaks. Baseline correction methods are used to subtract this noise, thereby improving the clarity of the spectral data and enhancing the detection of contaminants, even at low concentrations. Smoothing algorithms reduce random noise in the XRF spectra without distorting the signal, making it easier to identify and quantify contaminants accurately [10].

PCA is a widely used chemometric tool that reduces the dimensionality of the XRF data while retaining most of the variance. By transforming the original variables (elemental concentrations) into a set of uncorrelated principal components, PCA helps to visualize patterns and relationships within the data. In contamination detection, PCA can be used to differentiate between contaminated and uncontaminated samples or to identify clusters of samples with similar contamination profiles [62, 63].

The integration of XRF spectroscopy with chemometric tools significantly enhances the capabilities of XRF in contamination detection within agriculture. Chemometrics improves the quality of XRF data through preprocessing, enables sophisticated pattern recognition and quantitative analysis, and supports the development of classification and predictive models. By combining the strengths of XRF and chemometrics, researchers and agricultural professionals can achieve more accurate, reliable, and interpretable results, leading to better-informed decisions in contamination management and more sustainable agricultural practices. This integration represents a powerful approach to addressing the complex challenges of contamination in agriculture, ensuring the safety and productivity of agricultural systems [64].

10.9 Comparative Analysis: XRF Versus Other Analytical Techniques

Table 10.3 provides a summary of the key advantages and disadvantages of XRF compared to other common analytical techniques used in agriculture and environmental studies.

XRF is a non-destructive, rapid analytical technique ideal for on-site, multi-element detection with minimal sample preparation. However, it has limitations in sensitivity for light elements and trace analysis. In contrast, Inductively Coupled Plasma Mass Spectrometry (ICP-MS) offers high sensitivity for trace elements but requires extensive preparation and is more costly. AAS is simple and cost-effective for specific elements but slower and less versatile. FTIR excels in identifying organic compounds but is limited in elemental analysis. Each technique has distinct advantages depending on the analytical requirements [40].

10.10 Challenges and Future Prospects

XRF spectroscopy has become an invaluable tool in agricultural contamination analysis due to its ability to provide rapid, non-destructive, and multi-element detection. However, like any technology, XRF faces several challenges that limit its effectiveness in certain applications. One of the primary challenges with XRF is its limited sensitivity to light elements such as hydrogen, carbon, nitrogen, and oxygen, which are crucial in agricultural contexts. While XRF excels at detecting heavier elements like metals, its performance with lighter elements is less reliable. This limitation hampers its ability to fully characterize organic contamination or nutrient profiles in agricultural samples [5].

Ongoing advancements in XRF detector technology hold promise for overcoming some of the current limitations. Improved detector resolution and sensitivity

Table 10.3 Comparative analysis: XRF versus other analytical techniques

Analytical technique	Advantages	Disadvantages	Refs.
X-Ray fluorescence (XRF)	Non-destructive; rapid, on-site analysis; minimal sample preparation; capable of multi-element detection	Limited sensitivity to light elements (e.g., H, C, N); potential matrix effects; less accurate for trace elements	[47, 65]
Inductively coupled plasma mass spectrometry (ICP-MS)	Highly sensitive; capable of detecting trace elements and isotopes; wide dynamic range	Requires extensive sample preparation; more expensive and time-consuming; destructive analysis	[66]
Atomic absorption spectroscopy (AAS)	High sensitivity for specific elements; relatively simple and cost-effective for targeted analysis	Single-element analysis per run; slower compared to multi-element techniques; requires sample digestion	[66]
Fourier transform infrared spectroscopy (FTIR)	Effective for identifying organic compounds; non-destructive; minimal sample preparation	Limited to detecting functional groups; less effective for elemental analysis; lower sensitivity for metals	[67]
Ion chromatography (IC)	Excellent for analyzing anions and cations in solution; high precision and accuracy	Requires liquid samples; involves complex sample preparation; not suitable for solid samples directly	[68]
Neutron activation analysis (NAA)	Highly sensitive and accurate; non-destructive; capable of multi-element analysis	Requires a nuclear reactor; expensive and less accessible; complex data interpretation	[69]

could enhance the detection of light elements and trace contaminants. Silicon Drift Detectors (SDDs) and other advanced detectors are being developed to offer better energy resolution, allowing for more accurate analysis even in the presence of overlapping peaks [70].

The development of portable XRF devices is expanding the potential applications of XRF in agriculture. Portable XRF analyzers allow for in-situ analysis directly in the field, providing real-time data that can be used to make immediate decisions regarding contamination management. As these devices become more sophisticated, their accuracy and versatility will likely improve, making them indispensable tools for precision agriculture. Combining XRF with other analytical techniques, such as inductively coupled plasma mass spectrometry (ICP-MS) or near-infrared spectroscopy (NIR), could provide a more comprehensive analysis of agricultural samples. For instance, XRF could be used for rapid screening, followed by more detailed analysis using ICP-MS for elements that XRF struggles to detect. This integrated approach could offer a balanced solution, leveraging the strengths of multiple techniques [71].

The integration of artificial intelligence (AI) and machine learning (ML) into XRF analysis is a promising area for future development. AI and ML algorithms can be trained to recognize patterns in XRF spectra, improving the interpretation of complex data and reducing the impact of matrix effects and overlapping peaks. These technologies could also streamline the calibration process, making XRF more accessible for routine use in agricultural monitoring. As the agricultural sector increasingly adopts sustainable practices, XRF will play a crucial role in monitoring soil health and contamination in a non-invasive manner. The shift towards organic farming and the reduced use of chemical fertilizers and pesticides will necessitate precise monitoring techniques to ensure compliance with environmental standards. XRF, with its non-destructive nature and rapid analysis capabilities, is well-suited to support these sustainable agricultural practices [72].

Future research may explore expanding the scope of XRF applications in agriculture, such as monitoring nutrient uptake in plants or assessing the bioavailability of contaminants. This could involve developing new calibration methods or combining XRF with imaging techniques to provide spatial distribution maps of elements within a sample. Such innovations could offer deeper insights into plant-soil interactions and help optimize agricultural practices [73].

10.11 Conclusion

The integration of XRF spectroscopy into agricultural contamination analysis represents a significant advancement in the field, offering rapid, non-destructive, and multi-element detection capabilities. Despite its limitations, such as reduced sensitivity to light elements, matrix effects, and the need for precise calibration, XRF remains a valuable tool in monitoring soil health, detecting contaminants, and supporting sustainable agricultural practices. Future prospects, including advancements in detector technology, portable applications, and AI-driven data analysis, promise to address these challenges and expand XRF's applicability. As these innovations unfold, XRF is set to play an increasingly critical role in ensuring agricultural productivity, environmental safety, and the overall sustainability of farming practices. Through continued research and technological improvements, XRF will likely become an indispensable tool in the ongoing effort to balance agricultural demands with environmental stewardship.

References

1. Pawlak, K., Kołodziejczak, M.: The role of agriculture in ensuring food security in developing countries: considerations in the context of the problem of sustainable food production. Sustainability **12**, 5488 (2020). https://doi.org/10.3390/su12135488
2. Mathew, J.T., Musah, M., Azeh, Y., Muhammed, M.: Kinetic study of heavy metals removal from pharmaceutical wastewater using geopolymer/Fe_3O_4 nanocomposite. Bima J. Sci. Technol. **7**(4), 152–163 (2023b). https://doi.org/10.56892/bima.v7i4.539

3. El Hosry, L., Sok, N., Richa, R., Al Mashtoub, L., Cayot, P., Bou-Maroun, E.: Sample preparation and analytical techniques in the determination of trace elements in food: a review. Foods **12**, 895 (2023). https://doi.org/10.3390/foods12040895
4. Răcuşan Ghircoiaş, O., Tănăselia, C., Chintoanu, M., Crişan, I., Hoble, A., Ştefan, R., Dîrja, M. (2023). Relevance of soil heavy metal XRF screening for quality and landscaping of public playgrounds. Toxics **11**(6), 530. https://doi.org/10.3390/toxics11060530
5. Acquah, G.E., Hernandez-Allica, J., Thomas, C.L., Dunham, S.J., Towett, E.K., Drake, L.B., Shepherd, K.D., McGrath, S.P., Haefele, S.M.: Portable X-Ray fluorescence (pXRF) calibration for analysis of nutrient concentrations and trace element contaminants in fertilisers. PLoS ONE **17**(1), e0262460 (2022). https://doi.org/10.1371/journal.pone.0262460
6. Roberto de Alvarenga Junior, B., & Lajarim Carneiro, R. (2019). Chemometrics approaches in forced degradation studies of pharmaceutical drugs. Molecules (Basel, Switzerland) **24**(20), 3804. https://doi.org/10.3390/molecules24203804
7. Inobeme, A., Mathew, J.T., Jatto, E., Inobeme, J., Adetunji, C.O., Muniratu, M., Onyeachu, B.I., Adekoya, M.A., Ajai, A.I., Mann, A., Olori, E., Akhor, S.O., Eziukwu, C.A., Kelani, T., Omali, P.I.: Recent advances in instrumental techniques for heavy metal quantification. Environ. Monitor. Assess. **195**(4), 452 (2023b). https://doi.org/10.1007/s10661-023-11058-3
8. Gupta, A., Singh, U.B., Sahu, P.K., Paul, S., Kumar, A., Malviya, D., Singh, S., Kuppusamy, P., Singh, P., Paul, D., Rai, J.P., Singh, H.V., Manna, M.C., Crusberg, T.C., Kumar, A., Saxena, A.K.: Linking soil microbial diversity to modern agriculture practices: a review. Int. J. Environ. Res. Public Health **19**(5), 3141 (2022). https://doi.org/10.3390/ijerph19053141
9. Kharbach, M., Alaoui Mansouri, M., Taabouz, M., Yu, H. (2023). Current application of advancing spectroscopy techniques in food analysis: data handling with chemometric approaches. Foods **12**, 2753. https://doi.org/10.3390/foods12142753
10. Byers, H.L., McHenry, L.J., Grundl, T.J.: XRF techniques to quantify heavy metals in vegetables at low detection limits. Food Chem. X **1**, 100001 (2018). https://doi.org/10.1016/j.fochx.2018.100001
11. Mathew, J.T., Musah, M., Azeh, Y., Muhammed, M.: Adsorptive removal of selected toxic metals from pharmaceutical wastewater using Fe_3O_4/ZnO nanocomposite. Dutse J. Pure Appl. Sci. **9**(4a), 236–248 (2023a). https://doi.org/10.4314/dujopas.v9i4a.22
12. Sohn, S.I., Pandian, S., Oh, Y.J., Zaukuu, J.Z., Kang, H.J., Ryu, T.H., Cho, W.S., Cho, Y.S., Shin, E.K., Cho, B.K.: An overview of near infrared spectroscopy and its applications in the detection of genetically modified organisms. Int. J. Mol. Sci. **22**(18), 9940 (2021). https://doi.org/10.3390/ijms22189940
13. Kadadou, D., Tizani, L., Alsafar, H., Hasan, S.W.: Analytical methods for determining environmental contaminants of concern in water and wastewater. MethodsX **12**, 102582 (2024). https://doi.org/10.1016/j.mex.2024.102582
14. Zhou, J., Lu, X., Tian, B., Wang, C., Shi, H., Luo, C., Li, X.: A gas chromatography-flame ionization detection method for direct and rapid determination of small molecule volatile organic compounds in aqueous phase. 3 Biotech **10**(12), 520 (2020). https://doi.org/10.1007/s13205-020-02523-8
15. Inobeme, A., Adetunji, C.O., Mathew, J.T., Okonkwo, S., Bamigboye, M.O., Ajai, A.I., Afoso, E., Inobeme, J.: In: Shah, M. (ed.) Advanced Nanotechnology for the Degradation of Persistent Organic Pollutants. Microbial Degradation and Detoxification of Pollutants, pp. 51–72. De Gruyter, Berlin, Boston (2023b). https://doi.org/10.1515/9783110743623-003
16. Zhang, K., Banerjee, K.: A review: sample preparation and chromatographic technologies for detection of aflatoxins in foods. Toxins **12**(9), 539 (2020). https://doi.org/10.3390/toxins12090539
17. Mazarakioti, E.C., Zotos, A., Thomatou, A.A., Kontogeorgos, A., Patakas, A., Ladavos, A.: Inductively Coupled plasma-mass spectrometry (ICP-MS), a useful tool in authenticity of agricultural products' and foods' origin. Foods (Basel, Switzerland) **11**(22), 3705 (2022). https://doi.org/10.3390/foods11223705

18. Salem, M.A., Perez de Souza, L., Serag, A., Fernie, A.R., Farag, M.A., Ezzat, S.M., Alseekh, S.: Metabolomics in the context of plant natural products research: from sample preparation to metabolite analysis. Metabolites **10**(1), 37 (2020). https://doi.org/10.3390/metabo10010037
19. Alkhadra, M.A., Su, X., Suss, M.E., Tian, H., Guyes, E.N., Shocron, A.N., Conforti, K.M., de Souza, J.P., Kim, N., Tedesco, M., Khoiruddin, K., Wenten, I.G., Santiago, J.G., Hatton, T.A., Bazant, M.Z.: Electrochemical methods for water purification, ion separations, and energy conversion. Chem. Rev. **122**(16), 13547–13635 (2022). https://doi.org/10.1021/acs.chemrev.1c00396
20. Artika, I.M., Dewi, Y.P., Nainggolan, I.M., Siregar, J.E., Antonjaya, U.: Real-time polymerase chain reaction: current techniques, applications, and role in COVID-19 diagnosis. Genes **13**, 2387 (2022). https://doi.org/10.3390/genes13122387
21. Grigorieva, E., Livenets, A., Stelmakh, E.: Adaptation of agriculture to climate change: a scoping review. Climate **11**, 202 (2023). https://doi.org/10.3390/cli11100202
22. González-Domínguez, R., Sayago, A., Fernández-Recamales, Á.: An overview on the application of chemometrics tools in food authenticity and traceability. Foods (Basel, Switzerland) **11**(23), 3940 (2022). https://doi.org/10.3390/foods11233940
23. Abimbola, O.F., Adetunji, C.O., Dare, B.J., Mathew, J.T., Inobeme, A., Ghazanfar, S., Olaniyan, O.T, Ajiboye, M.D., Dauda, W.: Packaging of animal-based food products: the need for intelligent freshness sensors. In: Book: Sensing and Artificial Intelligence Solutions for Food Manufacturing, pp. 89–102. CRC Press. https://doi.org/10.1201/9781003207955-6.
24. Kim, J., Kim, D.-G., Ryu, K.H.: Enhancing response surface methodology through coefficient clipping based on prior knowledge. Processes **11**, 3392 (2023). https://doi.org/10.3390/pr11123392
25. Tobiszewski, M., Vakh, C.: Analytical applications of smartphones for agricultural soil analysis. Anal. Bioanal. Chem. Bioanal. Chem. **415**(18), 3703–3715 (2023). https://doi.org/10.1007/s00216-023-04558-1
26. Ahuja, V., Singh, A., Paul, D., Dasgupta, D., Urajová, P., Ghosh, S., Singh, R., Sahoo, G., Ewe, D., Saurav, K.: Recent advances in the detection of food toxins using mass spectrometry. Chem. Res. Toxicol. Toxicol. **36**(12), 1834–1863 (2023). https://doi.org/10.1021/acs.chemrestox.3c00241
27. Mashabela, M.D., Masamba, P., Kappo, A.P.: Metabolomics and chemoinformatics in agricultural biotechnology research: complementary probes in unravelling new metabolites for crop improvement. Biology **11**(8), 1156 (2022). https://doi.org/10.3390/biology11081156
28. Adetunji, C.O., Shariati, MA., Olugbenga, S.M., Osarenkhoe, O.O, Odoh, U.E., Olaniyan, O.T., Rebezov, M., Ajayi, O.O., Baibalinova, G., Bodunrinde, R.E., Adetunji, J.B., Adeniyi, M.J., Ugwu, P.N., Inobeme, A., Mathew, J.T., Elufisan, T.O., Omotayo, O.O.: Pineapple [*Ananas comosus* (L.) Merr.]: a biological and pharmacological active medicinal plant. In: Pullaiah, T. (ed.) Book: Phytochemical Composition and Pharmacy of Medicinal Plants, vol. 1, pp. 537–567. Apple Academic Press, Inc. Co-published with CRC Press (Taylor & Francis) (2023c)
29. Balaram, V.: Advances in analytical techniques and applications in exploration, mining, extraction, and metallurgical studies of rare earth elements. Minerals **13**, 1031 (2023). https://doi.org/10.3390/min13081031
30. Staufer, T., Grüner, F.: Review of development and recent advances in biomedical X-Ray fluorescence imaging. Int. J. Mol. Sci. **24**(13), 10990 (2023). https://doi.org/10.3390/ijms241310990
31. Croffie, M.E.T., Williams, P.N., Fenton, O., Fenelon, A., Metzger, K., Daly, K.: Optimising sample preparation and calibrations in EDXRF for quantitative soil analysis. Agronomy **10**, 1309 (2020). https://doi.org/10.3390/agronomy10091309
32. Mathew, J.T., Inobeme, A., Adetuyi, B.O., Adetunji, C.O., Popoola, O.A., Olaitan, F.Y., Akinbo, O., Shahnawaz, M., Oyewole, O.A., Eniola, K.I.T., Yerima, B.M.: General introduction of microplastic: uses, types, and generation. In: Book: Microplastic Pollution, vol. 1, pp. 3–21. Springer, Singapore (2024b). https://doi.org/10.1007/978-981-99-8357-5_1

33. Lu, J., Guo, J., Wei, Q., Tang, X., Lan, T., Hou, Y., Zhao, X.: A matrix effect correction method for portable X-Ray fluorescence data. Appl. Sci. **12**, 568 (2022). https://doi.org/10.3390/app12020568
34. Trojek, T., Trojková, D.: Uncertainty of quantitative X-Ray fluorescence micro-analysis of metallic artifacts caused by their curved shapes. Materials (Basel, Switzerland) **16**(3), 1133 (2023). https://doi.org/10.3390/ma16031133
35. Mondal, M.N., Horikawa, K., Seki, O., Nejigaki, K., Minami, H., Murayama, M., Okazaki, Y.: Investigation of adequate calibration methods for X-Ray fluorescence core scanning element count data: a case study of a marine sediment piston core from the Gulf of Alaska. J. Mar. Sci. Eng. **9**, 540 (2021). https://doi.org/10.3390/jmse9050540
36. Inobeme, A., Shahnawaz, M., Adetunji, C.O., Mathew, J.T., Adetuyi, B.O., Popoola, O.A., Olaitan, F.Y., Akinbo, O., Kolawole, O.M., Oyewole, O.A., Eniola, K.I.T., Yerima, B.M.: Microplastic as a pathogenic vector to cause diseases in marine biota. In: Book: Microplastic Pollution, vol. 1, pp. 321–349. Springer, Singapore (2024b). https://doi.org/10.1007/978-981-99-8357-5_19
37. Hou, Y., Huang, K.: A method for correcting the interference of overlapping absorption lines using second harmonic spectral reconstruction. Appl. Sci. **12**, 73 (2022). https://doi.org/10.3390/app12010073
38. Purwadi, I., Casey, L.W., Ryan, C.G., Erskine, P.D., van der Ent, A.: X-Ray fluorescence spectroscopy (XRF) for metallome analysis of herbarium specimens. Plant Methods **18**(1), 139 (2022). https://doi.org/10.1186/s13007-022-00958-z
39. Antonangelo, J., Zhang, H.: Soil and plant nutrient analysis with a portable XRF probe using a single calibration. Agronomy **11**, 2118 (2021). https://doi.org/10.3390/agronomy11112118
40. Terzano, R., Denecke, M.A., Falkenberg, G., Miller, B., Paterson, D., Janssens, K.: Recent advances in analysis of trace elements in environmental samples by X-Ray based techniques (IUPAC Technical Report). Pure and applied chemistry. Chim. Pure Appl. **91**(6), 1029–1063 (2019). https://doi.org/10.1515/pac-2018-0605
41. Adetunji, C.O., Egbuna, C., Ajiboye, A.A.A., Ajayi, O.O., Dauda, W.P., Ghazanfar, S., Ogundolie, F.A., Mathew, J.T., Adetunji, J.B., Inobeme, A., Olotu, T., Olaniyan, O.T., Ijabadeniyi, O.A., Aborode, A.T., Ficai, A., Akram, M., Bodunrinde, R.E., Laila, U., Olaniyan, O.P., Micheal, O.S., Aliyu, A., Akinbo, O., Adetuyi, B.O.: Biogenic nanomaterials with diverse biological activities in the food and biomedical industries. In: Adetunji, C.O., Hefft, D.I., Mbuge, D.O., Workneh, T.S. (eds.) Evaporative Coolers for the Postharvest Management of Fruits and Vegetables, pp. 395–420. Academic Press (2023a). https://doi.org/10.1016/B978-0-323-89864-5.00001-1.
42. Zhou, S., Wang, J., Wang, W., Liao, S.: Evaluation of portable X-Ray fluorescence analysis and its applicability as a tool in geochemical exploration. Minerals **13**, 166 (2023). https://doi.org/10.3390/min13020166
43. Pan, Y., Hu, L., Zhao, T.: Applications of chemical imaging techniques in paleontology. Natl. Sci. Rev. **6**(5), 1040–1053 (2019). https://doi.org/10.1093/nsr/nwy107
44. Mathew, J.T., Inobeme, A., Abdulkadir, A., Etsuyankpa, M.B., Shaba, E.Y., Musah, M., Azeh, Y., Adetunji, C.O., Ibrahim, A.M., Tanko, E., Mamman, A.: Properties and characteristics of nanocoatings for medicinal applications. In: Book: Sustainable Approach to Protective Nanocoatings. Publishing Tomorrow's Research Today, pp. 203–227. IGI Global (2024a). https://doi.org/10.4018/979-8-3693-3136-1.ch007
45. He, S., Niu, Y., Xing, L., Liang, Z., Song, X., Ding, M., Huang, W.: Research progress of the detection and analysis methods of heavy metals in plants. Front. Plant Sci. **15**, 1310328 (2024). https://doi.org/10.3389/fpls.2024.1310328
46. Mathew, J.T., Inobeme, A., Adetunji, C.O., Azeh, Y., Otori, A.A., Shaba, E.Y., Musah, M., Musa, S.T., Tanko, E., Mamman, A., Jibrin, N.A., Oyewole, O.A.: Structural elucidation, detection, and characterization of essential nutrients necessary for soil fertilization. In: Book: Handbook of Agricultural Biotechnology, vol. 5, pp. 117–147. Wiley (2024c). https://doi.org/10.1002/9781394211548.Ch5

47. Guimarães, D., Praamsma, M.L., Parsons, P.J.: Evaluation of a new optic-enabled portable XRF instrument for measuring toxic metals/metalloids in consumer goods and cultural products. Spectrochim. Acta Part B Atomic Spectrosc. **122**, 192–202 (2016). https://doi.org/10.1016/j.sab.2016.03.010
48. Adetunji, C.O., Olaniyan, O.T., Shariati, M.A., Bodunrinde, R.E., Ajayi, O.O., Adetunji, J.B., Hefft, D.I.,Olugbenga, S.M., Rebezov, M., Khayrullin, M., Adeniyi, M.J., Inobeme, A., Mathew, J.T.: Pharmacology, phytochemistry, and medicinal usefulness of *Basella alba* L. In: Pullaiah, T. (ed.) Book: Phytochemical Composition and Pharmacy of Medicinal Plants, vol. 2, pp. 499–451. Apple Academic Press, Inc. Co-published with CRC Press (Taylor & Francis) (2023b)
49. Borges, C.S., Weindorf, D.C., Carvalho, G.S., Guilherme, L.R.G., Takayama, T., Curi, N., Lima, G.J.E.O., Ribeiro, B.T.: Foliar elemental analysis of Brazilian crops via portable X-Ray fluorescence spectrometry. Sensors **20**, 2509 (2020). https://doi.org/10.3390/s20092509
50. Su, Q., Yu, H., Xu, X., Chen, B., Yang, L., Fu, T., Liu, W., Chen, G.: Using principal component analysis (PCA) combined with multivariate change-point analysis to identify brine layers based on the geochemistry of the core sediment. Water **15**, 1926 (2023). https://doi.org/10.3390/w15101926
51. Kandpal, L.M., Munnaf, M.A., Cruz, C., Mouazen, A.M.: Spectra fusion of mid-infrared (MIR) and X-Ray fluorescence (XRF) spectroscopy for estimation of selected soil fertility attributes. Sensors (Basel, Switzerland) **22**(9), 3459 (2022). https://doi.org/10.3390/s22093459
52. Milinovic, J., Santos, P., Sant'Ovaia, H., Futuro, A., Pereira, C.M., Murton, B.J., Flores, D., Azenha, M.: Multivariate analysis applied to X-Ray fluorescence to assess soil contamination pathways: case studies of mass magnetic susceptibility in soils near abandoned coal and W/Sn mines. Environ. Geochem. Health **46**(6), 202 (2024). https://doi.org/10.1007/s10653-024-01988-3
53. Wawire, A., Csorba, Á., Zein, M., Rotich, B., Phenson, J., Szegi, T., Tormáné Kovács, E., & Michéli, E. (2023). Farm household typology based on soil quality and influenced by socio-economic characteristics and fertility management practices in Eastern Kenya. Agronomy **13**, 1101. https://doi.org/10.3390/agronomy13041101
54. Trontelj Ml, J., Chambers, O.: Machine learning strategy for soil nutrients prediction using spectroscopic method. Sensors (Basel, Switzerland) **21**(12), 4208 (2021). https://doi.org/10.3390/s21124208
55. Inobeme, A., Mathew, J.T., Adetunji, C.O., Agbugui, M.O., Osarenren, E.J., Oti, C., Ngonso, B.F., Inobeme, J., Oyedolapo, M.B., Bernard, E., Jibrin, N.A.: Biocomposites with graphene derivatives. In: Book: Advances in Biocomposites and Their Applications, pp. 149–166. Elsevier Inc. (2024a). https://doi.org/10.1016/B978-0-443-19074-2.00005-8
56. Mouhamed, A.A., Nadim, A.H., Mostafa, N.M., Eltanany, B.M.: Application of smart chemometric models for spectra resolution and determination of challenging multi-action quaternary mixture: statistical comparison with greenness assessment. BMC Chem. **18**(1), 44 (2024). https://doi.org/10.1186/s13065-024-01148-9
57. Li, S., Shen, J., Bishop, T.F.A., Viscarra Rossel, R.A.: Assessment of the effect of soil sample preparation, water content and excitation time on proximal X-Ray fluorescence sensing. Sensors (Basel, Switzerland) **22**(12), 4572. https://doi.org/10.3390/s22124572
58. Priya, A.K., Muruganandam, M., Ali, S.S., Kornaros, M.: Clean-up of heavy metals from contaminated soil by phytoremediation: a multidisciplinary and eco-friendly approach. Toxics **11**(5), 422 (2023). https://doi.org/10.3390/toxics11050422
59. Pathak, V.M., Verma, V.K., Rawat, B.S., Kaur, B., Babu, N., Sharma, A., Dewali, S., Yadav, M., Kumari, R., Singh, S., Mohapatra, A., Pandey, V., Rana, N., Cunill, J.M.: Current status of pesticide effects on environment, human health and it's eco-friendly management as bioremediation: a comprehensive review. Front. Microbiol. Microbiol. **13**, 962619 (2022). https://doi.org/10.3389/fmicb.2022.962619
60. Bhardwaj, A.K., Arya, G., Kumar, R., Hamed, L., Pirasteh-Anosheh, H., Jasrotia, P., Kashyap, P.L., Singh, G.P.: Switching to nanonutrients for sustaining agroecosystems and environment:

the challenges and benefits in moving up from ionic to particle feeding. J. Nanobiotechnol. **20**(1), 19 (2022). https://doi.org/10.1186/s12951-021-01177-9
61. Woldetsadik, D., Llorent-Martínez, E.J., Gebrezgabher, S., Njenga, M., Mendum, R., Castillo-López, R., Fernández-de Córdova, M.L., Hailu, H., Evans, C.T., Madani, N., Mafika, T.P., Fleming, D.E.B.: Okra (Abelmoschus esculentus) in a refugee context in East Africa: kitchen gardening helps with mineral provision. SN Appl. Sci. **4**(1), 32 (2022). https://doi.org/10.1007/s42452-021-04898-6
62. Jolliffe, I.T., Cadima, J.: Principal component analysis: a review and recent developments. Philos. Trans. Ser. A Math. Physi. Eng. Sci. **374**(2065), 20150202 (2016). https://doi.org/10.1098/rsta.2015.0202
63. Mathew, J.T., Inobeme, A., Musah, M., Azeh, Y., Abdullahi, A., Shaba E.Y., Salihu, A.M., Muhammad, E.B., Josiah, J.G., Jibrin, N.A., Ismail, H., Muhammad, A.I., Maurice, J., Mamman, A., Ndamitso, M.M.: A critical review of green approach on wastewater treatment strategies. J. Appl. Sci. Environ. Manag. **28**(2), 363–391 (2024c). https://doi.org/10.4314/jasem.v28i2.9
64. Kharbach, M., Alaoui Mansouri, M., Taabouz, M., Yu, H.: Current Application of advancing spectroscopy techniques in food analysis: data handling with chemometric approaches. Foods (Basel, Switzerland) **12**(14), 2753 (2023). https://doi.org/10.3390/foods12142753
65. Adetunji, C.O., Akinbo, O., Mathew, J.T., Egbuna, C., Inobeme, A., Olotu, T.M., Ajayi, O.O., Dauda, W., Ghazanfar, S., Ogundolie, F.A., Adetunji, J.B., Adetuyi, B.O., Ajenifujah-Solebo, Shakirat, O., Aborode, A.T.: Relevance of nanomaterials derived from medicinal plants for marine and terrestrial environments: recent advances In: Book: Handbook of Agricultural Biotechnology, vol. 1, pp. 311–326. John Wiley & Sons (2024). https://doi.org/10.1002/9781119836483.ch12
66. Wilschefski, S.C., Baxter, M.R.: Inductively coupled plasma mass spectrometry: introduction to analytical aspects. Clin. Biochem. Rev. **40**(3), 115–133 (2019). https://doi.org/10.33176/AACB-19-00024
67. Barnes, M., Sulé-Suso, J., Millett, J., Roach, P.: Fourier transform infrared spectroscopy as a non-destructive method for analysing herbarium specimens. Biol. Let. **19**(3), 20220546 (2023). https://doi.org/10.1098/rsbl.2022.0546
68. Ngere, J.B., Ebrahimi, K.H., Williams, R., Pires, E., Walsby-Tickle, J., McCullagh, J.S.O.: Ion-Exchange chromatography coupled to mass spectrometry in life science, environmental, and medical research. Anal. Chem. **95**(1), 152–166 (2023). https://doi.org/10.1021/acs.analchem.2c04298
69. Kumpati, R., Skarka, W., Ontipuli, S.K.: Current trends in integration of nondestructive testing methods for engineered materials testing. Sensors (Basel, Switzerland) **21**(18), 6175 (2021). https://doi.org/10.3390/s21186175
70. Carlomagno, I., Antonelli, M., Aquilanti, G., Bellutti, P., Bertuccio, G., Borghi, G., Cautero, G., Cirrincione, D., de Giudici, G., Ficorella, F., Gandola, M., Giuressi, D., Medas, D., Mele, F., Menk, R.H., Olivi, L., Orzan, G., Picciotto, A., Podda, F., Rachevski, A., Meneghini, C.: Trace-element XAFS sensitivity: a stress test for a new XRF multi-detector. J. Synchrot. Radiat. **28**(Pt 6), 1811–1819 (2021). https://doi.org/10.1107/S1600577521008857
71. Balaram, V., Sawant, S.S.: Indicator minerals, pathfinder elements, and portable analytical instruments in mineral exploration studies. Minerals **12**, 394 (2022). https://doi.org/10.3390/min12040394
72. Webb, E.W., Scott, P.J.H.: Potential applications of artificial intelligence and machine learning in radiochemistry and radiochemical engineering. PET Clin. **16**(4), 525–532 (2021). https://doi.org/10.1016/j.cpet.2021.06.012
73. Kopittke, P.M., Lombi, E., van der Ent, A., Wang, P., Laird, J.S., Moore, K.L., Persson, D.P., Husted, S.: Methods to visualize elements in plants. Plant Physiol. **182**(4), 1869–1882 (2020). https://doi.org/10.1104/pp.19.01306

X-Ray Fluorescence Spectroscopy for Analyzing Food Products Coupled with Chemometrics

11

John Tsado Mathew, Abel Inobeme, Yakubu Azeh, Musah Monday, Elijah Yanda Shaba, Tanko Ezekiel, Etsuyankpa Muhammad Bini, Musa Safiyano Tanko, Jonathan Hussaini, Charles Oluwaseun Adetunji, Muhammad Aishetu Ibrahim, and Amos Mamman

Abstract

X-Ray fluorescence spectroscopy is a powerful analytical technique widely used for elemental analysis in various matrices, including food products. The complexity of food matrices and the need for precise quantification require advanced data analysis techniques, leading to the integration of XRF with chemometrics. Chemometrics, which involves the application of mathematical and statistical methods to chemical data, enhances the interpretative power of XRF by improving accuracy, precision, and sensitivity in detecting and quantifying elements in food products. This review explores the principles and applications of XRF Spectroscopy in food analysis, emphasizing its advantages and limitations. It delves into the fundamental aspects of chemometrics, including principal component analysis, partial least squares, and multiple linear regression, which are commonly employed to process and interpret complex spectral data. By coupling XRF with chemometric methods, researchers can overcome challenges

J. T. Mathew (✉) · Y. Azeh · M. Monday · J. Hussaini · M. A. Ibrahim · A. Mamman
Department of Chemistry, Ibrahim Badamasi Babangida University Lapai, Lapai, Niger State, Nigeria
e-mail: jmathew@ibbu.edu.ng

A. Inobeme
Department of Chemistry, Edo University Uzairue, Auchi, Edo State, Nigeria

E. Y. Shaba · T. Ezekiel
Department of Chemistry, Federal University of Technology Minna, Minna, Nigeria

E. M. Bini · M. S. Tanko
Department of Chemistry, Federal University of Lafia, Lafia, Nasarawa State, Nigeria

C. O. Adetunji
Applied Microbiology, Biotechnology and Nanotechnology Laboratory, Department of Microbiology, Edo University Uzairue, Auchi, Edo State, Nigeria

V. K. Singh (ed.), *X-Ray Fluorescence Spectroscopy and Chemometrics*,
https://doi.org/10.1007/978-3-031-98375-7_11

such as matrix effects, spectral overlaps, and interference, leading to more robust and reliable analytical results. The application of XRF in food analysis is examined through case studies that demonstrate its effectiveness in detecting elements like heavy metals, minerals, and trace elements in various food products. These studies highlight the potential of XRF in compliance with regulatory standards and ensuring consumer safety.

11.1 Introduction

Ensuring food safety and quality has become a critical concern worldwide, as the globalization of food supply chains and increasing consumer awareness demand stringent monitoring and analysis. Elemental analysis plays a significant role in this context, as it provides vital information on the presence of essential nutrients, contaminants, and trace elements in food products. Accurate determination of these elements is crucial for assessing nutritional content, identifying potential health risks, and ensuring compliance with regulatory standards. Among the various analytical techniques available, X-Ray fluorescence (XRF) spectroscopy has emerged as a powerful tool for elemental analysis due to its non-destructive nature, rapid analysis, and ability to detect multiple elements simultaneously [1, 2].

XRF spectroscopy works by exposing a sample to high-energy X-Rays, causing the atoms in the sample to emit secondary (or fluorescent) X-Rays that are characteristic of specific elements. The intensity and energy of these emitted X-Rays are measured and analyzed to determine the elemental composition of the sample. This technique has found widespread application in various fields, including geology, archaeology, environmental science, and, notably, food analysis. The non-destructive nature of XRF allows for the analysis of samples without altering their physical or chemical properties, making it ideal for food analysis where preserving the integrity of the sample is essential [3, 4].

Despite its many advantages, XRF spectroscopy faces challenges when applied to food analysis, primarily due to the complex and heterogeneous nature of food matrices. The presence of various organic and inorganic components in food can lead to matrix effects, spectral overlaps, and interferences, complicating the interpretation of XRF data. Additionally, while XRF provides qualitative and semi-quantitative information relatively easily, achieving accurate and precise quantitative analysis often requires advanced data processing techniques. This is where chemometrics comes into play [5].

Chemometrics is the application of mathematical and statistical techniques to chemical data, aiming to extract meaningful information and improve the interpretative power of analytical methods. In the context of XRF spectroscopy, chemometrics can be used to correct for matrix effects, resolve overlapping spectra, and enhance the accuracy and precision of elemental quantification. Techniques such as principal component analysis (PCA), partial least squares (PLS), and multiple linear regression (MLR) are commonly employed in chemometric analysis to model the complex relationships between spectral data and elemental concentrations [6].

The coupling of XRF spectroscopy with chemometrics has led to significant advancements in food analysis, enabling more reliable and robust detection and quantification of elements in various food products. This integrated approach has proven particularly valuable in analyzing complex food matrices where traditional methods may struggle to provide accurate results. For example, in cases where food products contain multiple elements that may interfere with each other's detection, chemometric techniques can help disentangle these effects, leading to more precise quantification [7].

Furthermore, the integration of chemometrics with XRF has opened new avenues for the development of predictive models and the classification of food products based on their elemental composition. This has important implications for food safety, as it allows for the rapid screening of large numbers of samples, the detection of adulteration, and the verification of food authenticity. The ability to quickly and accurately determine the elemental profile of food products is essential for ensuring that they meet regulatory standards and are safe for consumption [8].

11.2 Role of Chemometrics in Enhancing Analytical Techniques

At its core, chemometrics involves the application of algorithms and models to analyze chemical data, often derived from instruments that generate complex, multivariate datasets. Techniques like PCA, PLSR, and MLR are among the most widely used in chemometric analysis. Each of these techniques offers distinct advantages in dealing with specific challenges encountered in analytical chemistry [9].

PCA is a powerful tool for data reduction and pattern recognition. In analytical techniques where the data is multidimensional—such as spectra with hundreds or thousands of variables—PCA reduces the dimensionality of the data by identifying the principal components that capture the most variance. This reduction simplifies the data, making it easier to visualize and interpret, while also revealing patterns, trends, and outliers that may not be immediately apparent. For example, in spectroscopy, PCA can help distinguish between different sample types based on their spectral fingerprints, aiding in classification, quality control, and detection of adulteration [10, 11].

MLR is another common chemometric technique used for quantification, particularly when the relationship between the measured signals and analyte concentrations can be described by a linear model. MLR is effective in situations where the data is not highly collinear or where the number of variables is manageable. However, in cases of high multicollinearity or when dealing with highly complex data, techniques like PLS or ridge regression may be more appropriate [12, 13].

Improved Accuracy and Precision

By modeling the relationships between complex data and the concentrations of analytes, chemometrics reduces the impact of noise, interferences, and matrix effects. This leads to more accurate and precise measurements, which are crucial in fields like food analysis, where regulatory compliance and consumer safety depend on reliable data [7].

Data Interpretation and Visualization

Chemometrics simplifies the interpretation of complex datasets, making it easier for analysts to identify trends, patterns, and outliers. This is particularly useful in quality control, where rapid decision-making is required based on the data [6].

Enhanced Calibration and Quantification

Traditional calibration methods often struggle with complex matrices or overlapping signals. Chemometric models, such as PLS, can handle these challenges by incorporating information from the entire spectrum or chromatogram, leading to more robust and reliable quantification [14].

Chemometrics also plays a critical role in the design and optimization of analytical experiments. Techniques such as design of experiments (DOE) and response surface methodology (RSM) are used to optimize experimental conditions, reduce experimental errors, and maximize the information obtained from the data. By systematically varying experimental parameters, chemometrics allows researchers to identify the optimal conditions for analysis, ensuring that the analytical technique provides accurate and reproducible results. In food analysis, this can involve optimizing factors such as sample preparation methods, instrument settings, and calibration procedures to improve the accuracy and precision of techniques like XRF spectroscopy [14].

Another important application of chemometrics is in classification and pattern recognition. By analyzing multivariate data, chemometric techniques can classify samples into distinct groups based on their chemical composition. This is particularly useful in food authentication, where chemometrics can be used to verify the origin, quality, and authenticity of food products. For example, by analyzing the elemental profile of food samples using XRF and applying chemometric techniques, it is possible to distinguish between products from different geographical regions or identify adulterated samples [15].

11.3 Advantages and Limitations of XRF in Food Analysis

One of the most significant benefits of XRF is its non-destructive nature, allowing the analysis of food samples without altering their physical or chemical properties (Table 11.1). This feature is particularly valuable in food analysis, where preserving sample integrity is essential for further testing or consumption. XRF provides quick analysis, often delivering results within minutes. It can simultaneously detect and quantify multiple elements, making it highly efficient for comprehensive elemental analysis. This is especially beneficial for monitoring the nutritional content of food products or screening for contaminants such as heavy metals [7].

Table 11.1 Advantages and limitations of XRF in food analysis

Aspects	Advantages	Limitations	Details	Refs.
Sample preparation	Minimal or no sample preparation required	Requires homogeneity for accurate results	XRF's non-destructive nature preserves the sample integrity, but uneven samples can lead to inconsistent results	[19]
Analysis speed	Rapid analysis, providing results within minutes	Potentially slower for complex matrix corrections	Suitable for high-throughput environments; however, complex matrices may require additional data processing to correct matrix effects	[20, 21]
Non-destructive testing	Sample remains intact for further analysis or use	Matrix effects can complicate the interpretation of results	Ideal for food samples where preservation is necessary; however, matrix effects can distort results without proper calibration and correction	[22]
Portability	Availability of portable XRF devices for in-field analysis	May require calibration against laboratory standards for accuracy	Portable XRF devices are useful for on-site food testing, but field results might need to be validated against lab-based standards for precise quantification	[23, 24]
Regulatory compliance	Useful in monitoring compliance with food safety regulations	Detection limits may not meet all regulatory standards for trace elements	XRF is effective for routine screening of contaminants; however, trace elements at low concentrations might fall below detection limits required by regulations	[25]
Cost efficiency	Lower operational costs compared to some other analytical techniques	Initial equipment cost can be high	Cost-effective in long-term use due to minimal consumables, but the initial investment in XRF equipment can be substantia	[26]

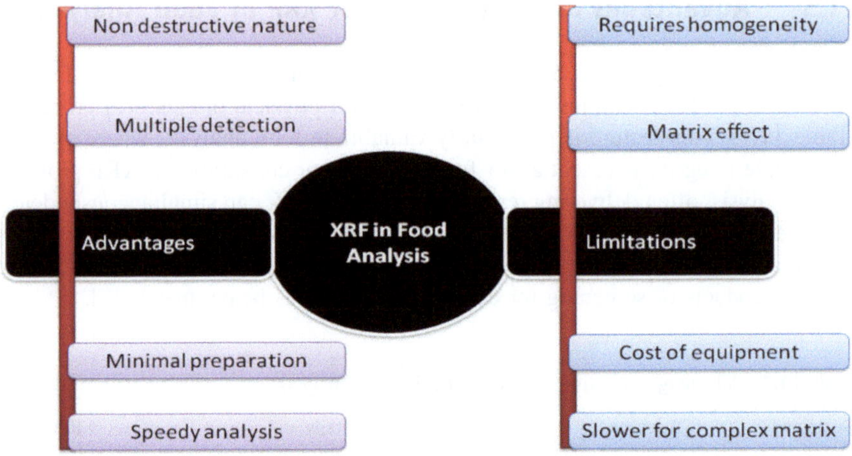

Fig. 11.1 Advantages and limitations of XRF in food analysis

XRF requires little to no sample preparation, reducing the risk of contamination and the time needed for analysis. This simplicity enhances the technique's utility in high-throughput environments, such as food processing plants or regulatory laboratories, where speed and efficiency are critical. The development of portable XRF devices has expanded the technique's applications to field-based food analysis. These portable spectrometers allow for on-site testing, providing real-time data that can be used for quality control, compliance checks, or rapid decision-making in various settings [16, 17] (Fig. 11.1).

The complex and heterogeneous nature of food matrices can lead to matrix effects, where the composition of the sample affects the absorption and emission of X-Rays. These effects can complicate the interpretation of results and may require advanced data processing techniques, such as chemometrics, to correct. XRF is generally less sensitive to light elements with low atomic numbers, such as sodium (Na) or magnesium (Mg). This limitation can be a challenge when analyzing food products that contain essential nutrients or contaminants at low concentrations, which may fall below the detection limits of XRF (Fig. 13.1) [18].

11.4 Chemometrics in Analytical Chemistry

Chemometrics, a discipline that merges chemistry, mathematics, and statistics, has revolutionized the field of analytical chemistry. It involves the application of mathematical and statistical techniques to design experiments, optimize procedures, and interpret complex data generated by analytical instruments. The integration of chemometrics into analytical chemistry enhances the accuracy, precision, and reliability of results, making it an indispensable tool in modern scientific research and industrial applications [27].

One of the fundamental roles of chemometrics in analytical chemistry is data preprocessing. Raw data obtained from analytical instruments often contain noise, baseline drifts, and other artifacts that can obscure the true signal. Chemometric techniques such as signal smoothing, baseline correction, and normalization are employed to preprocess data, eliminating these unwanted effects and enhancing the clarity of the analytical signal. This preprocessing step is crucial for ensuring the accuracy of subsequent data analysis and interpretation. For example, in spectroscopy, the presence of noise or baseline shifts can lead to incorrect identification or quantification of compounds. By applying chemometric techniques, these interferences can be minimized, allowing for more accurate and reliable results [28].

Analytical chemistry often deals with complex datasets involving multiple variables that are interrelated. Chemometrics provides a suite of multivariate data analysis tools to handle this complexity, enabling the extraction of meaningful information from large datasets. Techniques such as PCA, PLSR, and HCA are commonly used in chemometrics to identify patterns, trends, and relationships within the data [29, 30].

In chromatography, for instance, the choice of mobile phase composition, flow rate, and temperature can significantly impact the separation efficiency and resolution of analytes. Chemometric optimization helps in determining the best combination of these parameters to achieve the desired analytical performance [31].

Calibration is a crucial aspect of analytical chemistry, and chemometrics provides robust methods for developing calibration models. Techniques such as Multiple Linear Regression (MLR) and PLS regression are used to correlate the analytical signal (e.g., absorbance, intensity) with the concentration of analytes. Chemometrics also plays a key role in validating these models, ensuring that they are accurate, reliable, and applicable to unknown samples. Validation techniques such as cross-validation and bootstrapping are used to assess the robustness of the calibration models. These methods help in identifying any potential overfitting, ensuring that the models perform well not only on the training data but also on new, unseen data [32].

Chemometrics is instrumental in pattern recognition and classification of complex data. By applying multivariate analysis techniques, chemometrics can classify samples into distinct groups based on their chemical composition. This is particularly useful in quality control, where it is essential to verify the authenticity, origin, and quality of products [33].

For example, in the pharmaceutical industry, chemometrics can be used to distinguish between genuine and counterfeit drugs based on their spectral profiles. Similarly, in food analysis, chemometrics can identify adulterated products by comparing their chemical fingerprints with those of authentic samples [34].

In industrial settings, chemometrics is increasingly used for real-time process monitoring and control. By analyzing data from sensors and analytical instruments in real time, chemometric models can provide insights into the ongoing processes, allowing for timely adjustments and ensuring consistent product quality.

This application is particularly valuable in process industries such as pharmaceuticals, petrochemicals, and food processing, where maintaining product quality is critical [35].

11.5 Application of XRF Spectroscopy in Food Product Analysis

XRF spectroscopy has emerged as a powerful analytical tool in the food industry, offering several advantages for food product analysis. Its non-destructive nature, rapid analysis capabilities, and ability to detect multiple elements simultaneously make it particularly well-suited for a wide range of applications in food safety, quality control, and nutritional analysis [36, 37].

XRF spectroscopy is extensively used for the determination of essential nutrients and minerals in food products. It can accurately quantify elements like calcium (Ca), iron (Fe), zinc (Zn), and potassium (K), which are critical for human health. The technique's ability to analyze these nutrients without the need for complex sample preparation or destruction of the sample makes it ideal for routine nutritional labeling and quality assurance in food manufacturing. Food safety is a primary concern in the food industry, and XRF spectroscopy plays a crucial role in screening for toxic elements and contaminants. It is highly effective in detecting hazardous elements such as lead (Pb), cadmium (Cd), arsenic (As), and mercury (Hg) at trace levels. By ensuring that food products are free from these harmful substances, XRF helps manufacturers comply with stringent regulatory standards and protects public health [38].

XRF is also valuable for verifying the authenticity and quality of food products. By analyzing the elemental composition, it can detect adulteration and ensure that products meet their specified quality standards. For instance, the sodium (Na), magnesium (Mg), and phosphorus (P) content of a product can be measured to verify its authenticity and detect any potential substitutions or dilutions. This application is particularly important in high-value products, such as spices, dairy, and oils, where adulteration can have significant economic and health implications. The detection of trace elements is another critical application of XRF spectroscopy in food analysis. Elements such as selenium (Se), copper (Cu), and manganese (Mn) are essential in small quantities but can be harmful in excess. XRF allows for the precise monitoring of these elements in various food matrices, ensuring that their levels remain within safe and recommended limits [39].

In the context of food fortification, XRF is used to quantify added elements and fortifying agents. It ensures that elements like iodine (I) and fluorine (F) are present in the correct amounts, supporting the health claims of fortified foods and ensuring compliance with nutritional guidelines. XRF spectroscopy is also used to analyze food packaging materials to ensure they do not contaminate the food. Elements like aluminum (Al), silicon (Si), and tin (Sn) can be detected in packaging materials, helping to ensure that these materials are safe and compliant with food safety regulations [40] (Table 11.2).

Table 11.2 Application of XRF spectroscopy in food product analysis

Applications area	Description	Examples of analytes	Benefits	Refs.
Nutritional analysis	Determination of essential nutrients and minerals in food products	Calcium (Ca), Iron (Fe), Zinc (Zn), Potassium (K)	Provides quick, non-destructive analysis of key nutrients, aiding in accurate nutritional labeling	[41]
Contaminant detection	Screening for toxic elements and contaminants in food products	Lead (Pb), Cadmium (Cd), Arsenic (As), Mercury (Hg)	Ensures food safety by detecting harmful elements, complying with regulatory standards	[42]
Quality control and authenticity	Verifying the authenticity and quality of food products by elemental composition	Sodium (Na), Magnesium (Mg), Phosphorus (P)	Assists in identifying adulteration, ensuring product quality, and maintaining brand integrity	[43]
Trace element analysis	Detection of trace elements at low concentrations in various food matrices	Selenium (Se), Copper (Cu), Manganese (Mn)	Allows for the monitoring of trace elements important for health, supporting compliance with guidelines	[44]
Additive and fortification analysis	Quantification of added elements and fortifying agents in food products	Iodine (I), Fluorine (F), Fortified Minerals	Ensures the correct dosage of fortifying agents, supporting health claims and regulatory compliance	[45]
Packaging material analysis	Analysis of elements in food packaging materials to ensure safety and compliance	Aluminum (Al), Silicon (Si), Tin (Sn)	Helps in detecting contamination from packaging materials, ensuring food safety and quality	[46]

11.6 Challenges and Solutions in XRF-Chemometrics Applications

Table 11.3 presents combining X-Ray Fluorescence (XRF) spectroscopy with chemometrics offers powerful analytical capabilities, but it also presents several challenges. Addressing these challenges effectively is crucial for maximizing the benefits of this integrated approach in food product analysis and other applications [47].

Matrix effects occur when the composition of the sample influences the XRF signal, leading to inaccuracies in data interpretation. This can complicate the analysis, especially in heterogeneous or complex food matrices [48].

Advanced chemometric techniques such as Principal Component Analysis (PCA) and Partial Least Squares (PLS) regression can help mitigate matrix effects. PCA can reduce the dimensionality of data, isolating and correcting for variations caused by the matrix. PLS regression builds robust calibration models that account for these variations, improving accuracy in quantitative analysis [49].

Table 11.3 Challenges and solutions in XRF-chemometrics applications

Challenges	Description	Potential solutions	Details	Refs.
Matrix effects	Variations in sample composition can affect XRF signals and complicate data interpretation	Use of advanced chemometric techniques like PCA and PLS to correct for matrix effects	Applying principal component analysis (PCA) can help identify and adjust for matrix interferences, while Partial Least Squares (PLS) can model complex relationships between variables	[54]
Data complexity	XRF generates complex, multi-dimensional data that can be challenging to interpret	Implement multivariate analysis methods to manage data complexity	Techniques such as multivariate statistical analysis (MSA) and machine learning algorithms can simplify and extract meaningful patterns from complex datasets	[55, 56]
Calibration issues	Developing accurate calibration models can be difficult due to variability in sample matrices	Employ robust calibration techniques and regular model updates	Utilizing methods like cross-validation and bootstrapping can improve calibration models and adapt to variations in sample matrices	[57]
Sensitivity to light elements	XRF may have limited sensitivity for detecting elements with low atomic numbers	Combine XRF with complementary analytical techniques for enhanced detection	Integrating XRF with techniques like inductively coupled plasma (ICP) or atomic absorption spectroscopy (AAS) can enhance the detection of light elements	[58]
Sample homogeneity	Inhomogeneous samples can lead to inconsistent results and inaccuracies in analysis	Use preprocessing techniques to ensure sample homogeneity	Methods such as sample homogenization and standardization of sample preparation can help achieve more accurate and consistent results	[59, 60]

(continued)

Table 11.3 (continued)

Challenges	Description	Potential solutions	Details	Refs.
Data interpretation and validation	Interpreting and validating the results from chemometric models can be complex	Develop and apply rigorous validation protocols and cross-validation techniques	Regular validation of models through techniques like cross-validation and external validation against known standards can enhance the reliability of results	[61, 62]
Real time analysis challenges	Achieving real-time analysis with accurate results can be challenging due to data complexity	Optimize chemometric models for real-time processing and implement faster data acquisition techniques	Improving computational efficiency and using real-time data processing algorithms can enhance the capability to provide timely analysis	[63, 64]

Multivariate data analysis methods, including Multivariate Statistical Analysis (MSA) and machine learning algorithms, can simplify the interpretation of complex datasets. These methods help in identifying patterns, trends, and relationships within the data, facilitating more accurate analysis and decision-making [50, 51].

Developing accurate calibration models for XRF analysis can be challenging due to variability in sample matrices, which affects the calibration process. Robust calibration techniques, such as Cross-Validation and Bootstrapping, can improve model accuracy. Regular updates to calibration models and the use of diverse calibration standards help account for sample variability and ensure reliable results [52].

XRF spectroscopy is less sensitive to elements with low atomic numbers, such as sodium (Na) and magnesium (Mg), making it difficult to detect these light elements accurately. Combining XRF with complementary analytical techniques, such as Inductively Coupled Plasma (ICP) or Atomic Absorption Spectroscopy (AAS), enhances the detection of light elements. This multi-technique approach provides a more comprehensive analysis of the elemental composition [53].

11.7 Recent Advances and Innovations in XRF and Chemometrics

The fields of XRF spectroscopy and chemometrics are rapidly evolving, with recent advances enhancing their capabilities and broadening their applications. Innovations in these areas are transforming how we analyze and interpret complex data, leading to more accurate, efficient, and insightful analyses across various industries, including food safety, environmental monitoring, and material science [65].

- ***Enhanced XRF Instrumentation***

Recent advancements in XRF instrumentation include the development of high-resolution detectors that improve the sensitivity and precision of elemental analysis. These detectors allow for better differentiation between closely spaced X-Ray peaks, enhancing the ability to detect trace elements and providing more accurate quantitative results. The use of silicon drift detectors (SDDs) and advanced EDX has significantly increased the resolution and sensitivity of XRF systems. Portable XRF devices have seen significant advancements, making it possible to perform on-site analyses with greater accuracy and convenience. These handheld instruments are now equipped with improved calibration, increased sensitivity, and enhanced data processing capabilities. Portable XRF is increasingly used in field applications, such as soil and environmental monitoring, where real-time data collection is essential. Innovations in XRF technology have also led to the development of multi-elemental analysis capabilities. Modern XRF systems can simultaneously measure a broad range of elements, from light to heavy, with high accuracy. This capability is particularly valuable in applications requiring comprehensive elemental profiling, such as in food safety and material characterization [66].

- ***Advancements in Chemometric Techniques***

The integration of machine learning (ML) and artificial intelligence (AI) into chemometrics has revolutionized data analysis. Algorithms such as Support Vector Machines (SVM), Random Forests, and Neural Networks are being employed to handle complex datasets and improve predictive modeling. These AI-driven methods enhance the ability to classify and predict the quality of food products, detect anomalies, and optimize analytical procedures. Recent advancements in data fusion techniques allow for the integration of XRF data with information from other analytical methods, such as infrared spectroscopy (IR) or mass spectrometry (MS). Data fusion enhances the comprehensiveness of the analysis by combining complementary information from different sources. Chemometrics models can be developed to interpret fused data, leading to more accurate and holistic insights. The application of real-time chemometrics has improved the capability for dynamic and continuous monitoring of processes. Real-time chemometric models can be used to interpret XRF data on-the-fly, allowing for immediate feedback and adjustments during manufacturing or processing. This innovation is particularly useful in industrial applications where maintaining consistent quality is crucial [67, 68].

- ***Enhanced Calibration and Validation Methods***

Recent advancements in calibration techniques have led to the development of more robust models that account for sample variability and matrix effects. Techniques such as Partial Least Squares (PLS) with cross-validation and regularization methods improve the accuracy of calibration models. These advancements allow for better handling of complex matrices and enhance the reliability of quantitative analysis. Enhanced validation protocols, including Monte Carlo simulations and bootstrapping methods, provide more rigorous assessments of model performance. These techniques help to evaluate the robustness and reliability of chemometric models, ensuring that they perform well across different conditions and datasets. Improved validation helps in maintaining the accuracy and credibility of the analysis [69].

- ***Innovations in Data Interpretation***

The development of advanced visualization tools has transformed data interpretation. Interactive platforms allow researchers to explore and manipulate data in real-time, facilitating a deeper understanding of complex datasets. Visualization techniques such as heat maps, 3D plots, and interactive dashboards enable users to identify patterns, correlations, and outliers more effectively. Innovations in data mining and pattern recognition techniques enhance the ability to extract meaningful information from large datasets. Methods such as clustering, association rule mining, and anomaly detection are used to uncover hidden patterns and relationships within XRF data. These techniques provide valuable insights into the composition and quality of analyzed samples [70, 71].

11.8 Conclusion

The integration of XRF spectroscopy with chemometrics represents a significant advancement in analytical chemistry, offering a powerful toolkit for a wide range of applications. XRF provides a robust, non-destructive method for elemental analysis, while chemometrics enhances the interpretation and utilization of complex datasets. Together, they offer unparalleled capabilities for accurate, efficient, and insightful analysis of diverse samples. Recent advances in XRF technology, such as high-resolution detectors, portable devices, and multi-elemental analysis capabilities, have expanded the scope and applicability of this technique. These innovations have improved the precision, sensitivity, and convenience of XRF, making it a valuable tool for on-site and laboratory-based analyses. The development of high-resolution detectors has enhanced the ability to differentiate between closely spaced X-Ray peaks, while portable XRF devices enable real-time analysis in field settings. Simultaneously, the field of chemometrics has evolved with the incorporation of machine learning and artificial intelligence, advanced data fusion techniques, and real-time data processing capabilities. These advancements have significantly improved the ability to handle complex datasets, build robust calibration models, and provide real-time insights. Machine learning algorithms and AI-driven methods facilitate better classification, prediction, and optimization, while data fusion techniques enhance the comprehensiveness of analysis by integrating data from multiple sources. The combination of XRF and chemometrics has addressed several challenges, including matrix effects, data complexity, and calibration issues. By applying sophisticated chemometric techniques, such as PCA, PLSR, and advanced data validation methods, researchers can overcome these challenges and achieve more accurate and reliable results. Real-time chemometrics further enhances the capability for dynamic monitoring and immediate adjustments, particularly in industrial and manufacturing contexts.

References

1. Kruk, M.E., Gage, A.D., Arsenault, C., Jordan, K., Leslie, H.H., Roder-DeWan, S., Adeyi, O., Barker, P., Daelmans, B., Doubova, S.V., English, M., García-Elorrio, E., Guanais, F., Gureje, O., Hirschhorn, L.R., Jiang, L., Kelley, E., Lemango, E.T., Liljestrand, J., Malata, A., Pate, M.: High-quality health systems in the sustainable development goals era: time for a revolution. The Lancet Glob. Health **6**(11), e1196–e1252 (2018). https://doi.org/10.1016/S2214-109X(18)30386-3
2. Mathew, J.T., Musah, M., Azeh, Y., Muhammed, M.: Kinetic study of heavy metals removal from pharmaceutical wastewater using geopolymer/Fe_3O_4 nanocomposite. Bima J. Sci. Technol. **7**(4), 152–163 (2023b). https://doi.org/10.56892/bima.v7i4.539
3. Zhang, R., Li, L., Sultanbawa, Y., Xu, Z.P.: X-Ray fluorescence imaging of metals and metalloids in biological systems. Am. J. Nucl. Med. Mol. Imaging **8**(3), 169–188 (2018)
4. Varrà, M.O., Ghidini, S., Husáková, L., Ianieri, A., Zanardi, E.: Advances in troubleshooting fish and seafood authentication by inorganic elemental composition. Foods (Basel, Switzerland) **10**(2), 270 (2021). https://doi.org/10.3390/foods10020270
5. Zappi, A., Marassi, V., Giordani, S., Kassouf, N., Roda, B., Zattoni, A., Reschiglian, P., Melucci, D.: Extracting information and enhancing the quality of separation data: a review on chemometrics-assisted analysis of volatile, soluble and colloidal samples. Chemosensors **11**, 45 (2023). https://doi.org/10.3390/chemosensors11010045
6. Md Noh, M.F., Gunasegavan, R.D., Mustafa Khalid, N., Balasubramaniam, V., Mustar, S., Abd Rashed, A.: Recent techniques in nutrient analysis for food composition database. Molecules (Basel, Switzerland), 25(19), 4567 (2020). https://doi.org/10.3390/molecules25194567
7. Cottenet, G., Andrey, D., Dubascoux, S.: Evaluation of ED-XRF for the detection of inorganic adulterants in turmeric, paprika and oregano. Food Addit. Contamin. Part A Chem. Anal. Control Expos. Risk Assess **41**(5), 461–467 (2024). https://doi.org/10.1080/19440049.2024.2326425
8. Zhang, Y., Wang, Y.: Recent trends of machine learning applied to multi-source data of medicinal plants. J. Pharmaceut. Anal. **13**(12), 1388–1407 (2023). https://doi.org/10.1016/j.jpha.2023.07.012
9. Inobeme, A., Mathew, J.T., Jatto, E., Inobeme, J., Adetunji, C.O., Muniratu, M., Onyeachu, B.I., Adekoya, M.A., Ajai, A.I., Mann, A., Olori, E., Akhor, S.O., Eziukwu, C.A., Kelani, T., Omali, P.I.: Recent advances in instrumental techniques for heavy metal quantification. Environ. Monitor. Assess. **195**(4), 452 (2023b). https://doi.org/10.1007/s10661-023-11058-3
10. Jolliffe, I.T., Cadima, J.: Principal component analysis: a review and recent developments. Philos. Trans. Ser. A Math. Phys. Eng. Sci. **374**(2065), 20150202 (2016). https://doi.org/10.1098/rsta.2015.0202

11. Mathew, J.T., Musah, M., Azeh, Y., Muhammed, M.: Adsorptive removal of selected toxic metals from pharmaceutical wastewater using Fe_3O_4/ZnO nanocomposite. Dutse J. Pure Appl. Sci. **9**(4a), 236–248 (2023a). https://doi.org/10.4314/dujopas.v9i4a.22
12. Gadžurić, S., Kuzmanović, S.O.P., Vraneš, M.B., Petrin, M., Bugarski, T., Kovačević, S.Z.: Multivariate chemometrics with regression and classification analyses in heroin profiling based on the chromatographic data. Iran. J. Pharmaceut. Res. IJPR **15**(4), 725–734 (2016)
13. Inobeme, A., Adetunji, C.O., Mathew, J.T., Okonkwo, S., Bamigboye, M.O., Ajai, A.I., Afoso, E., Inobeme, J.: Advanced nanotechnology for the degradation of persistent organic pollutants. In: Shah, M. (ed.) Microbial Degradation and Detoxification of Pollutants, pp. 51–72. De Gruyter, Berlin, Boston (2023a). https://doi.org/10.1515/9783110743623-003
14. Zhang, Z., Li, Y., Zhao, S., Qie, M., Bai, L., Gao, Z., Liang, K., Zhao, Y.: Rapid analysis technologies with chemometrics for food authenticity field: a review. Curr. Res. Food Sci. **8**, 100676 (2024). https://doi.org/10.1016/j.crfs.2024.100676
15. Chiappini, F.A., Alcaraz, M.R., Escandar, G.M., Goicoechea, H.C., Olivieri, A.C.: Chromatographic applications in the multi-way calibration field. Molecules (Basel, Switzerland) **26**(21), 6357 (2021). https://doi.org/10.3390/molecules26216357
16. Kokkinofta, R., Yiannopoulos, S., Stylianou, M.A., Agapiou, A.: Use of chemometrics for correlating carobs nutritional compositional values with geographic origin. Metabolites **10**(2), 62 (2020). https://doi.org/10.3390/metabo10020062
17. Byers, H.L., McHenry, L.J., Grundl, T.J.: XRF techniques to quantify heavy metals in vegetables at low detection limits. Food Chem. X **1**, 100001 (2018). https://doi.org/10.1016/j.fochx.2018.100001
18. Abimbola, O.F., Adetunji, C.O., Dare, B.J., Mathew, J.T., Inobeme, A., Ghazanfar, S., Olaniyan, O.T, Ajiboye, M.D., Dauda, W.: Packaging of animal-based food products: the need for intelligent freshness sensors. In: Book: Sensing and Artificial Intelligence Solutions for Food Manufacturing, pp. 89–102. CRC Press (2023). https://doi.org/10.1201/9781003207955-6
19. Forde, C.G., Bolhuis, D.: Interrelations between food form, texture, and matrix influence energy intake and metabolic responses. Curr. Nutr. Rep. **11**(2), 124–132 (2022). https://doi.org/10.1007/s13668-022-00413-4
20. Povarov, V.G., Kopylova, T.N., Sinyakova, M.A., Rudko, V.A.: Quantitative determination of trace heavy metals and selected rock-forming elements in porous carbon materials by the X-Ray fluorescence method. ACS Omega **6**(38), 24595–24601 (2021). https://doi.org/10.1021/acsomega.1c03217
21. Panuwet, P., Hunter, R.E., Jr., D'Souza, P.E., Chen, X., Radford, S.A., Cohen, J.R., Marder, M.E., Kartavenka, K., Ryan, P.B., Barr, D.B.: Biological matrix effects in quantitative tandem mass spectrometry-based analytical methods: advancing biomonitoring. Crit. Rev. Anal. Chem. **46**(2), 93–105 (2016). https://doi.org/10.1080/10408347.2014.980775
22. Adetunji, C.O., Shariati, M.A., Olugbenga, S.M., Osarenkhoe, O.O, Odoh, U.E., Olaniyan, O.T., Rebezov, M., Ajayi, O.O., Baibalinova, G., Bodunrinde, R.E., Adetunji, J.B., Adeniyi, M.J., Ugwu, P.N., Inobeme, A., Mathew, J.T., Elufisan, T.O., Omotayo, O.O.: Pineapple [*Ananas comosus* (L.) Merr.]: a biological and pharmacological active medicinal plant. In: Pullaiah, T. (ed.) Book: Phytochemical Composition and Pharmacy of Medicinal Plants, vol. 1, pp. 537–567. Apple Academic Press, Inc. Co-published with CRC Press (Taylor & Francis) (2023c)
23. Wu, X., Liang, X., Wang, Y., Wu, B., Sun, J.: Non-destructive techniques for the analysis and evaluation of meat quality and safety: a review. Foods **11**, 3713 (2022). https://doi.org/10.3390/foods11223713
24. McIntosh, K.G., Guimarães, D., Cusack, M.J., Vershinin, A., Chen, Z.W., Yang, K., Parsons, P.J.: Evaluation of portable XRF instrumentation for assessing potential environmental exposure to toxic elements. Int. J. Environ. Anal. Chem. **96**(1), 15–37 (2016). https://doi.org/10.1080/03067319.2015.1114104

25. Mathew, J.T., Inobeme, A., Adetuyi, B.O., Adetunji, C.O., Popoola, O.A., Olaitan, F.Y., Akinbo, O., Shahnawaz, M., Oyewole, O.A., Eniola, K.I.T., Yerima, B.M.: General introduction of microplastic: uses, types, and generation. In: Book: Microplastic Pollution, vol. 1, pp. 3–21. Springer, Singapore (2024a). https://doi.org/10.1007/978-981-99-8357-5_1
26. Filippelli, G., Anenberg, S., Taylor, M., van Geen, A., Khreis, H.: New approaches to identifying and reducing the global burden of disease from pollution. GeoHealth **4**(4), e2018GH000167 (2020). https://doi.org/10.1029/2018GH000167
27. Brent, R.J.: Cost-benefit analysis versus cost-effectiveness analysis from a societal perspective in healthcare. Int. J. Environ. Res. Public Health **20**(5), 4637 (2023). https://doi.org/10.3390/ijerph20054637
28. Schmidt, J., Marques, M.R.G., Botti, S., et al.: Recent advances and applications of machine learning in solid-state materials science. npj Comput. Mater. **5**, 83 (2019). https://doi.org/10.1038/s41524-019-0221-0
29. Bos, T.S., Knol, W.C., Molenaar, S.R.A., Niezen, L.E., Schoenmakers, P.J., Somsen, G.W., Pirok, B.W.J.: Recent applications of chemometrics in one-and two-dimensional chromatography. J. Sep. Sci. **43**(9–10), 1678–1727 (2020). https://doi.org/10.1002/jssc.202000011
30. Truong, V.K., Dupont, M., Elbourne, A., Gangadoo, S., Rajapaksha Pathirannahalage, P., Cheeseman, S., Chapman, J., Cozzolino, D.: From academia to reality check: a theoretical framework on the use of chemometric in food sciences. Foods (Basel, Switzerland) **8**(5), 164 (2019). https://doi.org/10.3390/foods8050164
31. Kim, J., Kim, D.-G., Ryu, K.H.: Enhancing response surface methodology through coefficient clipping based on prior knowledge. Processes **11**, 3392 (2023). https://doi.org/10.3390/pr11123392
32. den Uijl, M.J., Schoenmakers, P.J., Pirok, B.W.J., van Bommel, M.R.: Recent applications of retention modelling in liquid chromatography. J. Sep. Sci. **44**(1), 88–114 (2021). https://doi.org/10.1002/jssc.202000905
33. Chen, H.Y., Chen, C.: Evaluation of calibration equations by using regression analysis: an example of chemical analysis. Sensors (Basel, Switzerland) **22**(2), 447 (2022). https://doi.org/10.3390/s22020447
34. Abraham, E.J., Kellogg, J.J.: Chemometric-guided approaches for profiling and authenticating botanical materials. Front. Nutr. **8**, 780228 (2021). https://doi.org/10.3389/fnut.2021.780228
35. Salim, M.R., Widodo, R.T., Noordin, M.I.: Proof-of-concept of detection of counterfeit medicine through polymeric materials analysis of plastics packaging. Polymers **13**(13), 2185 (2021). https://doi.org/10.3390/polym13132185
36. Dürauer, A., Jungbauer, A., Scharl, T.: Sensors and chemometrics in downstream processing. Biotechnol. Bioeng. Bioeng. **121**(8), 2347–2364 (2024). https://doi.org/10.1002/bit.28499
37. Kopittke, P.M., Punshon, T., Paterson, D.J., Tappero, R.V., Wang, P., Blamey, F.P.C., van der Ent, A., Lombi, E.: Synchrotron-based X-Ray fluorescence microscopy as a technique for imaging of elements in plants. Plant Physiol. **178**(2), 507–523 (2018). https://doi.org/10.1104/pp.18.00759
38. Inobeme, A., Shahnawaz, M., Adetunji, C.O., Mathew, J.T., Adetuyi, B.O., Popoola, O.A., Olaitan, F.Y., Akinbo, O., Kolawole, O.M., Oyewole, O.A., Eniola, K.I.T., Yerima, B.M.: Microplastic as a pathogenic vector to cause diseases in marine biota. In: Book: Microplastic Pollution, vol. 1, pp. 321–349. Springer, Singapore (2024b). https://doi.org/10.1007/978-981-99-8357-5_19
39. Dehayem-Kamadjeu, A., Okonda, J.: X-Ray fluorescence analysis of selected micronutrients in ten African indigenous leafy vegetables cultivated in Nairobi, Kenya. Pan Afric. Med. J. **33**, 296 (2019). https://doi.org/10.11604/pamj.2019.33.296.19501
40. Panebianco, S., Pellegriti, M.G., Finocchiaro, C., Musumarra, A., Barone, G., Caggiani, M.C., Cirvilleri, G., Lanzafame, G., Pulvirenti, A., Scordino, A., Mazzoleni, P.: XRF analysis searching for fingerprint elemental profile in south-eastern Sicily tomatoes. Sci. Rep. **13**(1), 13739 (2023). https://doi.org/10.1038/s41598-023-40124-6

41. Olson, R., Gavin-Smith, B., Ferraboschi, C., Kraemer, K.: Food fortification: the advantages, disadvantages and lessons from sight and life programs. Nutrients **13**(4), 1118 (2021). https://doi.org/10.3390/nu13041118
42. Tura, D.C., Belachew, T., Tamiru, D., Abate, K.H.: Optimization of a formula to develop iron-dense novel composite complementary flour with a reduced phytate/minerals molar ratio from dabi teff-field pea-based blends using a D-optimal mixture design. Front. Nutr. **10**, 1244571 (2023). https://doi.org/10.3389/fnut.2023.1244571
43. Mohamed, N.H., Yahya, G., El Bayoumi, R.M., Hussein, M.A.M., Cavalu, S., Dahshan, H., Alsayeqh, A.F., Darwish, W.S., Nasr, E.A.A.: Detection and health risk assessment of toxic heavy metals in chilled and frozen meat collected from Sharkia province in Egypt. Open Veter. J. **13**(12), 1729–1737 (2023). https://doi.org/10.5455/OVJ.2023.v13.i12.21
44. Żak, N., Wilczyńska, A.: The importance of testing the quality and authenticity of food products: the example of honey. Foods (Basel, Switzerland) **12**(17), 3210 (2023). https://doi.org/10.3390/foods12173210
45. Baudry, J., Kopp, J.F., Boeing, H., Kipp, A.P., Schwerdtle, T., Schulze, M.B.: Changes of trace element status during aging: results of the EPIC-Potsdam cohort study. Eur. J. Nutr. **59**(7), 3045–3058 (2020). https://doi.org/10.1007/s00394-019-02143-w
46. Santos, J.A.R., Christoforou, A., Trieu, K., McKenzie, B.L., Downs, S., Billot, L., Webster, J., Li, M.: Iodine fortification of foods and condiments, other than salt, for preventing iodine deficiency disorders. Cochr. Database Syst. Rev. **2**(2), CD010734 (2019). https://doi.org/10.1002/14651858.CD010734.pub2
47. Alamri, M.S., Qasem, A.A.A., Mohamed, A.A., Hussain, S., Ibraheem, M.A., Shamlan, G., Alqah, H.A., Qasha, A.S.: Food packaging's materials: a food safety perspective. Saudi J. Biol. Sci. **28**(8), 4490–4499 (2021). https://doi.org/10.1016/j.sjbs.2021.04.047
48. Selivanovs, Z., Panchuk, V., Kirsanov, D.: On the possibility of universal chemometric calibration in X-Ray fluorescence spectrometry: case study with ore and steel samples. Appl. Sci. **13**, 5415 (2023). https://doi.org/10.3390/app13095415
49. Tarakhovskaya, E., Marcillo, A., Davis, C., Milkovska-Stamenova, S., Hutschenreuther, A., Birkemeyer, C.: Matrix effects in GC-ms profiling of common metabolites after trimethylsilyl derivatization. Molecules (Basel, Switzerland) **28**(6), 2653 (2023). https://doi.org/10.3390/molecules28062653
50. Liu, C., Zhang, X., Nguyen, T.T., Liu, J., Wu, T., Lee, E., Tu, X.M.: Partial least squares regression and principal component analysis: similarity and differences between two popular variable reduction approaches. Gen. Psychiatry **35**(1), e100662 (2022). https://doi.org/10.1136/gpsych-2021-100662
51. Hazra, A., Gogtay, N.: Biostatistics series module 10: brief overview of multivariate methods. Indian J. Dermatol. **62**(4), 358–366 (2017). https://doi.org/10.4103/ijd.IJD_296_17
52. Adetunji, C.O., Egbuna, C., Ajiboye, A.A.A., Ajayi, O.O., Dauda, W.P., Ghazanfar, S., Ogundolie, F.A., Mathew, J.T., Adetunji, J.B., Inobeme, A., Olotu, T., Olaniyan, O.T., Ijabadeniyi, O.A., Aborode, A.T., Ficai, A., Akram, M., Bodunrinde, R.E., Laila, U., Olaniyan, O.P., Micheal, O.S., Aliyu, A., Akinbo, O., Adetuyi, B.O.: Biogenic nanomaterials with diverse biological activities in the food and biomedical industries. In: Adetunji, C.O., Hefft, D.I., Mbuge, D.O., Workneh, T.S. (eds.) Evaporative Coolers for the Postharvest Management of Fruits and Vegetables, pp. 395–420. Academic Press (2023a), https://doi.org/10.1016/B978-0-323-89864-5.00001-1.
53. Tavares, T.R., Mouazen, A.M., Alves, E.E.N., dos Santos, F.R., Melquiades, F.L., Pereira de Carvalho, H.W., Molin, J.P.: Assessing soil key fertility attributes using a portable X-Ray fluorescence: a simple method to overcome matrix effect. Agronomy **10**, 787 (2020). https://doi.org/10.3390/agronomy10060787
54. Mierzyńska, Z., Niemirska, M., Zgonina, K., Bieńkowski, T., Hryniów, K., Świder, P., Pawlak, K.: Multi-elemental analysis of hair and fingernails using energy-dispersive X-Ray fluorescence (ED XRF) method supported by inductively coupled plasma mass spectrometry (ICP MS). Molecules (Basel, Switzerland) **29**(4), 773 (2024). https://doi.org/10.3390/molecules29040773

55. Stanimirova, I., Daszykowski, M.: Assessment of the Kernel gram matrix representation of data in order to avoid the alignment of chromatographic signals. Molecules (Basel, Switzerland) **26**(3), 621 (2021). https://doi.org/10.3390/molecules26030621
56. Paliy, O., Shankar, V.: Application of multivariate statistical techniques in microbial ecology. Mol. Ecol. **25**(5), 1032–1057 (2016). https://doi.org/10.1111/mec.13536
57. Mathew, J.T., Inobeme, A., Abdulkadir, A., Etsuyankpa, M.B., Shaba, E.Y., Musah, M., Azeh, Y., Adetunji, C.O., Ibrahim, A.M., Tanko, E., Mamman, A.: Properties and characteristics of nanocoatings for medicinal applications. In: Book: Sustainable Approach to Protective Nanocoatings, publishing Tomorrow's Research Today, pp. 203–227. IGI Global (2024a). https://doi.org/10.4018/979-8-3693-3136-1.ch007
58. Baldoni, P.L., Sotres-Alvarez, D., Lumley, T., Shaw, P.A.: On the use of regression calibration in a complex sampling design with application to the hispanic community health study/study of Latinos. Am. J. Epidemiol. **190**(7), 1366–1376 (2021). https://doi.org/10.1093/aje/kwab008
59. El Hosry, L., Sok, N., Richa, R., Al Mashtoub, L., Cayot, P., Bou-Maroun, E.: Sample preparation and analytical techniques in the determination of trace elements in food: a review. Foods (Basel, Switzerland) **12**(4), 895 (2023). https://doi.org/10.3390/foods12040895
60. Rohde, A., Hammerl, J.A., Appel, B., Dieckmann, R., Al Dahouk, S.: Sampling and homogenization strategies significantly influence the detection of foodborne pathogens in meat. Biomed. Res. Int. **2015**, 145437 (2015). https://doi.org/10.1155/2015/145437
61. Adetunji, C.O., Olaniyan, O.T., Shariati, M.A., Bodunrinde, R.E., Ajayi, O.O., Adetunji, J.B., Hefft, D.I., Olugbenga, S.M., Rebezov, M., Khayrullin, M., Adeniyi, M.J., Inobeme, A., Mathew, J.T.: Pharmacology, phytochemistry, and medicinal usefulness of *Basella alba* L. In: Pullaiah, T. (ed.) Book: Phytochemical Composition and Pharmacy of Medicinal Plants, vol. 2, pp. 499–516. Apple Academic Press, Inc. Co-published with CRC Press (Taylor & Francis) (2023b)
62. Westad, F., Marini, F.: Validation of chemometric models–a tutorial. Anal. Chim. Acta **893**, 14–24 (2015). https://doi.org/10.1016/j.aca.2015.06.056
63. Inobeme, A., Mathew, J.T., Adetunji, C.O., Agbugui, M.O., Osarenren, E.J., Oti, C., Ngonso, B.F., Inobeme, J., Oyedolapo, M.B., Bernard, E., Jibrin, N.A.: Biocomposites with graphene derivatives. In: Book: Advances in Biocomposites and Their Applications, pp. 149–166. Elsevier Inc. (2024a). https://doi.org/10.1016/B978-0-443-19074-2.00005-8
64. Ahmed, Z., Mohamed, K., Zeeshan, S., Dong, X.: Artificial intelligence with multi-functional machine learning platform development for better healthcare and precision medicine. Database J. Biol. Databases Curat. **2020**, baaa010 (2020). https://doi.org/10.1093/database/baaa010
65. Mathew, J.T., Inobeme, A., Musah, M., Azeh, Y., Abdullahi, A., Shaba, E.Y., Salihu, A.M., Muhammad, E.B., Josiah, J.G., Jibrin, N.A., Ismail, H., Muhammad, A.I., Maurice, J., Mamman, A., Ndamitso, M.M.: A critical review of green approach on wastewater treatment strategies. J. Appl. Sci. Environ. Manag. **28**(2), 363–391 (2024d). https://doi.org/10.4314/jasem.v28i2.9
66. Kharbach, M., Alaoui Mansouri, M., Taabouz, M., Yu, H.: Current application of advancing spectroscopy techniques in food analysis: data handling with chemometric approaches. Foods (Basel, Switzerland) **12**(14), 2753 (2023). https://doi.org/10.3390/foods12142753
67. Moktan, H., Jayarathna, S., Cho, S.H.: Sensitivity enhancement of an experimental benchtop x-ray fluorescence imaging system by deploying a single crystal cadmium telluride detector system optimized for high flux x-ray operations. In: Nuclear Instruments & Methods in Physics Research. Section A, Accelerators, Spectrometers, Detectors and Associated Equipment, vol. 1025, p. 166198 (2022). https://doi.org/10.1016/j.nima.2021.166198
68. Williams, W.L., Zeng, L., Gensch, T., Sigman, M.S., Doyle, A.G., Anslyn, E.V.: The evolution of data-driven modeling in organic chemistry. ACS Cent. Sci. **7**(10), 1622–1637 (2021). https://doi.org/10.1021/acscentsci.1c00535
69. Adetunji, C.O., Akinbo, O., Mathew, J.T., Egbuna, C., Inobeme, A., Olotu, T.M., Ajayi, O.O., Dauda, W., Ghazanfar, S., Ogundolie, F.A., Adetunji, J.B., Adetuyi, B.O., Ajenifujah-Solebo, S.O., Aborode, A.T.: Relevance of nanomaterials derived from medicinal plants for marine and

terrestrial environments: recent advances. In: Book: Handbook of Agricultural Biotechnology, vol. 1, pp. 311–326. John Wiley & Sons (2024). https://doi.org/10.1002/9781119836483.ch12

70. Shah, R., Leigh, L., Kaewmanee, M., Pinto, C.T.: Validation of expanded trend-to-trend cross-calibration technique and its application to global scale. Remote Sens. **14**, 6216 (2022). https://doi.org/10.3390/rs14246216
71. Mathew, J.T., Inobeme, A., Adetunji, C.O., Azeh, Y., Otori, A.A., Musah, M., Shaba, E.Y., Badeggi, M.U., Mamman, A., Jemkur, M., Oyewole, O.A.: Characterization of nanoparticles used as nanobiofertilizers. In: Book: Handbook of Agricultural Biotechnology, vol. 5, pp. 175–199. Wiley (2024c). https://doi.org/10.1002/9781394211548.ch7

12 X-Ray Fluorescence for Geological Samples Combined with Chemometrics

Ankit Roy

Abstract

X-ray fluorescence (XRF) spectroscopy is one of the most fundamental and trusted analytical techniques for rapid analysis of geological samples. Geological samples are typical because of their inhomogeneous morphology and diversified matrices, unusual patterns of association of elements with soils and rocks, variability in grain size and composition. XRF spectra of a simple geological motif thus can produce very complex variables and possibilities. To handle these large quantum of data, one needs sophisticated data processing tools and leveraging the same in X-ray spectrometric experiments can indeed provide propitious results. Chemometric methods when applied to XRF can be a powerful tool to categorically differentiate samples into desired groups, to quantitatively characterize clustered, overlapping spectra and above all to provide veiled interrelation among variables. This chapter intends to delineate the multivariate approaches being used in the field of XRF analysis for geological samples. Concepts of spectral pre-processing, selection of variables, data dimensionality reduction and pattern recognition has also been addressed. Principal component analysis, hierarchical cluster analysis and partial least squares regression are the main components which seek attention. Other significant methods of chemometrics include factor analysis, linear discriminant analysis, support vector machines, artificial neural networks, multivariate cure resolution and data fusion.

A. Roy (✉)
Geological Survey of India, Eastern Region, Kolkata, India
e-mail: ankit.roy@gsi.gov.in

V. K. Singh (ed.), *X-Ray Fluorescence Spectroscopy and Chemometrics*,
https://doi.org/10.1007/978-3-031-98375-7_12

12.1 Introduction

X-ray fluorescence (XRF) is perhaps the most celebrated spectroscopic technique when it comes to the analysis of geological materials [1–3]. It is also a non-destructive, time saving and cost effective method for routine analysis [4]. The wide acceptance of XRF in geological domain is due to the fact that one can get quantitative measurements with very high precision over a short period of time for almost entire periodic table (from B to U) in a single scan [5]. Moreover, the method has no strenuous sample preparation steps, it can be used for all kinds of geological materials whether rocks, soil, sediments, metals and even for liquids like water and crude oil. Although this reliable technique has been in use in geological analyses for nearly six to seven decades, for the last few years XRF has reached new heights with the development in instrumentation (such as better X-ray optics, high resolution silicon based semiconductor detectors, sufficient intensity detection tools even from smaller surface areas) as well as in data processing mechanism [6–8]. Developers have made possible the conversion of overall x-ray analytical set up into a miniature version of portable handheld XRFs, alpha particle x-ray spectrometers etc. which further allow to determine the compositional attributes on a real-time scale by travelling across not only in earthen geological sites but also in distant planetary systems [8–10]. Though a majority of geochemical research still depend on two broad spectrum XRF categories viz. energy dispersive (ED) and wavelength dispersive (WD)-XRF, quite a good number of researches have been carried out to check the performance of micro-XRF, total reflection XRF (TXRF), polarized XRF, PIXE etc. [8]. The advantages of these new XRF techniques lie in their micro focused excitation sources, very small amount of sample consumption and improved sensitivity. Characterization and profiling of geological samples by XRF have also gained momentum in the recent past [11–15]. These tools facilitate transformation of geological data into specialized and thematic maps of a particular geological setting. Since XRF is applied to a very large number of sample points in almost every geological experiment, the data generated are often difficult to visualize, manage and interpret by the exact needs [16, 17].

Chemometrics, a recent development in the field of analytical chemistry where mathematical, statistical and computational methods are applied to discover sensible outcomes that are otherwise hidden in a given dataset, can be quite effective in easing out such complex problems [18–21]. The works related to XRF and chemometrics are gaining lightning importance in all fields of applied sciences and geological domain is no exception. Figure 12.1 shows yearly comparison of research papers published with the search title "XRF and chemometric studies for geological samples" of last 25 years (2025 excluded) which ascertains the aforementioned statement using the source "Science Direct".

This chapter, therefore, aims to portray the current global picture in respect of applicative XRF along with chemometrics that are employed in explaining geological observations.

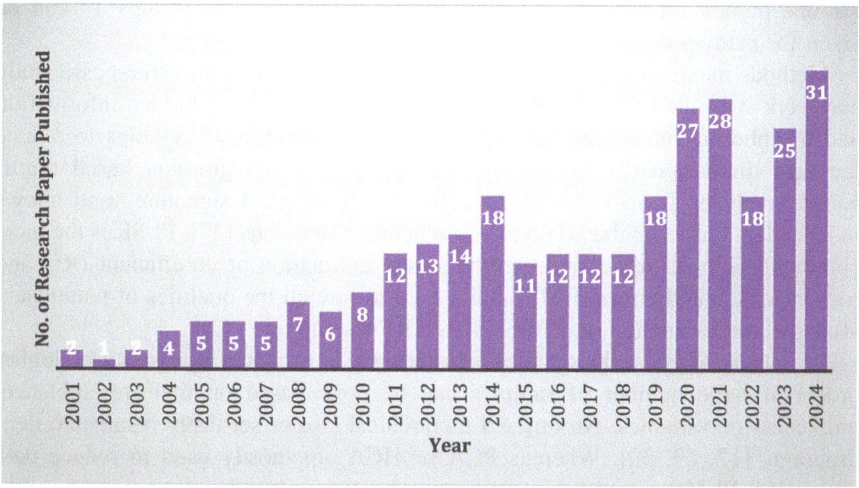

Fig. 12.1 Bar chart showing the number of research papers with search title "XRF and chemometric studies for geological samples" published from 2001 to 2024. The vertical axis represents the number of papers, while the horizontal axis shows the years. The data indicates a general upward trend, with notable increases in 2014, 2020, and 2024, reaching a peak of 31 papers in 2024

12.2 Necessity of Multivariate Approaches in Geology

Geological phenomena are multivariate. The timeline of occurrence for geological events is decently large and the process is slow but accumulative in nature [22]. Indicators from structural, geological and mineralogical factors those influence the physical characteristics seem to give multi-element patterns and anomalies as far as geochemical exploration is concerned. A knowledge of changes and classification of different microscopic as well as macroscopic variables can be beneficial for knowing earth crust, classifying soil patterns, targeting ores and apprehend economic significance. Geochemistry, in this regards, is helpful to identify lithology, alteration with respect to mineral abundance, host–guest interaction between rocks [23]. Hence to understand the inter-variable relationship in terms of geology, one needs to take multivariate approaches rather than classical univariate ones.

Chemometrics supports to establish such multivariate analytics in geochemistry. Its extensive practice in geochemical studies has encouraged scientists to use the term "Geochemometrics" in current literatures [24]. Incorporation of statistical tools like principal component analysis (PCA), hierarchical cluster analysis (HCA), partial least squares regression (PLSR), factor analysis (FA), linear discriminant aAnalysis (LDA), help in additional data generation [25]. A good way to simply classify these is to order them according to the number of population sampled rather than classify them solely based on theoretical statistical features [26]. Thus methods like PCA and multivariate normalization can be categorized for one population samples while discriminant analyses, Hotelling's T^2 can be taken for many population samples.

Methods mentioned above are the fundamentals of any multivariate case study and work with the principle of segregating batches, decoding hidden information and deciphering interesting sample features. For example, PCA helps to reduce the data dimensionality before spectral application in regression based multivariate analysis. It also mirrors local/regional geological signatures and allows independent clustering based on physicochemical functions [27]. PLSR is the most commonly applied regression method where correlation of co-efficient (R^2) and root mean squared errors (RMSEs) are used to foretell the qualities of parameters after spectral evaluation (e.g. VIS–NIR, XRF-Raman, etc.) [28].

By means of HCA, hierarchical clustering (i.e. each set is a subset of higher order) of large number of samples can be done based on their resemblance/unlikeness of variables. An ordered hierarchical cluster set lastly result into dendrograms [17, 29, 30]. Whereas PCA or HCA are mostly used to reduce data dimension, PLSR is used for quantification because it has the advantage of ability to make calibration against parameters [31]. Geochemical analyses by XRF often suffer the problem of matrix matching calibrators which can be addressed by using this technique. A more recent method involving least square regressions is uncertainty based weighted least square regression (UWLR) which can produce more reliable concentrations of elements in geochemical samples by XRF [32].

Factor analysis gathers sets of similarly behaving variables into groups called 'Factors'. In geochemical sense these factors signify cohesive effects of various geo-dynamic equilibria like sedimentation, secondary dispersion, weathering, changes in mineralization etc. [33]. LDA utilizes a linear combination of variables which can maximize the separation between a group of similar samples [34]. This discrimination is based on a linear equation or hyperspace when the boundaries of classes are linear [35]. For models where linearity between different set of observables lacks predominantly, Quadratic Discrimination with non-linear functions may be applied. A prerequisite of LDA is that the number of variables has to be fewer than the number of data points. However, in case of data points originating from geological landscape most of the time contains a higher number of observables, which may become a constraint to apply LDA extensively. In those cases, HCA or PCA can be applied first to reduce the number of variables and then LDA can be utilized on the score matrix derived from PCA [34].

While the fundamental statistical tools serve majority of chemometrics, machine learning algorithms like support vector machines, artificial neural networks (ANN), multivariate cure resolution (MCR), regression trees (RTs), random forest (RF) are also some advanced powerful data driven methods. However, they have come up very recently and still growing as far as geochemical applications are concerned. Support vector machine, a supervised learning model, has stand out to be an alternative method of lithology and sub-surface classification [36, 37]. This tool can also be employed in mapping regolith geology, mineral prospectively areas, measuring landslide susceptibility, and in slope stability analyses [38–43].

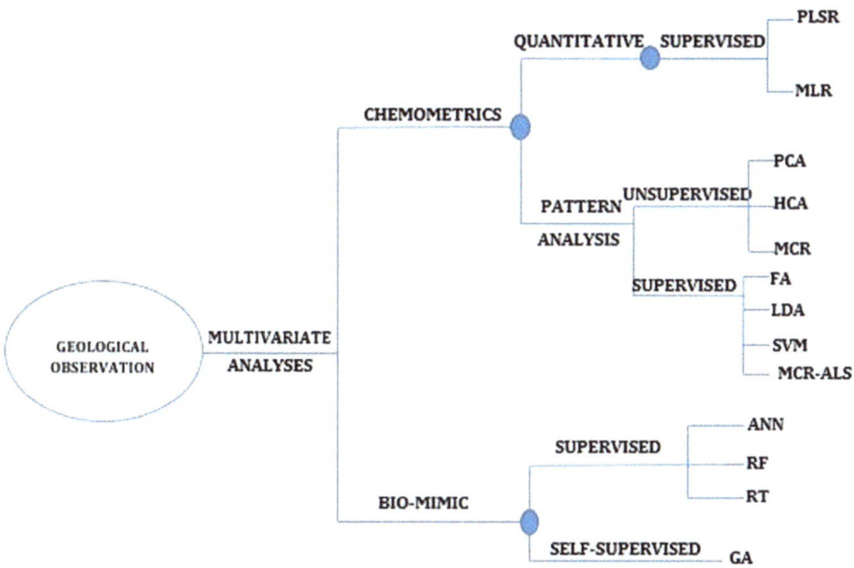

Fig. 12.2 Chemometric tools applied in geochemical studies

Artificial neural networks, on the other hand, are techniques derived from artificial intelligence replicating the structure and functioning of biological neuron systems [44]. Neural modelling of mining process are the significant contributors in decision making process while neural networks are also important for multi-dimensional anomaly identification in exploration geochemistry, for identification of hydrocarbon layers and physical characteristics of reservoir oil [45].

In the age of machine learning and artificial intelligence various types of algorithms are getting generated at a rapid pace. In this onset the major and minor ones applied in geochemical studies are snapped in Fig. 12.2 for a quick visit. An interesting fact which needs to be mentioned here that 66.14% of the published research articles with respect to the search item "multivariate analysis in geology" contains XRF data while searching in Google Scholar. This indeed emphasises the need to go through the application of XRF and chemometrics in geological samples.

12.3 Geochemical Variables and Their Selection

Geochemistry is vastly dependent on geochemical factors ranging from local mineral composition to large scale geology, type of matrix and topography. For such a large number of variables, no single probability function is suitable for all elements of interest and they seldom follow a normal Gaussian distribution [46]. This absence in normal relationship makes it more challenging to incorporate statistical tools in the domain of geochemistry. Prior understanding of probability features of geochemical variables is needed to process the data by fruitful chemometrics based application for geological samples.

Another persisting problem here is the abundance of noises around the dataset of actual interest. The end game is to find recognizable patterns which obviously need to transform the data and to remove the unwanted ones. Possible approaches to find geochemical patterns in large number of data include unsupervised and supervised learning [47]. Out of these two, the former approach is recommended widely and here data processing is done by simple to complex tools where the task of a geo-analyst remains in interpreting the possible sources so that patterns can be easily recognized [48, 49]. The second approach, however used less wisely, can also be useful when data from known sub-groups of rocks or soil are fitted into unknown sets to classify them in a systematic manner [50, 51]. It is worth mentioning in this regard that the sample density and survey area are two crucial aspects based on which one may use unsupervised or supervised approaches for any geochemical investigation. As the sample density decreases the chance of getting a mineralization or a deposit arbitrarily also decreases. Therefore, unsupervised approaches get advantage for large scale surveys.

For better understanding the geochemical variables let us take an example of the process called 'mineralization'. The process of minerals being deposited or chemically altered in rocks is very much different from the background geological changes. In this process some key elements can concentrate spatially giving rise to abundance in a particular set of minerals. This assemblage can be mapped in a toposheet by quantification of the clustered variables. Figure 12.3a and b show how mineralized entities can be segregated from background.

Mineral identification is also based on their mode of occurrence while identifying them in terms of population distribution. A cumulative distribution having two or more number of modes may be interpreted as possible occurrence of mineralization. Again for trace elements, if the standard deviation is fairly greater than 1 or the distribution is non-lognormal, chances of mineral association becomes higher. Sub-groups can also be identified out of multiple population by using scatter plot of a pair of variables.

A distinct cluster out of randomly spread data points is what a geochemist ideally seek for; the practical situation is rarely so specific. Once the dependence in terms of variables are identified, occurrence of deposits or mineralization can be specified by treating the data in a more filtered way by applying dimensionality reduction, data scaling and centring methods [17, 47, 52].

The following steps that should be considered when obtaining and compiling geochemical data prior to chemometric analyses and interpretation:

- **Appropriate Choice of Sample and Size Fraction**

Choice of sample media is one of the crucial aspects for any geochemical exploration. For example, sampling static rock gives geochemical dataset pertaining to the overall landmass of any region whereas sampling regolith resulting from weathered bed rocks can give information about the original unmetamorphosed

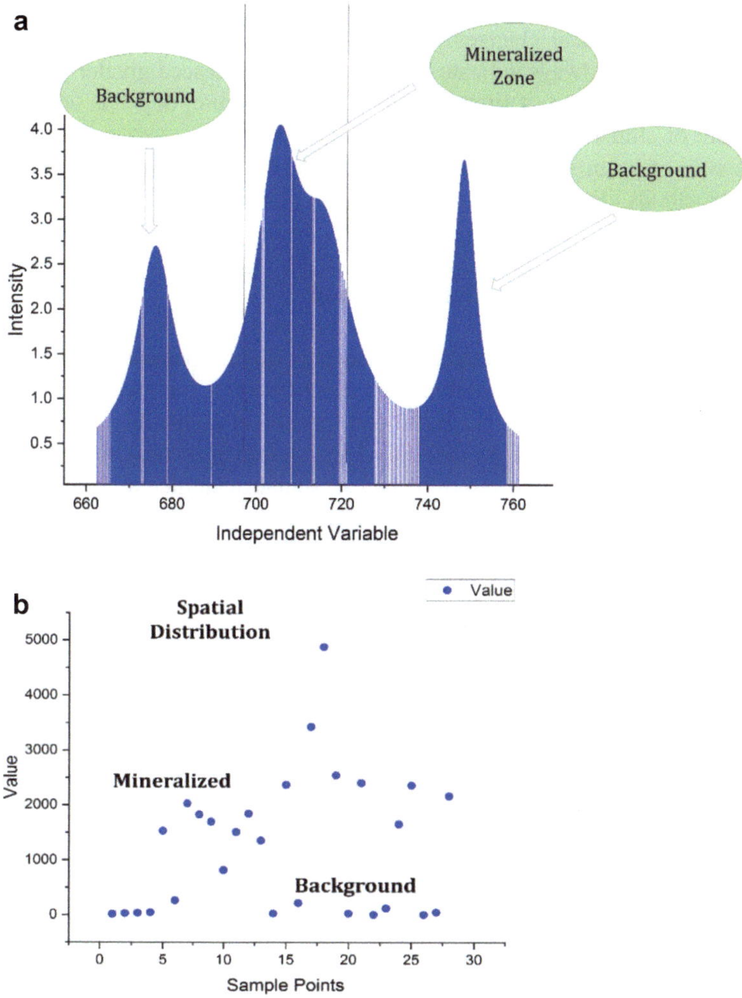

Fig. 12.3 **a** Imaginary mineral identification from background, **b** Imaginary mineral clustering in 2D space

rock and its weathered counterpart. On the other hand, sampling transported geomorphs like second or higher order stream sediments, alluvials etc. can indicate mixing of geogenic substances over the period of time. The underlying idea behind right choice of sample media is to understand capabilities as well as the limitations before data interpretation [53].

Size fraction of samples is also of great importance in differentiating different geochemical processes [54]. A formal distinction involves the grain size: samples with grain size greater than 63 microns but less than 2 mm are called 'coarse' while samples with grain size less than the above limit are fine grained. Coarse grained samples refer to locally derived ones or those which have not undergone weathering or chemical changes over a long period of time. A relatively finer geological samples always indicate that they have experienced greater geochemical alterations over a period of time but this, again in its turn, depends on the source and nature of climatic processes that occurred [55].

The role of grain size becomes significantly important especially for sample analysis methods like XRF because method like pressed pellet XRF which can give precise value of trace mineral concentrations, needs a finer particle size [56–58].

- **Appropriate Choice of Analytical Method**

The choice of appropriate analytical method is also important for the fact that some can be non-destructive while others can be destructive methods of analysis. If the sampling is very tough due to difficult location or access then non-destructive methods like XRF, NAA are preferred to deliver a true analysis. On the other hand, if organic matters are to be removed before actual quantification digestion techniques become important. Digestion methods also are useful for resistant minerals like zircon, monazite etc. When some of the important fractions tend to adsorb on mineral, extraction or leaching techniques are employed. Methods like total fusion come into play for actual quantification of major and minor oxides.

- **The Closure Problem**

The geochemical results are always reported as a part of overall composition of the matrix because the matrix itself may contain everything. Therefore, all the values are relative and should be normalized to a constant (e.g. 100% or 10,000 ppm). Thus the variables are always dependent to each other while in statistics every variable is independent to the other. One problem arises when the statistical methods are applied to real world geochemical perspectives. Statistical methods assume the possibility of any number of values while values in geochemical methods are limited to 0 to 100%. To overcome this closure or constant sum problem, use of ratios amongst variables, simple use of log-ratios is required while comparing the data points in two or more spaces [59–61]. Log-ratios can be additive, centred and isometric depending on the compositional choice of the study.

12.4 Chemometrics, XRF and Geology

The first application of chemometric data processing with respect to XRF was done in 1980s in which factor and cluster analyses were performed to XRF data to correlate various natural and external factors with elemental compositions of geological samples [62–64]. Since majority of geochemical analyses comprise XRF, nearly one third of XRF analyses combined with chemometrics are from geochemistry origin [17]. Recent researches also show this continuum [65–68].

Pre-processed XRF dataset may be subjected to unsupervised techniques like Principal Component Analysis and Cluster Analysis.

To date, PCA has been the most commonly used multivariate tool for this purpose in soil studies [69–76] and has been mainly applied for the purpose of identification. It lessens dimension of original data containing a larger number of possibly correlated variables to a fewer number of uncorrelated variables, known as principal components (PCs) [73]. Similar kind of samples form separate groups by virtue of PCA and moreover, samples those have abnormal variables, called outliers, remain apart in the scores plot [73]. A typical PCA operation converts a dataset matrix A into two matrices. The first one is the scores matrix and the second one is the loading. Plotting the principal components amongst each other provide a score plot while the loading plot is used to understand which input variables have greater weight in the actual data upon calculating the PCs. The importance of the first PC remains the most and weightage of consecutive PCs decreases gradually. The PCs do not divide the samples on any one particular property but use them all [75].

A few examples are illustrated where XRF data of geological/geochemical studies were subjected to pattern recognition by unsupervised chemometric techniques and researchers were able to grasp hidden information which cannot be done otherwise:

- In a case study from the Cadillac–Larder lake fault zone, Abitibi, Canada gold deposits hosted by various lithology were analysed by XRF for a plethora of elements including Si, Ti, Al, Fe, Mn, Mg, Ca, Na, K, P, Mn, C, S, Sc, V, Cr, Co, Ni, Cu, Zn, Ga, Ge, As, Se, Rb, Sr, Y, Zr, Nb, Mo, Ag, Sb, Ce, Ba, Nd, Eu, Ho, Er, Yb, Lu, Ta, W, Re, Pt, Au, Hg, Te, Pb, Bi, Th, and U [77]. Five PCs having eigen value greater than one were extracted from all the XRF data out of which the first three accounted for 68% variance. The correlation of elements and minerals in the loading plot governed the distribution of samples in the score plot. Plotting principal component scores and loadings in the same space allowed the identification of relationships among elements and minerals. The terms PC_E and PC_M were used to represent principal components based on elemental geochemistry and mineral abundance respectively.

Figure 12.4 shows results of PCA of centred log ratio transformed elemental concentrations. Element loadings for principal components having absolute values > 0.5 are considered statistically meaningful.

Figure 12.5 depicts the relationship between different PCs. The PC_{E1} component had positive Al, Fe, Mn, Mg, Ca, K, V, Ni, Zn, but negative Pt, W, Co, Hg, Re, and Au loadings whereas for PC_{E2} positive loadings for Cu, S, Si, Cr, Ni, and Fe, and negative loadings for Zr, P, and Ba were seen. The PC_{E3} component has positive loadings for Ti and K, and negative loadings for Mo. Albeit the fact that PC_{E3} had lower absolute loadings, it also positively correlated with Au, Re, and Hg, and negatively correlates with Co, Pt, and Mo.

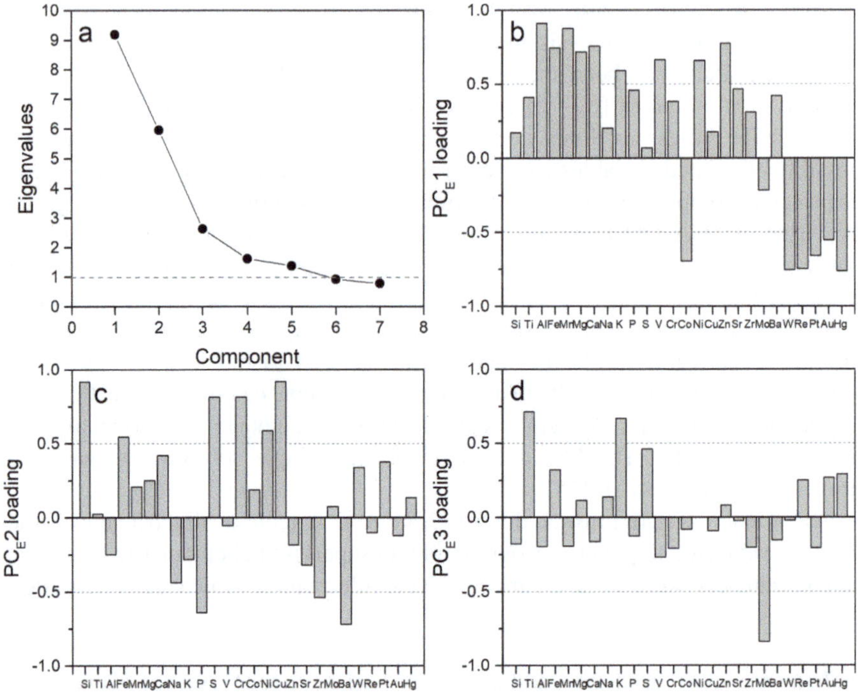

Fig. 12.4 Principal component analysis output of centre log-ratio transformed within mineralised rocks from orogenic gold deposits in Abitibi: **a** Eigen values of PCs, **b** Elemental loading of first three significant principal components. Figure adopted from Meng et al. [77] with the permission from Elsevier

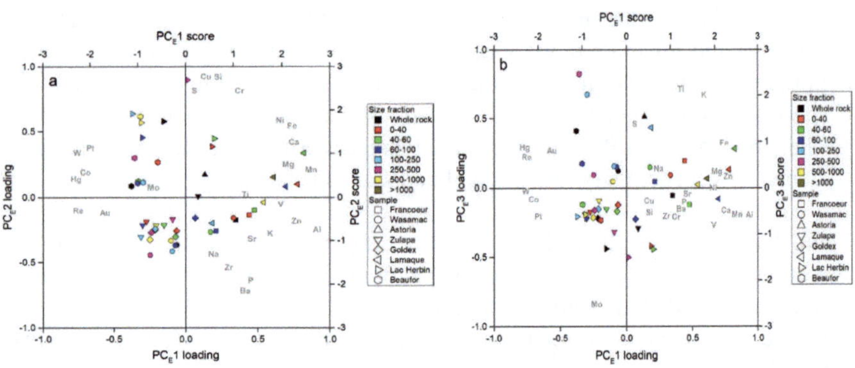

Fig. 12.5 Score plots of elemental concentrations within ore bearing rocks. Figure adopted from Meng et al. [77] with the permission from Elsevier

Using the first principal component along with elemental concentrations and their absolute loadings, the authors defined Element Mineralization Index (EMI):

$$\text{EMI} = \sum(C(X) * AL(X)) / \sum(C(Y) * AL(Y))$$

where C(i) is the elemental content for i^{th} element and AL(i) is the absolute loading of ith element.

X includes the ore-related elements Co, W, Re, Pt, Au, and Hg, and Y includes the rock forming elements Al, Fe, Mn, Mg, Ca, K, V, Ni, and Zn.

This study showcases how PCA can be used as a guided tool for effective differentiation between mineralised and barren rocks in gold deposits. A ready dependence of mineralization on principal components can be seen which helps to differentiate the mineralised zones.

PCA followed by PLSR study of XRF data can give information on the paleogeography of dried lake which are important for fossil preservation. XRF data of pragmatic earthen resources, macro and micro fossils, archaeological and anthropogenic events can be mapped to estimate past environmental, climatic, geographical and historical changes. An investigation carried out by M. Topuz et al. show that multivariate analysis of bottom sediments taken from dried lake Gavur, Turkiye can be beneficial for understanding the geological past of the region [78]. Sediment samples were taken from eleven depths (n = 11) and XRF analysis was carried out for non-oxidised and oxidised chemical elements. Radiocarbon dating of two sediment samples at depths 2.5 and 5.0 m showed the conventional ages were 4752 ± 33 BP and 5470 ± 35 BP (BP = before present). The basic statistical data comprised of 17 variables. These were taken as X variables in PCA analysis while in case of PLSR regression sampling depth (0–5 m) was used as Y variables with elements as X variables. Full cross validation (FCV) was done for both the treatments. The data were normalised to the factor of 1/SD for each variable to get equal chance to influence the model. The scores and loading plots, root means square errors and co-efficient of determination (R^2) were examined thereafter.

High negative correlations were observed for Sr, S + SO_3 and ZrO_2 while positive correlations among Ti + TiO_2, Al + Al_2O_3 and Cu + CuO. The correlation coefficients indicate that Sr, Zr, Ca, Mn follow a decreasing order of abundance with increasing depth whereas Si, Al, Ti, Cu follow the reverse order.

PCA analysis showed that the model had eight numbers of PCs accounting for a total of 98% and 75% variance in X variables in the calibration and FCV stages, respectively. The score plots with 51% PC1 variability and 22% PC2 variability showed a distinct grouping of samples. Loading plots revealed that three different clusters of elements viz. Si, Al, Mg, K, Ti, P, Cu, Rb in cluster-1 with positive loading, Fe, Mn, Cr, Ni in cluster-2 with negative loading and Ca, S, Sr, Zn, Zr in cluster-3 with almost neutral loading in PC1 axis.

Thus PCA study showed that samples can be grouped as deeper samples (4.0–5.0 m depth) and other samples (0–3.5 m depth). Their loading plot revealed that elements like Fe, Mn, Cr, Ni had least importance towards the geochemistry of the lake and surroundings.

The PLSR analysis was performed with these sets of samples to predict the sampling depth from concentrations of the elements. A good correlation with R^2cal = 0.95; R^2val = 0.67 and RMSEC = 0.37 m, RMSEP = 1.0 m was found in the PLSR model. A similar trend to PCA was also observed here. Elements Si, Al, Mg, K, Ti, P, Cu, Rb had positive, Ca, S, Sr, Zn, Zr had negative and Fe, Mn, Cr, Ni had neutral correlations with sampling depth. The overall outcome definitely indicated that variation of the elemental composition in the sediments had a strong reliance on the sample depth.

Strontium is known to be related with pterobod abundance and marker of biogenic origin, sulphur is related to pyrite and zirconium with tephra layers [79]. In this work, Sr had lower concentration in deeper samples while S and Zr had high negative correlation with sample depth. Availability of heavy metals in old lake sediments were indicative of anthropogenic activities in the past. High calcium content is linked with dry weather while high Ti concentration with rainy weather. For samples of lake Gavur the Ti + TiO_2 content was higher and Ca + CaO content was lower in deeper samples. It was also seen that samples at 2.5 m depth had twice the concentrations of Ca + CaO, Sr, Zn + ZnO and Zr + ZrO_2 when compared to samples at 5.0 m depth. These findings confirm rainy weather in deeper stages of sampling and dry weather in shallower region. For the fact that elements take a much longer period to gravitate the lake bottom, a short term climate variation can be inferred by this results.

HCA or simply CA can be employed to geological samples to trace elemental speciation in coal samples, [80, 81], heavy metals and pollutant estimation in soil [82, 83], reservoir characterization of natural geological formations [84], explain hydro-geochemistry in water–rock interaction zones [85], identify wash over beds in marine environment [86] etc.

In a study of iron mining trails in Paraopera riverbed, Brazil, researchers took borehole samples to perform X-ray analysis. HCA associated with PCA was employed which resulted into identification of six different groups with difference of one or the other mineral phases. HCA was used here to identify mineralogical fingerprints in iron ore deposits [87]. Satyamoorthy et al. investigated geochemical characteristics of fly ash samples from Neyveli thermal power station. Though the abundance in the earth's crust is low, several elements were found to be significantly concentrated in fly ash. Bottom ash consists of coarser particles which settles at the bottom while pond ash is a mixture of fly ash and bottom ash that is collected in a low-lying area. Harmful trace elements can get accumulated in such coal industry by-products. XRF concentrations of trace elements were determined first and then the results were subjected to cluster analysis to look into the correlation between bottom and pond ash components. Dendograms of two sections of fly ash indicate that As, Se, Hg, Pb, Mo, Cu, Mn, and Ni are strongly linked in the bottom ash whereas Hg, Mn, As, Se, Mo, Pb, Cu, Ni, Cr, and Zn are strongly linked in the pond ash [88].

Other than these vastly used chemometric techniques in geology and geochemistry, reports those use LDA, MCR, MLR are also available although they are not as profound as the other techniques discussed above [89–91].

12.5 Stepping into the Wild—Machine Learning in Geochemometrics

Machine learning (ML), a sub-discipline of artificial intelligence (AI), focuses on the development of algorithms and models with the capacity to learn from data and execute specific tasks without the need for explicit programming [92]. The objective of ML is to use known information to learn consequential patterns with which precise predictions can be made. This process of learning the known dataset is called 'training' [93]. ML is different from conventional computer programming in respect of extracting through training experiences to improve older capabilities of a system. Since the problems in earth science are often complex, traditional statistics may underperform when non-linear, higher order functions come into picture. Another big concern while dealing with geological samples is the cost estimates. This is where AI & ML has revolutionized the modern geosciences by modelling diverse and very large geological data in cohesive manner. The applications of AI and remote sensing ML algorithms in geological studies lie in characterization of stratigraphy, classifying ores, monitoring earthquakes, detecting geochemical anomalies and mineral prospecting [94–101].

Out of many unsupervised and supervised ML programs, random forests (RFs) and support vector machines (SVMs), artificial neural networks (ANNs) and decision trees (DTs) are the important categories as far as geological samples are concerned while other ML methods comprise of k-nearset neighbours (kNNs) and logistic regression (LR) [102].

ML methods are being used to improve the efficiency of many spectroscopic analyses as well, especially XRF and LIBS [103–105]. These techniques are also used in most of the geochemical researches.

The basic idea of ANNs is to mimic human neurons which receive signal and transmit to all other neurons in the vicinity after processing the data [106]. In a neural network, nodes in the core function as fundamental processing section and units in the form of layers are connected so that information may be freely passed uni-directionally. A neuron with the help of a non-linear operator performs linear regression. Weights are used to join neurons of more than one layer [107, 108].

Decision trees (DTs) accompanied with neural networks are one of the most popular ML algorithms in geosciences [109, 110]. Because they are simpler, readable graphically and can be done with low computation cost, their application is increasing very rapidly. A DT reorders limiting variables hierarchically which are constantly applied from the root to the end leaf of the tree. To close the DT from a dataset, an assessment measure of every ensemble is applied to maximize the inter-node difference. DTs can be categorized as classification trees and regression trees [111].

Random forests (RFs) are regression techniques combining utilities of plenty of DT algorithms to predict the value of a variable [112]. When (m) input is given containing different variables, RF builds n number of regression trees and medians the output. To reduce the anomalies in original data, RF bags to diversify the dataset by resampling some of the training points at random with no deletion from the parent set. This creates different sub sets within the original data and they are called random vectors and make the system more stable by acting robust with increased degree of accuracy.

Support vector machines (SVMs) are used to identify depositional microfacies and landslide susceptibility. The lines that are adjacent to the optimal hyperspace are known as support vectors as these lines run through the dataset which determine the maximal margin [113]. After deducing the vector new examples are presented into the same part in space and fit into the class. Figure 12.6 depicts all these supervised MLs in a simplified manner.

Figure 12.7 represents how ML algorithms can improve outcome of XRF and geochemical data and segregate various geological events by various modern day state of art computerized tools.

A few examples are now illustrated to show how geochemical data generated by XRF can be visualized in terms of ML algorithms which help researchers to characterize geological events.

A study done by Kirkwood et al. used quantile RFs to predict elemental concentrations, loss on ignition and soil p^H [114]. Quantile RFs are an elaboration of conventional RFs [115]. It takes specified quantiles from the outputs of ensembles of decision trees which gives a quantification of uncertainty associated with each possibility. 568 no. of random samples were collected during the summer season from a depth of 5–20 cm and XRF analysis of 48 major and trace elements were done. The major elements in this study were Si, Al, Fe, Ca, Mg, Na K, Ti and Zr. They were added with appropriate mass of oxygen considering being present as oxides. For most of the samples the elements did not sum to 100% and the remainder was included to see if this also can be modelled by RTs. For a relatively large number of predictions, the sum of element concentration and unmeasured remainder would fall very close to 100% in majority. The model predicted that this remainder had very close proximity with loss on ignition. To provide the quantile RFs with most information available, a 61,000 line-km of airborne geophysical and remote sensing data of magnetic and radiometric surveys were utilized. These were the auxiliary variables of the study. Each model used 101 decision trees which were grown until the terminal nodes contain 8 samples in order to reduce overfitting from outliers. The odd number of trees prevents possible ties in variable. The training was validated by a tenfold cross validation process. Cross validated R^2 values were also determined.

These quantile RTs could produce geochemical maps of each element present in the region. The results of tenfold cross validation could generate approximate acceptance of the model. R2 values of quantile RTs and ordinary kriging indicated that the former would give improved prediction accuracy for 37 out of 51 target variables. Any geochemical sampling strategy around this enhanced 37 sampling space rather than actual geographical space would improve the learning of distribution of transportable elements.

Another work reported by Kim et al. efficiently used micro-XRF and XRD data in a neural network, called sychrotron-based machine learning approach for RasTer (SMART) [116]. SMART could be used to provide 2D mineral maps.

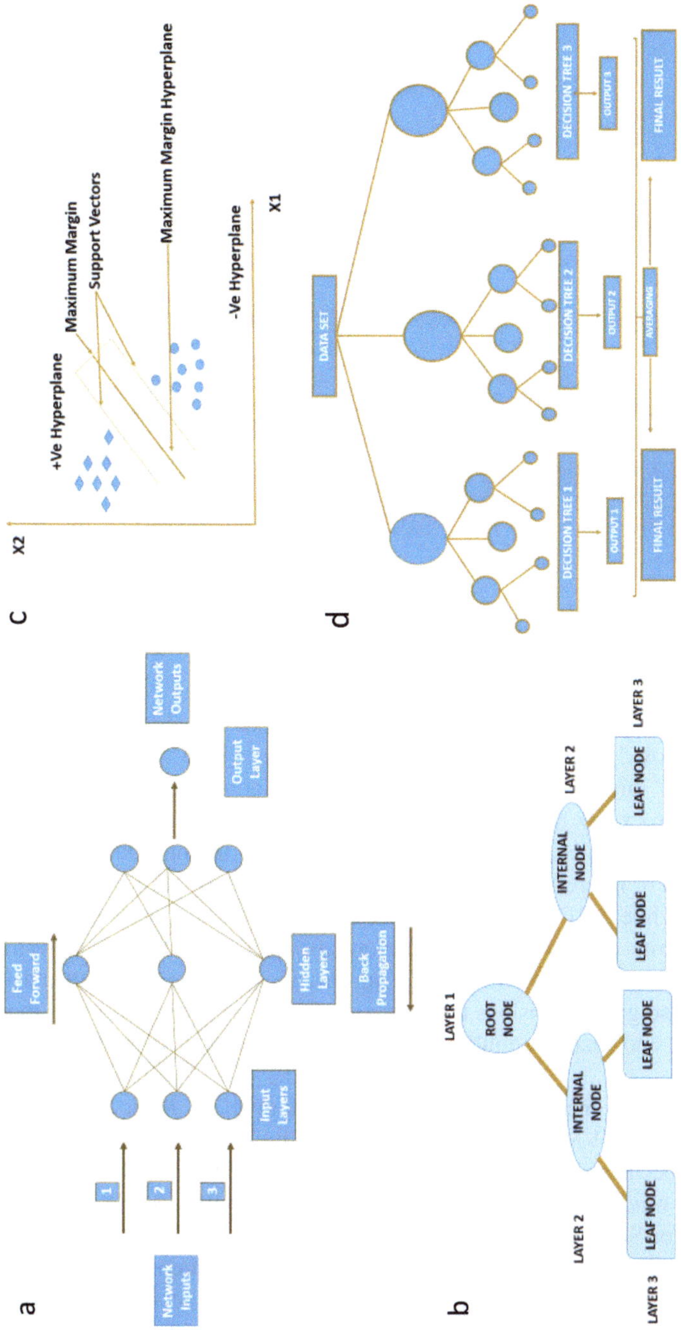

Fig. 12.6 Simplified depiction of machine learning techniques: **a** artificial neural networks, **b** random forests, **c** support vectors and **d** decision trees

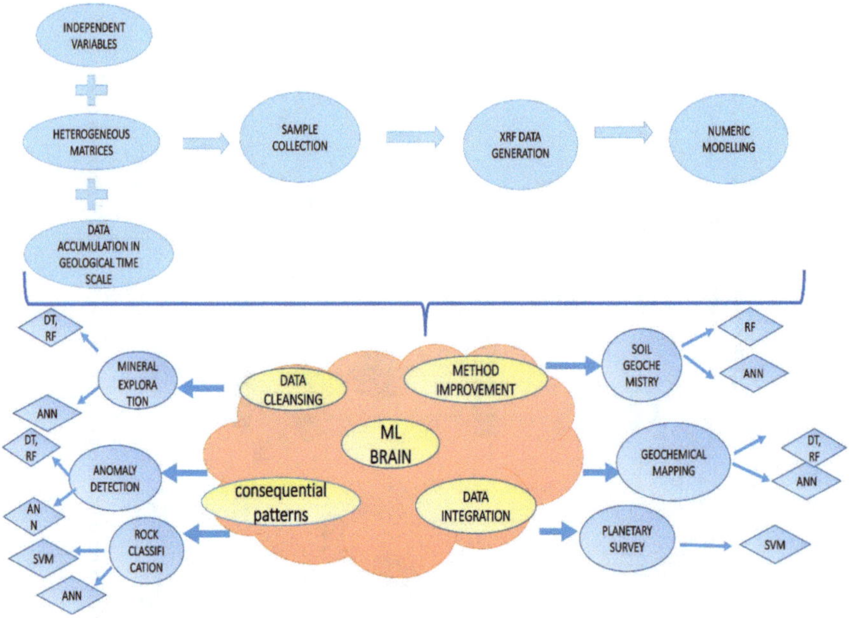

Fig. 12.7 Different methods of XRF data processing for geological events by means of machine learning

SMART classifier was applied to micro-XRF scans of five different sedimentary rocks which provided millimetre scale 2D mineral maps for a wide range of natural rocks. SMART mineral classifier was trained on a coupled micro x-ray fluorescence and micro x-ray diffraction dataset. Once trained, the classier can directly be applied to other micro-XRF systems. In a SMART training dataset, each data point consisted of fluorescence intensities of eight elements viz. Fe, Ca, K, Mn, Ti, As, S and Sr while XRD patterns indicated three dominant mineral phases including calcite, pyrite and quartz. When performances of SMART classifiers trained with different datasets were compared, a misclassification rate of less than 4% was observed. It also gave provisions to sub-pixel mineral characterization i.e. the ability to identify and quantify multiple mineral per pixel with insights of expected grain size.

MLs can also be incorporated in medical geology [117, 118]. Soft computing methods of ANN, SVM and Naive Bayes algorithms together can be utilised to estimate the occurrence of dental fluorosis arising from high fluoride contaminated ground water, as reported by Yetis et al. [119]. XRF and XRD analyses of 108 samples for SiO_2, Al_2O_3, Fe_2O_3, MgO, CaO, K_2O, TiO_2, P_2O_5, LOI, Cu, Ni, Sr, Zn, Zr, S, quartz and calcite were done. Each element/constituent was taken as input variables. For specifying output variables two aspects were considered. Firstly the utilization of fluoride concentration of 2.17 and 2.63 values as primary output and secondly to assign these values to dental fluorosis classes and then using those classes as the output value. Three MLs mentioned before were selected to evaluate the modalities of XRF and XRD analysis parameters modelled for the study. All the models were evaluated by the Naive-Bayes as a linear and ANN and SVM as non-linear classifiers. A tenfold cross validation was employed for all three MLs. Results indicated that ANN outperformed the other two algorithms. The study results revealed that for dental fluoride in groundwater, XRF variables had higher discrimination potential than XRD variables when MLs were considered with other calculations like fischer discrimination power (FDA) and normalized weighted voting map (NWVM).

12.6 Current Researches and Future Prospect

As far as this new branch of science is concerned, application of chemometrics which use XRF in geological domain is still in its early days but it is slowly trying to get mature through expanding researches increasing every year. Table 12.1 lists the references of recent research work on this field available in internet. This collection of references would definitely help a new researcher to delve into the broader domain of the current subject.

Use of AI driven chemometrics can reduce uncertainties and lessen cost. But the main problem that is persisting in modern data driven geological researches is the lack of large and high quality surface data. Remote sensing approaches can be a solution of this challenge because it can generate very large number of data within a shorter span of time as compared to physical survey and explorations. Also the algorithms available in the market are only few. Collaborative researches generating exciting AI and ML tools would help solving the

Table 12.1 References of application showing chemometrics using XRF data in geology

Branch of geology	Chemometric techniques	Refs.
Mineralogy and anomaly detection	PCA HCA PLSR RF	[120–128] [129–135] [136–139] [140]
Soil geochemistry and sedimentology	PCA HCA PLS RF ANN	[141–146] [147–151] [152, 153] [154, 155] [156–158]
Environmental geology	PCA PLS ANN	[142, 159–163] [164, 165] [166]
Palaeontology and historical geology	PCA	[167–170]
Planetary geology	ML	[171–174]
Petrology and stratigraphy	PCA	[175–177]

problems. Currently the XRF based chemometric researches are mainly centred in mineralogy, sedimentology, ore classification and environmental geology. But rapid growth of multivariate analyses can be seen in other geological fields like palaeontology, planetary geology etc. Researches those take care the data reduction approaches through deep learning in petrology, plate tectonics, oceanography, geothermal explorations, geochronology and physical geology have tremendous dormant potential. Further attention from global scientific community in these branches is necessary.

12.7 Conclusions

The book chapter discusses application of XRF and chemometrics in geological field which is definitely new and gaining popularity due to the fact that it best suits the models having large to very large dataset. Almost every geochemical sample is once analysed by XRF to get an insight of lithology. Since XRF scans are done on routine basis for many numbers of elements, application of chemometric tools always help reducing the data dimensionality for geochemical mapping. One can use both the raw XRF data and processed XRF data to extract information using multivariate analyses. Moreover, patterns predicted by chemometric setups assist modelling the geological exploration, mining strategies with greater confidence to geo-scientists. Lacunas those are otherwise unseen, can be easily identified with chemometric applications. Such accurate predictions also support any project economically. Major geological events like climate changes, alterations in plate tectonics, earthquakes and landslides etc. can be traced down to minimum error with the help of multivariate analyses associated with spectroscopy. Though these are the early days of using complex algorithms in geological data processing, the impact created by combined incorporation of chemometrics, artificial intelligence and machine learning is very much exciting and open new gateways with each new kind of research being reported. Further motivating developments are desired in this data driven science.

References

1. Kramar, U.: Advances in energy-dispersive X-ray fluorescence. J. Geochem. Explor. **58**, 73–80 (1997)
2. Ling, Y.-Y., Zhang, J.-J., Liu, K., Ge, M.-H., Wang, M., Wang, J.-M.: Geochemistry, geochronology, and tectonic setting of early cretaceous volcanic rocks in the northern segment of the Tan-Lu Fault region, northeast China. J. Asian Earth Sci. **144**, 303–322 (2017). https://doi.org/10.1016/j.jseaes.2016.12.025
3. Oyedotun, T.D.T.: X-ray fluorescence (XRF) in the investigation of the composition of earth materials: a review and an overview. Geol., Ecol., Landsc. **2**(2), 148–154 (2018). https://doi.org/10.1080/24749508.2018.1452459
4. Acquafredda, P.: XRF technique. Phys. Sci. Rev. **4**(8), 20180171 (2019). https://doi.org/10.1515/psr-2018-0171
5. Acquafredda, P.: XRF technique. Phys. Sci. Rev. **4**(8), 20180171 (2019)

6. De Pauw, E., Tack, P., Vincze, L.: A review of laboratory, commercially available, and facility based wavelength dispersive X-ray fluorescence spectrometers. J. Anal. At. Spectrom. (2024)
7. Vanhoof, C., Bacon, J.R., Fittschen, U.E., Vincze, L.: 2023 atomic spectrometry update–a review of advances in X-ray fluorescence spectrometry and its special applications. J. Anal. At. Spectrom. (2023)
8. Vanhoof, C., Bacon, J.R., Fittschen, U.E., Vincze, L.: 2020 atomic spectrometry update–a review of advances in X-ray fluorescence spectrometry and its special applications. J. Anal. At. Spectrom. **35**(9), 1704–1719 (2020)
9. Vanhoof, C., Bacon, J.R., Ellis, A.T., Fittschen, U.E., Vincze, L.: 2019 atomic spectrometry update–a review of advances in X-ray fluorescence spectrometry and its special applications. J. Anal. At. Spectrom. **34**(9), 1750–1767 (2019)
10. West, M., Ellis, A.T., Potts, P.J., Streli, C., Vanhoof, C., Wobrauschek, P.: 2015 Atomic spectrometry update–a review of advances in X-ray fluorescence spectrometry and their applications. J. Anal. At. Spectrom. **30**(9), 1839–1889 (2015)
11. Guimarães, C.C.B., Demattê, J.A., de Azevedo, A.C., Sayão, V.M., da Silva, R.C., Poppiel, R.R., dos Santos Teixeira, A.F.: Assessing soil mineralogy and weathering degree by a multi-range sensor synergistic approach: from parent rock to topsoil. J. S. Am. Earth Sci. **116**, 103855 (2022)
12. Silva, F.M., Weindorf, D.C., Silva, S.H.G., Silva, E.A., Ribeiro, B.T., Guilherme, L.R.G., Curi, N.: Tropical soil toposequence characterization via pXRF spectrometry. Soil Sci. Soc. Am. J. **83**(4), 1153–1166 (2019)
13. Stockmann, U., Cattle, S.R., Minasny, B., McBratney, A.B.: Utilizing portable X-ray fluorescence spectrometry for in-field investigation of pedogenesis. CATENA **139**, 220–231 (2016)
14. Gozukara, G., Akça, E., Dengiz, O., Kapur, S., Adak, A.: Soil particle size prediction using Vis-NIR and pXRF spectra in a semiarid agricultural ecosystem in central Anatolia of Türkiye. CATENA **217**, 106514 (2022)
15. Junior, C.R.P., Tavares, T.R., Pereira, M.G., Furquim, S.A.C., da Silva Terra, F., dos Anjos, L.H.C., de Oliveira, F.S.: Pedogenesis on Jurassic formations in the Araripe Basin, northeastern Brazil: Weathering and parent material. Catena. **223**, 106952 (2023)
16. Lemière, B., Harmon, R.S.: XRF and LIBS for field geology. Portable Spectrosc. Spectrom. 455–497 (2021)
17. Panchuk, V., Yaroshenko, I., Legin, A., Semenov, V., Kirsanov, D.: Application of chemometric methods to XRF-data–a tutorial review. Anal. Chim. Acta **1040**, 19–32 (2018)
18. Wehrens, R.: Chemometrics with R, vol. 3. Springer, New York (2011)
19. Kumar, N., Bansal, A., Sarma, G.S., Rawal, R.K.: Chemometrics tools used in analytical chemistry: an overview. Talanta **123**, 186–199 (2014)
20. Lavine, B., Workman, J.: Chemometrics. Anal. Chem. **80**(12), 4519–4531 (2008)
21. Santos, M.C., Nascimento, P.A.M., Guedes, W.N., Pereira Filho, E.R., Filletti, É.R., Pereira, F.M.V.: Chemometrics in analytical chemistry–an overview of applications from 2014 to 2018. Eclética Química **44**(2), 11–25 (2019)
22. Birks, H.J.B.: Multivariate analysis in geology and geochemistry: an introduction. Chemom. Intell. Lab. Syst. **2**(1–3), 15–28. ISSN 0169–7439 (1987). https://doi.org/10.1016/0169-7439(87)80082-5
23. Balaram, V., Sawant, S.S.: Indicator minerals, pathfinder elements, and portable analytical instruments in mineral exploration studies. Minerals **12**(4), 394 (2022)
24. Verma, S.P.: Road from Geochemistry to Geochemometrics, vol. 831. Springer, Singapore (2020)
25. Kokot, S., Grigg, M., Panayiotou, H., Phuong, T.D.: Data interpretation by some common chemometrics methods. Electroanal.: Int. J. Devoted Fundam. Pract. Asp. Electroanal. **10**(16), 1081–1088 (1998)
26. Savazzi, E., Reyment, R.A.: Aspects of Multivariate Statistical Analysis in Geology. Elsevier (1999)

27. Sadeghi, M., Casey, P., Carranza, E.J.M., Lynch, E.P.: Principal Components Analysis and K-means Clustering of Till Geochemical Data: Mapping and Targeting of Prospective Areas for Lithium Exploration in Västernorrland Region, p. 106002. Ore Geology Reviews, Sweden (2024)
28. Milinovic, J., Vale, C., Azenha, M.: Recent advances in multivariate analysis coupled with chemical analysis for soil surveys: a review. J. Soils Sediments **23**(3), 1085–1098 (2023)
29. Ghanbari, Y., Hezarkhani, A., Ataei, M., Pazand, K.: Regional geochemical pattern recognition with multivariate correspondence cluster analysis in the Ravar area, Iran. Appl. Earth Sci. **119**(4), 220−226 (2010)
30. Fresia, B., Ross, P.S., Gloaguen, E., Bourke, A.: Lithological discrimination based on statistical analysis of multi-sensor drill core logging data in the Matagami VMS district, Quebec, Canada. Ore Geol. Rev. **80**, 552–563 (2017)
31. Makvandi, S., Ghasemzadeh-Barvarz, M., Beaudoin, G., Grunsky, E.C., McClenaghan, M.B., Duchesne, C., Boutroy, E.: Partial least squares-discriminant analysis of trace element compositions of magnetite from various VMS deposit subtypes: application to mineral exploration. Ore Geol. Rev. **78**, 388–408 (2016)
32. Verma, S.P., Rosales-Rivera, M., Rivera-Gómez, M.A., Verma, S.K.: Comparison of matrix-effect corrections for ordinary and uncertainty weighted linear regressions and determination of major element mean concentrations and total uncertainties of sixty-two international geochemical reference materials from wavelength-dispersive X-ray fluorescence spectrometry. Spectrochim. Acta, Part B **162**, 105714 (2019)
33. Tripathi, V.S.: Factor analysis in geochemical exploration. J. Geochem. Explor. **11**(3), 263–275 (1979)
34. Hannon, J.S., Huff, W.D., Sturmer, D.M.: Geochemical relationships in Cretaceous bentonites as inferred from linear discriminant analysis. Sed. Geol. **390**, 1–14 (2019)
35. Coomans, D., Massart, D.L., Kaufman, L.: Optimization by statistical linear discriminant analysis in analytical chemistry. Anal. Chim. Acta **112**(2), 97–122 (1979)
36. Deng, C., Pan, H., Fang, S., Konaté, A.A., Qin, R.: Support vector machine as an alternative method for lithology classification of crystalline rocks. J. Geophys. Eng. **14**(2), 341–349 (2017)
37. Wohlberg, B., Tartakovsky, D.M., Guadagnini, A.: Subsurface characterization with support vector machines. IEEE Trans. Geosci. Remote Sens. **44**(1), 47–57 (2005)
38. De Boissieu, F., Sevin, B., Cudahy, T., Mangeas, M., Chevrel, S., Ong, C., Despinoy, M.: Regolith-geology mapping with support vector machine: a case study over weathered Ni-bearing peridotites, New Caledonia. Int. J. Appl. Earth Obs. Geoinformation. **64**, 377−385 (2018)
39. Abedi, M., Norouzi, G.H., Bahroudi, A.: Support vector machine for multi-classification of mineral prospectivity areas. Comput. Geosci. **46**, 272–283 (2012)
40. Chen, Y., Wu, W.: Mapping mineral prospectivity by using one-class support vector machine to identify multivariate geological anomalies from digital geological survey data. Aust. J. Earth Sci. **64**(5), 639–651 (2017)
41. Xu, C., Dai, F., Xu, X., Lee, Y.H.: GIS-based support vector machine modeling of earthquake-triggered landslide susceptibility in the Jianjiang River watershed, China. Geomorphol. **145**, 70–80 (2012)
42. Ballabio, C., Sterlacchini, S.: Support vector machines for landslide susceptibility mapping: the Staffora River Basin case study, Italy. Math. Geosci. **44**, 47–70 (2012)
43. Samui, P.: Slope stability analysis: a support vector machine approach. Environ. Geol. **56**, 255–267 (2008)
44. Tadeusiewicz, R.: Neural networks in mining sciences–general overview and some representative examples. Arch. Min. Sci. **60**(4), 971–984 (2015)
45. Chen, L., Guan, Q., Xiong, Y., Liang, J., Wang, Y., Xu, Y.: A spatially constrained multiautoencoder approach for multivariate geochemical anomaly recognition. Comput. Geosci. **125**, 43–54 (2019). https://doi.org/10.1016/j.cageo.2019.01.016

46. Shaw, D.M.: Element distribution laws in geochemistry. Geochim. Cosmochim. Acta **23**(1–2), 116–134 (1961)
47. Gredilla, A., de Vallejuelo, S.F.O., Elejoste, N., de Diego, A., Madariaga, J.M.: Non-destructive spectroscopy combined with chemometrics as a tool for green chemical analysis of environmental samples: a review. TrAC, Trends Anal. Chem. **76**, 30–39 (2016)
48. Sinclair, A.J.: Selection of thresholds in geochemical data using probability graphs: jour. Geochem. Exploration **3**(1), 129–149 (1974)
49. Harris, J.R., Wilkinson, L., Grunsky, E., Heather, K., Ayer, J.: Techniques for analysis and visualization of lithogeochemical data with applications to the Swayze greenstone belt, Ontario: Jour. Geochem. Exploration **67**(1–3), 301–334 (1999)
50. Marcotte, D., David, M.: Rtarget-definition of Kurokotype deposits in Abitibi by discriminant analysis of geochemical data. CIM Bull. **74**(828), 102−108 (1981)
51. Rehder, S., Van Den Boom, G.: Geochemical characterization of tin granites in Northern Thailand. In: Howarth, R.J. (ed.) 1983, Statistics and Data Analysis in Geochemical Prospecting: Handbook of exploration geochemistry, vol. 2, pp. 311–339. Elsevier, New York (1982)
52. Aitchison, J.: The Statistical Analysis of Compositional Data, p. 416. Chapman and Hall, London (1986). https://doi.org/10.1007/978-94-009-4109-0
53. Walling, D.E., Moorehead, P.W.: The particle size characteristics of fluvial suspended sediment: an overview. In: Sediment/Water Interactions: Proceedings of the Fourth International Symposium, pp. 125–149. Springer Netherlands (1989)
54. Folk, R.L.: A review of grain-size parameters. Sedimentol. **6**(2), 73–93 (1966)
55. Le Roux, J.P., Rojas, E.M.: Sediment transport patterns determined from grain size parameters: overview and state of the art. Sed. Geol. **202**(3), 473–488 (2007)
56. Ichikawa, S., Nakamura, T.: Approaches to solid sample preparation based on analytical depth for reliable X-ray fluorescence analysis. X-Ray Spectrom. **45**(6), 302–307 (2016)
57. Markowicz, A.: An overview of quantification methods in energy-dispersive X-ray fluorescence analysis. Pramana **76**, 321–329 (2011)
58. Ikumapayi, O.M., Akinlabi, E.T., Adedeji, P.A., Akinlabi, S.A.: Preparation, characterization, image segmentation and particle size analysis of cow bone powder for composite applications. Trends Manuf. Eng. Manag.: Sel. Proc. ICMechD **2019**, 273–283 (2021)
59. Egozcue, J.J., Pawlowsky-Glahn, V., Mateu-Figueras, G., et al.: Isometric logratio transformations for compositional data analysis. Math. Geol. **35**, 279–300 (2003). https://doi.org/10.1023/A:1023818214614
60. Buccianti, A., Grunsky, E.: Compositional data analysis in geochemistry: are we sure to see what really occurs during natural processes? J. Geochem. Explor. **141**, 1–5 (2014)
61. Pawlowsky-Glahn, V., Buccianti, A.: Compositional Data Analysis. Wiley, Chichester (2011)
62. Hahn, J.U., Schöler, H.F.: Determination of heavy metals in sediments by X-ray fluorescence in SEM (qualitative and quantitative evaluation) Part 2: application. Fresenius' Z. Für Anal. Chem. **315**, 683–686 (1983)
63. Selinus, O.: Factor and discriminant analysis to lithogeochemical prospecting in an area of central Sweden. J. Geochem. Explor. **19**(1–3), 619–642 (1983)
64. Duchesne, J.C., Rouhart, A., Schoumacher, C., Dillen, H.: Thallium, nickel, cobalt and other trace elements in iron sulfides from Belgian lead-zinc vein deposits. Miner. Deposita **18**, 303–313 (1983)
65. Custo, G., Boeykens, S., Cicerone, D., Vázquez, C.: Combining XRF analysis and chemometric tools for a preliminary classification of argentine soils. X-Ray Spectrom.: Int. J. **31**(2), 132–135 (2002)
66. Singh, V., Agrawal, H.M., Joshi, G.C., Sudershan, M., Sinha, A.K.: Elemental profile of agricultural soil by the EDXRF technique and use of the principal component analysis (PCA) method to interpret the complex data. Appl. Radiat. Isot. **69**(7), 969–974 (2011)
67. Dempster, M., Dunlop, P., Scheib, A., Cooper, M.: Principal component analysis of the geochemistry of soil developed on till in Northern Ireland. J. Maps **9**(3), 373–389 (2013)
68. Schneider, A.R., Morvan, X., Saby, N.P., Cancès, B., Ponthieu, M., Gommeaux, M., Marin, B.: Multivariate spatial analyses of the distribution and origin of trace and major elements

in soils surrounding a secondary lead smelter. Environ. Sci. Pollut. Res. **23**, 15164–15174 (2016)

69. Fang, G.C., Chang, C.N., Wu, Y.S., Fu, P.P.C., Yang, I.L., Chen, M.H.: Characterization, identification of ambient air and road dust polycyclic aromatic hydrocarbons in central Taiwan. Taichung. Sci. Total. Environ. **327**, 135–146 (2004)
70. Wang, X.J., Liu, R.M., Wang, K.Y., Hu, J.D., Ye, Y.B., Zhang, S.C., Xu, F.L., Tao, S.: Application of multivariate spatial analysis in scale-based distribution and source study of PAHs in the topsoil: an example from Tianjin, China. Environ. Geol. **49**, 1208–1216
71. Wang, X.S., Qin, Y., Sang, S.X.: Accumulation and sources of heavy metals in urban topsoils: a case study from the city of Xuzhou. China Environ. Geol. **48**(1), 101–107 (2005)
72. Navarro, A., Tauler, R., Lacorte, S., Barcelo, D.: Chemometrical investigation of the presence and distribution of organochlorine and polyaromatic compounds in sediments of the Ebro river Basin. Anal. Bioanal. Chem. **385**(6), 1020–1030 (2006)
73. Heinrich, A.: The application of multivariate statistical methods for the evaluation of soil profiles. J. Soils Sediments **7**(1), 45–52 (2007)
74. Facchinelli, A., Sacchi, E., Mallen, L.: Multivariate statistical and GIS-based approach to identify heavy metal sources in soils. Environ. Pollut. **114**(3), 313–324 (2001)
75. Kokot, S., Grigg, M., Panayiotou, H., Phuong, T.D.: Data interpretation by some common chemometrics methods. Electroanalysis **10**(16), 1081–1088 (1999)
76. Yu, X.Z., Gao, Y., Wu, S.C., Zhang, H.B., Cheung, K.C., Wong, M.H.: Distribution of polycyclic aromatic hydrocarbons in soils at Guiyu area of China, affected by recycling of electronic waste using primitive technologies. Chemosphere **65**, 1500–1509 (2006)
77. Meng, Y.M., Jébrak, M., Sasseville, C., Huang, X.W.: Geochemical and mineralogical diagnosis on gold ores: a case study from the Cadillac-Larder Lake Fault Zone, Abitibi, Canada. Ore Geol. Rev. **127**, 103840 (2020)
78. Topuz, M., Keskin, M., Karabulut, M., KILIÇ, N.K.: An investigation of the chemical composition of bottom sediments from dried lake Gavur of Turkiye by using XRF and multivariate data analysis. Gazi Univ. J. Sci. 1–1 (2024)
79. Rothwell, R.G., Croudace, I.W.: Twenty years of XRF core scanning marine sediments: what do geochemical proxies tell us?. In: Micro-XRF Studies of Sediment Cores: Applications of a Non-Destructive Tool for the Environmental Sciences, pp 25–102 (2015)
80. Srikanth, S., Naga Raju, G.J., Gowri Naidu, B., Tiwari, M.K., Sarita, P.: Chemometric speciation of trace elements in Indonesian coal imported by Indian thermal power stations-a comparative study with Indian coals using SRXRF. J. Geol. Soc. India **100**(7), 964–970 (2024)
81. Srikanth, S., Naidu, B.G., Tiwari, M.K., Sarita, P., Raju, G.N.: A comparative evaluation of enrichment of trace elements in South African and Australian coals using SRXRF. Nucl. Part. Phys. Proc. **339**, 10–17 (2023)
82. Usman, H., Maigari, A.U., Zanna, H.K., Mohammed, I., Muktari, M.: Chemometric analysis for distribution of physicochemical parameters and heavy metals of soil. Bima J. Sci. Technol. (2536-6041). **8**(2A), 135–147 (2024)
83. Vakanjac, B., Naunovic, Z., Ristić Vakanjac, V., Đumić, T., Bakrač, S., Štrbački, J., Gajić, V., Alzarog, T.M.: Distribution of potentially toxic elements in the city of Zintan and its surroundings (Northwestern Libya) by surface soil sampling. Minerals **13**(8), 1048 (2023)
84. Al-Morakhi, R., Verma, N., Long Nguyen, K., Nelson, N., Hayes, D., Rowe, H., Reed, J., Ghiniwa, I., Arrington, S., Stemberger, D., Narasimhan, S.: Reservoir characterization of the middle/upper Jurassic Sargelu and Najmah formations, West Kuwait: key workflows and findings for future development of unconventionals. In: Abu Dhabi International Petroleum Exhibition and Conference, p. D032S179R001. SPE (2022)
85. Mohamed, A., Asmoay, A., Alshehri, F., Abdelrady, A., Othman, A.: Hydro-geochemical applications and multivariate analysis to assess the water–rock interaction in arid environments. Appl. Sci. **12**(13), 6340 (2022)

86. Williams, H.F., Beaubouef, C.E., Liu, K.B., Culligan, N., Riedlinger, L.: Testing XRF identification of marine washover sediment beds in a Coastal Lake in Southeastern Texas, USA. Mar. Geol. **443**, 106705 (2022)
87. Laureano, F.V., Kwitko-Ribeiro, R., Guimarães, L., Leão, L.P.: Mineralogical fingerprint of iron ore tailings in Paraopeba River bedload sediments after the B1 dam failure in Brumadinho, MG (Brazil). Minerals **12**(6), 716 (2022)
88. Sathiyamoorthy, G., Vasudevan, S., Balamurugan, P., Selvaganapathi, R.: Geochemical characteristics of the Bottom and Pond Fly ash, Neyveli, Cuddalore District, Tamil Nadu, India. J. Geol. Soc. India **99**(8), 1121–1130 (2023)
89. Cui, H., Deng, Y., Zhong, R., Li, W., Yu, C., Danyushevsky, L.V., Belousov, I., Li, Z., Wang, H.: Determining the ore-forming processes of Dongshengmiao Zn-Pb-Cu deposit: evidence from the linear discriminant analysis of pyrite geochemistry. Ore Geol. Rev.105782 (2023)
90. Miller, J.D., Nicholson, S.W., Easton, R.M., Ripley, E.M., Feinberg, J.M.: Geology and mineral deposits of the 1.1 Ga midcontinent Rift in the lake superior region—an overview. In: Field Guide to the Copper-Nickel-Platinum Group Element Deposits of the Lake Superior Region, Duluth, MN, pp. 2013–01 (2013)
91. Ding, T., Gao, J., Tian, S., Shi, G., Chen, F., Wang, C., Luo, X., Han, D.: Chemical and isotopic characteristics of the water and suspended particulate materials in the Yangtze River and their geological and environmental implications. Acta Geol. Sin.-Engl. Ed. **88**(1), 276–360 (2014)
92. Zhou, J., Gandomi, A.H., Chen, F., Holzinger, A.: Evaluating the quality of machine learning explanations: a survey on methods and metrics. Electron. **10**(5), 593 (2021)
93. Jooshaki, M., Nad, A., Michaux, S.: A systematic review on the application of machine learning in exploiting mineralogical data in mining and mineral industry. Mineral. **11**(8), 816 (2021)
94. El-Omairi, M.A., El Garouani, A. (2023). A review on advancements in lithological mapping utilizing machine learning algorithms and remote sensing data. Heliyon.
95. Peyghambari, S., Zhang, Y.: Hyperspectral remote sensing in lithological mapping, mineral exploration, and environmental geology: an updated review. J. Appl. Remote. Sens. **15**(3), 031501–031501 (2021)
96. Upadhyay, R.K., Kishore, N., Sharma, M.: Delineation and mapping of palaeochannels using remote sensing, geophysical, and sedimentological techniques: a comprehensive approach. Water Sci. **35**(1), 100–108 (2021)
97. Saha, S.K.: Remote sensing and geographic information system applications in hydrocarbon exploration: a review. J. Indian Soc. Remote. Sens. **50**(8), 1457–1475 (2022)
98. Han, W., Zhang, X., Wang, Y., Wang, L., Huang, X., Li, J., Wang, S., Chen, W., Li, X., Feng, R., Fan, R.: A survey of machine learning and deep learning in remote sensing of geological environment: challenges, advances, and opportunities. ISPRS J. Photogramm. Remote Sens. **202**, 87–113 (2023)
99. Bhadauria, P.K.S.: Comprehensive review of AI and ML tools for earthquake damage assessment and retrofitting strategies. Earth Sci. Inf. **17**(5), 3945–3962 (2024)
100. Pwavodi, J., Ibrahim, A.U., Pwavodi, P.C., Al-Turjman, F., Mohand-Said, A. (2024). The role of artificial intelligence and IoT in prediction of earthquakes. Artif. Intell. Geosci. 100075
101. Shirmard, H., Farahbakhsh, E., Müller, R.D., Chandra, R.: A review of machine learning in processing remote sensing data for mineral exploration. Remote Sens. Environ. **268**, 112750 (2022)
102. Binetti, M.S., Massarelli, C., Uricchio, V.F.: Machine learning in geosciences: a review of complex environmental monitoring applications. Mach. Learn. Knowl. Extr. **6**(2), 1263–1280 (2024)
103. Rao, A.P., Jenkins, P.R., Pinson, R.E., Auxier, J.D., II., Shattan, M.B., Patnaik, A.K.: Machine learning in analytical spectroscopy for nuclear diagnostics. Appl. Opt. **62**(6), A83–A109 (2023)

104. de Carvalho, G.G.A., Guerra, M.B.B., Adame, A., Nomura, C.S., Oliveira, P.V., de Carvalho, H.W.P., Santos, D., Nunes, L.C., Krug, F.J.: Recent advances in LIBS and XRF for the analysis of plants. J. Anal. At. Spectrom. **33**(6), 919–944 (2018)
105. Pathak, A.K., Kumar, R., Singh, V.K., Agrawal, R., Rai, S., Rai, A.K.: Assessment of LIBS for spectrochemical analysis: a review. Appl. Spectrosc. Rev. **47**(1), 14–40 (2012)
106. Li, L.N., Liu, X.F., Yang, F., Xu, W.M., Wang, J.Y., Shu, R.: A review of artificial neural network based chemometrics applied in laser-induced breakdown spectroscopy analysis. Spectrochim. Acta, Part B **180**, 106183 (2021)
107. Meza Ramirez, C.A., Greenop, M., Ashton, L., Rehman, I.U.: Applications of machine learning in spectroscopy. Appl. Spectrosc. Rev. **56**(8–10), 733–763 (2021)
108. Lu, T., Lerner, J.: Spectroscopy and hybrid neural network analysis. Proc. IEEE **84**(6), 895–905 (1996)
109. Rodriguez-Galiano, V., Sanchez-Castillo, M., Chica-Olmo, M., Chica-Rivas, M.J.O.G.R.: Machine learning predictive models for mineral prospectivity: an evaluation of neural networks, random forest, regression trees and support vector machines. Ore Geol. Rev. **71**, 804–818 (2015)
110. Collins, A.L., Pulley, S., Foster, I.D., Gellis, A., Porto, P., Horowitz, A.J.: Sediment source fingerprinting as an aid to catchment management: a review of the current state of knowledge and a methodological decision-tree for end-users. J. Environ. Manage. **194**, 86–108 (2017)
111. Milanović, M., Stamenković, M.: CHAID decision tree: methodological frame and application. Econ. Themes **54**(4), 563–586 (2016)
112. Schnitzler, N., Ross, P.S., Gloaguen, E.: Using machine learning to estimate a key missing geochemical variable in mining exploration: application of the Random Forest algorithm to multi-sensor core logging data. J. Geochem. Explor. **205**, 106344 (2019)
113. Kecman, V.: Support vector machines–an introduction. In Support vector machines: theory and applications, pp. 1–47. Springer Berlin Heidelberg, Berlin, Heidelberg (2005)
114. Kirkwood, C., Cave, M., Beamish, D., Grebby, S., Ferreira, A.: A machine learning approach to geochemical mapping. J. Geochem. Explor. **167**, 49–61 (2016)
115. Breiman, L.: Random forests. Mach. Learn. **45**(5), 32 (2001)
116. Kim, J.J., Ling, F.T., Plattenberger, D.A., Clarens, A.F., Peters, C.A.: Quantification of mineral reactivity using machine learning interpretation of micro-XRF data. Appl. Geochem. **136**, 105162 (2022)
117. Zhang, C., Zuo, R., Xiong, Y., Shi, X., Donnelly, C.: GIS, geostatistics, and machine learning in medical geology. Pract. Appl. Med. Geol. 215–234 (2021)
118. Odoh, B.I., Ahaneku, C.V., Madu, F.M., Muogbo, C.D., Ugwu, F.O., Ozoemena, G.O., Chukwu, D.C., Aseh, P., Osuagwu, J., Elomba, U.F.: Applications of geospatial techniques in medical geology: a review
119. Yetis, A.D., Yesilnacar, M.I., Atas, M.: A machine learning approach to dental fluorosis classification. Arab. J. Geosci. **14**(2), 95 (2021)
120. Spaleta, K.J., Hayes, S.M., Newberry, R.J., Piatak, N.M.: Evaluating the utility of principal component analysis on EDS X-ray maps to determine bulk mineralogy. Geostand. Geoanal. Res. **44**(4), 821–843 (2020)
121. Singh, V., Agrawal, H.: Qualitative soil mineral analysis by EDXRF, XRD and AAS probes. Radiat. Phys. Chem. **81**(12), 1796–1803 (2012)
122. Germinario, L., Cossio, R., Maritan, L., Borghi, A., Mazzoli, C.: Textural and mineralogical analysis of volcanic rocks by μ-XRF mapping. Microsc. Microanal. **22**(3), 690–697 (2016)
123. Gazley, M.F., Collins, K.S., Roberston, J., Hines, B.R., Fisher, L.A., McFarlane, A.: Application of principal component analysis and cluster analysis to mineral exploration and mine geology. In: AusIMM New Zealand Branch Annual Conference, vol. 2015, pp. 131–139. Dunedin New Zealand (2015)
124. Emami, M., Eslami, M., Fadaei, H., Karami, H.R., Ahmadi, K.: Mineralogical–geochemical characterization and provenance of the stones used at the Pasargadae complex in Iran: a new perspective. Archaeometry **60**(6), 1184–1201 (2018)

125. Plantz, J.B., Borghi, L.: Unsupervised learning for identifying silicified and dolomitized zones in pre-salt using portable XRF data. J. S. Am. Earth Sci. **128**, 104421 (2023)
126. Holcomb, J.A., Karkanas, P.: Elemental mapping of micromorphological block samples using portable X-ray fluorescence spectrometry (pXRF): integrating a geochemical line of evidence. Geoarchaeol. **34**(5), 613–624 (2019)
127. Mohamed, N., Rifai, K., Selmani, S., Constantin, M., Doucet, F.R., Özcan, L.Ç., Sabsabi, M., Vidal, F.: Chemical and mineralogical mapping of platinum-group element ore samples using laser-induced breakdown spectroscopy and micro-X-ray fluorescence. Geostand. Geoanal. Res. **45**(3), 539–550 (2021)
128. Saleh, M., Bonizzoni, L., Orsilli, J., Samela, S., Gargano, M., Gallo, S., Galli, A.: Application of statistical analyses for lapis lazuli stone provenance determination by XRL and XRF. Microchem. J. **154**, 104655 (2020)
129. Worobiec, A., Stefaniak, E.A., Potgieter-Vermaak, S., Sawlowicz, Z., Spolnik, Z., Van Grieken, R.: Characterisation of concentrates of heavy mineral sands by micro-Raman spectrometry and CC-SEM/EDX with HCA. Appl. Geochem. **22**(9), 2078–2085 (2007)
130. Hart, N., Dix, M., Rowe, H., Morrell, A., Torrez, G.: Applications of high-resolution X-ray fluorescence (XRF) elemental data in mineral modeling, brittleness indices, and chemostratigraphic correlation in the STACK play of central Oklahoma. In: 2019 AAPG Annual Convention and Exhibition (2019)
131. Guarino, V., De Bonis, A., Peña, J.T., Verde, M., Morra, V.: Multianalytical investigation of wasters from the Tower 8/Porta di Nola refuse middens in Pompeii: Sr–Nd isotopic, chemical, petrographic, and mineralogical analyses. Geoarchaeol. **36**(5), 712–739 (2021)
132. Guo, H., Wang, P., Li, Y., Hu, C., Zheng, J., Mei, H., Zhu, J., Fan, S., Zhong, Q.: Mineralogical and elemental data for soil discriminating and geolocation tracing. Sci. Justice **62**(1), 76–85 (2022)
133. Khelfaoui, M., Medjram, M.S., Kabir, A., Zouied, D., Mehri, K., Chikha, O., Trabelsi, M.A.: Chemical and mineralogical characterization of weathering products in mine wastes, soil, and sediment from the abandoned Pb/Zn mine in Skikda, Algeria. Environ. Earth Sci. **79**, 1–15 (2020)
134. Islam, N., Rabha, S., Subramanyam, K.S.V., Saikia, B.K.: Geochemistry and mineralogy of coal mine overburden (waste): a study towards their environmental implications. Chemosphere **274**, 129736 (2021)
135. Hammon, H., Prather, T., Rowe, H., Mainali, P., Matheny, M., Krumm, R.: Geochemical, mineralogical, and lithological linkages in a thick, early permian, siliciclastic succession, midland Basin, West Texas, USA. In: Unconventional Resources Technology Conference, Denver, Colorado, 22–24 July 2019, pp. 3843–3851. Unconventional Resources Technology Conference (URTeC); Society of Exploration Geophysicists (2019)
136. Breen, C., Clegg, F., Herron, M.M., Hild, G.P., Hillier, S., Hughes, T.L., Jones, T.G.J., Matteson, A., Yarwood, J.: Bulk mineralogical characterisation of oilfield reservoir rocks and sandstones using diffuse reflectance infrared fourier transform spectroscopy and partial least squares analysis. J. Petrol. Sci. Eng. **60**(1), 1–17 (2008)
137. Seo, C., Jo, H.Y., Byun, Y., Ryu, J.H., Joo, Y.: Montmorillonite content prediction in bentonite using Vis–NIR spectroscopy and PLSR analysis: effects of humidity and mineralogical variability. Geoderma **448**, 116980 (2024)
138. Burley, L.L., Barnes, S.J., Laukamp, C., Mole, D.R., Le Vaillant, M., Fiorentini, M.L.: Rapid mineralogical and geochemical characterisation of the fisher east nickel sulphide prospects, western Australia, using hyperspectral and pXRF data. Ore Geol. Rev. **90**, 371–387 (2017)
139. Atheaux, I., Medvedev, A., Makarychev-Mikhailov, S.: Analysis of minerals in drilling fluids using synergy in X-ray fluorescence and fourier transform infrared spectra realized with machine learning. In: SPE Gas & Oil Technology Showcase and Conference, p. D031S041R001. SPE (2024)
140. Nawal, M., Shekar, B., Jaiswal, P.: Integration of sparse and continuous data sets using machine learning for core mineralogy interpretation. Lead. Edge **42**(6), 421–432 (2023)

141. Srivastava, D., Dubey, C.P., Banerji, U.S., Joshi, K.B.: Geochemical trends in sedimentary environments using PCA approach. J. Earth Syst. Sci. **133**(3), 122 (2024)
142. Wen, B., Jiang, H., Gao, Y., Zhou, Q., Qie, H.: Source analysis and bioavailability of soil cadmium in Poyang lake plain of China based on principal component analysis and positive definite matrix factor. Mineral. **14**(5), 514 (2024)
143. Milinovic, J., Santos, P., Sant'Ovaia, H., Futuro, A., Pereira, C.M., Murton, B.J., Flores, D., Azenha, M.: Multivariate analysis applied to X-ray fluorescence to assess soil contamination pathways: case studies of mass magnetic susceptibility in soils near abandoned coal and W/Sn mines. Environ. Geochem. Health. **46**(6), 202 (2024)
144. Pan, Y., Ren, L., Huo, J., Xiang, X., Meng, D., Wang, Y., Yu, C., Liu, Y., Suo, J., Huang, Y.: Soil geochemistry prevails over root functional traits in controlling soil organic carbon fractions of the alpine meadow on the Qinghai-Tibet Plateau, China. CATENA **237**, 107814 (2024)
145. Lee, S., Song, S.Y., Lee, N.R., Kim, M.J., Ahn, Y., Choi, J.: Development and validation of a statistical prediction model for estimating soil lead contamination based on geochemical and geophysical data. Available at SSRN 4960882
146. Saiano, F., Buttafuoco, G., Fantappié, M., Priori, S., Costantini, E.A., Scalenghe, R.: Spatial distribution of soil rare earth elements in Sicily (Italy). Geoderma Reg. **38**, 00845 (2024)
147. Papazotos, P., Liakopoulos, A., Kontodimos, K., Koukoulis, A.: Integrated geochemical analysis of urban and peri-urban soils: a case study of Lamia City, Greece. Environ. Monit. Assess. **196**(11), 1–31 (2024)
148. Sangwan, P., Nandal, A., Nain, T., Sharma, N., Yadav, S.: Discrimination of soil samples using X-ray fluorescence combined with multivariate statistical tools. Available at SSRN 4910442
149. Lin, C.Y., Lam, S.S., Hasnan, H.K., Yue, F.J., El Hidayah, N., Govindaraju, K.: Deconvolving geochemical micro-spatial variability of an unconsolidated aquifer through chemometric and geostatistical techniques. Environ. Earth Sci. **83**(6), 177 (2024)
150. Sunkari, E.D., Nkansah, J., Adams, S.J.: Geochemical exploration for tantalum in coltan-rich pegmatites at Bewadze-Mankoadze area of the Kibi-Winneba belt, southern Ghana: constraints from exploratory data analysis. Heliyon. **10**(18) (2024)
151. Milinovic, J., Vale, C., Futuro, A., Pereira, C., Flores, D., Azenha, M.: Improved X-ray fluorescence spectroscopic monitoring of potentially toxic elements by multivariate analysis: a case study of soils near abandoned coal mines (NW Portugal). J. Environ. Chem. Eng. 114437 (2024)
152. Lucena, P.G., Aquino, R.V., Sousa, J.E., Júnior, V.S.S., Pacheco Filho, J.G., Pereira, C.F.: Mineral and particle-size chemometric classification using handheld near-infrared instruments for soil in Northeast Brazil. Geoderma Reg. **38**, 00819 (2024)
153. da Silva Brochado, M.G., de Noronha, B.G., da Costa Lima, A., Guedes, A.G., da Silva, R.C., dos Santos Dias, D.C.F., Mendes, K.F.: What is the most effective analytical method for quantification and identification of microplastics in contaminated soils? Environ. Geochem. Health **46**(7), 260 (2024)
154. Wang, Y., Gan, T., Zhao, N., Yin, G., Ye, Z., Sheng, R., Li, T., Liang, T., Jia, R., Fang, L., Hu, X.: Identification study of soil types based on feature factors of XRF spectrum combining with machine learning. Spectrochim. Acta, Part B **219**, 107001 (2024)
155. Lu, L., Lei, M., Zhou, Y., Cui, H., Du, H.: In vitro tungsten bioaccessibility in Chinese residential soils: Implications for human health risk assessments and soil screening level derivation. J. Hazard. Mater. **477**, 135368 (2024)
156. Lv, K.N., Huang, Y., Yuan, G.L., Sun, Y.C., Li, J., Li, H., Zhang, B.: Uptake of zinc from the soil to the wheat grain: nonlinear process prediction based on artificial neural network and geochemical data. Sci. Total. Environ. **947**, 174582 (2024)
157. Kodikara, G.R., McHenry, L.J., Stanistreet, I.G., Stollhofen, H., Njau, J.K., Toth, N., Schick, K.: Wide & deep learning for predicting relative mineral compositions of sediment cores solely based on XRF scans, a case study from Pleistocene Paleolake Olduvai, Tanzania. Artif. Intell. Geosci. **5**, 100088 (2024)

158. Liu, B., Ge, J., Li, S., Du, H., Liang, X., Jin, H., Jin, J., Zhao, H., Chen, F.: Quantification of middle to late holocene precipitation in the Gonghe Basin, northeastern Qinghai-Tibetan Plateau, from the geochemistry of aeolian surface soil. Quatern. Sci. Rev. **343**, 108940 (2024)
159. Tennent, E.R., Maxbauer, D.P., Umbanhowar, C.E.: Shallow lake response to Holocene climate variation in south-central Minnesota, USA. Holocene **34**(11), 1521–1533 (2024)
160. Boumaza, B., Kechiched, R., Chekushina, T.V., Benabdeslam, N., Senouci, K., Merzeg, F.A., Rezgui, W., Rebouh, N.Y., Harizi, K.: Geochemical distribution and environmental assessment of potentially toxic elements in farmland soils, sediments, and tailings from phosphate industrial area (NE Algeria). J. Hazard. Mater. **465**, 133110 (2024)
161. Gerivani, H., Beni, A.N., Leroy, S.A., Lahijani, H.K.: Environmental history of the South Caspian Sea basin and its relation to the Siberian high dynamics since 14,000 years ago. Mar. Geol. **475**, 107356 (2024)
162. Medeiros Filho, L.C., Asp, N.E., Lafon, J.M., Souza, T.P., Berredo, J.F., Salomão, G.N.: Geochemical dynamics and evolutionary implications of sediments at the Xingu-Amazon rivers' confluence: proxies for mixing, mobility and weathering. Minerals **14**(11), 1101 (2024)
163. Bouabid, N., Srarfi, F., Mnasri, H., Tagorti, M.A.: Geochemical, mineralogical records, and statistical approaches in establishing sedimentary in the environment of a Mediterranean coastal system: case of Sebkha El-Guettiate (southeastern Tunisia). Environ. Monit. Assess. **196**(4), 398 (2024)
164. Bacon, J.R., Butler, O.T., Cairns, W.R., Cavoura, O., Cook, J.M., Davidson, C.M., Mertz-Kraus, R.: Atomic spectrometry update–a review of advances in environmental analysis. J. Anal. At. Spectrom (2024)
165. Yuda, A.P., Petrus, H.T., Astuti, W.: Effect of pH neutral on the separation of nickel and cobalt from laterite leaching solution using cyanex 272. In: IOP Conference Series: Earth and Environmental Science, vol. 1388, no. 1, p. 012027. IOP Publishing (2024)
166. Na, S., Jeong, H., Kim, I., Hong, S.M., Shim, J., Yoon, I.H., Cho, K.H.: Distribution coefficient prediction using multimodal machine learning based on soil adsorption factors, XRF, and XRD spectrum data. J. Hazard. Mater. **478**, 135285 (2024)
167. Kechiched, R., Sinisi, R., Kocsis, L., Bruguier, O., Mongelli, G., Ferhaoui, S., Bosch, D., Ameur-Zaimeche, O., Buccione, R., Laouar, R.: Phosphate deposits in North Africa: an overview of Algerian-Tunisian deposits, geochemistry and paleoenvironments. Geol. North Afr. 471–499 (2024)
168. Lewis, R.J., Tibby, J., Arnold, L.J., Barr, C., Marshall, J., McGregor, G., Gadd, P., Yokoyama, Y.: Insights into subtropical Australian aridity from Welsby Lagoon, north Stradbroke Island, over the past 80,000 years. Quatern. Sci. Rev. **234**, 106262 (2020)
169. Kapur, V.V., Kumar, K., Morthekai, P., Chaddha, A.S.: Palaeodiet of Miocene producers and depositional environments: inferences from the first evidence of microcoprolites from India. Acta Geol. Sin.-Engl. Ed. **94**(5), 1574–1590 (2020)
170. Zimmerman, H.A., Meizel-Lambert, C.J., Schultz, J.J., Sigman, M.E.: Chemical differentiation of osseous, dental, and non-skeletal materials in forensic anthropology using elemental analysis. Sci. Justice **55**(2), 131–138 (2015)
171. Veneranda, M., Manrique, J.A., Sanz-Arranz, A., Julve Gonzalez, S., Prieto Garcia, C., Pascual Sanchez, E., Konstantinidis, M., Charro, E., Lopez, J.M., Gonzalez, M.A., Rull, F.: Application of chemometrics on Raman spectra from mars: recent advances and future perspectives. J. Chemom. **37**(9), 3438 (2023)
172. Chen, T., Zhang, T., Li, H.: Applications of laser-induced breakdown spectroscopy (LIBS) combined with machine learning in geochemical and environmental resources exploration. TrAC, Trends Anal. Chem. **133**, 116113 (2020)
173. Manrique, J.A., Lopez-Reyes, G., Cousin, A., Rull, F., Maurice, S., Wiens, R.C., Madsen, M.B., Madariaga, J.M., Gasnault, O., Aramendia, J., Arana, G.: SuperCam calibration targets: design and development. Space Sci. Rev. **216**, 1–27 (2020)
174. Aramendia, J., Gomez-Nubla, L., Castro, K., de Vallejuelo, S.F.O., Arana, G., Maguregui, M., Baonza, V.G., Medina, J., Rull, F., Madariaga, J.M.: Overview of the techniques used for

the study of non-terrestrial bodies: proposition of novel non-destructive methodology. TrAC, Trends Anal. Chem. **98**, 36–46 (2018)

175. Ragland, P.C., Conley, J.F., Parker, W.C., Von Orman, J.A.: Use of principal components analysis in petrology: an example from the martinsville igneous complex, Virginia, USA. Mineral. Petrol. **60**(3–4), 165 (1997)
176. Caggiani, M.C., Colomban, P.: Advanced procedures in Raman forensic, natural, and cultural heritage studies: mobile set-up, optics, and data treatment—state of the art and perspectives. J. Raman Spectrosc. **55**(2), 116–124 (2024)
177. Szefer, P., Grembecka, M.: Chemometric assessment of chemical element distribution in bottom sediments of the southern Baltic sea including Vistula and Szczecin lagoons-an overview. Pol. J. Environ. Stud. **18**(1) (2009)

Total Reflection X-Ray Fluorescence: Technological Developments and Expanding Applications

13

John Tsado Mathew, Abel Inobeme, Yakubu Azeh, Musah Monday, Elijah Yanda Shaba, Tanko Ezekiel, Etsuyankpa Muhammad Bini, Muhammad Aishetu Ibrahim, Musa Safiyano Tanko, Jibrin Mohammed Ndejiko, Jonathan Hussaini, Charles Oluwaseun Adetunji, and Amos Mamman

Abstract

Total reflection X-Ray fluorescence (TXRF) spectroscopy has emerged as a powerful analytical technique due to its sensitivity, non-destructive nature, and ability to analyze trace elements. Recent developments have significantly expanded the application potential of TXRF across various fields. One major advancement is the enhancement in detection limits, achieving sub-ppb (parts per billion) levels for many elements. This improvement is crucial for environmental monitoring, where minute concentrations of pollutants can be accurately detected. Furthermore, developments in instrumentation have led to increased

J. T. Mathew (✉) · Y. Azeh · M. Monday · M. A. Ibrahim · J. Hussaini · A. Mamman
Department of Chemistry, Ibrahim Badamasi Babangida University Lapai, Lapai, Niger State, Nigeria
e-mail: jmathew@ibbu.edu.ng

A. Inobeme
Department of Chemistry, Edo State University Uzairue, Auchi, Edo State, Nigeria

E. Y. Shaba · T. Ezekiel
Department of Chemistry, Federal University of Technology Minna, Minna, Nigeria

E. M. Bini · M. S. Tanko
Department of Chemistry, Federal University of Lafia, Lafia, Nasarawa State, Nigeria

J. M. Ndejiko
Department ofMicrobiology, Ibrahim Badamasi Babangida University Lapai, Lapai, Niger State, Nigeria

C. O. Adetunji
Applied Microbiology, Biotechnology and Nanotechnology Laboratory, Department of Microbiology, Edo State University Uzairue, Auchi, Edo State, Nigeria

V. K. Singh (ed.), *X-Ray Fluorescence Spectroscopy and Chemometrics*,
https://doi.org/10.1007/978-3-031-98375-7_13

automation and throughput, reducing analysis times and improving data reliability. This is particularly advantageous in industrial applications such as semiconductor manufacturing and quality control of materials, where rapid elemental analysis is essential. TXRF has also found application in archaeometry and cultural heritage preservation, allowing non-destructive elemental analysis of artifacts and historical materials with high precision. This aids in conservation efforts and understanding the composition of ancient artifacts without damaging them. In addition, recent advancements in TXRF spectroscopy have significantly broadened its applicability across scientific and industrial disciplines. Continued research and innovation promise further enhancements in sensitivity, speed, and versatility, making TXRF an indispensable tool for elemental analysis in diverse fields ranging from environmental monitoring to cultural heritage preservation.

13.1 Introduction

Total reflection X-Ray fluorescence (TXRF) spectroscopy has become an invaluable analytical tool across numerous scientific and industrial domains, thanks to its exceptional ability to detect and quantify trace elements with high sensitivity and precision. By combining the principles of XRF with total external reflection, TXRF allows for the effective analysis of ultra-thin samples on smooth substrates, minimizing background interference and enhancing signal clarity. Since its inception, TXRF has evolved from a specialized research tool into a widely adopted method across various fields, including environmental science, where it monitors trace pollutants; biological research, where it analyzes essential and toxic elements in tissues and fluids; materials science, for quality control and surface analysis; and cultural heritage preservation, where it helps study and conserve artifacts without causing damage. Ongoing advancements in TXRF technology-such as improved X-Ray sources, detectors, and data processing capabilities-have significantly broadened its application scope, making it a versatile and indispensable technique in modern analytical chemistry. Its continued evolution ensures that TXRF remains a critical tool for addressing complex analytical challenges in diverse settings [1, 2].

13.1.1 Historical Background and Evolution

The development of TXRF can be traced back to the mid-20th century when researchers began exploring the potential of XRF for elemental analysis. It was during this period that the concept of total reflection was recognized as a means to significantly enhance the sensitivity of XRF. By reflecting incident X-Rays off a surface at very shallow angles, it became possible to confine the interaction to a thin surface layer, thereby reducing background noise and improving detection limits. In the 1970s, the first TXRF instruments were developed, primarily aimed at surface and thin film analysis. These early systems, though relatively simple compared to modern standards, demonstrated the technique's capability to detect trace elements at very low concentrations. The success of these initial applications laid the groundwork for the broader adoption of TXRF, as researchers and industry professionals recognized its potential for high-sensitivity analysis. Over time, TXRF evolved beyond surface analysis to encompass a wide range of applications, driven by continuous advancements in technology and an increasing demand for precise elemental analysis in various fields [1].

Over the decades, technological innovations have significantly enhanced the performance of TXRF, transforming it into one of the most sensitive analytical methods available today. Improvements in X-Ray sources, such as high-brightness microfocus tubes and synchrotron radiation, have increased the intensity and focus of the incident X-Ray beam, leading to better excitation of the sample and improved detection capabilities. Additionally, advances in detector technology, particularly the use of silicon drift detectors (SDDs) and advanced multichannel analyzers, have provided greater energy resolution and faster data acquisition, further refining the technique's precision. Moreover, the development of sophisticated data processing algorithms has enabled more accurate analysis and interpretation of the fluorescence signals, pushing detection limits down to the parts per trillion (ppt) range for certain elements. These advancements have expanded the applicability of TXRF, allowing it to detect and quantify trace elements at ultra-low concentrations with unparalleled sensitivity. As a result, TXRF has gained widespread acceptance in both research and industry, particularly in fields like environmental monitoring, biomedical research, and semiconductor manufacturing, where the analysis of trace elements is crucial [3].

13.1.2 Principles of Operation

TXRF operates on the principle of total external reflection of X-Rays, which occurs when X-Rays strike a smooth surface at angles below the critical angle for total reflection. In a typical TXRF setup, a collimated X-Ray beam is directed onto a sample that has been carefully deposited on a flat, polished substrate, such as a quartz disc. When the angle of incidence is less than the critical angle, the X-Rays are completely reflected off the surface, rather than penetrating the substrate. This reflection confines the interaction volume of the X-Rays to just a few nanometers at the sample's surface. The confinement of the X-Ray interaction to such a shallow depth significantly reduces the background signal originating from the substrate, which in turn enhances the sensitivity of the fluorescence signal emitted by the sample. This enhanced sensitivity is crucial for detecting trace elements, as it allows for clearer differentiation between the sample's fluorescence and any potential background interference. The ability of TXRF to minimize background noise while maximizing signal strength makes it a powerful tool for high-sensitivity elemental analysis [4].

The XRF emitted by a sample during TXRF analysis is characteristic of the specific elements present, allowing both qualitative and quantitative analysis. The intensity of this fluorescence signal is directly proportional to the concentration of each element, enabling precise determination of trace elements within the sample [5].

This proportional relationship allows for accurate quantification, even at very low concentrations. Moreover, TXRF requires only a small sample volume and exhibits minimal matrix effects, making it particularly well-suited for analyzing small and complex samples. This includes biological fluids, where only minute quantities may be available, as well as nanoparticles and environmental particulates, which can present analytical challenges due to their size and heterogeneity. The minimal matrix effects mean that the sample's composition does not significantly interfere with the analysis, ensuring reliable and reproducible results. These features make TXRF an ideal tool for applications requiring high sensitivity and precision in the analysis of challenging samples [1, 6].

13.1.3 Advantages and Unique Features

TXRF offers distinct advantages over conventional XRF and other elemental analysis techniques (Fig. 13.1), particularly in its ability to achieve extremely low detection limits, often in the sub-ppb (parts per billion) range. This heightened sensitivity is primarily due to the reduced background signal achieved through the total reflection of X-Rays, which significantly enhances the detection of trace elements. As a result, TXRF is highly effective for analyzing samples where elemental concentrations are exceedingly low, such as in environmental monitoring, where detecting trace pollutants is crucial, or in semiconductor manufacturing, where even minute impurities can impact product performance. The technique's ability to detect such low concentrations makes it an invaluable tool in applications where precision and sensitivity are paramount. By providing accurate, low-level detection without the need for extensive sample preparation or complex correction procedures, TXRF ensures reliable results, making it a preferred choice in industries and research areas that require high-sensitivity elemental analysis [7].

A significant advantage of TXRF is the minimal sample preparation it requires, streamlining the analytical process. Unlike other techniques that may demand complex chemical treatments or matrix corrections, TXRF often allows samples to be analyzed directly after deposition on the substrate. This straightforward approach not only accelerates the preparation time but also reduces the potential for contamination or loss of analytes, which can be critical when dealing with trace element analysis. By minimizing the steps involved in sample preparation, TXRF reduces the likelihood of introducing external variables that could affect the accuracy of the results. This is particularly beneficial when working with sensitive or limited samples, where preserving the original composition is crucial. The simplicity of

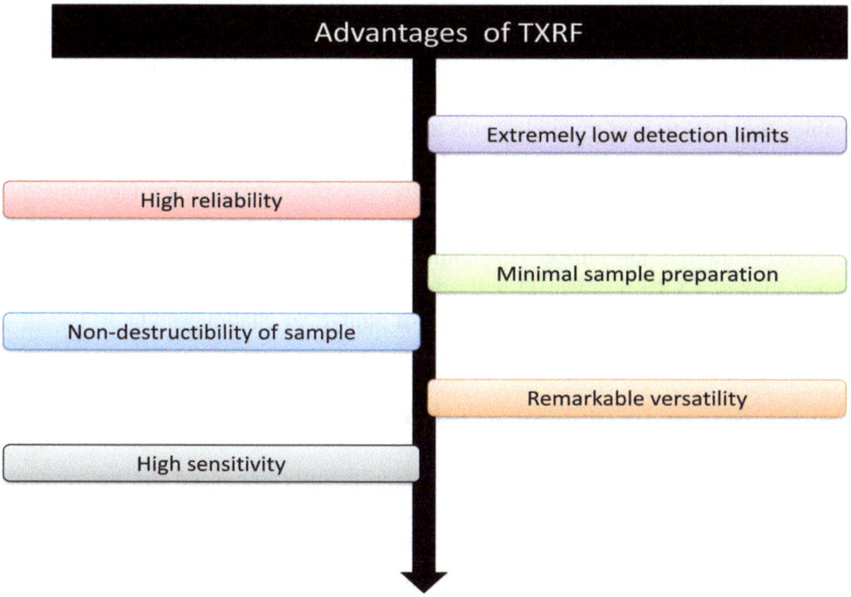

Fig. 13.1 Advantages of TXRF

the preparation process also enhances the reproducibility of the analysis, making TXRF a reliable choice for routine quality control, environmental monitoring, and research applications. Overall, the minimal sample preparation required by TXRF contributes to its efficiency, reliability, and widespread applicability across various fields [8].

TXRF stands out as a non-destructive analytical technique, ensuring that the integrity of the sample is maintained for further analysis or archival purposes. This feature is particularly crucial in fields like archaeometry and cultural heritage preservation, where the analysis of valuable or irreplaceable artifacts must be conducted with the utmost care. The ability of TXRF to perform precise elemental analysis without altering or damaging the sample makes it an invaluable tool for conservators and researchers. In the context of cultural heritage, TXRF allows for the detailed examination of artifacts, such as ancient ceramics, metals, or paintings, revealing their composition and providing insights into their origin, manufacturing techniques, and historical context. Since the technique does not require extensive sample preparation or extraction, it preserves the artifacts' physical and chemical integrity, allowing them to remain intact for future study or display. This non-invasive approach ensures that the cultural and historical value of these objects is fully preserved, while still providing critical analytical information [9].

TXRF is exceptionally versatile due to its ability to analyze a wide range of sample types, including solids, liquids, and aerosols. This broad applicability allows TXRF to be employed across various disciplines, from environmental monitoring and industrial quality control to biomedical research and cultural heritage preservation. Its capacity to handle different sample forms without extensive preparation is a significant advantage, simplifying the analytical process and reducing the risk of contamination or sample loss. Additionally, TXRF's high sensitivity enables the detection of trace elements at very low concentrations, making it ideal for applications where precision is critical, such as detecting pollutants in environmental samples or monitoring impurities in semiconductor manufacturing. The minimal sample requirements also mean that even small or precious samples can be analyzed without compromise. This combination of versatility, sensitivity, and efficiency makes TXRF a powerful and attractive tool for researchers and industries alike, capable of addressing a wide array of analytical challenges [1].

13.1.4 Recent Technological Advancements

Recent technological advancements have significantly enhanced the performance and usability of TXRF. Innovations in X-Ray sources, such as high-brightness microfocus X-Ray tubes and synchrotron radiation sources, have boosted the intensity and collimation of the incident X-Ray beam. These improvements have led to better detection limits, allowing for the identification of trace elements at even lower concentrations, and have reduced analysis times, making the process more efficient. In parallel, advances in detector technologies, particularly the use of silicon drift detectors (SDDs) and sophisticated multichannel analyzers, have further refined TXRF's capabilities. These enhanced detectors provide better energy resolution, enabling more precise elemental identification, and facilitate faster data acquisition, improving overall throughput. Together, these technological improvements have made TXRF a more powerful and versatile tool, expanding its application potential across various scientific and industrial fields [3].

The integration of automation and high-throughput capabilities in modern TXRF instruments has significantly expanded their use in industrial settings. Automated sample handling systems, paired with advanced software for data analysis and interpretation, enable the rapid and reliable processing of large sample batches. This automation reduces manual intervention, minimizing errors and improving consistency across analyses. These advancements make TXRF an efficient tool for routine quality control and process monitoring, particularly in industries where high precision and quick turnaround times are essential, such as semiconductor manufacturing and pharmaceuticals. The ability to swiftly analyze numerous samples with minimal operator input not only enhances productivity but also ensures that products meet stringent quality standards. As a result, TXRF has become increasingly valuable in maintaining the quality and reliability of industrial processes, driving greater efficiency and accuracy in elemental analysis [10, 11].

The development of portable TXRF instruments has revolutionized in situ analysis, especially in environmental and field-based research. These compact and mobile systems allow for on-site elemental analysis, significantly reducing the need for transporting samples to laboratories. By enabling immediate analysis at the point of sample collection, portable TXRF instruments facilitate real-time decision-making, which is critical in time-sensitive field studies such as environmental monitoring, pollution assessment, and disaster response. This capability is particularly advantageous in remote or challenging locations where traditional laboratory access may be limited or impractical. The portability of these instruments ensures that researchers can obtain accurate and reliable elemental data on-site, enhancing the efficiency and effectiveness of field investigations. Additionally, the non-destructive nature of TXRF allows for the preservation of samples for further analysis if needed. Overall, the advent of portable TXRF technology has expanded the possibilities for rapid, on-the-ground analysis across a wide range of environmental and research applications [12, 13].

13.1.5 Expanding Applications

The continuous advancements in TXRF technology have expanded its adoption across various fields. In environmental science, TXRF plays a crucial role in detecting trace metals in water, soil, and air, aiding in the monitoring and management of pollution. Its high sensitivity and minimal sample preparation requirements make it ideal for analyzing environmental pollutants at trace levels. In the biomedical field, TXRF is employed to analyze trace elements in biological fluids, tissues, and pharmaceutical products. This application supports research in toxicology, where understanding metal exposure is essential, as well as in nutrition and drug development, where elemental composition plays a critical role. TXRF's ability to provide precise, non-destructive analysis with minimal sample volume makes it an invaluable tool for advancing research and ensuring quality control in these critical areas [14].

In the industrial sector, TXRF is essential for quality control, especially in the semiconductor industry, where even trace impurities can significantly affect product performance. TXRF's ability to detect and quantify these minute contaminants with high sensitivity ensures the production of high-purity materials critical for advanced electronics. The technique's precision and non-destructive nature make it ideal for continuous monitoring and quality assurance in manufacturing processes. Additionally, TXRF is a valuable tool in the analysis of cultural heritage objects, where preserving the artifact's integrity is crucial. Its non-invasive approach allows for detailed elemental analysis without damaging delicate or irreplaceable historical items. This capability aids in conservation efforts and enhances the understanding of the composition and provenance of artifacts, supporting informed decisions in preservation and restoration. TXRF's role in both industry and cultural heritage underscores its versatility and importance in maintaining the quality and authenticity of materials [1].

As TXRF technology continues to advance, new applications and capabilities are expected to emerge. Ongoing research is centered on further improving detection limits, allowing for even more precise trace element analysis, and expanding the range of elements that can be accurately detected. Additionally, efforts are being made to integrate TXRF with complementary analytical techniques, creating more comprehensive and versatile analytical platforms. These developments are set to significantly enhance the versatility and utility of TXRF, enabling it to address increasingly complex analytical challenges across various fields. As a result, TXRF is poised to become an even more integral tool for elemental analysis, supporting breakthroughs in scientific research and industrial processes. The continued evolution of TXRF technology promises to solidify its position as a leading technique in the field of analytical chemistry, driving innovation and enabling new discoveries for years to come [7].

Recent developments in TXRF have greatly expanded its application potential, establishing it as a vital tool across numerous fields. Enhanced detection limits, automation, and integration with other techniques have made TXRF indispensable for high-sensitivity elemental analysis in areas such as environmental monitoring, biomedical research, and industrial quality control. The ability to detect trace elements at sub-ppb levels has enabled more accurate monitoring of pollutants, precise analysis of biological samples, and stringent quality control in industries like semiconductor manufacturing. As technological advancements continue to enhance TXRF's performance—improving sensitivity, speed, and ease of use—the technique is poised to play an increasingly critical role in scientific research and industrial applications. TXRF's versatility and precision will drive innovation in elemental analysis, enabling new discoveries and supporting the development of safer, more effective products and processes. The ongoing evolution of TXRF technology promises to unlock new applications and capabilities, further solidifying its importance in modern analytical chemistry [3].

13.2 Recent Technological Developments on TXRF

Recent technological developments in TXRF have significantly enhanced its capabilities. Advances include high-brightness X-Ray sources, such as microfocus tubes and synchrotron radiation, which increase intensity and collimation, improving detection limits. Silicon drift detectors (SDDs) have enhanced sensitivity and data acquisition speed, enabling more precise analysis. Portable TXRF instruments allow for on-site, real-time analysis in field research, expanding the technique's applicability. Automation and high-throughput capabilities in TXRF systems streamline sample processing and reduce human error, making it more efficient for routine industrial use. Additionally, sophisticated data processing algorithms have refined accuracy and lowered detection limits, advancing TXRF's role in environmental monitoring, biomedical research, and materials science (Table 13.1) [7, 15].

Table 13.1 An overview of the latest technological advancements in TXRF

Technological development	Description	Impact on performance	Application area	Future potential	Refs.
High-brightness X-Ray sources	Development of high-brightness microfocus X-Ray tubes and synchrotron radiation sources	Increased X-Ray intensity and collimation	Semiconductor industry, environmental monitoring	Enhanced detection limits, reduced analysis times	[16]
Silicon drift detectors (SDDs)	Advanced detector technology with improved energy resolution and faster data acquisition	Better sensitivity and precision	Biomedical research, materials science	Broader application in high-throughput settings	[17]
Portable TXRF instruments	Compact systems for in situ analysis in field research	On-site analysis, real-time data	Environmental science, field-based studies	Expanded use in remote or difficult-to-access areas	[18]
Automated sample handling	Integration of automation in TXRF systems for high-throughput analysis	Increased efficiency, reduced human error	Industrial quality control, pharmaceutical industry	Widespread adoption in routine industrial processes	[19]
Advanced data processing algorithms	Development of sophisticated algorithms for more accurate data interpretation	Lower detection limits, improved accuracy	Research and development, environmental monitoring	Continued enhancement of precision in trace element analysis	[17]

13.3 Applications in Environmental Monitoring

TXRF is highly effective in environmental monitoring due to its sensitivity and precision. In water quality monitoring, TXRF detects trace metals and pollutants at very low concentrations, ensuring accurate assessments of surface and groundwater. For soil analysis, TXRF quantifies heavy metals and trace elements, providing crucial data on contamination levels in agricultural and industrial soils. In air quality assessment, TXRF measures trace metals in airborne particulates, aiding in the evaluation of pollution levels. It is also used in sediment analysis to track environmental pollution history by analyzing trace elements in river and lake sediments. Additionally, TXRF helps identify pollution sources by detecting specific elemental signatures, supporting targeted remediation efforts. These applications underscore TXRF's role in maintaining environmental health and safety (Table 13.2) [14].

Table 13.2 Diverse applications of TXRF in environmental monitoring

Application	Description	Benefits	Typical samples	Challenges	Refs.
Water quality monitoring	Detection of trace metals and pollutants in water	High sensitivity for low concentration analysis	Surface and groundwater, drinking water	Sample preparation and matrix effects	[20, 21]
Soil analysis	Analysis of trace elements and heavy metals in soil	Accurate detection of contaminants in soil	Agricultural and industrial soils	Soil matrix complexity	[22, 23]
Air quality assessment	Measurement of trace metals in airborne particulates	Detection of airborne pollutants	Ambient air, particulate matter	Collection of air samples and particle size	[24, 25]
Sediment analysis	Quantification of trace elements in sediment samples	Insight into environmental pollution history	River and lake sediments	Sample homogenization and sediment variability	[26]
Pollution source identification	Identification of elemental signatures to trace pollution sources	Pinpointing pollution sources for remediation	Environmental samples from various sources	Differentiating between multiple contamination sources	[27, 28]

13.4 Industrial Applications of TXRF

TXRF plays a pivotal role in various industrial applications due to its high sensitivity and precision. In semiconductor manufacturing, TXRF is essential for analyzing trace contaminants in ultra-pure materials, such as wafers and thin films. This ensures the purity of components and prevents device failures caused by impurities. In materials science, TXRF characterizes and controls the quality of alloys, coatings, and nanomaterials, providing accurate elemental composition data critical for maintaining desired material properties and performance [29].

In the pharmaceutical industry, TXRF ensures the safety and quality of drug products by detecting trace elements in raw materials, formulations, and packaging. This helps prevent contamination and ensures compliance with stringent regulatory standards. TXRF also serves a vital role in quality control and assurance across various industries, including electronics and automotive, by monitoring product surfaces and process parameters to maintain high standards and regulatory compliance. Additionally, TXRF is used in environmental impact assessments to analyze waste products and emissions, aiding in the evaluation of environmental compliance and minimizing industrial impacts. Despite its advantages, challenges such as sample complexity and matrix effects must be managed to fully leverage TXRF's capabilities in industrial settings [30] (Table 13.3).

Table 13.3 The key industrial applications of TXRF

Application	Description	Benefits	Typical samples	Challenges	Refs.
Semiconductor manufacturing	Analysis of trace contaminants in ultra-pure materials	Ensures high purity, prevents device failure	Wafers, cleanroom dust, thin films	High sensitivity required for trace levels	[31, 32]
Materials science	Characterization and quality control of materials	Accurate elemental analysis, ensures material properties	Alloys, coatings, nanomaterials	Sample complexity, matrix effects	[33]
Pharmaceuticals	Analysis of trace elements in drugs and ingredients	Ensures safety and compliance, prevents contamination	Raw materials, formulations, packaging	Detection limits and sample preparation	[8]
Quality control and assurance	Routine monitoring of product quality and process parameters	Maintains product standards, ensures regulatory compliance	Surfaces, industrial waste, emissions	Variability in sample types and conditions	[17]
Environmental impact assessment	Analysis of waste and emissions to assess environmental impact	Ensures compliance with regulations, minimizes environmental footprint	Waste products, emissions, pollutants	Sample collection and matrix effects	[34]

13.5 Archaeometry and Cultural Heritage Preservation

This Table 13.4 outlines the diverse applications of TXRF in archaeometry and cultural heritage preservation, emphasizing its benefits, typical samples analyzed, and challenges encountered in each application. TXRF is instrumental in archaeometry and cultural heritage preservation. It enables non-destructive analysis of archaeological artifacts, revealing material provenance, manufacturing techniques, and authenticity. TXRF helps trace the origin of materials like pottery and metals, providing insights into historical trade and technology. In cultural heritage, it assesses the condition of artifacts by detecting surface corrosion and contaminants, guiding effective conservation and restoration efforts. TXRF also aids in authenticating artworks by analyzing pigments and materials, distinguishing original elements from modern additions. Its high sensitivity and minimal sample preparation make it ideal for preserving valuable and irreplaceable objects while ensuring their historical and artistic integrity [35].

Table 13.4 The diverse applications of TXRF in archaeometry

Application	Description	Benefits	Typical samples	Challenges	Refs.
Material sourcing and provenance	Determines the origin and source of materials in artifacts	Reveals trade routes and cultural exchanges	Pottery, metals, glass	Requires comparison with known material databases	[36, 37]
Technology and manufacturing techniques	Analyzes composition to understand ancient manufacturing methods	Provides insights into technological practices	Metal alloys, ceramics	Interpretation of complex compositional data	[38]
Dating and authenticity	Assists in dating artifacts and confirming authenticity	Validates the age and origin of artifacts	Artifacts with historical significance	Combining with other dating methods for accuracy	[39]
Condition assessment	Assesses the condition of artifacts by analyzing surface layers	Identifies corrosion products and contaminants	Ancient metals, sculptures, and ceramics	Surface preparation and detection of subtle changes	[40]
Restoration and conservation	Guides selection of materials and techniques for restoration	Ensures compatibility and effectiveness of restoration efforts	Restored and conserved artifacts	Matching original composition with restoration materials	[41]
Preventive conservation	Monitors environmental factors affecting artifacts	Helps protect artifacts from environmental damage	Storage environment dust and particulate matter	Accurate sampling and environmental monitoring	[42]
Art authentication	Analyzes pigments and materials in artworks to verify authenticity	Detects modern or non-original materials	Paintings, sculptures, and historical artifacts	Differentiating between historical and modern materials	[43]

13.6 Challenges and Limitations of TXRF

TXRF offers numerous advantages but also faces several challenges and limitations:

Matrix Effects: TXRF can be sensitive to the matrix of the sample, which may impact the accuracy of measurements. Variations in the composition and surface conditions of the sample can introduce errors, as the matrix effects can alter the

X-Ray fluorescence signal. For example, differences in sample density, thickness, or the presence of interfering elements can affect the intensity and shape of the fluorescence peaks. To address these issues, sophisticated calibration and data processing techniques are required. Accurate calibration involves using standards with similar matrix properties to the sample and applying corrections for matrix effects during analysis. Advanced data processing algorithms help deconvolute complex spectra and account for variations, ensuring precise quantification of trace elements. Despite these measures, matrix effects remain a challenge and necessitate ongoing refinement of TXRF methodologies [44].

- ***Sample Size and Complexity***

TXRF requires a flat, smooth sample surface to achieve optimal results. The technique relies on the total external reflection of X-Rays at very shallow angles, which is best achieved on a well-prepared, uniform surface. Analyzing complex or irregularly shaped samples can be challenging due to difficulties in maintaining the precise incident angle and ensuring uniform X-Ray reflection. For very small or intricately detailed objects, achieving a suitable sample surface and managing the spatial resolution can be particularly difficult. Additionally, irregularities or deviations in the sample surface can affect the accuracy of the measurements. This limitation necessitates careful sample preparation and sometimes limits the types of samples that can be effectively analyzed using TXRF. Continued advancements in sample handling and instrumentation may help address these challenges [45].

- ***Detection Limits***

Although TXRF is renowned for its high sensitivity, achieving detection limits in the parts per trillion (ppt) range for certain elements can still be challenging. The technique's sensitivity is influenced by factors such as the X-Ray source intensity, detector resolution, and sample matrix effects. For elements with very low concentrations, the inherent background noise and signal-to-noise ratio can limit detection capabilities. To extend detection limits further, advancements in TXRF technology are necessary, including improvements in X-Ray sources, such as higher-brightness microfocus tubes or synchrotron radiation, and enhancements in detector technology, such as more sensitive silicon drift detectors. Additionally, innovations in data processing algorithms and sample preparation techniques will play a crucial role in achieving even lower detection limits [44].

- ***Elemental Overlaps***

Elements with similar X-Ray fluorescence lines can lead to spectral overlaps, complicating the identification and quantification of individual elements in TXRF analysis. When elements emit X-Rays at similar energies, their fluorescence signals can overlap in the spectrum, making it challenging to distinguish and accurately measure each element. This spectral interference requires advanced data analysis techniques to deconvolute the overlapping signals and accurately quantify the elements present. Methods such as peak fitting, background correction, and mathematical deconvolution are employed to address these overlaps. Despite these solutions, resolving spectral interference remains complex and may impact the precision of the analysis. Continued development of improved algorithms and detector technologies is essential for enhancing the accuracy and reliability of TXRF measurements in the presence of spectral overlaps [46, 47].

13.7 Future Prospects and Research Directions

TXRF has already established itself as a powerful tool for trace elemental analysis across various fields. However, ongoing research and development promise to further enhance its capabilities and broaden its applications. The future prospects for TXRF lie in several key areas [1].

- ***Enhanced Sensitivity and Detection Limits***

A key goal for the future development of TXRF is to achieve even lower detection limits, extending into the sub-parts per trillion (ppt) range for a broader array of elements. Achieving these ultra-sensitive detection levels will likely require significant advancements in several areas of TXRF technology. One crucial area is the development of more powerful X-Ray sources, such as enhanced microfocus X-Ray tubes and the use of synchrotron radiation, which can provide higher intensity and better collimation of the incident X-Ray beam. Additionally, improvements in detector technology are essential. The next generation of silicon drift detectors (SDDs) will need to offer higher energy resolution and faster data acquisition rates to accurately capture the faint signals associated with trace elements at these low concentrations. These advancements will not only push the boundaries of TXRF's sensitivity but also expand its applicability in fields where detecting ultra-low concentrations of elements is critical, such as environmental monitoring, semiconductor manufacturing, and biomedical research [44, 48].

- ***Integration with Complementary Techniques***

Combining TXRF with other analytical techniques, such as mass spectrometry or Raman spectroscopy, represents a promising direction for future research. This integration can provide a more comprehensive understanding of samples by enabling simultaneous elemental and molecular analysis. For example, while TXRF excels at detecting trace elements with high sensitivity, mass spectrometry can offer detailed information about isotopic composition, and Raman spectroscopy can reveal molecular structure and chemical bonds. In environmental science, this combination could be invaluable for studying pollutants by identifying both their elemental makeup and chemical forms. In materials research, it could allow for the precise characterization of complex materials, from their elemental distribution to their molecular architecture. Similarly, in biomedical studies, such integrated approaches could enhance research into disease mechanisms, drug interactions, or nutritional studies by offering insights into both elemental trace analysis and molecular pathways. The synergy between TXRF and other methods promises to open new frontiers in analytical chemistry, making it a powerful tool for multidisciplinary research [49].

- ***Portable and In Situ Analysis***

The development of portable TXRF instruments has significantly expanded the possibilities for in situ analysis, especially in environmental monitoring and field-based research. These portable systems allow for on-site elemental analysis, providing immediate results without the need to transport samples to a laboratory. Future advancements are expected to make these portable TXRF instruments even more robust and user-friendly, with enhanced durability for use in remote or challenging environments. Improvements in sensitivity and detection limits will enable high-precision analysis in the field, making it possible to detect trace elements in real time. This capability will be particularly valuable in environmental studies, where rapid decision-making is crucial for assessing contamination levels, monitoring pollution, or managing natural resources. As portable TXRF technology continues to evolve, it will become an increasingly essential tool for researchers and professionals working in diverse, field-based settings [50, 51].

- ***Automated and High-Throughput Systems***

As the industrial applications of TXRF expand, there is an increasing demand for automated, high-throughput systems capable of processing large volumes of samples with minimal human intervention. This demand is particularly strong in sectors like quality control, semiconductor manufacturing, and pharmaceuticals, where precision and efficiency are paramount. Future developments in TXRF technology will likely focus on enhancing automated sample handling systems, enabling faster and more reliable preparation and analysis of multiple samples simultaneously. Advances in data processing will also play a critical role, with improved algorithms and software designed to quickly and accurately interpret complex datasets. Additionally, streamlined reporting systems will be developed to ensure that results are easily accessible and actionable, further reducing the need for manual oversight. These innovations will make TXRF even more efficient and reliable for routine industrial analysis, helping to meet the growing demands of high-volume production environments and ensuring consistent quality across various manufacturing processes [52, 53].

- ***New Applications and Expanding Elemental Range***

Researchers are actively exploring new applications for TXRF, particularly in cutting-edge fields like nanotechnology, advanced materials, and biomedical research. These areas demand precise and sensitive elemental analysis, making TXRF an ideal tool due to its high sensitivity and minimal sample preparation. In nanotechnology, TXRF is used to characterize nanoparticles and detect trace contaminants that could affect performance. In advanced materials science, the technique helps in analyzing the composition of complex alloys and thin films. Biomedical research benefits from TXRF's ability to analyze trace elements in biological samples, aiding in studies on nutrition, drug development, and disease diagnosis. A significant focus is also on expanding the range of elements that TXRF can accurately analyze. This includes optimizing the technique for light elements, which are often challenging due to their low fluorescence yields, and developing methods to resolve overlapping fluorescence lines for elements with similar emission spectra. These advancements will broaden TXRF's applicability, making it even more versatile for diverse scientific and industrial applications [54].

- ***Sustainable and Green Analytical Chemistry***

As sustainability becomes a central focus in analytical chemistry, future developments in TXRF are expected to prioritize reducing the technique's environmental impact. Researchers will likely work on minimizing the amount of sample and reagents required for analysis, thereby decreasing waste and conserving resources. Efforts to improve the energy efficiency of TXRF instruments will also be a key area of innovation, aiming to reduce the overall energy consumption of the technique. Additionally, the development of greener methods that rely less on hazardous materials, such as toxic solvents or complex chemical treatments, will be crucial in making TXRF more environmentally friendly. These advancements will not only align TXRF with the principles of green chemistry but also enhance its appeal as a sustainable, efficient, and responsible analytical tool across various scientific and industrial applications [55].

The future of TXRF is bright, with ongoing research focused on enhancing its sensitivity, expanding its applications, and improving accessibility. Key developments include achieving lower detection limits, integrating TXRF with complementary techniques, and advancing portable and high-throughput systems for in situ and industrial analysis. Additionally, the exploration of new applications in fields like nanotechnology and biomedical research, along with efforts to expand the range of analyzable elements, will further solidify TXRF's role. As these innovations materialize, TXRF will remain a critical tool for elemental analysis, driving progress and discovery in diverse scientific and industrial domains, and reinforcing its position as a versatile and indispensable analytical technique [56, 57].

13.8 Conclusion

In general, TXRF has evolved significantly since its inception, becoming a crucial tool for trace elemental analysis across diverse scientific and industrial fields. Recent developments, including advancements in X-Ray sources, detector technologies, and the integration of automation, have greatly enhanced the sensitivity, efficiency, and versatility of TXRF. These innovations have expanded its applications, making TXRF indispensable in areas such as environmental monitoring, biomedical research, materials science, and cultural heritage preservation.

As TXRF technology continues to advance, its future prospects are promising. Ongoing research aims to achieve even lower detection limits, improve portability for in situ analysis, and integrate TXRF with complementary analytical techniques. Additionally, there is a growing focus on sustainability, with efforts to minimize the environmental impact of the technique. The continued evolution of TXRF will undoubtedly drive further innovation and discovery, solidifying its role as a vital analytical tool for addressing the complex challenges of modern science and industry. As these developments unfold, TXRF is poised to remain at the forefront of elemental analysis, offering unmatched sensitivity and precision for a wide range of applications.

References

1. Gruber, A., Müller, R., Wagner, A., Colucci, S., Spasić, M.V., Leopold, K.: Total reflection X-Ray fluorescence spectrometry for trace determination of iron and some additional elements in biological samples. Anal. Bioanal. Chem. **412**(24), 6419–6429 (2020). https://doi.org/10.1007/s00216-020-02614-8
2. Mathew, J.T., Adetunji, C.O., Inobeme, A., Musah, M., Azeh, Y., Otori, A.A., Shaba, E.Y., Mamman, A., Tanko, E.: Removal of heavy metals using bio-remedial techniques. In: Shah, M.P. (ed.), Modern Approaches in Waste Bioremediation, pp. 117–130. Springer Nature Switzerland AG 2023 (2023a). https://doi.org/10.1007/978-3-031-24086-7_6.
3. Staufer, T., Grüner, F.: Review of development and recent advances in biomedical X-Ray fluorescence imaging. Int. J. Mol. Sci. **24**(13), 10990 (2023). https://doi.org/10.3390/ijms241310990
4. Mathew, J. T., Musah, M., Azeh, Y., Muhammed, M.: Kinetic study of heavy metals removal from pharmaceutical wastewater using geopolymer/Fe_3O_4 nanocomposite. Bima J. Sci. Technol. **7**(4), 152–163 (2023g). https://doi.org/10.56892/bima.v7i4.539
5. Vassholz, M., Salditt, T.: Observation of electron-induced characteristic x-ray and bremsstrahlung radiation from a waveguide cavity. Sci. Adv. **7**(4), eabd5677 (2021). https://doi.org/10.1126/sciadv.abd5677
6. Mathew, J.T., Musah, M., Azeh, Y., Muhammed, M.: Adsorptive removal of selected toxic metals from pharmaceutical wastewater using Fe_3O_4/ZnO nanocomposite, Dutse J. Pure Appl. Sci. **9**(4a), 236–248 (2023b). https://doi.org/10.4314/dujopas.v9i4a.22
7. Terzano, R., Denecke, M.A., Falkenberg, G., Miller, B., Paterson, D., Janssens, K.: Recent advances in analysis of trace elements in environmental samples by X-Ray based techniques (IUPAC Technical Report). Pure Appl. Chem. Chim. Pure Appl. **91**(6), 1029–1063 (2019). https://doi.org/10.1515/pac-2018-0605

8. El Hosry, L., Sok, N., Richa, R., Al Mashtoub, L., Cayot, P., Bou-Maroun, E.: Sample preparation and analytical techniques in the determination of trace elements in food: a review. Foods (Basel, Switzerland) **12**(4), 895 (2023). https://doi.org/10.3390/foods12040895
9. Orsilli, J., Galli, A. (2024). Angle-dependent XRF analyses: Pros and Cons of a novel technique in the field of cultural heritage. Sustainability **16**, 1460 (2024). https://doi.org/10.3390/su16041460
10. Upadhya, R., Kosuri, S., Tamasi, M., Meyer, T.A., Atta, S., Webb, M.A., Gormley, A.J.: Automation and data-driven design of polymer therapeutics. Adv. Drug Deliv. Rev. **171**, 1–28 (2021). https://doi.org/10.1016/j.addr.2020.11.009
11. Inobeme, A., Adetunji, C.O., Ajai, A.I., Inobeme, J., Mathew, J.T., Obar, A., Maliki, M., Nwakife, N., Eziukwu, C.: Chemical nanosensors for monitoring environmental pollution. In: Book: Advanced Application of Nanotechnology to Industrial Wastewater, pp. 93–103. Springer Nature Singapore (2023). https://doi.org/10.1007/978-981-99-3292-4_6
12. McIntosh, K.G., Guimarães, D., Cusack, M.J., Vershinin, A., Chen, Z.W., Yang, K., Parsons, P.J.: Evaluation of portable XRF instrumentation for assessing potential environmental exposure to toxic elements. Int. J. Environ. Anal. Chem. **96**(1), 15–37 (2016). https://doi.org/10.1080/03067319.2015.1114104
13. Ogundolie, F.A., Adetunji, C.O., Mathew, J.T., Inobeme, A., Ghazanfar, S., Olaniyan, O.T., Ajiboye, M.D., Dauda, W.P., Adetunji, J.B., Akinbo, O., Adetuyi, B.O.: Chapter Twenty-two-intelligent freshness indicators in packaging of plant-based food products. In: Adetunji, C.O., Hefft, D.I., Mbuge, D.O., Workneh, T.S. (eds.) Evaporative Coolers for the Postharvest Management of Fruits and Vegetables, pp. 323–331. Academic Press (2023). https://doi.org/10.1016/B978-0-323-89864-5.00008-4
14. Garg, V.K., Tanta, A., Lal Srivastav, A., Tiwari, M.K., Sharma, A., Kanwar, V.S.: Water quality assessment using synchrotron-based TXRF. Water Environ. Res. Res. Publ. Water Environ. Feder. **94**(7), e10759 (2022). https://doi.org/10.1002/wer.10759
15. Mathew, J.T., Inobeme, A., Musah, M., Azeh, Y., Abdullahi, A., Shaba E.Y., Salihu, A.M., Muhammad, E.B., Josiah, J.G., Jibrin, N.A., Ismail, H., Muhammad, A.I., Maurice, J., Mamman, A., Ndamitso, M.M.: A critical review of green approach on wastewater treatment strategies. J. Appl. Sci. Environ. Manag. **28**(2), 363–391 (2024e). https://doi.org/10.4314/jasem.v28i2.9
16. Gambari, M., Clady, R., Stolidi, A., Utéza, O., Sentis, M., Ferré, A.: Exploring phase contrast imaging with a laser-based Kα x-ray source up to relativistic laser intensity. Sci. Rep. **10**(1), 6766 (2020). https://doi.org/10.1038/s41598-020-63614-3
17. Wang, C., He, T., Zhou, H., Zhang, Z., Lee, C.: Artificial intelligence enhanced sensors - enabling technologies to next-generation healthcare and biomedical platform. Bioelectron. Med. **9**(1), 17 (2023). https://doi.org/10.1186/s42234-023-00118-1
18. Tiihonen, T.E., Nissinen, T.J., Turhanen, P.A., Vepsäläinen, J.J., Riikonen, J., Lehto, V.P.: Real-time on-site multielement analysis of environmental waters with a portable X-Ray fluorescence (pXRF) system. Anal. Chem. **94**(34), 11739–11744 (2022). https://doi.org/10.1021/acs.analchem.2c01490
19. Casian, T., Nagy, B., Kovács, B., Galata, D.L., Hirsch, E., Farkas, A.: Challenges and opportunities of implementing data fusion in process analytical technology-a review. Molecules (Basel, Switzerland) **27**(15), 4846 (2022). https://doi.org/10.3390/molecules27154846
20. Zainurin, S.N., Wan Ismail, W.Z., Mahamud, S.N.I., Ismail, I., Jamaludin, J., Ariffin, K.N.Z., Wan Ahmad Kamil, W.M.: Advancements in monitoring water quality based on various sensing methods: a systematic review. Int. J. Environ. Res. Public Health **19**(21), 14080 (2022). https://doi.org/10.3390/ijerph192114080
21. Mathew, J.T., Adetunji, C.O., Inobeme, A., Musah, M., Shaba, E.Y., Azeh, Y., Abulude, F.O., Mamman, A.: Organic Compounds in Atmospheric Aerosols: Atmospheric Aerosols Properties, Sources and Detection, pp. 43–59. Nova Science Publishers, Inc. (2022)
22. Kabir, E., Ray, S., Kim, K.H., Yoon, H.O., Jeon, E.C., Kim, Y.S., Cho, Y.S., Yun, S.T., Brown, R.J.: Current status of trace metal pollution in soils affected by industrial activities. Sci. World J. **2012**, 916705 (2012). https://doi.org/10.1100/2012/916705

23. Olorunsola, A., Adetunji, C.O., Olaniyan, O., Adetunji, J.B., Akinbo, K., Adetuyi, B.O., Inobeme, A., Mathew, J.T., Ajenifujah-Solebo, S.O.: Application of nanotechnology for the production of biopesticides, bioinsecticides, bioherbicides, mosquitoe repellants and biofungicides. In: Book: Handbook of Agricultural Biotechnology, pp. 293-309. John Wiley & Sons (2024). https://doi.org/10.1002/9781119836483.ch11
24. Jia, Y., Dong, H., Zheng, J., Sun, H.: Portable detection of trace metals in airborne particulates and sediments via μPADs and smartphone. Biomicrofluidics **11**(6), 064101 (2017). https://doi.org/10.1063/1.5003308
25. Adetuyi, B.O., Adetunji, C.O., Olajide, P.A., Ogunlana, O.O., Mathew, J.T., Inobeme, A., Popoola, O.A., Olaitan, F.Y., Akinbo, O., Oyewole, O.A., Eniola, K.I.T., Yerima, M.B.: Removal of microplastic from wastewater treatment plants. In: Book: Microplastic Pollution, VOL. 1, PP. 271–286. Springer, Singapore (2024). https://doi.org/10.1007/978-981-99-8357-5_16
26. Leão, L.P., da Costa, R.D.V.F., Leite, M.G.P., Nalini Júnior, H.A., Fonseca, R.M.F.: Distribution and assessment of trace elements contamination in sediments of Conceição River Basin, Brazil. Geosciences **11**, 236 (2021). https://doi.org/10.3390/geosciences11060236
27. Nazarpour, A., Watts, M.J., Madhani, A., Elahi, S.: Source, spatial distribution and pollution assessment of Pb, Zn, Cu, and Pb, isotopes in urban soils of Ahvaz City, a semi-arid metropolis in southwest Iran. Sci. Rep. **9**(1), 5349 (2019). https://doi.org/10.1038/s41598-019-41787-w
28. Mathew, J.T., Musah, M., Azeh, Y., Muhammed, M.: Development of Fe_3O_4 Nanoparticles for the removal of some toxic metals from pharmaceutical wastewater. Caliphate J. Sci. Technol. (CaJoST) **6**(1), 26–4 (2024f). https://doi.org/10.4314/cajost.v6i1.4
29. Sulka, G.D.: Electrochemistry of thin films and nanostructured materials. Molecules (Basel, Switzerland) **28**(10), 4040 (2023). https://doi.org/10.3390/molecules28104040
30. Manghani, K.: Quality assurance: importance of systems and standard operating procedures. Perspect. Clin. Res. **2**(1), 34–37 (2011). https://doi.org/10.4103/2229-3485.76288
31. Han, C., Hwang, H., Kang, J.-H., Hong, S.-B., Han, Y., Lee, K., Hur, S.D., Hong, S.: Reliable ultra trace analysis of Cd, U and Zn concentrations in greenland snow and ice by using ultra-clean methods for contamination control. Molecules **25**, 2519 (2020). https://doi.org/10.3390/molecules25112519
32. Inobeme, A., Shahnawaz, M., Adetunji, C.O., Mathew, J.T., Adetuyi, B.O., Popoola, O.A., Olaitan, F.Y., Akinbo, O., Kolawole, O.M., Oyewole, O.A. and KIT, E.B.: Primary microplastic: source, uses, transportation from land to marine environment. In: Book: Microplastic Pollution, vol. 1, PP. 23–33. Springer, Singapore (2024). https://doi.org/10.1007/978-981-99-8357-5_2
33. Saleh, H.M., Hassan, A.I.: Synthesis and characterization of nanomaterials for application in cost-effective electrochemical devices. Sustainability **15**, 10891 (2023). https://doi.org/10.3390/su151410891
34. Abubakar, I.R., Maniruzzaman, K.M., Dano, U.L., AlShihri, F.S., AlShammari, M.S., Ahmed, S.M.S., Al-Gehlani, W.A.G., Alrawaf, T.I.: Environmental sustainability impacts of solid waste management practices in the global south. Int. J. Environ. Res. Public Health **19**(19), 12717 (2022). https://doi.org/10.3390/ijerph191912717
35. Giuntini, L., Castelli, L., Massi, M., Fedi, M., Czelusniak, C., Gelli, N., Liccioli, L., Giambi, F., Ruberto, C., Mazzinghi, A., et al.: Detectors and cultural heritage: the INFN-CHNet experience. Appl. Sci. **11**, 3462 (2021). https://doi.org/10.3390/app11083462
36. Lazzari, M., Pereyra Domingorena, L., Stoner, W.D., Scattolin, M.C., Korstanje, M.A., Glascock, M.D.: Compositional data supports decentralized model of production and circulation of artifacts in the pre-Columbian south-central Andes. Proc. Natl. Acad. Sci. USA **114**(20), E3917–E3926 (2017). https://doi.org/10.1073/pnas.1610494114
37. Mathew, J.T., Inobeme, A., Abdulkadir, A., Etsuyankpa, M.B., Shaba, E.Y., Musah, M., Azeh, Y., Adetunji, C.O., Ibrahim, A.M., Tanko, E., Mamman, A.: Properties and Characteristics of Nanocoatings for Medicinal Applications. In book: Sustainable Approach to Protective Nanocoatings; publishing Tomorrow's Research Today, pp. 203–227. IGI Global (2024c). https://doi.org/10.4018/979-8-3693-3136-1.ch007

38. Alkunte, S., Fidan, I., Naikwadi, V., Gudavasov, S., Ali, M.A., Mahmudov, M., Hasanov, S., Cheepu, M.: Advancements and challenges in additively manufactured functionally graded materials: a comprehensive review. J. Manuf. Mater. Process. **8**, 23 (2024). https://doi.org/10.3390/jmmp8010023
39. Megahed, M.M., Elashery, N.H., Saleh, S.M., El-Shamy, A.M.: Flawless polyaniline coating for preservation and corrosion protection of ancient steel spearheads: an archaeological study from military museum, Al-Qala, Egypt. Sci. Rep. **14**(1), 7215 (2024). https://doi.org/10.1038/s41598-024-57184-x
40. Sedig, J.W., Olade, I., Patterson, N., Harney, É., Reich, D.: Combining ancient dna and radiocarbon dating data to increase chronological accuracy. J. Archaeol. Sci. **133**, 105452 (2021). https://doi.org/10.1016/j.jas.2021.105452
41. Rosa, E.D.A.R., Silva, L.F.V.D., Silva, P.F.D., Silva, A.L.F.E.: Color matching and color recovery in large composite restorations using single-shade or universal composites. Brazilian Dent. J. **35**, e245665 (2024). https://doi.org/10.1590/0103-6440202405665
42. Vilcassim, R., Thurston, G.D.: Gaps and future directions in research on health effects of air pollution. EBioMedicine **93**, 104668 (2023). https://doi.org/10.1016/j.ebiom.2023.104668
43. Pelowski, M., Gerger, G., Chetouani, Y., Markey, P.S., Leder, H.: But is it really art? The classification of images as "Art"/"Not Art" and correlation with appraisal and viewer interpersonal differences. Front. Psychol. **8**, 1729 (2017). https://doi.org/10.3389/fpsyg.2017.01729
44. Wilschefski, S.C., Baxter, M.R.: Inductively coupled plasma mass spectrometry: introduction to analytical aspects. Clin. Biochem. Rev. **40**(3), 115–133 (2019). https://doi.org/10.33176/AACB-19-00024
45. Li, P., Phillips, N.W., Leake, S., Allain, M., Hofmann, F., Chamard, V.: Revealing nano-scale lattice distortions in implanted material with 3D Bragg ptychography. Nat. Commun. **12**(1), 7059 (2021). https://doi.org/10.1038/s41467-021-27224-5
46. Pushie, M.J., Pickering, I.J., Korbas, M., Hackett, M.J., George, G.N.: Elemental and chemically specific X-Ray fluorescence imaging of biological systems. Chem. Rev. **114**(17), 8499–8541 (2014). https://doi.org/10.1021/cr4007297
47. Mathew, J.T., Adetunji, C.O., Inobeme, A., Musah, M., Azeh, Y., Otori, A.A. Mamman, A.: Nutritional qualities of agricultural crops after application of nanobioherbicides. In: Book: Handbook of Agricultural Biotechnology, vol. 3, pp. 189–204. John Wiley & Sons (2024a). https://doi.org/10.1002/9781119836513.ch9
48. Mathew, J.T., Inobeme, A., Adetuyi, B.O., Falana, Y.O., Adetunji, C.O., Shahnawaz, M.: Application of microplastics in toiletry products. In: Shahnawaz, M., Adetunji, C.O., Dar, M.A., Zhu, D. (eds.) Microplastic Pollution, pp. 73–84. Springer, Singapore (2024c). https://doi.org/10.1007/978-981-99-8357-5_5
49. Fan, X., Sun, A.R., Young, R.S.E., Afara, I.O., Hamilton, B.R., Ong, L.J.Y., Crawford, R., Prasadam, I.: Spatial analysis of the osteoarthritis microenvironment: techniques, insights, and applications. Bone Res. **12**(1), 7 (2024). https://doi.org/10.1038/s41413-023-00304-6
50. Guimarães, D., Praamsma, M.L., Parsons, P.J.: Evaluation of a new optic-enabled portable XRF instrument for measuring toxic metals/metalloids in consumer goods and cultural products. Spectrochim. Acta Part B Atom. Spectrosc. **122**, 192–202 (2016). https://doi.org/10.1016/j.sab.2016.03.010
51. Adetunji, C.O., Akinbo, O. Mathew, J.T., Egbuna, C., Inobeme, A., Olotu, T.M., Ajayi, O.O., Dauda, W., Ghazanfar, S., Ogundolie, F.A., Adetunji, J.B., Adetuyi, B.O., Ajenifujah-Solebo, S.O., Aborode, A.T.: Nanotechnology: history, trends and modern applications. In: Book: Handbook of Agricultural Biotechnology, pp. 1–177. John Wiley & Sons (2024). https://doi.org/10.1002/9781119836513.ch1
52. Oyewole, O.A., Chimbekujwo, K.I., Oniha, M., Omoregie, I.P., Ayanda, O.I., Adetunji, C.O.,Mathew, J.T.: Application of nanobiofertilization for bioremediation and ecorestoration of polluted soil/farmland. In: Book: Handbook of Agricultural Biotechnology, vol. 5, pp. 1–36. Wiley (2024). https://doi.org/10.1002/9781394211548.ch1

53. Mathew, J.T., Inobeme, A., Adetuyi, B.O., Adetunji, C.O., Popoola, O.A., Olaitan, F.Y., Akinbo, O., Shahnawaz, M., Oyewole, O.A., Eniola, K.I.T., Yerima, B.M.: General introduction of microplastic: uses, types, and generation. In: Book: Microplastic Pollution, vol. 1, pp. 3–21. Springer, Singapore (2024d). https://doi.org/10.1007/978-981-99-8357-5_1
54. Altammar, K.A.: A review on nanoparticles: characteristics, synthesis, applications, and challenges. Front. Microbiol. **14**, 1155622 (2023). https://doi.org/10.3389/fmicb.2023.1155622
55. Sajid, M., Płotka-Wasylka, J.: Green analytical chemistry metrics: a review. Talanta **238**(Pt 2), 123046 (2022). https://doi.org/10.1016/j.talanta.2021.123046
56. Sena-Torralba, A., Álvarez-Diduk, R., Parolo, C., Piper, A., Merkoçi, A.: Toward next generation lateral flow assays: integration of nanomaterials. Chem. Rev. **122**(18), 14881–14910 (2022). https://doi.org/10.1021/acs.chemrev.1c01012
57. Mathew, J.T., Inobeme, A., Shaba, E.Y., Musah, M., Azeh, Y., Etsuyankpa, M.B., Adetunji, C.O., Mamman, A., Tanko, M.S.: Application of diatoms in the treatment of industrial effluent. In: Rafatullah, M., Siddiqui, M.R., Khan, M.A., Kapoor, R.T. (eds) Multidisciplinary Applications of Marine Resources. Springer, Singapore, 47–62 (2024d). https://doi.org/10.1007/978-981-97-5057-3_3

14

X-Ray Fluorescence Detection of Fluorine, Chromium, Arsenic, Cadmium, Mercury, Lead and Uranium Measuring Drinking Water Toxicity

Kaushik Sanyal and Sangita Dhara

Abstract

Contamination of drinking water is one of the most serious concerns as it affects the lives of millions of people annually throughout the world. Both natural and anthropogenic activities are responsible for the increasing water pollution. Amongst all the contaminants, the heavy metal pollution in drinking water is the most critical. The toxicity level of these heavy metals depends on its chemical form, solubility and biological role. These hazardous metals in drinking water, which are referred to as poisoning, include As, Cr, Cd, Hg, Pb, U, etc. Nevertheless, low Z elements such as F and Cl toxicity are also well known. Detection and accurate determination of these hazardous elements in water samples requires highly sensitive analytical techniques. X-ray fluorescence is the most versatile technique having reasonably good accuracy and precision for trace elemental analysis in water samples. Development of preconcentration methodologies using novel ligands, activated charcoal, nano-materials, metal organic framework (MOF), etc. have drastically improved the detection limits and sensitivity for trace analysis by XRF. The combination of total reflection-XRF, an advance variant of energy dispersive-XRF, with synchrotron radiation's (SR) excellent characteristics have made SR-TXRF, an ideal analytical tool especially for ultra-trace elemental analysis in drinking water.

K. Sanyal · S. Dhara (✉)
Fuel Chemistry Division, Bhabha Atomic Research Centre, Mumbai, India
e-mail: sdhara@barc.gov.in

Homi Bhabha National Institute, Anushakti Nagar, Mumbai, India

V. K. Singh (ed.), *X-Ray Fluorescence Spectroscopy and Chemometrics*,
https://doi.org/10.1007/978-3-031-98375-7_14

14.1 Introduction

Environmental quality assessment is of paramount importance to all living organisms and among them water, the most basic of human necessities, requires continuous monitoring. Water is used in food production, industries, sanitization and most essentially, for drinking by humans. With increase in industrialization, a great amount of contaminates in form of trace and heavy elements are been dissipated into the environments [1–3]. Presence of these contaminants, heavy metals and other pollutants, in environmental samples especially drinking water is a serious problem and therefore, there has been a growing demand globally, for the qualitative as well as quantitative determination of these environmentally hazardous substances in drinking water. Though trace elements are present ubiquitously in very small quantities however, if they exceed the bio-available toxic limits, they can be very dangerous to human health. Further, their presence is not only of concern due to their toxicity but also for aesthetic impact with respect to their taste, colour, smell, etc. The World Health Organization (WHO) produces the norms on water quality in form of guidelines which forms the basis for their regulation in countries worldwide. These Guidelines for Drinking Water Quality (GDWQ) are built over 60 years of guidance [4]. These guidelines provide safety and health protection of public associated with the quality of the drinking water through the control of the hazardous constituents of water. In these guidelines the concentration limits for various contaminants are defined, which vary from elements to elements, considering its speciation, distribution and mobility as well as the regional environment, social, economic and cultural conditions. These guideline values signify the concentration of a physical, chemical, radioactive and microbial constituent that does not exceed the acceptable nominal risk to health of public over a lifetime of consumption.

The sources of drinking water can be groundwater, spring water, lakes, river, ponds, artificial reservoirs, etc. Though the groundwater forms only a very small fraction (0.6%) of the total water resource on earth, it is the major source of drinking water in both rural and urban areas [5]. It is very important to provide safe drinking water to each individual and therefore, before delivering of the water for drinking purpose it must comply with the physical, chemical and biological guidelines. The potential drinking water contaminants can be categorized as physical, chemical, biological and radiological [6]. The physical substances that are potential contaminants include the particles that cause turbidity, sediments and suspended solids. The chemical contaminants include both organic and inorganic substances. In organic contaminants, human-sourced pollutants are pesticides, pharmaceuticals, surfactants, personal care products, food dyes and preservatives, Per- and Poly Fluoro Alkyl Substances (PFAS), etc. These compounds have undesirable and detrimental effects and are collectively termed as Emerging Organic Contaminants (EOCs) [7–9]. Presence of EOCs in drinking water can affect the overall health of humans consuming it. The inorganic contaminants are heavy metals, metalloids, toxic elements and various fluorides, nitrates, nitrites and sulphate, which pose health risk to environment. The biological contaminants are the organisms referred as microbes and microbiological contaminants. These include bacteria, virus, protozoa, parasites, etc. [10]. The radiological contaminants are Ra, Cs, U and Pu, which gets accumulated in drinking water by both natural and anthropogenic activities. These contaminants are of major concern due to their known health effects [11]. Weathering, erosion, volcanic activity, etc. are the natural process that can lead to occasional contaminants in the drinking water sources. However, the ground water is less susceptible to these contaminants in comparison to the other sources of drinking water. But with increasing industrialization, agricultural, mining activities as well as over use of toxic chemicals, all the sources of drinking water are under potential threat of contamination with hazardous elements. The various anthropogenic sources which are responsible for the hazardous elements in drinking water are shown in Fig. 14.1. Industries play a major role in release of variety of harmful chemicals into the water sources. Annually 1500 km^3 wastewater is discharged [12]. Excessive use of fertilizers, pesticides, etc. has lead to heavy metal contamination in soil which is eventually carried to various nearby water bodies. The harmful chemicals in the landfills from electronics and pharmaceuticals wastes, with time, find its way into the ground water. Overuse and depletion of water levels are also responsible for the increase of toxic elements into these sources. Before being distributed the water is given a kind of treatment which includes, disinfection using Cl, ozone or UV treatment. Sometimes, these treatments can lead to addition of contaminants into drinking water. These hazardous elements are directly responsible for the physiological and biological health issues. Continuous exposure to these elements over a long period of time can cause adverse health effects such as anemia, low blood platelets, skin diseases, chronic lung issues, cardiovascular diseases, cancer and even death.

With ever increasing societal awareness as well as pollution, together with the issues related to increasing diseases in both urban and rural regions, toxicological risk assessment of contaminants is of utmost concern. Therefore, monitoring of trace elements having potential health risks in drinking water has become very necessary [13]. A substantial amount of financial assistance is provided by the government to the research related to the assessment of drinking water quality. To meet the WHO specification limits for hazardous elements and develop remediation strategies, rapid, reliable, sensitive and cost effective analytical detection systems are required. Figure 14.2 gives the impact of consumption of contaminated drinking water on human body for a prolonged time.

14.2 Potentially Toxic Trace Contaminants in Drinking Water

The effect of elemental pollution in drinking water is becoming more and more critical due to the lethal health effects caused by these elements. Due to their persistence, bio-accumulation and toxicity they are causing a serious threat to human lives. After absorption they generally get accumulated in the vital organs like lungs, liver, kidney, skin, bone, brain etc. as shown in Fig. 14.2. Fluorine

Fig. 14.1 Various anthropogenic sources which are responsible for the hazardous elements in drinking water

is a very common element that occurs in nature and accounts for 0.3 ppm of the earth crust and 1.3 ppm in seawater, approximately. It exists in a form of fluorspar, cryolite and fluorapatite with oxidation state of –1. Fluorides may enter the water sources as a result of industrial discharge as it is used in several industries including aluminum production, steel, fertilizers, tiles, ceramics and glass fibre industries. Further, sodium fluoride, fluorosilicic acid and sodium hexafluorosilicate are used in municipal water for fluoridation. In ground water fluoride concentration is higher due to the presence of fluoride –bearing minerals. The safe limit of fluoride in drinking water is 1 mg/L [5, 14]. Fluoride is required in minute quantities, 0.5–1 mg/L, for normal mineralization of bones and formation of dental enamel. The water soluble fluorides after uptake get completely absorbed in the gastrointestinal tract. It gets incorporated into teeth and bones. Long term exposure effects skeletal tissues and causes tooth decay or dental flurosis.

Chromium is a colourless and odorless metallic element found naturally in rocks, plants, soil, etc. The three thermodynamically stable states of Cr are Cr (0), Cr (III) and Cr (VI). In natural waters, it generally occurs in Cr (III) or Cr (VI) depending on the redox potential, pH or presence of oxidizing or reducing compounds. The hexavalent Cr is more toxic than trivalent, Cr (VI) is believed to be carcinogenic whereas, Cr (III) is a micronutrient having an important role in the metabolism of glucose and fats [15, 16]. Further, Cr (VI) has high environmental mobility due to the solubility of its salts in water. The causes of anthropogenic

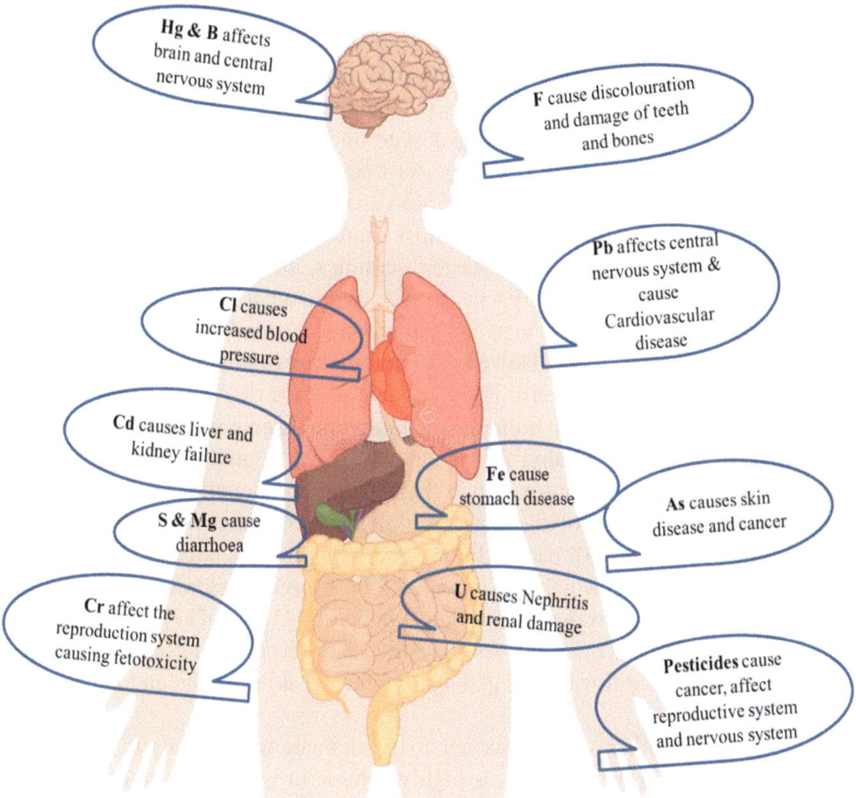

Fig. 14.2 The impact of consumption of contaminated drinking water on human body for a prolonged time

contamination of Cr in drinking water includes disposal of incompletely processed chromite ores, disposal of toxics from industries, etc. The absorption of Cr in human body also depends on its speciation. Cr (VI) gets absorbed in the gastrointestinal tract to a larger extent than Cr (III). After absorption Cr(VI) penetrates readily into the cell membrane and found in both erythrocytes and plasma. Inside the cell Cr (VI) is reduced to Cr (III), which binds to the macromolecules. Cr was found to accumulate mainly in liver, kidneys, spleen and bone marrow, the distribution depending on the speciation. Cr is known to affect the reproduction system causing fetotoxicity. Long term exposure of Cr (VI) causes cancer [15]. The WHO recommends a guideline value of 50 μg/L for total Cr in drinking water irrespective of Cr(III) or Cr (VI).

Arsenic is one of the most potent toxicant having a very complex metabolic pathway which is affecting millions of people worldwide. It is estimated that worldwide 4.01×10^{16} kg of As is present in the earth crust and it is mainly available as sulphide mineral. It is equally potential hazard for humans as well as animals. It is a metalloid as it possesses the properties of both metals as well as non-metals. It is known to exist in several organic as well as inorganic forms with varying toxicity. The common valence states of naturally occurring As are As (−III), As (0), As(III) and As (V) known as arsine, arsenic, arsenite and arsenate, respectively, which exists in a particular state under the certain conditions of pH, redox conditions, temperature, etc. The As finds its way into the drinking water sources, through several anthropogenic activities. The occupational, environmental and medicinal activities are the foremost sources of both organic and inorganic forms of As. Pesticides, electronic wastes, ceramics, medical supplements, etc. are the other sources. It mainly exists in arsenite and arsenate form in water. In animals the main source of As is ingestion of drinking water. The WHO limit for As in drinking water is 10 μg/L. However, contaminated food is another major route of As contamination in humans. The inorganic species of As is considered more toxic than its organic form. In humans, the inorganic arsenic metabolizes to methylated metabolites Mono Methyl Arsonous acid (MMA) and Di Methyl Arsinous acid (DMA) which are more toxic than their parent compounds. As contamination in drinking water is one of the major problems across the world [16]. Arsenic concentration in the eastern part of India is quite high, especially the Murshidabad district of West Bengal, which has recorded 54.6% higher arsenic than the recommended value in groundwater with average value of 141 μg/L [17]. There has been a strong correlation between the As ingestion and several acute and chronic disorders like cancer, cardiovascular diseases, diabetes, skin lesions and neurological effects.

Cadmium is another very important toxic element which finds its way to the environment through the ores as it is highly soluble in water. The other factors are discharge of metals from refineries, waste from Ni–Cd batteries, plastic, pigment and paint industry. In drinking water presence of Cd is as a result of galvanized pipes carrying water which may be contaminated by Cd or from solders in fittings containing Cd. When ingested it is absorbed in the gastrointestinal tract, the absorption is more in case of anemic person (more then 50%) [18]. The absorbed Cd gets stored in both kidney and liver, through the bloodstream. The short term effects on human health includes nausea, vomiting, diarrhea, muscular cramp, sensory disturbances, liver damage and renal failure. Long term Cd poisoning in humans lead to cancer, kidney dysfunction, acute gastroenteritis and osteoporosis. The biological half-life of Cd in humans range from 10 to 35 years and with age, the tissue cadmium concentration increases [19]. The WHO guidelines recommended values for Cd in drinking water is 3.0 μg/L, however, the Environmental Protection Agency (EPA) allowed the maximum contaminant level of 5.0 μg/L in drinking water [19].

Lead is another toxic element which can greatly harm the ecological environment and human health and therefore, strict control of its emission and accurate monitoring is important. It is a stable chemical element used widely in batteries, paints, dyes and several industries. Further Pb is also present in the cooking ceramics, cosmetics, inks as well as automobile exhaust. The source of Pb in drinking water is due to the plumbing systems in which the pipes, solder and fitting may contain Pb. The plastics used for storage may also contain Pb. It enters and easily gets absorbed into the body mainly thorough digestive tract and skin, then it reacts with biological systems, affecting different organs of organisms with acute or chronic toxic effects. This heavy metal cannot be decomposed by organisms and is extremely difficult to discharge, so it accumulates in living organisms. It is a neuro-toxic substance, which affects the central nervous system, causes anemia, cardiovascular diseases, lung dysfunction, etc. including carcinogenic [20]. Further, in human body Pb is stored in the bone as phosphate and a small amount also gets stored in liver, spleen, cells of blood and brain [21]. According to the WHO and US environmental protection agency guidelines, the maximum allowable concentration of Pb in drinking water are 10 μg/L and 15 μg/L, respectively [22].

Another highly toxic ubiquitous element, which gets distributed into the environment through natural processes like volcanic eruption is mercury. The increasing use of Hg in industrial processes such as in electrochemical processes, control instruments like switches, thermometers, barometers, etc., in electrical appliances such as lamps, arc rectifiers, mercury cells, in fungicides, preservatives, antiseptics, etc. has lead to its increase in the environment due to these anthropogenic activities. It exits as elemental mercury which is insoluble. Mercury (II) is readily soluble and mercury (I) is less soluble. In sea and fresh water, the methylation of inorganic mercury is a common process. Methylmercury or monomethylmercury enters the food chain through rapid diffusion and tight binding to proteins. In the intestinal tract the extent of transport of inorganic mercury depends on the intestinal pH, solubility, and the ease of dissociation. The main target of organ is kidney where Hg get accumulated. It also has harmful effects on the growing fetus, kidney and brain. Mercury (II) chloride has evidence of carcinogenic activity. Acute poisoning can affect neurological and renal activities. Moreover, it can cause cardiovascular collapse, acute renal failure and severe gastrointestinal damage. Acute toxication also includes pharyngitis, dysphagia, abdominal pain, nausea and vomiting and diarrhoea. Later, swelling of the salivary glands, stomatitis, loosening of the teeth, nephritis, anuria and hepatitis may also occur. The guideline value of inorganic mercury is 6 μg/L. Monitoring and determination of uranium is an important task for the safety, regulatory compliance and its disposal, due to its radio and chemical toxicity. Its chemical toxicity supersedes the radiological toxicity. The average U concentration in the earth‘s crust is 2.8 μg/g and in seawater it is 3.3 μg/L [23]. Uranium is a radionuclide which decays by both alpha and gamma emission. It occurs naturally in +2, +3, +4, +5, and +6 oxidation states, but in nature it usually exists in its hexavalent form as uranyl ion UO_2^{2+}. It is mainly used as fuel in nuclear reactors. In the environment, U finds its path through release of runoff water from mill tailings, combustion of coal, use of phosphate fertilizers, etc. In water U exists in hydroxyl, biphosphate and carbonate complexes in neutral pH solutions under ambient conditions. The dioxouranate (UO_2^{2+}) ion exists in equilibrium with carbonate and bicarbonate ligands in the presence of dissolved CO_2 in water. In ground water conditions the complexes of uranium are dioxouranate, UO_2^{2+}, tris(carbonato)dioxouranate,

$UO_2(CO_3)_3^{4-}$, and hexakis(carbonato)tris[dioxouranate], $(UO_2)_3(CO_3)_6^{6-}$ [24]. In humans the primary effect of uranium intake is Nephritis and renal damage. Long exposure of high specific activity compounds of U can cause cancer. Table 14.1 gives the summary of the hazardous elements with their occurrence, health effects and specification limits.

14.3 Analytical Techniques for Trace Determination in Drinking Water

In order to ascertain the water quality for consumption of drinking water, investigation of ultra-trace, trace, minor and major elements is essential. The minerals present in drinking water are necessary for human life and play an essential role in the metabolic functions, however if their presence is beyond certain specific limits they act as toxics and can cause harmful effect. Monitoring of major and minor elements such as Ca, K, Mg, Na (Major) and Ni, Cu, Zn, Mn, Fe (Minor) in drinking water is necessary to ensure that their quality meets the regulatory guidelines. Flow injection analysis with fluorescence detection, was used to determine Zn, Al, Mg, Cu, Fe and Cl^- ion in tap water samples [25, 26] potentiometric ion-selective electrode detectors in ion chromatography have advantage of online analysis of water for chloride, bromide, iodide, fluorideions [27]. Traditionally these elements in drinking water are determined by analytical techniques such as atomic absorption spectrometry (AAS) [28, 29] and atomic emission spectrometry (AES) [30]. Though analytical methodologies based on AAS techniques are used routinely to perform analysis of numerous laboratory samples, the main drawback of these methodologies is that they have narrow ranges of calibration and sequentially single-element determination is carried out [29, 31]. This can be quite time consuming. Therefore, Inductively Coupled Plasma (ICP) based AES or Optical Atomic Spectrometry (OES) are preferred which shows high sensitivity and allow multi-elemental analysis [32]. Moreover, ICP-Mass Spectrometric systems allows achieving the lowest detection limits in ng/L level [33].

Trace element as defined by IUPAC is any element having an average concentration less than 100 ppm or μg/mL [34]. With rapid increase of detection capabilities of analytical techniques, another new term was coined as ultra-trace elements. Ultra-trace concentration refers to elemental concentrations below 1 μg/mL. Nowadays a large number of analytical techniques are available for trace and ultra-trace determinations in a variety of matrix. However, proper precautions should be taken during sampling, storage, sample preparation and analysis of samples for trace and ultra-trace determinations. Routinely, the determination of trace metals has been carried out by inductively coupled plasma atomic emission spectrometry (ICP-AES), inductively coupled plasma mass spectrometry (ICP-MS) [30, 35], electrothermal atomic absorption spectrometry (ET-AAS) [36], and flame atomic absorption spectrometry (FAAS) [29].

XRF is another powerful analytical technique which complies with the desired feature of multi-elemental capability. XRF is a very simple technique which allows

Table 14.1 Hazardous elements with their occurrence, health effects and specification limits

S. no.	Contaminant	Occurrence in water	Adverse health effects	WHO limit/ EPA limit
1.	Fluoride	F^-	Effects skeletal tissues and causes tooth decay or dental flurosis	1.5 mg/L/ 4 mg/L
2.	Chromium	Cr (VI) as CrO_4^{2-} or $HCrO_4^-$ and Cr (III) as $Cr(OH)_n^{(3-n)+}$	Affect the reproduction system causing fetotoxicity and Cancer	50 μg/L/ 0.1 mg/L
3.	Arsenic	**Arsenite**: [H_3AsO_3 and $H_2AsO_3^{1-}$] **Arsenate**: [$(H_2AsO4)^{1-}$ and $(HAsO_4)^{2-}$]	Severe problems in the skin, bladder, lung, kidney and liver, endocrine disruptor Cancer in skin, Cardiovascular Problems Neurological problems	10 μg/mL/ 10 μg/mL
4.	Cadmium	Cd^{2+}	Short term effects are diarrhea, muscular cramp, sensory disturbances, liver damage, and renal failure. Long term effects include kidney dysfunction, acute gastroenteritis, osteoporosis and cancer	3.0 μg/mL/ 5.0 μg/mL
5.	Lead	$PbCO_3$, Pb_2O and $Pb(OH)_2$	Neuro-toxic substance, which effects the central nervous system, causes anemia, cardiovascular diseases, lung dysfunction	10 μg/L / 15 μg/L

(continued)

Table 14.1 (continued)

S. no.	Contaminant	Occurrence in water	Adverse health effects	WHO limit/ EPA limit
6.	Mercury	Mercuric (Hg^{2+}) Methyl mercury mercurous (Hg^{1+})	Gets accumulated in Kidney, liver, brain. Affects growing foetus. Acute poisoning causes cardiovascular collapse, acute renal failure and severe gastrointestinal damage. Further symptoms include pharyngitis, dysphagia, abdominal pain, nausea and vomiting and diarrhoea	6 μg/L/ 2 μg/L
7.	Uranium	U (IV): UO_2^{2+}U (VI): $UO_2(CO_3)_3^{4-}$ & $(UO_2)_3(CO_3)_6^{6-}$		30 μg/L/ 20 μg/L

direct sample analysis of solids and liquids. To perform XRF analysis a few ml or mg of the sample is taken in the sample cup and directly measured by exciting the sample using incident X-rays. The primary X-rays interact with the analyte atoms resulting in the emission of characteristic secondary X-rays. The characteristic X-rays emitted from the atoms of the sample are detected by a detector. XRF can be operated in two modes: wavelength dispersive XRF (WDXRF) and energy dispersive XRF (EDXRF), with sequential and simultaneous detection systems, respectively. Total reflection XRF (TXRF) is an advanced variant of EDXRF, with low glancing angles of incident beam. In Fig. 14.3 the geometrical difference between EDXRF and TXRF is shown. In conventional XRF the sample amount required is high and so is the scattering. Hence, the typical LODs are in the range of mg/L [37]. Further, it also suffers from severe matrix effect. TXRF offers better sensitivity, improved detection limits in the range of several μg/L and negligible matrix effect [38]. In TXRF analysis the sample required is very less in the range of a few micro liters. A schematic diagram of sample preparation in TXRF analysis is shown in Fig. 14.4.

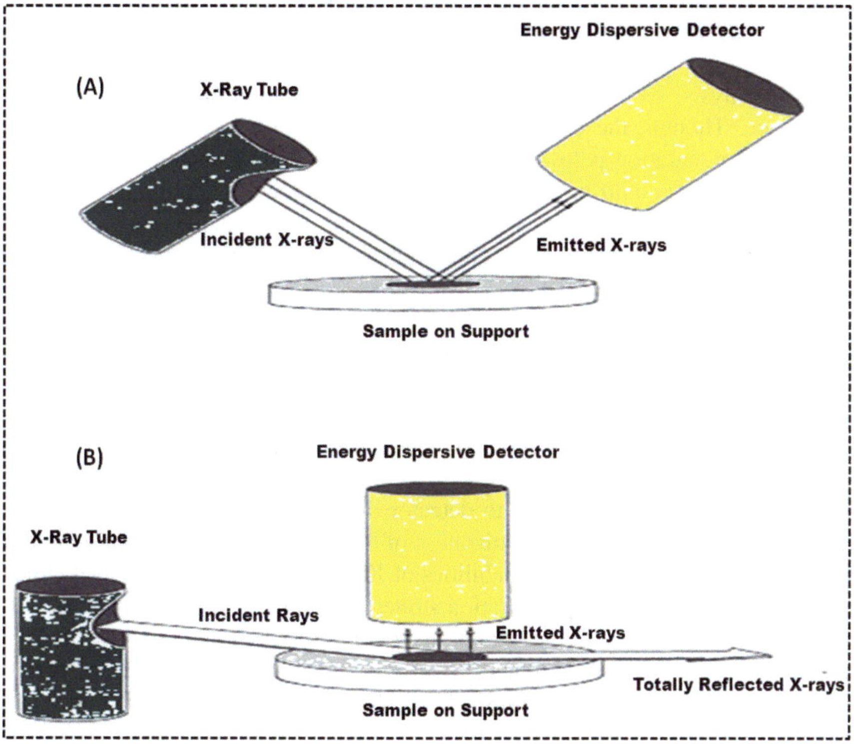

Fig. 14.3 Geometric difference between **a** EDXRF and **b** TXRF

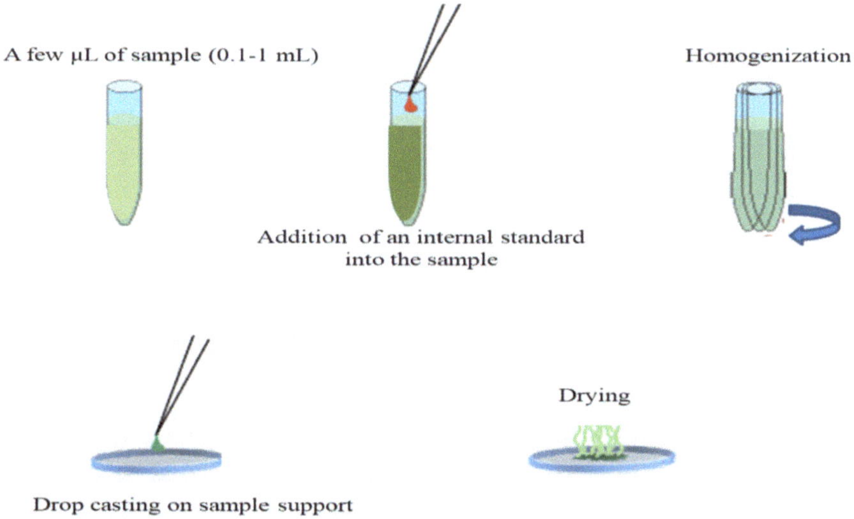

Fig. 14.4 Schematic diagram of sample preparation in TXRF

14.4 Development of Sample Preparation Methodologies for Improved Sensitivity and Detection Limits in XRF

In XRF analysis, employing the right sample preparation strategy is of utmost importance. Though, the detection limit obtained by EDXRF or TXRF without any sample preparation is low [39], but the WHO recommended limits for different toxic elements could be as low as a few ng/mL. Direct determination of such extremely low concentrations is difficult because of the matrix interferences or poor sensitivity and it may not be achieved by EDXRF. The variation in the detection limits of elements excited by TXRF and EDXRF is shown Fig. 14.5. In order to analyze these toxic elements in natural water down to such low concentration levels requires different sample preparation strategies.

With recent advancement in the instrumentation and development of sample pre-concentration strategies, the analytical capabilities and applications of XRF has expanded. Improvements in electronics signal processing systems, with thin window / windowless X-ray excitation system and compactness in instrumentations with less X-ray tube- to sample- to detector distances, have drastically improved the instrumental sensitivity. The utilization of Synchrotron sources has also significantly enhanced the analytical capabilities of XRF. In addition to this, development of novel and efficient pre-concentration strategies and methodologies such as solid phase extraction (SPE), liquid liquid extraction (LLE), miniature mode micro-extraction, such as dispersive liquid liquid micro-extraction (DLLME), dispersive micro solid phase extraction (DMSPE), etc. for high recoveries, enrichments and

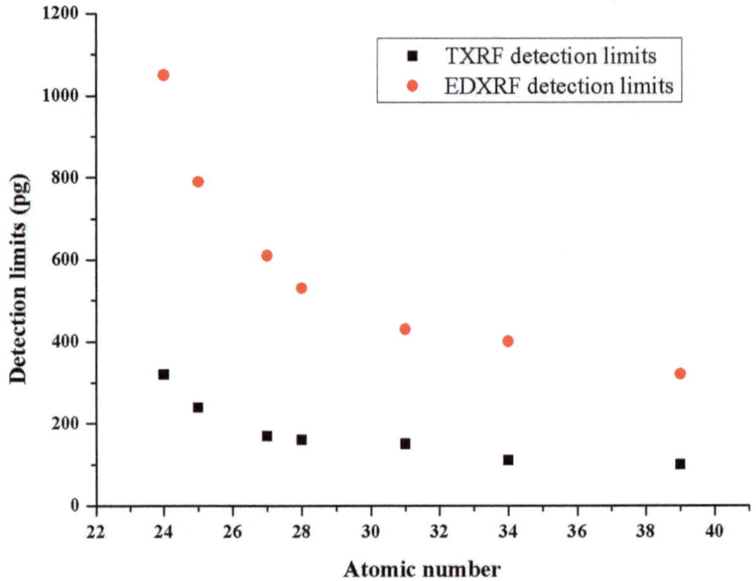

Fig. 14.5 Variation in the detection limits of elements excited by TXRF and EDXRF

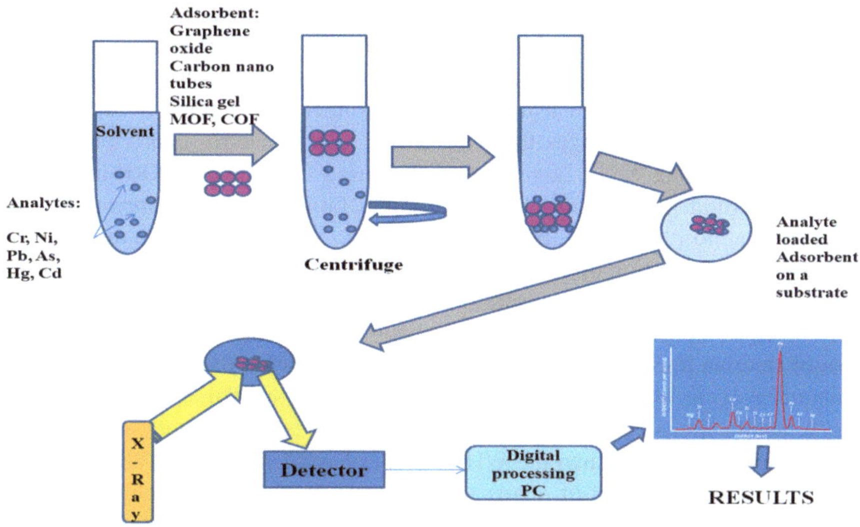

Fig. 14.6 The sample preparation strategy using SPE followed by XRF determination

low sample consumption. SPE is a widely used sample preparation technique for pre-concentration and selective isolation of analytes from a liquid matrices. A solid-phase sorbent material selectively binds with the analyte and selective isolated from the rest of the matrix. Figure 14.6 shows the sample preparation strategy using SPE followed by XRF determination. LLE, known as solvent extraction and partitioning, is a separation technique used to transfer a solute containing the analyte from one liquid phase to another. This method exploits differences in solubility or chemical affinity between two immiscible liquids, typically a polar and an organic solvent. The analyte rich phase is then selectively isolated and analysed by XRF.

In literature there are different types of sample preparation methodologies for pre-concentration and separation are discussed for different elements using nanoparticles, nanotubes, carbon fullerenes, membranes, grapheme oxide, activated charcoal, etc. In this chapter we will discuss about the techniques and methods employed for determination of toxic elements like F, Cr, As, Cd, Hg, Pb and U in drinking water samples.

14.5 Fluorine Determination

There are many challenges associated with the analysis of fluorine in water sample. First fluorine is a low Z element ($Z = 9$), so the fluorescent yield is very low. Moreover, the F Kα line has very low energy (677 eV), which could be easily absorbed by the air molecules present in the environment or by the window of the detector. So to overcome these problems the excitation energy should be as close as possible to the absorption edge of F. Further, the experiment has to be carried out in vacuum environment and the window of the detector should be as thin as possible to avoid/reduce any possible absorption of F Kα line. Another problem associated with fluorine determination using X-ray spectrometric technique is its volatility loss. Due to these challenges there are a very few literature reports on fluorine determination using XRF. Tarsolet et al. had used Cr targeted TXRF instrument for the determination fluorine in water samples [40]. They have reported the use of vacuum atmosphere and Silicon Drift Detector (SDD) having ultra thin window to reduce the absorption of F Kα X-rays. For the TXRF analysis, internal standard plays a very crucial role for quantification [41]. Most of the internal standards used in TXRF spectrometry are acidic in nature, this would increase the volatility loss of fluorine during the drying step in TXRF analysis. To avoid this problem, a neutral reagent like NH_4VO_3 is added into the sample. This however increases the detection limits. The detection limits achieved using this methodology was 5 mg/L which is higher than the WHO limit (1.5 mg/L). Sanyal et al. had shown that silicon wafer instead of quartz sample supports could drastically reduce the detection limit of fluorine down to 0.375 mg/L [42]. In another report the authors had further improved the detection limit of fluorine using lower energy synchrotron excitation source having excitation energy of 1.9 keV [43]. In these studies it was observed that the sensitivity for fluorine determination deteriorates as the matrix concentration increases. So, although it could be possible to determine fluorine in water samples filtered by Reverse Osmosis (RO) using TXRF, but the situation become very difficult when it is required to analyze fluorine in water samples containing high matrices like tap water or ground water. In such case selective separation of F will be required prior to analysis.

14.5.1 Chromium Determination

Different pre-concentration methodologies were adapted in combination with EDXRF and TXRF to improve the sensitivity for the analyses of chromium down to a level of a few ppb. Orescanin et al. used ammonium pyrolidine dithiocarbamate (APDC) based pre-concentration followed by its determination using EDXRF [44]. This APDC forms complex with chromium species and forms a type of suspension which can be filtered and analysed by XRF. The detection limit achieved using this methodology was 2.95 μg/L. Membrane based pre-concentration techniques are attractive way of sample pre-concentration. Membranes can be attached on a solid substrate like the TXRF quartz sample support (Fig. 14.4) and the analytes can be directly pre-concentrated on these supports. The main advantage of this method is that they do not require further separation of the extractant from the analyte as the extractant is directly attached on the quartz sample support. Kallithrakas et al. initially developed this methodology for the detection of chromium using TXRF [45]. The TXRF support with the membrane can be directly immersed into the solution and equilibrated for a few hours. After equilibration these supports can be taken out from the solution and dried. After drying they can be directly measured by TXRF. In this work Kallithrakas et all have used various types of membranes for the pre-concentration purpose. For the preparation of these membranes basically four components were required which are polymer matrix, solvent, extractant and plasticizer. Finally, it was observed that Poly Vinyl Chloride (PVC) polymer matrix in combination with Aliquat 336 extractant gave the best results for the detection of Cr (VI) species. Aliquat 336 is basically a type of anion exchanger which has high affinity towards chromate species [46, 47]. On the other hand membrane with nafion as main component shows selectivity towards Cr(III) species. The whole process of sample preparation is shown schematically in Fig. 14.7. The detection limits obtained using this methodology is in the range of 0.33–0.6 ng/mL for Cr. The quartz sample supports could be easily cleaned and reused.

Many Solid Phase Extraction (SPE) techniques have been combined with TXRF/EDXRF for the determination of Cr in water samples at ultra trace level. This offers a powerful approach to enhance the sensitivity and selectivity for trace elemental analysis. In this method depending on the elements of interest, different solid phase extractant can be chosen such as graphene oxide, carbon nanotube, nano particles, different nano composites, metal organic framework (MOF), Covalent Organic Framework (COF), ion exchange resins, etc. They can also be functionalized properly to enhance their selectivity towards particular element/species. These extractants can be dispersed into the solution which needs to be analysed. When the equilibrium is reached the extractant could be separated and directly measured by TXRF/EDXRF. Graphene based nano materials are widely used for the preconcentration of different toxic elements followed by its detection using EDXRF/TXRF technique. A graphene sheet is a single layer of carbon atoms arranged in a two-dimensional honeycomb lattice. Different forms of graphene e.g. graphene sheet, graphene oxide, graphene nano composite, etc. are applied as solid phase absorbent for the pre-concentration purpose. Both graphene oxide and graphene has large surface area and they also have many sites where functional modification can be done [48]. These types of absorbent mainly consist of low Z elements which produce minimum matrix effect as well as interference during the analysis. Hence, these type of materials are ideal absorbent for the pre-concentration purpose when XRF is used as a probe. Katarzyana has reported that graphene oxide nano particles can be used to pre-concentrate Cr (III) species

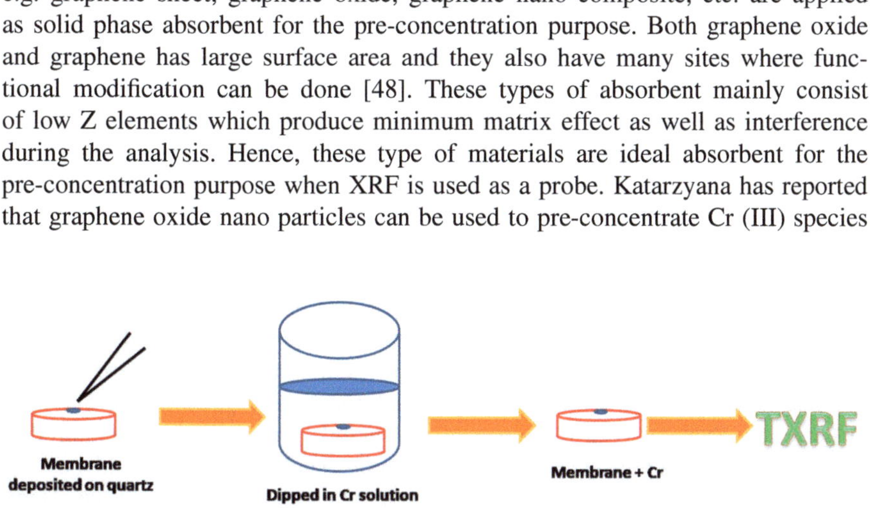

Fig. 14.7 Schematic representation of sample preparation using membrane preconcentration methodology

followed by its analysis using EDXRF [49]. The graphene oxide has negatively charged surface so, Cr(III) species are selectively absorbed on these surfaces. The total Cr concentration was determined by simply reducing all Cr (VI) species into Cr(III) species. Therefore, by knowing the total Cr concentration and Cr(III) concentration one can determine the concentration of Cr(VI) species. Hence, the developed methodology not only provided the information about the concentration but also the data for each species. This methodology could detect chromium concentration down to 0.06 μg/L, and can analyse chromium concentration ranging from 1–150 ng/mL. In another work, Vanesha et al. have reported the usage of graphene membrane for determination of Cr. In this work it was shown that absorption sites on graphene membrane are uniformly distributed and these sites can bind selectively with Cr(VI) species [50]. By combining graphene membrane pre-concentration technique and TXRF, the detection limit achieved was 0.08 μg/L for Cr. A novel absorbent was reported by Baranik et al. for preconcentration of Cr using aluminium oxide coated with nano-graphite (Al_2O_3-nano-G) [51, 52]. In this work 200μL of the aliquot suspension containing Al_2O_3-nano-G was added in 25 mL of water sample. Under certain pH conditions, Al_2O_3-nano-G displayed a very selectivity towards Cr (III) species. After equilibrium was reached this suspension was filtered and measured directly by EDXRF. This type of preconcentration technique is also called Dispersive Micro Solid Phase Extraction (DMSPE) as the extractant volume is only a few microlitres. The detection limits obtained using this methodology was 0.04 μg/L for Cr. Activated Carbon (AC) is another excellent candidate for the environmental pollution control. This AC has large surface area along with porous structure [53]. It has high thermal stability, adsorptive capacity, chemical inertness and easy regeneration properties. It is finding a lot of applications in wastewater treatment, water purification, catalyst, food processing, air pollution control etc. Activated Carbon functionalized with extactant is widely used for pre-concentration and separation of trace elements in water samples. Aliquat 336 fuctionalised AC was used for the selective absorption followed by the detection of Cr (VI) in drinking water samples [54]. Carbon Nanotubes (CNTs) are proposed as the novel solid phase extractants having high surface area, good chemical and mechanical stability. CNTs are classified as single –walled and multi-walled. Multi Walled Carbon Nanotubes (MWCNTs) is also a well known solid absorbent frequently used in DMSPE. This MWCNTs could also be functionalized or loaded for the selective absorption of different toxic elements like Ni, Sr, Pb, Cu, Ar, etc. [55–57]. Bahadir et al. have modified MWCNTs with anionic exchanger Aliquat 336 and used it for the preconcentration of Cr(VI) followed by its detection using TXRF [58]. After the absorption of Cr(VI) species by this modified MWCNTs the solution was centrifuged to separate out them and directly analyzed by a bench top TXRF spectrometer. The limit of detection obtained is in the range of 2–3 ng/mL. Sometimes it is required to analyse toxic elements on-site. In that situation hand held XRF could be very much useful to instantly get information about the elemental concentrations. Hagiwara et al. have reported a rapid, simple, on-site determination methodology for Cr (VI), As and Se in drinking water by combining solid phase extraction method

Table 14.2 Summary of different pre-concentration techniques used in combination with XRF for the determination of ultra trace level of Cr in drinking water

Sorbent	Analyte	Pre-concentration process	Detection Limit (ng/mL)	X-ray based analytical technique used	References
Ammonium pyrolidine dithio-carbamate (APDC)	Cr (III) & Cr(VI)	Liquid–liquid extraction	2.95	EDXRF	[44]
Activated Carbon (AC) funtionalized with Aliquat 336	Cr(VI)	Solid phase extraction (SPE)	1	EDXRF	[54]
Aliquat 336 in PVC matrix	Cr(III) & Cr(VI)	Membrane based pre-concentration	0.6	TXRF	[45]
Multi Walled Carbon Nanotubes (MWCNTs) funtionalized with Aliquat 336	Cr(VI)	DMSPE	2–3	TXRF	[58]
Graphene oxide nano paticles	Cr(III)	Dispersive Micro Solid Phase Extraction (DMSPE)	0.06	EDXRF	[49]
Graphene membrane	Cr(VI)	Membrane based pre-concentration	0.08	TXRF	[50]
Aluminium oxide coated nano-graphite (Al_2O_3-nano-G) nanocomposite	Cr(III)	DMSPE	0.04	EDXRF	[52]
Ti-loaded anion exchange disk	Cr(VI)	SPE	1	Hand held EDXRF	[59]

in combination with hand held XRF technique [59]. The analyte preconcentration was carried out on a Ti-loaded anion-exchange disk (Ti-AED).The detection limit reported was 1 ng/mL foe As, Se and Cr. Table 14.2 summarizes different pre-concentration techniques uses for the determination of chromium in combination with XRF.

14.5.2 Arsenic Determination

Similar to chromium, arsenic exists in many species forms. It can be divided into two parts one is the organic form of As: MMA and DMA. Another is inorganic form where As remain in +5 and +3 oxidation states in arsenate (AsO_4^{3-}) and arsenite (AsO_3^{3-}), respectively. Among them inorganic form of arsenic especially inorganic arsenic in +3 oxidation state is highly toxic in nature. This species can bind with sulfhydryls or thiols groups which are present at the active sites of different enzymes [60]. The structural details of different species of arsenic is shown in Fig. 14.8. Similar to chromium different types of sample preparation strategies were deployed in combination with EDXRF/TXRF for the determination as well as speciation of arsenic at ultra trace level. Barros et al. have used alumina as an absorbing reagent for the preconcentration of arsenic [61]. About 50 mL of sample were mixed with 10 mg of alumina. The pH, time and temperature were optimized for maximum enrichment. The alumina loaded with As was separated from the slurry by centrifugation, then it was washed with de-ionized water and analyzed directly by TXRF on the sample holder. A pre-concentration factor of 100 was observed, with detection limit of 0.7 μg/L for As.

During the sample collection for speciation studies of arsenic species, utmost care should be given to ensure that the species should be retained in its original form during transport and storage process. There are some guidelines recommended by International Standardization Organization (ISO) for the collection and preservation of water samples. During the transportation of inorganic arsenic species use of glass or polyethylene container is recommended. All these environmental samples should be stored at very low temperature (around 253 K) in a airtight container in dark so that photo-oxidation can be avoided [62]. Now to avoid such problems associated with transport and preservation of samples, usage of hand-held XRF is another good option where one could do on the spot analysis.. Hagiwara et al. have reported a rapid method of quantifying arsenic using

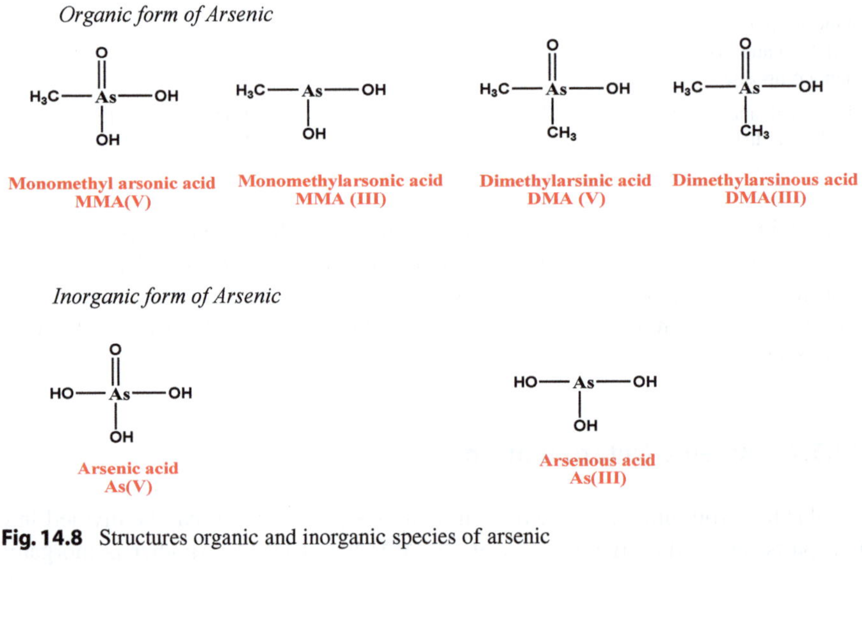

Fig. 14.8 Structures organic and inorganic species of arsenic

hand-held XRF spectrometer in combination with SPE based pre-concentration technique [63]. In this study Ti and Zr loaded carbon disk (TiZr-CD) has been used for the absorption and preconcentration of inorganic (As (III) and AS (V)) as well as organic As (as methyl, phenyl and aromatic arsenic). These disks after pre-concentration were analyzed by the mobile XRF spectrometer. The detection limit achieved using this methodology was 2.0 ng/mL. Sitko et al. had reported a novel DMSPE based pre-concentration method for ultrarace determination of As(III) species. They have functionalized graphene oxide nano sheets by grafting 3-mercaptopropyl trimethoxy silane [64]. This mercapto group can selectively absorb As(III) species along with Co, Ni, Cd and Pb ions. After pre-concentration TXRF was used to determine As concentration. An enrichment factor of 150 was obtained with detection limits of 0.064 ng/mL for As (III). Aranda et al. have used funtionalized multiwalled carbon nano tube as solid phase extractant for As (V) species [65]. In this work ALIQUAT 336 has been functionalized on multiwalled carbon nanotubes (ALIQUAT 336-MWCNTs), which was found to be a promising solid absorbent. This funtionalized carbon nanotubes were finally loaded on membrane filter papers. WDXRF method was utilized to analyze the samples. A detection limit of 0.1 ng/mL was achieved using this methodology. Mihucz et al. have reported a method based on SPE for on-site separation of arsenic using micro-columns of strong anion exchanger DOWEX 1-X8 followed by detection of the residual using TXRF [66]. A new concept for arsenic speciation as well as determination was developed by functionalization of the quartz sample supports with N-Methyl-D-Glucamine (NMDG) followed by its separation. The NMDG forms a very thin membrane on quartz sample supports. It has mild reducing property and it can specifically bind with As (V) species. During TXRF analysis usage of internal standard is very common. In this work a novel approach was developed for introducing the internal standard. The NMDG modified quartz sample supports were dipped in a solution containing $AuCl_4^-$ ions with a fixed concentration of Au. The reducing agent NMDG present in the quartz sample supports reduced $AuCl_4^-$ ions into Au nano-particles which served as internal standard during TXRF measurements. This Au loaded NMDG grafted quartz supports were simply dipped in water samples in normal pH conditions (pH 6–7) for 6 h. The arsenate species got absorbed by the NMDG via hydrogen bonding as well as electrostatic force. The details of this work could be found in the literature [67]. Au loaded NMDG grafted quartz supports before and after dipping in As (V) solution is shown in Fig. 14.9. The total arsenic concentration could be obtained by oxidizing all As(III) species into As(V). The detection limit of arsenic obtained using this methodology is 0.05 ng/mL. The pH of the solution plays a crucial role for the speciation studies as well as pre-concentration of arsenic. Inorganic As (V) species has three pK_a values in three different pH range e.g. 2.16, 6.94 and 11.5. This means in natural pH condition (pH 5–7) the predominating species of As(V) are $H_2AsO_4^-$ and $HAsO_4^{2-}$. On the contrary As(III) has only one pK_a value which is 9.1. So under natural pH condition As(III) will remain in un-dissociated conditions [68]. All these factors have to be considered while designing the extractant. In recent time Metal Organic Framework (MOF) has gained great popularity as a potential absorbent material for the separation and pre-concentration of different pollutants [69–71]. Metal–Organic Frameworks (MOFs) are a class of materials, having diverse analytical applications, that consist of metal ions or clusters coordinated to organic molecules to form a porous, crystalline structure. The major advantages of MOF are its tunable pore sizes, high surface area, possibility of various types of functionalization, etc. However, there are very limited studies on the usage of MOF based SPE technique coupled with XRF for the ultra trace elemental analysis in water samples. Aslan et al. have reported usage of mixed matrix membrane formed by using Fe based MOF for the pre-concentration of As(V) followed by its detection employed by EDXRF technique [72]. In this study the a novel miniaturized MIL-101(Fe) based MOF/ Poly Vinyl Dene Fluoride (PVDF) mixed matrix membrane was developed as a new SPE material for selective extraction of As (V). The MOF suspension was drop-casted on holytex non-woven fabric supports as shown in Fig. 14.10. This MOF based Mixed Matrix Membrane or MOF-MMA is equilibrated with the solution containing the analytes. After the equilibration was done the whole MOF-MMA system could be directly measured in EDXRF spectrometer. The detection limit of arsenic obtained using this methodology is 0.094 ng/mL.

Polymer Inclusion Membrane (PIM) were initially used as the sensing component in ion- selective electrodes, however PIM can also be used for the extraction/ pre-concentration purpose. It is an attractive greener alternative to solvent extraction. PIMs are a type of liquid membranes consisting of an liquid phase extracting agent and a base polymer. The polymer forms the skeleton of the membrane. PIM consisting of cellulose triacetate as polymer and Aliquat 336 as extractant have been prepared to preconcentrate As(V) species [73, 74]. Recently Chillé et al.

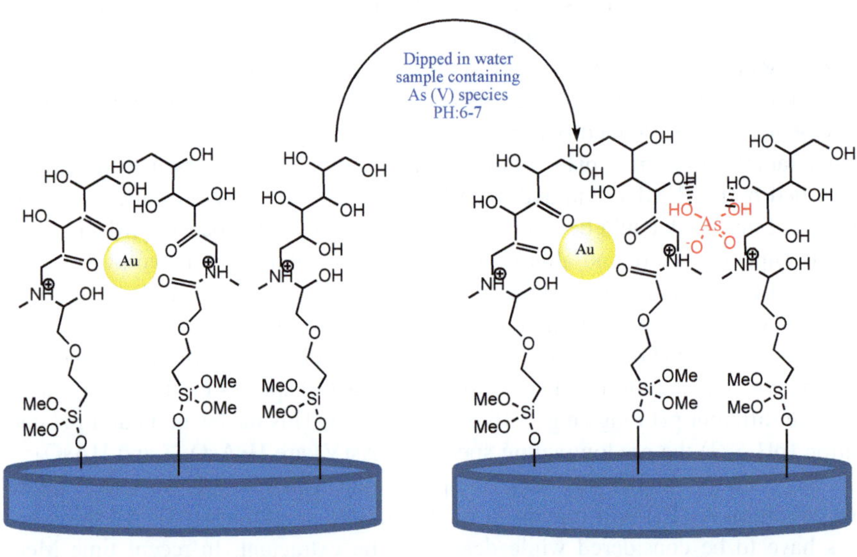

Fig. 14.9 Au loaded NMDG grafted quartz supports before and after dipping in As (V) solution

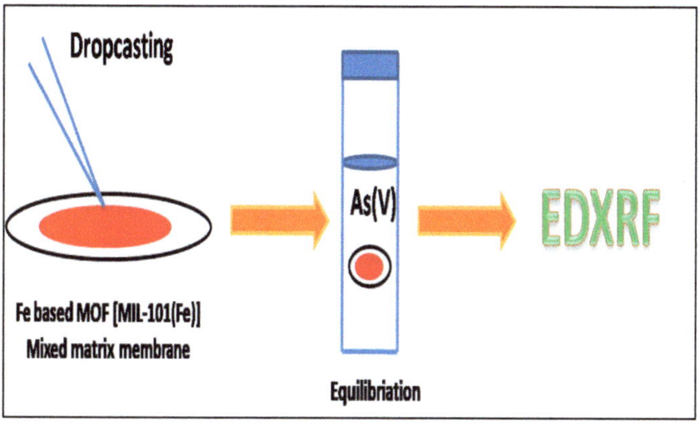

Fig. 14.10 Schematic of MOF based SPE technique coupled with EDXRF

have reported usage of PIM based pre-concentration coupled with EDXRF based detection for the ultra trace analysis as well as speciation of arsenic in water samples [75]. This PIM is made of cellulose triacetate as polymer and Cyanex 301 (bis(2,4,4-trimethylpentyl)dithiophosphinic acid) as extractant for As(III) species. Here, the thiol group present in Cyanex 301 has high affinity towards As(III) species. This PIM after pre-concentration can be directly measured by EDXRF. The As(V) species could be reduced to As(III) and one can get information about both As(III) as well as As(V) species. The detection limit obtained using this methodology is 0.6 ng/mL. Till now the methodologies employed for arsenic speciation involves separation of one particular species of arsenic either As (III) or As (V). However, if both the species of arsenic are absorbed then the information about the speciation will be lost. This is because in EDXRF/TXRF geometry typically a semiconductor (like Si(Li) or SDD) is being used to directly measure the energy of incoming X-rays. Though this methodis faster, it lacks the ability to finely resolve X-rays with very close energies, leading to lower resolution. On the other hand, in case of Wave Length Dispersive X-Ray Fluorescence (WDXRF) a crystal is used as a dispersive element that separates X-rays based on their wavelengths. This method allows for more precise differentiation of X-rays with very close energies thus, improving the resolution. The system scans over different wavelengths, providing detailed spectra with clear separation between closely spaced peaks. Deb Barman et al. have showed that As$K\beta_{2,5}$ line energy is sensitive to oxidation state of arsenic [76]. Now As$K\beta_{1,3}$ (11.72 keV) and As$K\beta_{2,5}$ (11.86 keV) lines are very closely spaced which is almost impossible to resolved by EDXRF geometry. So WDXRF was employed to resolve these two lines. The calibration curve was constructed which is As(III) fraction Vs intensity ratio of As$K\beta_{2,5}$ and As$K\beta_{1,3}$. It was seen that as As(III) species fraction increased the ratio also increased. The pre-concentration was done initially using activated alumina pellets. The total As concentration was determined from As$K\alpha_{1,2}$ line (10.54 keV). The detection limit obtained for the total arsenic determination

Table 14.3 Summary of different pre-concentration techniques used in combination with XRF for the determination of ultra trace level of As in drinking water

Sorbent	Analyte	Pre-concentration process	Detection limit (ng/mL)	X-ray based analytical technique used	References
Sodiumdibenzylthiocarbamate	As (III) & As(V)	Absorption followed by elution	10	TXRF	[77]
Alumina	As (III) & As(V)	Solid Phase Extraction (SPE)	0.7	TXRF	[61]
Ti and Zr loaded carbon disk	As (III) & As(V)	SPE	0.10	Hand-held XRF spectrometer	[63]
Graphene oxide nano sheets funtionalized with mercaptopropyl trimethoxy silane	As(III)	Dispersive Micro Solid Phase Extraction (DMSPE)	0.064	TXRF	[58]
Funtionalized carbon nano tube	As(V)	Membrane based pre-concentration	0.10	EDXRF	[78]
Quartz sample supports were funtionalized with N-Methyl-D-Glucamine (NMDG)	As(V)	Membrane based direct pre-concentration	0.05	TXRF	[67]
Mixed matrix membrane formed by using Fe based MOF	As(V)	Membrane based pre-concentration	0.094	EDXRF	[72]
Polymer Inclusion Membrane (PIM)	As(III)	Membrane based pre-concentration	1	EDXRF	[75]

using this methodology is 0.23 ng/mL. Table 14.3 summarizes the different pre-concentration techniques uses for the determination of arsenic in combination with XRF.

14.5.3 Cadmium Determination

There are few reports on the determination of Cd in water samples at ultra trace level using XRF technique due to several limitations. For example it is difficult to excite Cd K line due to very high absorption edge energy (26.11 keV). So, one has to excite Cd L lines to get concentration information of Cd. Here Cd Lα (3.31 keV) has low energy and its fluorescent yield is quite low. Moreover, in this energy region there are severe interference of K Kα, Ar Kα, etc. So it is always necessary to pre-concentrate Cd before its analysis. Earlier, liquid–liquid extraction technique has been used for the preconcentration/separation purpose. However this technique requires lots of organic solvent. To avoid this problem people have used Liquid Phase Micro Extraction (LPME) technique which is a green alternative requiring few microlitres of organic solvent [79]. Now DLLME is slightly modified version of LPME technique where a certain mixture of extraction solvent and disperser solvent are injected into the solution being analyzed. The fine droplet of extraction solvent is dispersed into the aqueous phase and these organic extractant solvent efficiently interacts with the analyte. The organic extractant binds along with the analyte which is then separated by using centrifugation method and the separated phase has a few microlitres of volume. So this results in high preconcentration factor. TXRF technique requires only few microlitres of sample to be deposited on quartz sample supports. This is the reason why DLLME types of technique is well suited for TXRF. Eva Margui et al. have reported the usage of Dispersive Liquid–Liquid Micro-Extraction (DLLME) methodology combined with TXRF for the ultra trace determination of Cd in water samples [80]. The dispersed phase consisted of organic solvent and a complexing agent, Ammonium Pyrrolidine Dithio-Carbamate (APDC). The detection limit of Cd obtained using this methodology was 0.04 ng/mL. Similar to As and Cr, Cd could also be pre-concentrated using DMSPE method as discussed earlier. Kocot et al. have reported a DMSPE based pre-concentration method combined with TXRF for the determination of both Pb and Cd in water samples at ultra trace level [76]. Here APDC was used as complexing agent which interact with Cd (II) and forms an ahydrophobic complex which can be absorbed on the surface of MWCNT by Vander Waals forces [81]. After absorption is completed these MWCNTs can be separated through centrifugation, followed by adition of an internal standard and directly depositing on quartz sample supports. A detection limit of 1 ng/mL was achieved using this methodology. As mentioned earlier Ar Kα and K Kα lines severely interfere with Cd Lα line. To overcome this problem Kunimura et al. proposed a methodology where they have used a portable TXRF instrument and carried out the measurement in vacuum atmosphere. Moreover, SPE based method was employed to remove the potassium. The detection limit achieved using this methodology is 20 ng/mL [82]. Medina González et al. has solved this interference problem using multivariate calibration method [83].

14.5.4 Mercury Determination

Mercury is an extremely toxic element which can be found in different forms in the environment. It can be roughly categorized into two types: inorganic and organic form. Generally, while using XRF based techniques, few microlitres of the sample volume is placed in the support and dried. Howeve,r this approach cannot be used in case of Hg due to its volatile nature. It was observed that after depositing the mercury sample on a sample support, within 15 min the signal intensity reduced to 50%. This was a clear indication how rapidly Hg vaporize in ambient air atmosphere [84]. Several researchers have developed different types of methodologies to tackle this volatilization problem which will be discussed briefly. Greaves et al. have suggested trapping of mercury using amalgamation technique. In this work

50 nm thick silver coating was given on a quartz sample support/reflector which was given contact with the Hg solution. The Hg vapor generated will form amalgam with Ag and the reflector could be directly analyzed by TXRF [85]. The whole setup is depicted in Fig. 14.11. The detection limit for Hg achieved using this methodology was 15 ng/mL. However, there are several disadvantages associated with this technique. One is each time one has to prepare the special coated sample supports. Moreover, there will be strong spectral interferences between Ag–L and Hg-L lines which will complicate the quantification part. Koulouridakis et al. have utilized membrane based complexation technique for the determination of Hg at ultra trace level. The PVC based membrane along with dithizone complexing agent has good selectivity for Hg (II) species. The quartz sample supports along with the membrane is directly dipped into the solution for 1–2 days and after that these supports could be taken out and directly measured by TXRF. The detection limit of Hg achieved using this methodology is 0.2 ng/mL [86]. Later, this detection limit was further improved to 0.3 ng/mL by improving the extraction parameter and using low power TXRF instrumentation (50W), which resulted better signal to noise ratio [87] (Fig. 14.12).

Similar type of immobilization technique has been reported by Aretaki et al. In this work, first the quartz sample supports were activated by 1,1-carbonyldiimidazole (CDI). This CDI activated quartz supports could be combined with different types of ligands which has affinity for Hg (II) species. The whole process of immobilization of ligand on quartz sample support is depicted in Fig. 14.12. These reflector immobilized with ligands could be simply dipped into the solution containing mercury for 10 min to preconcentrate Hg on the quartz sample supports which could be directly analysed by TXRF. Among various types of ligand tested it was observed that 4-(2-pyridylazo)-resorcinol gives the best results with detection limit achieved 2.5 ng/mL [88]. E. Marguí et al. have reported that reducing the viscosity of the solution as well as trapping of Hg by using

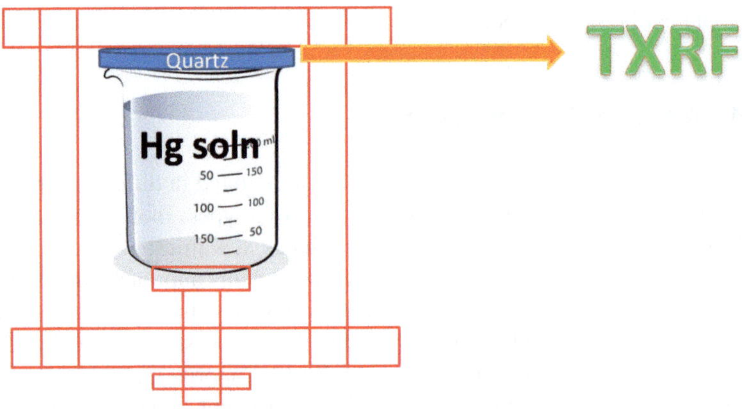

Fig. 14.11 Determination of Hg using amalgamation technique followed by TXRF measurements

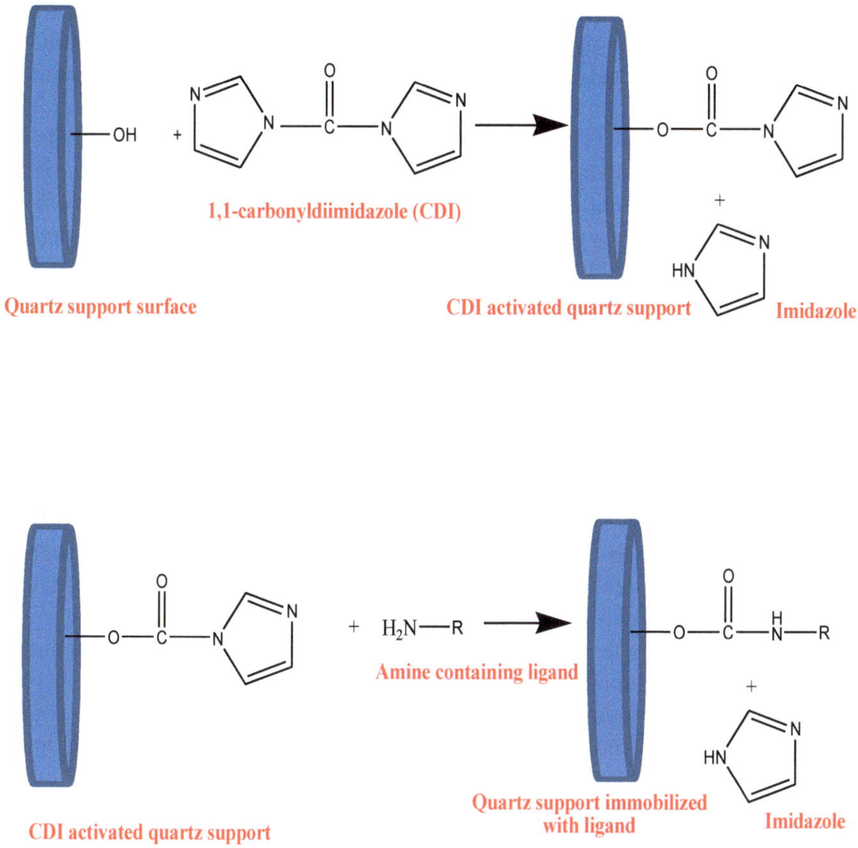

Fig. 14.12 Procedure of immobilization of ligand on quartz sample support

thiourea as complexing agent could be a possible way to reduce the volatization loss of mercury [89]. The detection limit achieved using this methodology could be as low as 0.7 ng/mL for Hg. Aranda et al. have shown that Aliquat 336 activated carbon could be useful for the preconcentration of Hg in HCl medium. Mercury remains in different forms e.g. Hg^{2+}, $HgCl^{+}$, $HgCl_2$, $HgCl_3^{-}$ and $HgCl_4^{2-}$ in HCl medium [90]. The amine group present in Aliquat 336 can form complex with these anionic chloride species. There is no requirement of elution steps and the activated carbon can be directly analyzed by XRF. The detection limit of Hg achieved using this methodology in drinking water is 0.001 ng/mL [91]. In the last few years, nanomaterials have been largely used for the pre-concentration of different toxic elements [92–94]. The nano particles can be functionalized to get desired selective absorption properties. Quartz sample supports immobilized with these nano particles could be useful for the direct pre-concentration of different toxic elements. First the quartz sample supports are silanized followed by the immobilization of nano-particles. Then the quartz sample supports immobilized with the nano particles can be directly dipped into the aqueous solution which needed to be analyzed. The analyte will be directly pre-concentrated on the quartz surface. This type of pre-concentration methodology is known as Thin Film Micro Extraction (TFME). Romero et al. have utilized this TFME based pre-concentration methodology for the pre-concentration of Hg(0) species. Here, a continuous flow generation method has been developed for the generation of Hg vapor which was pre-concentrated on the quartz reflector [95]. This quartz supports can be directly presented for TXRF measurements. The detection limit obtained for Hg using this methodology is 0.18 ng/mL. LLME based technique could also be used for the pre-concentration of Hg. K. V. Oskolok has reported that in presence of KI, HgI_2 will be formed which could be extracted by using benzene. Benzene forms a molecular complex with HgI_2. This benzene along with the molecular complex has lower density than water so it would form a localized droplet which could be removed and can be directly deposited on the surface of quartz reflector and dried for the TXRF measurements [96]. The detection limit obtained using this methodology was 8 ng/mL. Recently Musielak et al. have utilized DMSPE based pre-concentration methodology in combination with TXRF for Hg determination [97]. In this work Carbon Nano Tubes (CNTs) were functionalized with thiosemicarbazide group which has good affinity towards Hg (II) species at certain pH condition. After pre-concentration these functionalized CNTs could be directly presented on quartz reflectors for TXRF measurements. The detection limit obtained using this methodology was 0.0026 ng/mL for Hg. The schematic of the pre-concentration process is presented in Fig. 14.13. Table 14.4 gives Summary of different pre-concentration techniques used in combination with XRF for the determination of ultra trace level of Hg in drinking water.

14.5.5 Lead Determination

The major challenges arises during the determination of Pb using XRF based method is the spectral interference between AsKα (10.54 keV) and PbLα lines (10.55 keV). Noemi Nagata et al. have proposed a Partial Least-Squares regression (PLSR) based multivariate technique to overcome this problem [98]. Another possible way to avoid the spectral interference is selective pre-concentration of Pb from water samples. Koulouridakis has shown that membranes containing complexing reagents could be used for the selective pre-concentration of Pb from water samples [99]. It was observed that dithizone and citric acid are the best complexing reagent in combination with Nafion. A detection limit of 0.5 ng/mL is achieved for Pb in combination with TXRF using this methodology. Karina Kocot et al. have employed DMSPE based pre-concentration in combination with TXRF for the selective preconcentration of Pb [76]. In this work Multi Walled Carbon Nano Tubes (MWCNTs) functionalized with ammonium pyrrolidine dithiocarbamate (APDC) was used as sorbent material. The pre-concentration procedure was similar as mentioned before.

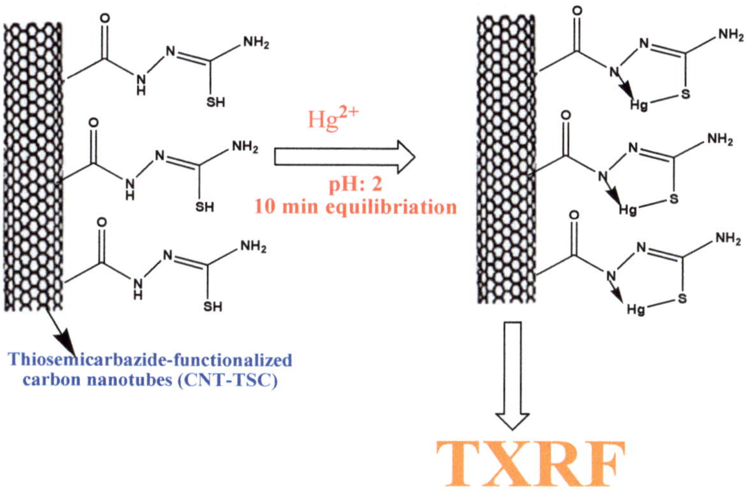

Fig. 14.13 Schematic of DMSPE based pre-concentration of Hg using thiosemicarbazide-functionalized carbon nanotubes (CNT-TSC)

Table 14.4 Summary of different pre-concentration techniques used in combination with XRF for the determination of ultra trace level of Hg in drinking water

Sorbent	Pre-concentration process	Detection limit (ng/mL)	X-ray based analytical technique used	Reference
Au thin film	Amalgamation	15	TXRF	[85]
PVC based membrane along with dithizone complexing agent	Membrane based pre-concentration	0.2	TXRF	[86]
4-(2-pyridylazo)-resorcinol immobilized on quartz sample supports	Thin Film Micro Extraction (TFME)	2.5	TXRF	[88]
Aliquat 336 activated carbon	SPE	0.001	EDXRF	[90]
Au Nps immobilized on quartz supports	TFME	0.18	EDXRF	[95]
Benzene	LLME	8	TXRF	[96]
Carbon Nano Tubes (CNTs) are functionalized with thiosemicarbazide group	DMSPE	0.0026	TXRF	[97]

14.5.6 Uranium Determination

Different strategies were adopted for the determination of uranium using XRF technique. Intially, evaporation methodology was adopted for the determination of uranium in water samples at ultra trace level. Here the sample is either multiple time deposited and dried on quartz sample supports or large volume of the sample is being evaporated to reduce the total volume and preconcentrate uranium in it [100–102]. The detection limit achieved using this procedure were in the range of 20–32 ng/mL. This evaporation technique could be combined with liquid–liquid extraction process for getting better detection limit of uranium. Dhara et al. have utilized DLLME process for the pre-concentration of uranium from water samples followed by its detection using TXRF [103]. Tri n Butyl Phosphate (TBP) was used as an extracting agent which has excellent affinity for UO_2^{2+} species under 3–4 M HNO_3 medium. Here uranium will be extracted as $UO_2(NO_3)_2.2TBP$ complex into the organic phase. The uranium present in the organic phase needed to be stripped out into the aqueous phase for TXRF measurements. The detection limit of uranium achieved using the following methodology was 0.39 ng/mL. This developed methodology was quite helpful to not only pre-concentrate uranium but also to avoid the spectral interference of Rb Kα (13.39 keV) and Br Kβ (13.29 keV) with U Lα (13.61 keV). Many SPE based pre-concentration method has been tried in combination with XRF to improve the detection limit of uranium while analyzing water samples. Takashi et hal have used Graphene Oxide (GO) suspended in water samples to absorb uranium under certain pH condition followed by its separation [104]. The separated GO along with internal standard was presented on quartz sample supports for TXRF measurements. The detection limit reported using this methodology was 0.042 ng/mL for U. However the major problem associated with this methodology was that GO is not selective for U. It can absorb other elements which could affect the pre-concentration process of uranium. To incorporate selectivity, these GO could be functionalized with certain groups which are highly selective towards uranium. Kocot et al. have reported β-cyclodextrin modified graphene oxide for the pre-concentration of uranium by DMSPE [105]. The detection limit of uranium achieved using this methodology is 0.40 ng/mL by EDXRF and 0.014 ng/mL by TXRF. Cloud Point Extraction (CPE) is another type of SPE based technique which is well suited in combination with TXRF for the ultra trace determination of different elements in water samples. In CPE, a non-ionic surfactant (such as Triton X-100 or similar compounds) is added to the solution containing the analyte. These surfactants self-assemble into small aggregates called micelles when they reach a certain concentration (called the critical micelle concentration or CMC). Micelles have a hydrophobic core and a hydrophilic outer surface, making them useful for trapping hydrophobic substances. At a specific temperature, known as the cloud point, the surfactant solution becomes turbid, and the micelles undergo phase separation. This separates the solution into two phases, e.g. Surfactant Rich Phase (SRP) and aqueous phase. The analyte will be in the SRP which could be separated by centrifugation [106–108]. As the volume of the SRP is only a few microlitres, it could be directly presented on quartz sample supports for TXRF measurements. CPE based method was also utilized for the determination of uranium in water samples at ultra trace level [109]. The detection limit of uranium achieved with this methodology is 0.02 ng/mL. Hatzistavros et al. have utilized membrane based pre-concentration for the direct pre-concentration of uranium on quartz sample supports [110]. Different types of complexing agents were used to prepare the membrane and it was observed that the combination of dithizone and thiourea gives the best result. This methodology can achieve a detection limit of 0.8 ng/mL for uranium. Although membrane based pre-concentration technique is very simple straight forward which do not require any separation steps; still the major hurdle associated with this method is the introduction of internal standard. Generally, in XRF method an internal standard is added to avoid the error associated with the changes in the X-ray flux, alignment condition or sample position. In a membrane based preconcentration methodology for uranium Compton scattered peak of Rh Kα was utilized to construct the calibration curve. This acted as an internal standard for normalizing all the errors as discussed above. The quartz sample support was immobilized with amidoxime group ($-CH_2-CH-C(NH_2) = N-OH$) which has excellent selectivity towards UO_2^{2+} species under natural pH condition (pH 6–7). The details of this methodology could be found in the literature [111]. A detection limit of 0.013 ng/mL was achieved using the following methodology. A similar type of amidoxime based membrane has been utilized by S A. Kumar et al. for the pre-concentration of uranium followed by its analysis using EDXRF [112]. In this work polyhydroxamic acid was attached on microporous polypropylene membrane. The structure of the complex formed between polyhydroxamic acid and uranyl ion is shown in Fig. 14.14.

Poly Urathane Foam (PUF) also has excellent affinity towards uranium. S. Carvalho et al. have utilized this PUF powder as solid absorbent for the pre-concentration followed by EDXRF based detection of uranium in water samples [113]. The detection limit achieved using this methodology is 5.5 ng/mL. Recently, Carvalho et al. have proposed another SPE based method where 1-(-2-thiazolylazo)-2-napthol (TAN) was used as extractant which forms complex with uranium [114]. In this work the solution was pumped through a narrow channel and the complex was retained by C18 disk. This disk could be directly measured by EDXRF for getting uranium concentration. The detection limit of uranium achieved using this above methodology is 1.5 ng/mL. Table 14.5 gives the summary of different preconcentration method employed along with XRF technique for the ultra trace determination of uranium in water samples.

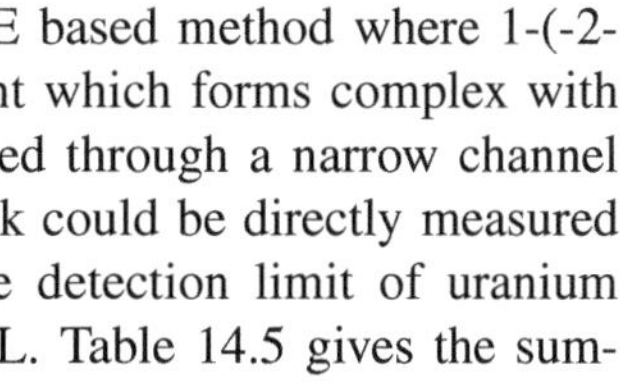

Fig. 14.14 The structure of complex formed by uranyl ion and polyhydroxamic acid

Table 14.5 Different preconcentration method employed along with XRF technique for the ultra trace determination of uranium in water samples

Sorbent	Pre-concentration process	Detection limit (ng/mL)	X-ray based analytical technique used	References
Poly Urathane Foam (PUF)	SPE	5.5	EDXRF	[113]
Quartz sample support was immobilized with amidoxime group	TFME	8	TXRF	[111]
Task Specific Ionic Liquid (TSIL) named as phamdNMe3] [NTf2]	Cloud point extraction (CPE)	0.02	TXRF	[109]
Graphene Oxide (GO)	SPE	0.042	TXRF	[104]
Tri n Butyl Phosphate (TBP)	DLLME	0.39	TXRF	[103]
β-cyclodextrin modified graphene oxide	SPE	0.014	TXRF	[105]
Dithizone and thiourea	Membrane based preconcentration	0.013	TXRF	[102]
1-(-2-thiazolylazo)-2-napthol (TAN)	SPE	1.5	EDXRF	[114]

14.6 Future Prospects

We have seen that XRF has significant potential and its capabilities have enhanced for determination of various hazardous elements in drinking water owing to its precision, accuracy, speed, ease, versatility and non-destructive in nature. Different sample preparation methodologies were adopted and developed time to time by various researchers in the field of XRF, which results in improvement of detection limit as well as other analytical parameters drastically. Synchrotron radiation has been the most important development in the field of science and technology which has upgraded the level analytical research. However, the limitation of SR is that it cannot be made available for routine analysis. An important scope includes development of real time online monitoring and quantification systems in remote areas. Development of small handheld portable systems with improved detection limits and sensitivity can assist significantly. Extension of elemental range is another important scope as it has low sensitivity for the detection of lighter elements. Development of high flux low energy X-ray tubes with windowless/ ultrathin window detectors will help to overcome this challenge. These advancements will broaden the appeal of XRF spectroscopy for cutting-edge analytical research for science, industry, and environmental applications.

References

1. Alagan, M., Kishore, S.C., Perumal, S., Manoj, D., Raji, A., Kumar, R.S., et al.: Narrative of hazardous chemicals in water: Its potential removal approach and health effects. Chemosphere **335**, 139178 (2023)
2. Rathi, B.S., Kumar, P.S., Vo, D.-V.N.: Critical review on hazardous pollutants in water environment: occurrence, monitoring, fate, removal technologies and risk assessment. Sci. Total. Environ. **797**, 149134 (2021)
3. Zamora-Ledezma, C., Negrete-Bolagay, D., Figueroa, F., Zamora-Ledezma, E., Ni, M., Alexis, F., et al.: Heavy metal water pollution: A fresh look about hazards, novel and conventional remediation methods. Environ. Technol. Innov. **22**, 101504 (2021)
4. Edition, F.: Guidelines for drinking-water quality. WHO chronicle **38**(4), 104–108 (2011)
5. Maheshwari, R.: Fluoride in drinking water and its removal. J. Hazard. Mater. **137**(1), 456–463 (2006)
6. Shah, A., Arjunan, A., Baroutaji, A., Zakharova, J.: A review of physicochemical and biological contaminants in drinking water and their impacts on human health. Water Sci. Eng. **16**(4), 333–344 (2023)
7. Sharma, S., Bhattacharya, A.: Drinking water contamination and treatment techniques. Appl Water Sci **7**(3), 1043–1067 (2017)
8. Mukhopadhyay, A., Duttagupta, S., Mukherjee, A.: Emerging organic contaminants in global community drinking water sources and supply: a review of occurrence, processes and remediation. J. Environ. Chem. Eng. **10**(3), 107560 (2022)
9. Capdeville, M., Budzinski, H.: Trace-level analysis of organic contaminants in drinking waters and groundwaters. TrAC, Trends Anal. Chem. **30**(4), 586–606 (2011)
10. Kristanti, R.A., Hadibarata, T., Syafrudin, M., Yılmaz, M., Abdullah, S.: Microbiological contaminants in drinking water: current status and challenges. Water Air Soil Pollut. **233**(8), 299 (2022)
11. Lytle, D.A., Sorg, T., Wang, L., Chen, A.: The accumulation of radioactive contaminants in drinking water distribution systems. Water Res. **50**, 396–407 (2014)
12. Kumar, R., Qureshi, M., Vishwakarma, D.K., Al-Ansari, N., Kuriqi, A., Elbeltagi, A., et al.: A review on emerging water contaminants and the application of sustainable removal technologies. Case Stud. Chem. Environ. Eng. **6**, 100219 (2022)
13. Baken, K.A., Sjerps, R.M., Schriks, M., van Wezel, A.P.: Toxicological risk assessment and prioritization of drinking water relevant contaminants of emerging concern. Environ. Int. **118**, 293–303 (2018)
14. Kashyap, S.J., Sankannavar, R., Madhu, G.: Fluoride sources, toxicity and fluorosis management techniques–a brief review. J. Hazard. Mater. Lett. **2**, 100033 (2021)
15. Zhitkovich, A.: Chromium in drinking water: sources, metabolism, and cancer risks. Chem. Res. Toxicol. **24**(10), 1617–1629 (2011)
16. Parks, J.L., McNeill, L., Frey, M., Eaton, A.D., Haghani, A., Ramirez, L., et al.: Determination of total chromium in environmental water samples. Water Res. **38**(12), 2827–2838 (2004)
17. Das, A., Joardar, M., De, A., Mridha, D., Chowdhury, N.R., Khan, M.T.B.K., et al.: Pollution index and health risk assessment of arsenic through different groundwater sources and its load on soil-paddy-rice system in a part of Murshidabad district of West Bengal, India. Groundw. Sustain. Dev. **15**, 100652 (2021)
18. Krajnc, E.: Integrated Criteria Document. Cadmium—Effects. Appendix. Bilthoven, National Institute of Public Health and Environmental Protection (Report No. 758476004) (1987)
19. Organization, W.H.: Cadmium in drinking-water: background document for development of WHO guidelines for drinking-water quality. World Health Organization (2004)
20. Kumar, S.: Occupational and environmental exposure to lead and reproductive health impairment: an overview. Indian J. Occup. Environ. Med. **22**(3), 128–137 (2018)

21. Wu, D., Hu, Y., Cheng, H., Ye, X.: Detection techniques for lead ions in water: a review. Molecules **28**(8), 3601 (2023)
22. Organization, W.H.: Guidelines for Drinking-water Quality: First Addendum to, vol. 1. World Health Organization, Recomendations (2006)
23. Jha, S.K., Patra, A.C., Verma, G.P., Jha, V., Aswal, D.K.: Uranium standards in drinking water: An examination from scientific and socio-economic standpoints of India. Environ. Sci. Pollut. Res. **31**(35), 47461–47474 (2024)
24. Marchenko, A., Truflandier, L.A., Autschbach, J.: Uranyl carbonate complexes in aqueous solution and their ligand NMR chemical shifts and 17O quadrupolar relaxation studied by ab initio molecular dynamics. Inorg. Chem. **56**(13), 7384–7396 (2017)
25. Porter, N., Hart, B., Morrison, R., Hamilton, I.: Determination of metal ion mixtures using pH gradient flow-injection analysis with fluorescence detection. Anal. Chim. Acta **308**(1–3), 313–328 (1995)
26. Haj-Hussein, A.T.: Ultraviolet determination of chloride in water by flow injection analysis. Anal. Lett. **29**(5), 793–806 (1996)
27. Isildak, I., Covington, A.K.: Ion-selective electrode potentiometric detection in ion-chromatography. Electroanalysis **5**(9–10), 815–824 (1993)
28. Barroso, M.F., Silva, A., Ramos, S., Oliva-Teles, M.T., Delerue-Matos, C., Sales, M.G.F., et al.: Flavoured versus natural waters: macromineral (Ca, Mg, K, Na) and micromineral (Fe, Cu, Zn) contents. Food Chem. **116**(2), 580–589 (2009)
29. Gómez-Nieto, B., Gismera, M.J., Sevilla, M.T., Procopio, J.R.: Fast sequential multi-element determination of major and minor elements in environmental samples and drinking waters by high-resolution continuum source flame atomic absorption spectrometry. Anal. Chim. Acta **854**, 13–19 (2015)
30. Birke, M., Reimann, C., Demetriades, A., Rauch, U., Lorenz, H., Harazim, B., et al.: Determination of major and trace elements in European bottled mineral water—analytical methods. J. Geochem. Explor. **107**(3), 217–226 (2010)
31. Boschetti, W., Rampazzo, R.T., Dessuy, M.B., Vale, M.G.R., de Oliveira Rios, A., Hertz, P., et al.: Detection of the origin of Brazilian wines based on the determination of only four elements using high-resolution continuum source flame AAS. Talanta **111**, 147–155 (2013)
32. Nadeau, K., Mester, Z., Yang, L.: The direct and accurate determination of major elements Ca, K, Mg and Na in water by HR-ICPMS. Sci. Rep. **8**(1), 17750 (2018)
33. Krachler, M.: Environmental applications of single collector high resolution ICP-MS. J. Environ. Monit. **9**(8), 790–804 (2007)
34. McNaught, A.D., Wilkinson, A.: Compendium of Chemical Terminology (Vol. 1669): Blackwell Science Oxford (1997)
35. Miller-Ihli, N.J., Baker, S.A.: Trace element composition of municipal waters in the United States: a comparison of ICP-AES and ICP-MS methods. J. Food Compos. Anal. **14**(6), 619–629 (2001)
36. Hudnik, V., Gomiščck, S., Gorenc, B.: The determination of trace metals in mineral waters: Part I. Atomic absorption spectrometric determination of Cd, Co, Cr, Cu, Ni and Pb by electrothermal atomization after concentration by co-precipitation. Analytica chimica acta **98**(1), 39–46 (1978)
37. Marguí, E., Zawisza, B., Sitko, R.: Trace and ultratrace analysis of liquid samples by X-ray fluorescence spectrometry. TrAC, Trends Anal. Chem. **53**, 73–83 (2014)
38. Streli, C.: Recent advances in TXRF. Appl. Spectrosc. Rev. **41**(5), 473–489 (2006)
39. Kanrar, B., Sanyal, K., Misra, N., Aggarwal, S.: Improvements in energy dispersive X-ray fluorescence detection limits with thin specimens deposited on thin transparent adhesive tape supports. Spectrochim. Acta, Part B **101**, 130–133 (2014)
40. Tarsoly, G., Ovari, M., Zaray, G.: Determination of fluorine by total reflection X-ray fluorescence spectrometry. Spectrochim. Acta, Part B **65**(4), 287–290 (2010)
41. Klockenkamper, R.: Total-reflection X-ray fluorescence analysis. Institut fur Spectrochemie und Angewandte Spektroskopie 140 (1997)

42. Sanyal, K., Misra, N.: Direct determination of fluorine in high-purity water samples using vacuum sample chamber total reflection X-ray fluorescence spectrometry. J. Anal. At. Spectrom. **33**(5), 876–882 (2018)
43. Sanyal, K., Kanrar, B., Misra, N., Czyzycki, M., Migliori, A., Karydas, A.: A comparative study on the total reflection X-ray fluorescence determination of low Z elements using X-ray tube and synchrotron radiation as excitation sources. X-Ray Spectrom. **46**(3), 164–170 (2017)
44. Orescanin, V., Mikelic, L., Lulic, S., Rubcic, M.: Determination of Cr (III) and Cr (VI) in industrial and environmental liquid samples by EDXRF method. Anal. Chim. Acta **527**(2), 125–129 (2004)
45. Kallithrakas-Kontos, N., Koulouridakis, P., Hatzistavros, V., Aretaki, I.: Chromium speciation by TXRF analysis. X-Ray Spectrom.: Int. J. **38**(2), 152–156 (2009)
46. Fontàs, C., Queralt, I., Hidalgo, M.: Novel and selective procedure for Cr (VI) determination by X-ray fluorescence analysis after membrane concentration. Spectrochim. Acta, Part B **61**(4), 407–413 (2006)
47. Lo, S., Shiue, S.: Recovery of Cr (VI) by quaternary ammonium compounds. Water Res. **32**(1), 174–178 (1998)
48. Kuila, T., Bose, S., Mishra, A.K., Khanra, P., Kim, N.H., Lee, J.H.: Chemical functionalization of graphene and its applications. Prog. Mater. Sci. **57**(7), 1061–1105 (2012)
49. Pytlakowska, K.: Speciation of inorganic chromium in water samples by energy dispersive X-ray fluorescence spectrometry. J. Anal. At. Spectrom. **31**(4), 968–974 (2016)
50. Romero, V., Costas-Mora, I., Lavilla, I., Bendicho, C.: Graphene membranes as novel preconcentration platforms for chromium speciation by total reflection X-ray fluorescence. RSC Adv. **6**(1), 669–676 (2016)
51. Baranik, A., Sitko, R., Gagor, A., Zawisza, B.: Alumina/nano-graphite composite as a new nanosorbent for the selective adsorption, preconcentration, and determination of chromium in water samples by EDXRF. Anal. Bioanal. Chem. **410**, 7793–7802 (2018)
52. Baranik, A., Gagor, A., Queralt, I., Marguí, E., Sitko, R., Zawisza, B.: Determination and speciation of ultratrace arsenic and chromium species using aluminium oxide supported on graphene oxide. Talanta **185**, 264–274 (2018)
53. Lemos, V.A., Teixeira, L.S.G., Bezerra, M.d.A., Costa, A.C.S., Castro, J.T., Cardoso, L.A.M., et al.: New materials for solid-phase extraction of trace elements. Appl. Spectrosc. Rev. **43**(4), 303–334 (2008)
54. Aranda, P.R., Moyano, S., Martinez, L.D., De Vito, I.E.: Determination of trace chromium (VI) in drinking water using X-ray fluorescence spectrometry after solid-phase extraction. Anal. Bioanal. Chem. **398**, 1043–1048 (2010)
55. Sitko, R., Zawisza, B., Malicka, E.: Graphene as a new sorbent in analytical chemistry. TrAC, Trends Anal. Chem. **51**, 33–43 (2013)
56. Tarigh, G.D., Shemirani, F.: Magnetic multi-wall carbon nanotube nanocomposite as an adsorbent for preconcentration and determination of lead (II) and manganese (II) in various matrices. Talanta **115**, 744–750 (2013)
57. Islam, A., Rais, S.: A facile approach for grafting ion imprinted polymer onto magnetic multi-walled carbon nanotubes for selective removal and preconcentration of cadmium in food and wastewater samples prior to atomic spectrometric determination. Food Chem. **405**, 134751 (2023)
58. Bahadir, Z., Bulut, V., Hidalgo, M., Soylak, M., Marguí, E.: Determination of trace amounts of hexavalent chromium in drinking waters by dispersive microsolid-phase extraction using modified multiwalled carbon nanotubes combined with total reflection X-ray fluorescence spectrometry. Spectrochim. Acta, Part B **107**, 170–177 (2015)
59. Hagiwara, K., Koike, Y., Aizawa, M., Nakamura, T.: On-site determination of arsenic, selenium, and chromium (VI) in drinking water using a solid-phase extraction disk/handheld x-ray fluorescence spectrometer. Anal. Sci. **34**(11), 1309–1315 (2018)
60. Nearing, M.M., Koch, I., Reimer, K.J.: Complementary arsenic speciation methods: a review. Spectrochim. Acta, Part B **99**, 150–162 (2014)

61. Barros, H., Parra, L.-M.M., Bennun, L., Greaves, E.D.: Determination of arsenic in water samples by Total Reflection X-Ray Fluorescence using pre-concentration with alumina. Spectrochim. Acta, Part B **65**(6), 489–492 (2010)
62. Radke, B., Jewell, L., Namieśnik, J.: Analysis of arsenic species in environmental samples. Crit. Rev. Anal. Chem. **42**(2), 162–183 (2012)
63. Hagiwara, K., Koike, Y., Aizawa, M., Nakamura, T.: On-site quantitation of arsenic in drinking water by disk solid-phase extraction/mobile X-ray fluorescence spectrometry. Talanta **144**, 788–792 (2015)
64. Sitko, R., Janik, P., Zawisza, B., Talik, E., Margui, E., Queralt, I.: Green approach for ultratrace determination of divalent metal ions and arsenic species using total-reflection X-ray fluorescence spectrometry and mercapto-modified graphene oxide nanosheets as a novel adsorbent. Anal. Chem. **87**(6), 3535–3542 (2015)
65. Aranda, P.R., Llorens, I., Perino, E., De Vito, I., Raba, J.: Environmental Nanotechnology, Monitoring & Management (2016)
66. Mihucz, V.G., Enesei, D., Veszely, Á., Bencs, L., Pap-Balázs, T., Óvári, M., et al.: A simple method for monitoring of removal of arsenic species from drinking water applying on-site separation with solid phase extraction and detection by atomic absorption and X-ray fluorescence based techniques. Microchem. J. **135**, 105–113 (2017)
67. Sanyal, K., Chappa, S., Bahadur, J., Pandey, A.K., Mishra, N.L.: Arsenic quantification and speciation at trace levels in natural water samples by total reflection X-ray fluorescence after pre-concentration with N-methyl-d-glucamine functionalized quartz supports. J. Anal. At. Spectrom. **35**(11), 2770–2778 (2020)
68. Gupta, R., Gamare, J.S., Pandey, A.K., Tyagi, D., Kamat, J.V.: Highly sensitive detection of arsenite based on its affinity toward ruthenium nanoparticles decorated on glassy carbon electrode. Anal. Chem. **88**(4), 2459–2465 (2016)
69. Ma, J., Yao, Z., Hou, L., Lu, W., Yang, Q., Li, J., et al.: Metal organic frameworks (MOFs) for magnetic solid-phase extraction of pyrazole/pyrrole pesticides in environmental water samples followed by HPLC-DAD determination. Talanta **161**, 686–692 (2016)
70. Wang, Y., Rui, M., Lu, G.: Recent applications of metal–organic frameworks in sample pretreatment. J. Sep. Sci. **41**(1), 180–194 (2018)
71. Gumber, N., Pai, R.V., Sanyal, K., Dutta, B., Hassan, P.: Synthesis and uranium adsorption studies of UiO-66 (Ce) based metal organic frameworks from aqueous solutions. Microporous Mesoporous Mater. **341**, 112108 (2022)
72. Aslan, F., Tor, A.: Determination and speciation of trace inorganic arsenic species in water samples by using metal organic framework mixed-matrix membrane and EDXRF spectrometry. Chemosphere **307**, 135661 (2022)
73. Almeida, M.I.G., Cattrall, R.W., Kolev, S.D.: Polymer inclusion membranes (PIMs) in chemical analysis-a review. Anal. Chim. Acta **987**, 1–14 (2017)
74. Güell, R., Anticó, E., Kolev, S.D., Benavente, J., Salvadó, V., Fontàs, C.: Development and characterization of polymer inclusion membranes for the separation and speciation of inorganic As species. J. Membr. Sci. **383**(1–2), 88–95 (2011)
75. Chillè, D., Marguí, E., Anticó, E., Foti, C., Fontàs, C.: Facilitating inorganic arsenic speciation and quantification in waters: Polymer inclusion membrane preconcentration and X-ray fluorescence detection. Anal. Chim. Acta **1324**, 343098 (2024)
76. Kocot, K., Zawisza, B., Marguí, E., Queralt, I., Hidalgo, M., Sitko, R.: Dispersive micro solid-phase extraction using multiwalled carbon nanotubes combined with portable total-reflection X-ray fluorescence spectrometry for the determination of trace amounts of Pb and Cd in water samples. J. Anal. At. Spectrom. **28**(5), 736–742 (2013)
77. Staniszewski, B., Freimann, P.: A solid phase extraction procedure for the simultaneous determination of total inorganic arsenic and trace metals in seawater: Sample preparation for total-reflection X-ray fluorescence. Spectrochim. Acta, Part B **63**(11), 1333–1337 (2008)
78. Aranda, P.R., Llorens, I., Perino, E., De Vito, I., Raba, J.: Removal of arsenic (V) ions from aqueous media by adsorption on multiwall carbon nanotubes thin film using XRF technique. Environ. Nanotechnol., Monit. & Manag. **5**, 21–26 (2016)

79. Dadfarnia, S., Shabani, A.M.H.: Recent development in liquid phase microextraction for determination of trace level concentration of metals—A review. Anal. Chim. Acta **658**(2), 107–119 (2010)
80. Marguí, E., Queralt, I., Hidalgo, M.: Determination of cadmium at ultratrace levels in environmental water samples by means of total reflection X-ray spectrometry after dispersive liquid–liquid microextraction. J. Anal. At. Spectrom. **28**(2), 266–273 (2013)
81. Sitko, R., Zawisza, B., Malicka, E.: Modification of carbon nanotubes for preconcentration, separation and determination of trace-metal ions. TrAC, Trends Anal. Chem. **37**, 22–31 (2012)
82. Kunimura, S., Amagasu, K.: Measurement of trace amount of cadmium using a portable total reflection X-ray fluorescence spectrometer. ISIJ Int. **55**(12), 2697–2699 (2015)
83. Medina González, G., Castillo, R.d.P., Neira, J.Y.: Multivariate calibration for the improvement of the quantification of cadmium in the presence of potassium as interferent by total reflection X-ray fluorescence. X-Ray Spectrom. **48**(6), 700–707 (2019)
84. Holtkamp, M., Elseberg, T., Wehe, C.A., Sperling, M., Karst, U.: Complexation and oxidation strategies for improved TXRF determination of mercury in vaccines. J. Anal. At. Spectrom. **28**(5), 719–723 (2013)
85. Greaves, E., Sosa, J.A., Sajo-Bohus, L., Alvarez, M., Wobrauschek, P., Streli, C.: Trace element determination of mercury by total-reflection X-ray fluorescence. Spectrochim. Acta, Part B **52**(7), 945–951 (1997)
86. Koulouridakis, P.E., Kallithrakas-Kontos, N.G.: Selective mercury determination after membrane complexation and total reflection X-ray fluorescence analysis. Anal. Chem. **76**(15), 4315–4319 (2004)
87. Marguí, E., Queralt, I., Guerra, M., Kallithrakas-Kontos, N.: Mercury determination at trace levels using membrane preconcentration and benchtop total reflection X-ray fluorescence analysis. Spectrochim. Acta, Part B **149**, 84–90 (2018)
88. Aretaki, I., Koulouridakis, P., Kallithrakas-Kontos, N.: Total reflection X-ray fluorescence mercury analysis after immobilization on quartz surfaces. Anal. Chim. Acta **562**(2), 252–257 (2006)
89. Marguí, E., Kregsamer, P., Hidalgo, M., Tapias, J., Queralt, I., Streli, C.: Analytical approaches for Hg determination in wastewater samples by means of total reflection X-ray fluorescence spectrometry. Talanta **82**(2), 821–827 (2010)
90. Rice, N.M., Smith, M.R.: Recovery of zinc, cadmium and mercury (II) from chloride and sulphate media by solvent extraction. J. Appl. Chem. Biotech. **25**(5), 379–402 (1975)
91. Aranda, P.R., Colombo, L., Perino, E., De Vito, I.E., Raba, J.: Solid-phase preconcentration and determination of mercury (II) using activated carbon in drinking water by X-ray fluorescence spectrometry. X-Ray Spectrom. **42**(2), 100–104 (2013)
92. Rezaei, A., Khani, H., Masteri-Farahani, M., Rofouei, M.K.: A novel extraction and preconcentration of ultra-trace levels of uranium ions in natural water samples using functionalized magnetic-nanoparticles prior to their determination by inductively coupled plasma-optical emission spectrometry. Anal. Methods **4**(12), 4107–4114 (2012)
93. Filik, H., Avan, A.A.: Magnetic nanostructures for preconcentration, speciation and determination of chromium ions: a review. Talanta **203**, 168–177 (2019)
94. Zou, Z., Wang, S., Jia, J., Xu, F., Long, Z., Hou, X.: Ultrasensitive determination of inorganic arsenic by hydride generation-atomic fluorescence spectrometry using Fe3O4@ ZIF-8 nanoparticles for preconcentration. Microchem. J. **124**, 578–583 (2016)
95. Romero, V., Costas-Mora, I., Lavilla, I., Bendicho, C.: Silver nanoparticle-assisted preconcentration of selenium and mercury on quartz reflectors for total reflection X-ray fluorescence analysis. J. Anal. At. Spectrom. **29**(4), 696–706 (2014)
96. Oskolok, K., Monogarova, O., Alov, N.: TXRF determination of mercury (II) in water in combination with liquid–liquid microextraction. Mosc. Univ. Chem. Bull. **72**, 174–177 (2017)
97. Musielak, M., Serda, M., Talik, E., Gagor, A., Korzuch, J., Sitko, R.: Highly selective and sensitive determination of mercury ions by total-reflection X-ray fluorescence spectrometry. J. Anal. At. Spectrom. **36**(7), 1533–1543 (2021)

98. Nagata, N., Peralta-Zamora, P.G., Poppi, R.J., Perez, C.A., Bueno, M.I.M.: Multivariate calibrations for the SR-TXRF determination of trace concentrations of lead and arsenic in the presence of bromine. X-Ray Spectrom.: Int. J. **35**(1), 79–84 (2006)
99. Koulouridakis, P.E., Domazos, E.A., Galani-Nikolakaki, S.M., Kallithrakas-Kontos, N.G.: Low level lead determination after membrane complexation and TXRF analysis. Microchim. Acta **146**, 245–250 (2004)
100. Kunimura, S., Kawai, J.: Trace elemental analysis of commercial bottled drinking water by a portable total reflection X-ray fluorescence spectrometer. Anal. Sci. **23**(10), 1185–1188 (2007)
101. Matsuyama, T., Izumoto, Y., Ishii, K., Sakai, Y., Yoshii, H.: Development of methods to evaluate several levels of uranium concentrations in drainage water using total reflection x-ray fluorescence technique. Front. Chem. **7**, 152 (2019)
102. Sanyal, K., Dhara, S.: Recent advances in ultra-trace determination of uranium in natural water using total reflection X-ray fluorescence (TXRF) spectrometry. X-Ray Spectrom. **53**(5), 326–339 (2024)
103. Shukla, V.K., Dhara, S., Mishra, N.: Total reflection X-ray fluorescence spectrometric determination of ultra-trace uranium in natural water samples using a dispersive liquid–liquid micro-extraction method. J. Anal. At. Spectrom. **35**(8), 1632–1640 (2020)
104. Takahashi, H., Izumoto, Y., Matsuyama, T., Yoshii, H.: Trace determination of uranium preconcentrated using graphene oxide by total reflection X-ray fluorescence spectrometry. X-Ray Spectrom. **48**(5), 366–374 (2019)
105. Kocot, K., Pytlakowska, K., Talik, E., Krafft, C., Sitko, R.: Sensitive determination of uranium using β-cyclodextrin modified graphene oxide and X-ray fluorescence techniques: EDXRF and TXRF. Talanta **246**, 123501 (2022)
106. Mortada, W.I.: Recent developments and applications of cloud point extraction: A critical review. Microchem. J. **157**, 105055 (2020)
107. Labrecque, C., Potvin, S., Whitty-Léveillé, L., Larivière, D.: Cloud point extraction of uranium using H2DEH [MDP] in acidic conditions. Talanta **107**, 284–291 (2013)
108. Saha, A., Deb, S.B., Sarkar, A., Saxena, M.K., Tomar, B.: Simultaneous preconcentration of uranium and thorium in aqueous samples using cloud point extraction. RSC Adv. **6**(24), 20109–20119 (2016)
109. Saha, A., Sanyal, K., Rawat, N., Deb, S.B., Saxena, M.K., Tomar, B.S.: Selective micellar extraction of Ultratrace levels of uranium in aqueous samples by task specific ionic liquid followed by its detection employing Total reflection X-ray fluorescence spectrometry. Anal. Chem. **89**(19), 10422–10430 (2017)
110. Hatzistavros, V., Koulouridakis, P., Kallithrakas-Kontos, N.: Complexing membrane for uranium detection by total reflection X-ray fluorescence. Anal. Sci. **21**(7), 823–826 (2005)
111. Sanyal, K., Dhara, S., Gumber, N., Pai, R.V.: A highly sensitive method for uranium quantification in water samples at ultra-trace level by total reflection X-ray fluorescence, after its direct pre-concentration on the surface of amidoxime functionalized quartz sample supports. Talanta **254**, 124129 (2023)
112. Kumar, S.A., Pandey, S.P., Thakur, N., Parab, H., Shinde, R.N., Pandey, A.K., et al.: Synthesis and application of a unified sorbent for simultaneous preconcentration and determination of trace metal pollutants in natural waters. J. Hazard. Mater. **262**, 265–273 (2013)
113. Carvalho, M.S., Maria de Lourdes, F.D., Mantovano, J.L., S Filho, E.Q.: Uranium determination at ppb levels by X-ray fluorescence after its preconcentration on polyurethane foam. Spectrochim. Acta Part B: At.Ic Spectrosc. **53**(14), 1945–1949 (1998)
114. Carvalho, R.N., Anunciação, T.A., Dantas, A.F., de S. Dias, F., Teixeira, L.S.: Determination of uranium in water samples by energy-dispersive X-ray fluorescence spectrometry after solid-phase extraction. J. Iran. Chem. Soc. 1–8 (2024)

Different XRF Methods for Detection of Chemical Hazards in Drinking Water

15

Nuray Kup Aylikci, Volkan Aylikci, and Abdelhalim Kahoul

Abstract

Contaminated water is not only a threat in itself, but also has negative effects on the surrounding soil and other living habitats. Protecting water resources, developing strategies to improve water quality, and assessing and responding to the risks that threaten water resources are essential for a healthy society. One of the possible risks is the presence of pollutants in the water that are a threat to human health. Arsenic, nitrate salts, copper, lead, and radon are the most important of these chemical pollutants. Such pollutants are known to be present at levels that threaten human health when consumed with water. The XRF technique provides fast, convenient and reliable results for the determination of these types of contaminants. For quantitative analysis of these contaminants, ICP method was reported to be more expensive and inaccessible to viable uses. Quantitative analysis of chemical hazards in a faster, more practical and low-cost way is accomplished with XRF technique. This study will be an in-depth look at the next generation of trends in detecting chemical hazards in water and will be an introduction to possible new recommendations in light of the available literature including artificial intelligence and machine learning.

N. K. Aylikci (✉)
Department of Engineering Sciences, Iskenderun Technical University–ISTE, Hatay, Turkey
e-mail: nuray.aylikci@iste.edu.tr

V. Aylikci
Department of Metallurgical and Material Science Engineering, Iskenderun Technical University–ISTE, Hatay, Turkey

A. Kahoul
Department of Materials Science, Faculty of Sciences and Technology, Mohamed El Bachir El Ibrahimi University, Bordj-Bou-Arreridj, Algeria

Laboratory of Materials Physics, Radiation and Nanostructures (LPMRN), Mohamed El Bachir El Ibrahimi University, Bordj-Bou-Arreridj, Algeria

V. K. Singh (ed.), *X-Ray Fluorescence Spectroscopy and Chemometrics*,
https://doi.org/10.1007/978-3-031-98375-7_15

15.1 Introduction

The demand for fresh water is increasing due to the rapid growth of the world population. This growth, combined with factors such as uncontrolled urbanization, unplanned industrialization and unsustainable use of energy resources, leads to pollution of water, air, soil, and ultimately endangers the ecosystem. The amount of oxygen dissolved in water is reduced by chemicals and organic compounds in polluted water. This leads to an increase in the mortality rate of plants and animals living in the water. It is essential to protect water resources, develop strategies to improve water quality, and mitigate the risks that threaten water resources for a healthy society. For chemical analysis and the water quality control, different analytical tools are available such as ICP, AAS, ICP-MS, and ICP-OES. Depending on the conditions required by the analyst, each method has its own advantages and disadvantages. When the required detection limits of quantitative analysis are above 1 ppm, XRF technique will be very attractive method to consider, especially for analysis of solids, powders, slurries, filters and oil owing to its non-destructive capability. In contrast to ICP and AAS, XRF spectroscopy requires no sample dissolution or digestion and so, this advantage is a sign of eliminating the potential inaccuracies of incomplete dissolution and large dilutions. Apart from its non-destructive characteristic, sample preparation is simple, fast and safe as no chemical waste is generated. Other advantages can be described as low operating costs, analysis at the point of manufacture and eliminating the daily recalibration.

Different types of XRF instrument systems are commercially available in which demonstrated as Fig. 15.1, which is based on the information provided by Margui et al. [1]. In this figure, the analyzing conditions were summarized according to the instruments capabilities. According to this figure, it is clear that only the TXRF analysis can be used to measure the chemical hazards in drinking water. However, the analysis of water samples can also be possible using different filtration or preconcentration system for obtaining solid samples. This section provides detailed information about the new trends about the analysis chemical hazards in drinking water and other water samples by using X-ray fluorescence.

15.1.1 Chemical Hazard Analysis by Portable Energy Dispersive X-Ray Fluorescence (p-EDXRF)

The advantages and disadvantages of portable energy dispersive XRF method (p-EDXRF) are summarized as Fig. 15.1 and it can be seen that water samples only can be analyzed using TXRF method. Although TXRF is widely used in water analysis, scientists have analyzed water samples using other XRF instruments due to the lack of specific standard and their accuracy should be tested by round-robin (inter-laboratory comparison) test. In-situ monitoring of toxic metals in mining water was investigated by tri-axial portable X-ray fluorescence method (3pXRF). The reason of choosing tri-axial geometry (3pXRF) was explained as for improving peak-to-background ratios compared to conventional pXRF systems. The water

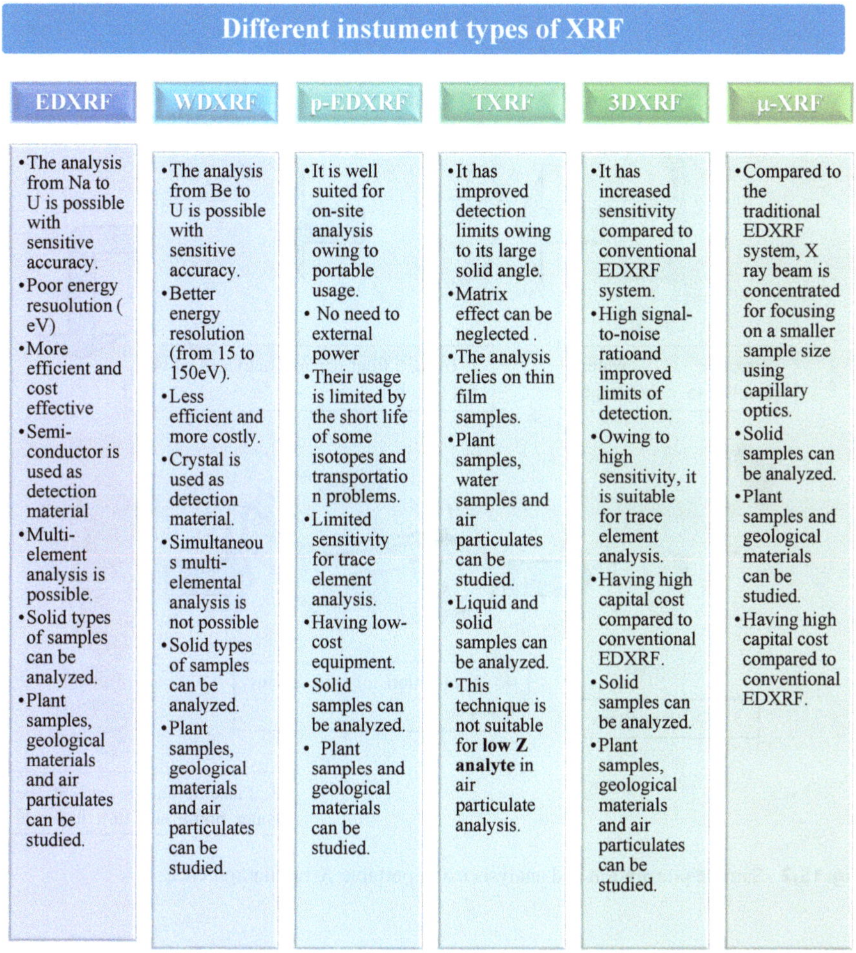

Fig. 15.1 Different XRF techniques according to the different instrumentation types

samples were deposited onto a commercial filter paper and after drying process the filter papers were analyzed [2]. Despite all the restrictions of pXRF as summarized in Fig. 15.1, Pb analysis was performed using activated carbon felt [3]. According to this new sensitive screening method, activated carbon felt was used as a trapping agent for Pb elements.

Figure 15.2 shows the water sample preparation and analysis process as explained in Ref. [3]. Filter cap, activated carbon felt and washer were screwed onto the bottle including water samples and turned upside down until the bottle has been allowed to drain completely. The study reported that XRF signals of dried filters were 7.5 fold stronger than that of damp ones. The attenuation of X-rays in water was explained as the reason for the weak XRF signals of wet filters.

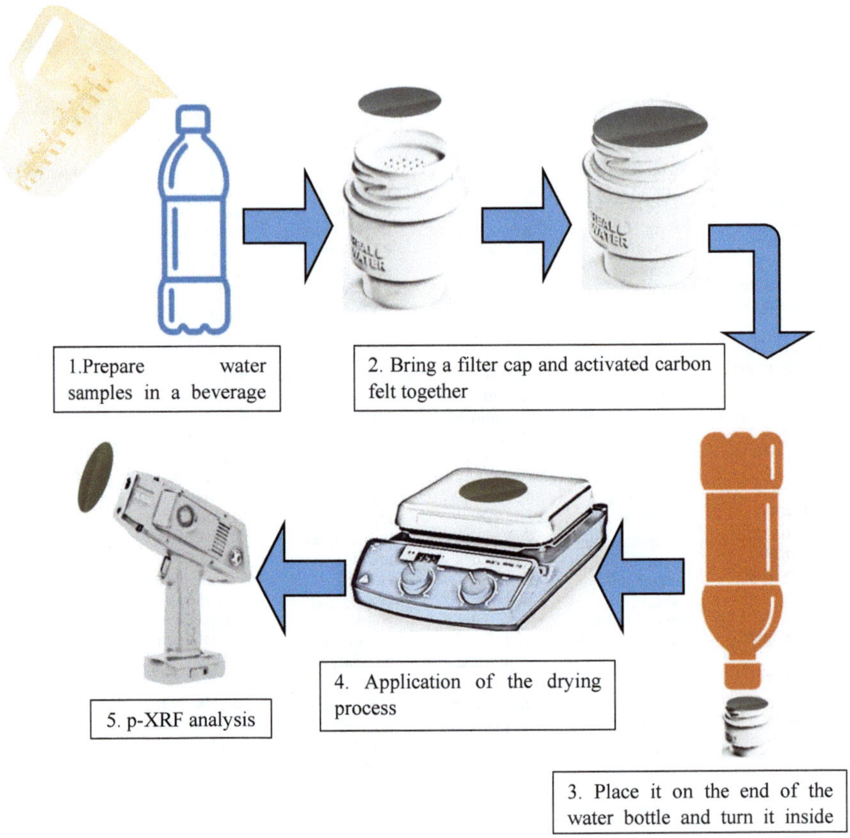

Fig. 15.2 Sample preparation and analysis using portable X-ray fluorescence

On site determination process was applied to the analysis of arsenic, selenium and chromium (VI) in drinking water by using solid phase extraction for sample preparation. In addition to filtration of water samples, it was noted that pretreatments such as evaporation, co-precipitation, and extraction should be used in sample preparation for p-XRF analysis. Especially, solid phase extraction (SPE) disk is a competitive candidate in sample preparation for XRF analysis without using acid–base solutions and volatile organic compounds. A combination of membrane disk type SPE and p-XRF system was designed as a new analytical technique for the determination of As (arsenic), Se (selenium) and Cr(VI) (chromium) in drinking water [4]. The pollution of the drinking water is getting worse due to the industrial wastes, soil pollution, and the irresponsible usage of ground water. Arsenic and other toxic elements can result in dangerous consequences for public health. ICP-MS and ICP-AES analytical methods are generally defined as a sensitive tool for the water analysis but not practical. Portable X-ray fluorescence equipments do not require any extra power systems owing to its rechargeable property. Despite the drawbacks outlined in Fig. 15.1, the ability to simultaneously survey the surrounding environment (plants, soils, rocks) for drinking water analysis may make it the preferred method. For water analysis, solid phase extraction is explained as a rapid and stable pre-concentration way to prepare water samples. Based on this knowledge, a Ti and Zr loaded carbon disk was used for pre-concentration of water samples. 1 ml of titanium isopropoxide and 0.5 ml of 70% zirconium propoxide solution were prepared in a 15 ml of ethanol and then, carbon disks were stirred in the prepared solution for 5 min. After the process was finished, the modified disks were washed with deonized water and used within 1 day. Water samples were prepared in solution form by adjusting pH = 3 using aqueous HCl and passed through the modified carbon disks to pre-concentration of As [5].

It is common knowledge that portable XRF systems have limited sensitivity for trace element analysis (Fig. 15.1). However, an analysis of trace metal contaminations in ageing storm water constructed wetlands was performed with pXRF. The storage and use of rainwater is on the current global priority list in worldwide due to the over-consumption of drinking water. In addition, stormwater can have devastating effect in urban areas and it can be a source of hazardous contaminations that is harmful to the nature. For the stated reasons, water management should be a crucial issue. That is why metal mapping distribution in reed-bed filters was performed for stormwater management adjustement using portable X-ray [6]. The study focus on developing a robust analyzing, compution a distrbution mapping of metals in stromwater lands and finally quick analysis of hotspots with calibrated pXRF. Moreover, it was explained that shooting time, sample water content, homogenity, texture and organic matter content could influence the main parameters of XRF. The linearity, limit of detection, limit of quantification and working range of quantitative pXRF analysis were performed for metals in liquid matrises [7]. For the preparation of calibration curves, the concentrations in %mg/L should be converted to %m/m written below as Eq. (15.1).

$$C(\%mg/L) = C(\%m/m) \times 10000 \times d \tag{15.1}$$

The sample holder of the X-ray fluorescence (XRF) instrument for liquids was designed as a stack of three polypropylene films (4 μm each) to prevent any leakage. This design was explained as effective for sample containment during analysis and for maintenance of the integrity of the experimental setup [7]. A valid limit of detection and limit of quantification values for liquid matrices were calculated as Eqs. (15.2) and (15.3).

$$\begin{pmatrix} Limit \\ of \\ detection \end{pmatrix} = 3.66 \times \frac{s_0}{\sqrt{n}} \tag{15.2}$$

$$\begin{pmatrix} Limit \\ of \\ quantification \end{pmatrix} = k_Q \times \frac{s_0}{\sqrt{n}} \tag{15.3}$$

In which n is the number of measurements, s_0 stands for the estimated standard deviation and finally k_Q is a value of 10 determined by IUPAC [7].

In summary, the detection of chemical hazards for water samples using different filters [8–10] or pre-concentration methods will be getting more important owing to the practical usage of pXRF.

15.1.2 Chemical Hazard Analysis by Micro X-ray Fluorescence (μ-XRF) and X-Ray Fluorescence with Polarized Beam (3D-XRF)

Micro X-ray fluorescence (μ-XRF) is defined as an elemental analytical technique focusing X-rays on very small sample areas. Polycapillary lenses and doubly curved crystal focusing X-ray optics provide a high X-ray flux creating small focus spots. Reaching a small X-ray beam size can be achived by using a pinhole collimator which reduces the input intensity of beam. Policapillary lenses are the best choice for the higher proportion of transmitted radiation than is obtained with a pinhole collimator due to the total reflection process. Besides, these optical materials overcome the limitation of square dependence of X-ray intensity with distance, thus providing the ability to design small size, low power XRF systems. Owing to the focusing of X-ray beams, micro X-ray fluorescence systems are capable of small feature evaluation, elemental mapping, film and plating thickness measurement, micro-contamination detection, small particle analysis and forensics. Furthermore, focusing device of X-rays makes micro X-ray fluorescence analysis possible for studying small particles and also for complex and inhomogeneous materials. X-ray radiation from source will be transported by reflection process in policapillary lenses and other optical materials for use as focus device. A smooth surface is required for obtaining a small reflection angle than a critical value written as Eq. (15.4) [11];

$$\theta_{cr} \approx 20\sqrt{\rho}/E \tag{15.4}$$

In which E denotes photon energy and ρ is the lenses density. Equation (15.4) demonstrates that the greater photon energy means smaller critical angle.

Micro X-ray fluorescence technique is generally preferred for studies in material science, archaeological works, biological and geological samples. Notably, the elemental composition of rocks and minerals can give information about the chemical composition of surrounding water samples [11]. Besides, confocal micro X-ray fluorescence method is applicable for the monitoring of chemical reaction in solution. Corrosion and deposition process on sample surface occur as a result of chemical reaction process. Especially for drinking water, any corrosive reactions can be monitored through the piping system [12]. The products from chemical reactions can be a source of metallic contaminations along the drinking water pipe line. Another way to investigate chemical hazards in water can possible using a pearl in freshwater [13, 14]. Sediments can also give critical information about surrounding water resources and soils however, it is debated whether reliable measurements can be made for elemental analysis due to the moisture content of the samples. Traditional XRF methods can analyze dry sediment samples but new generation scanning XRF instruments are generally presented with wet sediment cores. Conversion of wet core XRF data to dry mass concentrations will aid improved palaeo-environmental interpretation, as the methodology used is reliable and avoids loss of data quality [15]. The corrections for obtaining micro-XRF data of wet sediments can be achieved with two steps. First procedure is the estimation of water content in the sediment and the second is the correction of water content effects on X-ray intensity [16]. In addition to sediment samples, the presence of chemical hazards in water (lake, river) samples can be determined by analyzing water–rock interactions in fractures but further studies were recommended to estimate error of sorption process in microfluidic test [17]. The other way of determining chemical hazards in drinking water may be the aquatic plants with a laboratory scale phytofiltration system. *Salvinia biloba* Raddi was defined as an efficient accumulator of Cu and Zn elements. After the accumulation process on aquatic plant, in vivo quantitative analysis Cu and Zn metal were performed using micro-XRF [18]. Owing to the advantage of direct verification of accuracy, Monte Carlo iterative simulation approach was applied by comparing the experimental and simulated data even though it was time consuming and a requirement of specific skill.

Up to this point of the literature review, it has been emphasized that the indirect analysis of chemical hazards using micro X-ray fluorescence in water resources has been presumed by using the surrounding natural samples or aquatic plants without any sample preparation process. Using the chips and automated printing technology, new type sample preparation process was described for the direct analysis of liquid samples. The first method was developed by Miller et al. in which nano-liter dried spot sample preparation process onto AP1 and Kapton film substrates was performed for liquid samples. The study proposed the method as a rapid and minimizing the human error and operator skills [19]. And the second technique was related the using of chip technology for analyzing the Pu content in spent nuclear fuel [20].

Another confocal excitation/detection experimental set up is used for chemical imaging techniques called XRF computed tomography (XRF-CT) and 3D-XRF spectrometry [21]. To the best of our knowledge from the literature review, there are no studies on the determination of chemical hazards in drinking water using the 3D-XRF method. However, the non-destructive depth profile analysis for industrial plastics, chemical microchip and biological samples [22], the identification of oil emulsion species [23], analysis of ancient paintings [24] and paint layers [25, 26] and archeometric studies [27].

15.1.3 Chemical Hazard Analysis by Energy Dispersive X-Ray Fluorescence (EDXRF)

The chemical hazard analysis by means of EDXRF is a traditional X-ray fluorescence technique which uses semi-conductor detection material. Considering all the features of XRF, EDXRF's lower price, improved portability and simultaneous multi-element capabilities make the energy dispersive system more desirable than WDXRF in environmental applications [1]. The analyzing results with minimized errors can be possible with proper sample preparation process. Sample preparation methods best suited for EDXRF can be announced as pressed pellet, loose powder and fused bead for solid samples but however, direct analysis of solid samples is possible for flat and smooth surface, and signal is thickness dependent. The direct analysis for liquid samples is only be probable with TXRF method due to the detection limit criteria. EDXRF method cannot be the best choice for the direct analysis of liquid samples with a loe detection limit. Instead of the direct analysis, the preconcentration method can improve the detection limit providing the removal of matrix effects.

The preconcentration technique is a process of concentration of liquid samples for analyte detection with the use of an absorbent material. Membranes or disks, liquid phase micro-extraction and using nanomaterials are generally explained as preconcentrating methods to eliminate the matrix effects and increase the detection limit. The commercial filter papers, extraction disks membranes and other retaining materials can supply a preconcebtration of analyte in liquid samples. Furthermore, the origami-type folding of filter paper can increase the sorbent surface for multifold analyte preconcentration [28]. In recent times, biopolymer nano-fiber based sorbents are recommended for extraction media, filter membranes and sensors owing to its accessibility, scalability, high surface areaand ease of handling properties [29]. In addition, some researchers have decleared that "Green Analytical Methodologies" should be applied to perform analytical determinations considering encironmental benefits. Environmental benefits mean maximizing operator safety and environmentally friendly analytical method. Consequently, reducing agent, energy, equipment consumption and reducing waste will be the most important arguments for determining the analysis technique [30]. Besides these preconcentartion methods, coprecipitation is an another technique for the determination of elemental analysis of water. The precipitated water samples were collected on a Nucleopore membrane using vacuum filtration system shown as Fig. 15.3.

The aforementioned work used an EDXRF spectrometers with X-ray tube having oil cooled Rh anode and X rays were detected by a Si(Li) detector with a 143 eV resolution at 5.9 keV [31]. Over the past few decades, new nanocomposites with the sorption properties of metallic contaminants have gain more importance not only from the environmental point of view but also for also developing eco-friendly methods for chemical hazard determination by X-ray fluorescence methods. An aluminum oxide coated nano-graphite (Al_2O_3/nano-G) composite

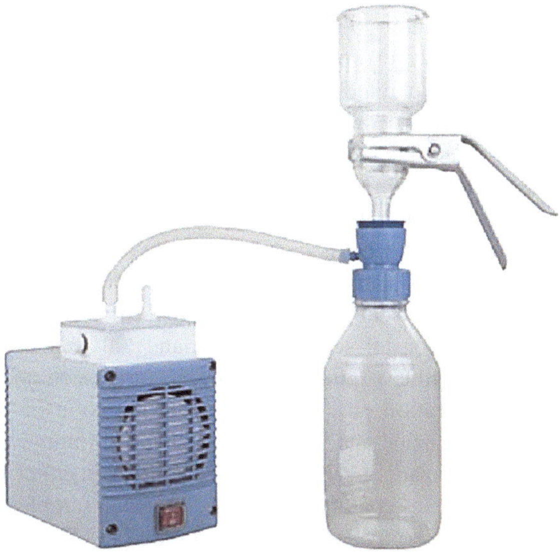

Fig. 15.3 A vacuum filtration system

was evolved for the sorption of chromium species in water with preconcentration method [32]. The molybdenum disulfide decorated reduced graphene oxide composite was used as a solid sorbend for chromium, cobalt, nickel, copper, zinc and lead ions in water for analysis using EDXRF. The calibration curves showed a linear relationship between the concentration of the analytes and the intensity of the fluorescent radiation in which optimal conditions were described as pH = 6.5, 50 mL of sample volume and 10 min of sonication time [33].

Nowadays, the removal of mercury ions is a challenging case by using practical ways with nano absorbents. A functionalized graphdiyne oxide (GDYO-3 M) was desingned for the absorbance of mercury ions and it was reported that an exceeding 97% removal efficiency was obtained. Due to the—SH groups, the functionalized graphydine oxides showed better removal efficiency for mercury ions and any other metallic samples in drinking water [34]. A dispersive liguid-liquid microextraction technique was applied as a new sample preparation method using Zn and Cu complexes for trace element determination with EDXRF. However, the optimized contiditions should be determined for this method explained as extraction and disperser solvent type, volume of solvent, pH, concentration of chelating agent for complex and salt effect. And also, the calibration graph must show a linear calibration to detect trace amount of complex in drinking water [35]. For the determination of inorganic chromium species in water samples, a combination of simple and practical micro solid phase extraction (DMSPE) and energy dispersive X-ray fluorescence spectrometry (EDXRF) method was developed. The preconcentration of trivalent chromium ion process was performed using the graphene oxides nanoparticles as a solid sorbent. Whereas, some parameters have dominant effect on the sorprtion capacity of graphene oxide nanomaterials. In acidic conditions, the low degree of dissolution of the graphene oxide (GO) functional groups and inter-competing of hydrogen and metallic ions for the same sorption sites reduces its sorption process. For the enhanced cation-exchange capacity, pH value was explained as to be higher than 3–4 [36–38]. In addition, the same process was applied to the determination of trace and ultra trace selenium ion in mineral, tap, lake and sea water [39]. The coprecipitation process is also another pre-concentration method for EDXRF analysis of metallic contaminants in water samples [40–42]. In contrast, for selenium ion determination, solid phase extraction process was defined as the best choice for the concentration process owing to its inexpensive ability compared to the liquid–liquid extraction systems.

The analysis of chemical hazards will also be possible by the indirect analysis using coastal sediments [43] and Salvinia natans [44]. However, in recent years, a number of different materials have been under investigation for use in the preconcentration process for the determination of mercury. [45], chromium [46, 47], copper, zinc and lead [48], arsenic, cadmium, cobalt, iron, gallium, and other chemical hazards [49] by EDXRF. While pre-concentartion is the most commonly used sample preparation method for water samples, in the future, glass fiber filter [50], adsorptive membrane systems [51, 52], metal organic framework based mixed-matrix membrane materials [53] and nano metal fixed on cellulose filter paper which has multivariate properties (photocatalyst, solid phase adsorbent, filter membrane, and chemical sensor) [54] may also be preferred for eliminating diverse matrix effects in the case of analysis of water by means of EDXRF.

15.1.4 Chemical Hazard Analysis by Wavelength Dispersive X-Ray Fluorescence (WDXRF)

Environmental investigations [55, 56] and advances in X-ray spectrometry [57–62] have been the subject of various studies. This means that the chemical contamination analysis using practical, green and non-destructive methods are gaining more importance due to the increase of pollution factors. In many industrial applications, both of EDXRF (energy dispersive) and WDXRF (wavelength dispersive) are generally used for the qualitative and quantitative analysis. To compare two methods, EDXRF is best for providing fast and compherensive analysis of a wide sample range but beyond that, WDXRF enables detailed and high-resolution analysis of selected elements in complex matrix.

The preconcentration of liquid samples are commonly used for chemical hazards determinationin water samples using X-ray fluorescence. Although, preconcentration of liquid samples for XRF analysis has a variety of applications amoung themselves, solid-phase extraction and liquid-phase extraction are the most preferred concentrating methods [59]. A solid phase extraction disk was implemented for the simultaneous collection of As(III) and As(V) in water. However, it is important to note that the X-ray peaks of PbL_α and AsK_α will overlap with each other. The best quantification of arsenic contents will possible with the removal of Pb^{+2} ions from water samples [63]. Another X-ray-convenient water analysis was performed by using iminodiacetate extraction disk for the determination of Mn, Fe, Co, Ni, Cu, Zn, Cd, Hg and Pb. To prevent the disk from strong X-ray energy (50 kV), iminodiacetate extraction disk was coated with commercially available laminate film including both faces [60]. Ion (cation/anion) exchange disks were also practiced for As [64], Cr [65] and Br [66] quantification with WDXRF.

Recent studies have shown that carbon and its derivatives have been improved with the synthesis of various types of energy storage materials. However, these materials may be good candidates for determining chemical hazards in water samples. As an exception to the use of disks, carbon derivatives defined activated carbon, carbon nanotubes and multiwalled carbon nanotubes showed exhilarating results for the preconcentrated water samples. The activated carbon Aliquat 336-AC, were placed in a shaker with solutions including Hg(II) ions. In that study, it was reported that the best adsorption was obtained with a shaking time of 5 min and a pH value of 1.5. Moreover, the linearity between X-ray intensity and Hg content was obtained at values of 20–30 ppm Hg(II) [67]. WDXRF analysis for trace amounts of some metals (V, Cr, Mn, Fe, Co, Ni, Cu, Zn, Ga, Se, Pb and Cd) were investigated by the preconcentration of liquid samples using carbon nanotubes. Carbon nanotubes were explained as a new adsorbent which showed the better adsorption properties than activated carbon. Besides carbon nanotubes, its oxidized form was emphasized as better adsorption capacity than non-oxidized form due to carboxyl and hydroxyl groups in CNTs surface [68]. The oxidized form of multi-walled carbon nanotubes (MWCNTs) were dispersed in an aquatic solution for the adsorption of trace metals. After the adsorption process, MWCNTs@metallic ions were collected onto a filter paper and the analysis of trace amounts of metals were performed with WDXRF. The effect of solution pH on the adsorption capacity were related with the surface charge of the adsorbent and the degree of ionization [69].

The preconcentration of thorium (Th) ions in natural water was applied by using polyurethane foam loaded with EXPA (2-etilhexylphosphonic acid) [70], and silica gels [71] for the metallic ion adsorption for trace amount of WDXRF analysis. In contrast to the preconcentration process with using an absorber, droplet method onto a porous filter [72, 73] and dried residue [74, 75] were studied as different sample preparation methods before WDXRF analysis. However, the effectiveness of activated carbon disks was studied by Lee and Yoon et al. for repeatable measurements. An activated carbon disk was used as an adsorber for the preconcentration process for iodine analysis in water using WDXRF [76] described as simple, rapid, efficiently and environmentally friendly. According to the information provided by the authors, carbon and its derivatives may be the future effective adsorbers owing to their repeatable usage.

15.1.5 Chemical Hazard Analysis by Total Reflection X-Ray Fluorescence (TXRF)

The chemical hazards are generally released from pharmaceutical, battery, steel, paper and other industries which have harmfull effects on natural water and other ecosystem. By the development industrial applications, the issue of access to clean, fresh and potable water is becoming increasingly important. X-ray fluorescence analysis are generally defined as non-destructive, repeatable and cost-effective compared to the other destructive systems but sampling process is the most important step avoiding the matrix effects and also other contaminations. Time span between the process of storing, transporting and analysis water samples may be an example of this situation which can have a negative impact on the results of the analysis.

Water sampling should be performed with high purity HNO_3 and HCl not greater than pH 2 to avoid precipitation of metal hydroxides or interaction of water with metal ions in the sampling vessel. However, HCl is generally not the preferred choice because of the high chlorine content that can be the result. For avoiding negative impacts on analysis results, water collection vessels were chosen from the polypropylene, polyethylene and fluorinated ethylene polymers, but high density polyethylene was explained as not suitable for As element included water samples [77]. Generally, the recent studies are related by using solid phase extraction methods for all x-ray fluorescence analysis of water samples. The sensitive mercury determination in water, environmental and biological samples was performed by using both of dispersive micro solid phase extraction (DMSPE) + EDXRF and DMSPE + TXRF owing to the low detection limit and the results was approved with certified reference samples [78].

The chemical hazards and types of industries were reviewed by a study of Sharma et al. [79]. It was emphasized that the combination of TXRF equipment with a synchrotron radiation was the best choice for ultra-trace elemental analysis due to the excellent characteristics. However, in spite of the allowing to the sensitive analysis, a standardization process is not available for TXRF emthod for environmental, biological and water samples. Generally, a Round-Robin test methodology [80] is applied in which the analyzing results are compared with the other certified laboratuaries. The Round-Robin test method is defined as the measurements, analysis and experiments which are performed in different laboratuaries for several times for comparing the analysis results. In TXRF method, the analyzing results are compared with the the data of other certified laboratuary results as can be seen in the diagram provided in the Fig. 15.4. Besides, the verification process can also be accomplished with the other analytical methods. However, for all laboratuaries, the accuracy measurements will be achieved by avoiding from the errors resulted from data analysis, contaminations, spectral deconvolution, instrumental calibration process and finally sample preparation.

In contrast to Round-Robin test method, the validation of chemical method for proposing as a standart to the International Standart Organization was explored by defining the linear calibration range, sensitivies, limits of detection and accuracy.

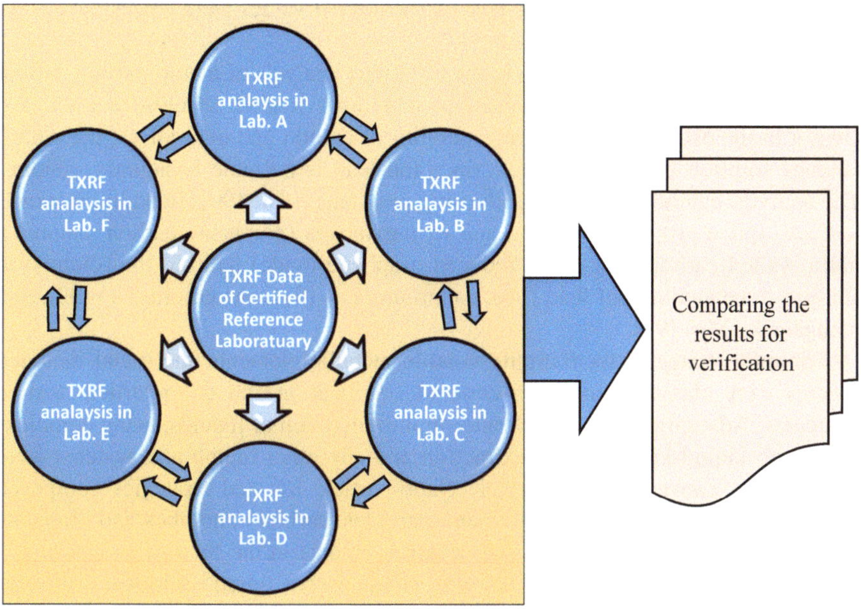

Fig. 15.4 The data verification process of TXRF analysis

For each element of analyzed, a linear least squares regression was individually applied to estimate sensitivities with all statistical tests included [81]. One of the key advantages of TXRF is that it allows water samples to be analysed directly without requiring sample preparation. However, Hg analysis of water samples needs some complexing agent such as ammonium pyrrolidine dithiocarbamate (APDC) solution [82], ethylendiaminetetraacetic acid (EDTA) at pH = 10, thiourea at pH = 10 and sodium thiosulphate at pH = 6 [68] avoiding volatility losses. Despite allowing direct analysis of water samples without sample preparation, there have also been studies in the literature applying pre-concentration methodologies. Generally, it was emphasized that the most common sample pre-concentration method was drop-casting defined as liquid drop deposition and drying on the quartz and other supporting material. In the case of high matrix content, drop casting method was not suitable for accurate analysis [83]. For avoiding matrix effects, micro-extraction is the most preferred method. Dispersive liquid–liquid micrextraction for chromium [84], selenium [85], uranium [86], lanthanum, cerium, praseodymium, neodymium, and samarium in mineral water [87] with the combination of total X-ray fluorescence analysis. Using the hollow fiber liquid phase micro-extraction, trace level determination of arsenic (As) [88] and antimony (Sb) [89] were achieved in water species. In extraction processes, the optimum parameters should be determined for types of extractant, organic solvent, agitation speed, solution pH, extraction time and any other conditions. For arsenic (As) species in water, the best extraction pH values were determined as 13 for As(III), 8.5 for As(V) [88].

For a green chemistry approach, the ultratrace analysis of metal ions was investigated by using graphene oxide/carbon nanotubes loaded cellulose membranes. Owing to the excellent adsorptive capability of carbon and derivatives, the simultaneous sorption of cobalt, nickel, zinc, and lead is possible in aquatic samples. The analytes can be eliminated with using graphene oxide [90]. In addition, excellent adsorptive propery of carbon and its derivatives (graphene, carbon nanotube, multi walled carbon nano tube, reduced graphene oxide) had led the scientist to analyze trace amounts of lead (Pb), cadmium (Cd) [91], chromium (Cr) [92], and manganese (Mn) [93].

Common strategies for trapping volatile metallic elements and metal included species were classified into conventional stabilization and nano-particle assited strategies. Adsorption onto membranes, immobilization of ligands, using complexing agent, kiquid-kiquid micro extraction and any other membrane systems were described as conventional strategies. Nano-particle assisted strategies composed of the magnetic and non magnetic nano-particles, carbon nanotubes and graphene based materials [94]. Promising recent developments in the field of materials science give metal organic frameworks [95], metal–carbon based composites [96] and other materials a future for their ability to absorb heavy metals in water. Finally, algae technology may be a promising future candidate for water chemical hazard analysis because of its accessibility and ease of production [97].

References

1. Marguí, E., Queralt, I., de Almeida, E.: X-ray fluorescence spectrometry for environmental analysis: basic principles, instrumentation, applications and recent trends. Chemosphere **303**(P1), 135006 (2022). https://doi.org/10.1016/j.chemosphere.2022.135006
2. Pessanha, S., Marguí, E., Carvalho, M.L., Queralt, I.: A simple and sustainable portable triaxial energy dispersive X-ray fluorescence method for in situ multielemental analysis of mining water samples. Spectrochim. Acta-Part B At.Ic Spectrosc. **164**(November 2019) (2020). https://doi.org/10.1016/j.sab.2019.105762
3. Tighe, M., Bielski, M., Wilson, M., Ruscio-Atkinson, G., Peaslee, G.F., Lieberman, M.: A sensitive XRF screening method for lead in drinking water. Anal. Chem. **92**(7), 4949–4953 (2020). https://doi.org/10.1021/acs.analchem.9b05058
4. Hagiwara, K., Koike, Y., Aizawa, M., Nakamura, T.: On-site determination of arsenic, selenium, and chromium(VI) in drinking water using a solid-phase extraction disk/handheld X-ray fluorescence spectrometer. Anal. Sci. **34**(11), 1309–1315 (2018). https://doi.org/10.2116/analsci.18P217
5. Hagiwara, K., Koike, Y., Aizawa, M., Nakamura, T.: On-site quantitation of arsenic in drinking water by disk solid-phase extraction/mobile X-ray fluorescence spectrometry. Talanta **144**, 788–792 (2015). https://doi.org/10.1016/j.talanta.2015.07.002
6. Lenormand, É., Kustner, C., Combroux, I., Bois, P., Wanko, A.: Diagnosing trace metals contamination in ageing stormwater constructed wetlands by portable X-ray Fluorescence Analyzer (pXRF). Sci. Total. Environ. **844**(June) (2022). https://doi.org/10.1016/j.scitotenv.2022.157097

7. Krummenauer, A., Lapazin, L.H., Cenci, M.P., Veit, H.M.: Portable EDXRF spectrometer: Determination of linearity, LOD, LOQ, and working range for the analysis of base and precious metals in liquid matrices. J. Phys.: Conf. Ser. **2606**(1) (2023). https://doi.org/10.1088/1742-6596/2606/1/012001
8. Pearson, D., Chakraborty, S., Duda, B., Li, B., Weindorf, D.C., Deb, S., Brevik, E., Ray, D.P.: Water analysis via portable X-ray fluorescence spectrometry. J. Hydrol. **544**, 172–179 (2017). https://doi.org/10.1016/j.jhydrol.2016.11.018
9. Pearson, D., Weindorf, D.C., Chakraborty, S., Li, B., Koch, J., Van Deventer, P., de Wet, J., Kusi, N.Y.: Analysis of metal-laden water via portable X-ray fluorescence spectrometry. J. Hydrol. **561**(April), 267–276 (2018). https://doi.org/10.1016/j.jhydrol.2018.04.014
10. Tiihonen, T.E., Nissinen, T.J., Turhanen, P.A., Vepsäläinen, J.J., Riikonen, J., Lehto, V.P.: Real-time on-site multielement analysis of environmental waters with a portable X-ray fluorescence (pXRF) system. Anal. Chem. **94**(34), 11739–11744 (2022). https://doi.org/10.1021/acs.analchem.2c01490
11. Heimler, K., Gottschalk, C., Vogt, C.: Confocal micro X-ray fluorescence analysis for the non-destructive investigation of structured and inhomogeneous samples. Anal. Bioanal. Chem. **415**(21), 5083–5100 (2023). https://doi.org/10.1007/s00216-023-04829-x
12. Tsuji, K., Matsuno, T., Takimoto, Y., Yamanashi, M., Kometani, N., Sasaki, Y.C., Hasegawa, T., Kato, S., Yamada, T., Shoji, T., Kawahara, N.: New developments of X-ray fluorescence imaging techniques in laboratory. Spectrochim. Acta-Part B At.Ic Spectrosc. **113**(November), 43–53 (2015). https://doi.org/10.1016/j.sab.2015.09.001
13. Gao, J., Zhang, J., Wu, W., Yen, C.K., Su, W.: Synchrotron radiation μ-XRF imaging reveals mn zoning in freshwater pearls. J. Phys. Chem. C **126**(50), 21381–21389 (2022). https://doi.org/10.1021/acs.jpcc.2c05988
14. Leppänen, J.J., Saarinen, T., Jilbert, T., Oulasvirta, P.: The analysis of freshwater pearl mussel shells using μ-XRF (micro-x-ray fluorescence) and the applicability for environmental reconstruction. SN Appl. Sci. **3**(1), 1–8 (2021). https://doi.org/10.1007/s42452-020-03978-3
15. Boyle, J.F., Chiverrell, R.C., Schillereff, D.: Approaches to water content correction and calibration for μXRF core scanning: comparing X-ray scattering with simple regression of elemental concentrations 373–390 (2015). https://doi.org/10.1007/978-94-017-9849-5_14
16. Katsuta, N., Takano, M., Sano, N., Tani, Y., Ochiai, S., Naito, S., Murakami, T., Niwa, M., Kawakamiichi, S.: Quantitative micro-X-ray fluorescence scanning spectroscopy of wet sediment based on the X-ray absorption and emission theories: Its application to freshwater lake sedimentary sequences. Sedimentology **66**(6), 2490–2510 (2019). https://doi.org/10.1111/sed.12603
17. Oh, Y.S., Jo, H.Y., Ryu, J.H., Kim, G.Y.: A microfluidic approach to water-rock interactions using thin rock sections: Pb and U sorption onto thin shale and granite sections. J. Hazard. Mater. **324**, 373–381 (2017). https://doi.org/10.1016/j.jhazmat.2016.10.071
18. Sbarato, V.M., Falchini, G.E., Sánchez, H.J., Perez, R.D.: Monitoring of metal phytofiltration performance by micro-XRF methodology. Anal. Methods **13**(24), 2701–2709 (2021). https://doi.org/10.1039/d1ay00360g
19. Miller, T.C., Hastings, E.P., Havrilla, G.J.: Automated printing technology as a new tool for liquid sample preparation for micro x-ray fluorescence (MXRF). X-Ray Spectrom. **35**(2), 131–136 (2006). https://doi.org/10.1002/xrs.882
20. Mcintosh, K.G., Neal, J.A., Nath, P., Havrilla, G.J.: Microfluidic sample preparation for elemental analysis in liquid samples using micro X-ray fluorescence spectrometry. X-Ray Spectrom. **43**(6), 332–337 (2014). https://doi.org/10.1002/xrs.2559
21. Vanhoof, C., Bacon, J.R., Fittschen, U.E.A., Vincze, L.: Atomic spectrometry update—a review of advances in X-ray fluorescence spectrometry and its special applications. J. Anal. At. Spectrom. **38**(9), 1730–1743 (2023). https://doi.org/10.1039/d3ja90026f
22. Nakano, K., Tsuji, K.: Development of laboratory confocal 3D-XRF spectrometer and non-destructive depth profiling. J. Anal. At. Spectrom. **25**(4), 562–569 (2010). https://doi.org/10.1039/b916974a

23. Wang, Z., Zhao, Y., Kong, D.: Application of 3D fluorescence spectroscopy and a convolutional neural network for oil emulsion species identification. Meas.: J. Int. Meas. Confed. **237**(October 2023) (2024). https://doi.org/10.1016/j.measurement.2024.115177
24. Seim, C., Laurenze-Landsberg, C., Schröder-Smeibidl, B., Mantouvalou, I., De Boer, C., Kanngießer, B.: Old traces, read anew-'The Reading Hermit' painting in the light of X-ray fluorescence. J. Anal. At. Spectrom. **29**(8), 1354–1360 (2014). https://doi.org/10.1039/c3ja50301a
25. Kanngießer, B., Malzer, W., Rodriguez, A.F., Reiche, I.: Three-dimensional micro-XRF investigations of paint layers with a tabletop setup. Spectrochim. Acta-Part B At.Ic Spectrosc. **60**(1), 41–47 (2005). https://doi.org/10.1016/j.sab.2004.10.012
26. Musílek, L., Novotný, P., Trojek, T.: Possibilities of various arrangements of X-ray fluorescence analysis in the study of paintings. Radiat. Phys. Chem. **220**(December 2023) (2024). https://doi.org/10.1016/j.radphyschem.2024.111731
27. Kanngießer, B., Malzer, W., Reiche, I.: A new 3D micro X-ray fluorescence analysis set-up–first archaeometric applications. Nucl. Instrum. Methods Phys. Res., Sect. B **211**(2), 259–264 (2003). https://doi.org/10.1016/S0168-583X(03)01321-1
28. Kalligosfyri, P.M., Cinti, S.: 3D paper-based origami device for programmable multifold analyte preconcentration. Anal. Chem. **96**(24), 9773–9779 (2024). https://doi.org/10.1021/acs.analchem.4c02032
29. Lhotská, I., Kholová, A., Švec, F., Šatínský, D.: Nanofibers prepared from synthetic polymers and biopolymers as advanced extraction materials for sample preparation prior to liquid chromatography. TrAC-Trends Anal. Chem. **180**(July) (2024). https://doi.org/10.1016/j.trac.2024.117912
30. Esteve-Turrillas, F.A., Garrigues, S., de la Guardia, M.: Green extraction techniques in green analytical chemistry: a 2019–2023 up-date. TrAC-Trends Anal. Chem. **170**(September 2023) (2024). https://doi.org/10.1016/j.trac.2023.117464
31. Joshi, G.C., Agrawal, H.M., Mohanta, B., Sudarshan, M., Sinha, A.K.: Elemental study of nainital lake water by EDXRF. Nucl. Instrum. Methods Phys. Res., Sect. B **251**(1), 223–226 (2006). https://doi.org/10.1016/j.nimb.2006.06.001
32. Baranik, A., Sitko, R., Gagor, A., Zawisza, B.: Alumina/nano-graphite composite as a new nanosorbent for the selective adsorption, preconcentration, and determination of chromium in water samples by EDXRF. Anal. Bioanal. Chem. **410**(29), 7793–7802 (2018). https://doi.org/10.1007/s00216-018-1397-8
33. Pytlakowska, K., Kocot, K., Hachuła, B., Pilch, M., Wrzalik, R., Zubko, M.: Determination of heavy metal ions by energy dispersive X-ray fluorescence spectrometry using reduced graphene oxide decorated with molybdenum disulfide as solid adsorbent. Spectrochim. Acta-Part B At.Ic Spectrosc. **167**(March), 105846 (2020). https://doi.org/10.1016/j.sab.2020.105846
34. He, Q., Li, X., Chai, W., Chen, L., Mao, X.: A novel functionalized graphdiyne oxide membrane for efficient removal and rapid detection of mercury in water. J. Hazard. Mater. **467**(February), 133711 (2024). https://doi.org/10.1016/j.jhazmat.2024.133711
35. Paquin, F., Rivnay, J., Salleo, A., Stingelin, N., Silva, C.: Multi-phase semicrystalline microstructures drive exciton dissociation in neat plastic semiconductors. J. Mater. Chem. C **3**(207890), 10715–10722 (2015). https://doi.org/10.1039/b000000x
36. Pytlakowska, K.: Speciation of inorganic chromium in water samples by energy dispersive X-ray fluorescence spectrometry. J. Anal. At. Spectrom. **31**(4), 968–974 (2016). https://doi.org/10.1039/c5ja00495k
37. Menéndez-Alonso, E., Hill, S.J., Foulkes, M.E., Crighton, J.S.: Speciation and preconcentration of CrIII and CrVI in waters by retention on ion exchange media and determination by EDXRF. J. Anal. At. Spectrom. **14**(2), 187–192 (1999). https://doi.org/10.1039/A806002I
38. Musielak, M., Serda, M., Gagor, A., Talik, E., Sitko, R.: Ultratrace determination and speciation of hexavalent chromium by EDXRF and TXRF using dispersive micro-solid phase extraction and tetraethylenepentamine graphene oxide. Spectrochim. Acta-Part B At.Ic Spectrosc. **213**(January), 106863 (2024). https://doi.org/10.1016/j.sab.2024.106863

39. Kocot, K., Leardi, R., Walczak, B., Sitko, R.: Determination and speciation of trace and ultra-trace selenium ions by energy-dispersive X-ray fluorescence spectrometry using graphene as solid adsorbent in dispersive micro-solid phase extraction. Talanta **134**, 360–365 (2015). https://doi.org/10.1016/j.talanta.2014.11.036
40. Kaplan, D.I., Hunter, D.B., Bertsch, P.M., Bajt, S., Adriano, D.C.: Research communications: application of synchrotron X-ray fluorescence spectroscopy and energy dispersive X-ray analysis to identify contaminant metals on groundwater colloids. Environ. Sci. Technol. **28**(6), 1186–1189 (1994). https://doi.org/10.1021/es00055a033
41. Bhalla, A., Singh, G., Kumar, S., Shahi, J.S., Mehta, D.: Elemental analysis of groundwater from different regions of Punjab State (India) using EDXRF technique and the sources of water contamination. Proc. Int. Conf. Environ. Comput. Sci. **19**, 156–164 (2011)
42. Singh, et al.: Elemental Analysis of Satluj River Water Using EDXRF Nature and Science, Elemental Analysis Elemental Analysis of Satluj River Water Using EDXRF 2014. October (2010)
43. Ravisankar, R., Sivakumar, S., Chandrasekaran, A., Kanagasabapathy, K.V., Prasad, M.V.R., Satapathy, K.K.: Statistical assessment of heavy metal pollution in sediments of east coast of Tamilnadu using Energy Dispersive X-ray Fluorescence Spectroscopy (EDXRF). Appl. Radiat. Isot. **102**, 42–47 (2015). https://doi.org/10.1016/j.apradiso.2015.03.018
44. Shan, Q., Liu, Y., Zhang, X., Shao, J., Hei, D., Ling, Y., Jia, W.: EDXRF analysis of TiO2 nanoparticles bioaccumulation in aquatic plant, Salvinia natans. Microchem. J. **155**(November 2019), 104784 (2020). https://doi.org/10.1016/j.microc.2020.104784
45. Pytlakowska, K., Malicka, E., Talik, E., Gągor, A.: Nano-bismuth sulfide based dispersive micro-solid phase extraction combined with energy dispersive X-ray fluorescence spectrometry for determination of mercury ions in waters. J. Anal. At. Spectrom. **36**(4), 786–795 (2021). https://doi.org/10.1039/d0ja00477d
46. Orescanin, V., Mikelic, L., Lulic, S., Rubcic, M.: Determination of Cr(III) and Cr(VI) in industrial and environmental liquid samples by EDXRF method. Anal. Chim. Acta **527**(2), 125–129 (2004). https://doi.org/10.1016/j.aca.2004.09.027
47. Pytlakowska, K., Kocot, K., Pilch, M., Zubko, M. (2020). Ultrasound-assisted dispersive micro-solid phase extraction using molybdenum disulfide supported on reduced graphene oxide for energy dispersive X-ray fluorescence spectrometric determination of chromium species in water. Microchim. Acta **187**(9). https://doi.org/10.1007/s00604-020-04476-4
48. Pytlakowska, K., Pilch, M., Hachuła, B., Nycz, J.E., Kornaus, K., Pisarski, W.A.: Energy dispersive X-ray fluorescence spectrometric determination of copper, zinc, lead and chromium species after preconcentration on graphene oxide chemically modified with mercapto-groups. J. Anal. At. Spectrom. **34**(7), 1416–1425 (2019). https://doi.org/10.1039/c9ja00081j
49. Rathod, T.D., Tiwari, M., Maity, S., Sahu, S.K., Pandit, G.G.: Multi-element detection in sea water using preconcentration procedure and EDXRF technique. Appl. Radiat. Isot. **135**(May 2017), 57–60 (2018). https://doi.org/10.1016/j.apradiso.2018.01.014
50. Li, X., Zhang, X., Qin, J., Zhang, Y., Chen, X., Jia, J., Feng, F., Wu, H., Bai, Y.: Facile and fast elemental analysis using glass fiber filter by X-ray fluorescence. Spectrochim. Acta-Part B At.Ic Spectrosc. **216**(February), 106951 (2024). https://doi.org/10.1016/j.sab.2024.106951
51. Bacon, J.R., Butler, O.T., Cairns, W.R.L., Cavoura, O., Cook, J.M., Davidson, C.M., Mertz-Kraus, R.: Atomic spectrometry update - a review of advances in environmental analysis. J. Anal. At. Spectrom. **37**(1), 10–56 (2022). https://doi.org/10.1039/d2ja90055f
52. Zheng, F., Lin, X., Yu, H., Li, S., Huang, X.: Visible-light photoreduction, adsorption, matrix conversion and membrane separation for ultrasensitive chromium determination in natural water by X-ray fluorescence. Sens. Actuators, B Chem. **226**, 500–505 (2016). https://doi.org/10.1016/j.snb.2015.12.022
53. Aslan, F., Tor, A.: Determination and speciation of trace inorganic arsenic species in water samples by using metal organic framework mixed-matrix membrane and EDXRF spectrometry. Chemosphere **307**(May) (2022). https://doi.org/10.1016/j.chemosphere.2022.135661

54. Kumar, S.A., Pandey, S.P., Thakur, N., Parab, H., Shinde, R.N., Pandey, A.K., Wagh, D.N., Kumar, S.D., Reddy, A.V.R.: Synthesis and application of a unified sorbent for simultaneous preconcentration and determination of trace metal pollutants in natural waters. J. Hazard. Mater. **262**, 265–273 (2013). https://doi.org/10.1016/j.jhazmat.2013.08.053
55. Vo, T.S., Hossain, M.M., Jeong, H.M., Kim, K.: Heavy metal removal applications using adsorptive membranes. Nano Converg. **7**(1) (2020). https://doi.org/10.1186/s40580-020-00245-4
56. Bacon, J.R., Butler, O.T., Cairns, W.R.L., Cavoura, O., Cook, J.M., Davidson, C.M., Mertz-Kraus, R.: Atomic spectrometry update—a review of advances in environmental analysis. J. Anal. At. Spectrom. **39**(1), 11–65 (2023). https://doi.org/10.1039/d3ja90044d
57. West, M., Ellis, A.T., Potts, P.J., Streli, C., Vanhoof, C., Wobrauschek, P.: Atomic spectrometry update-a review of advances in X-ray fluorescence spectrometry and its applications. J. Anal. At. Spectrom. **31**(9), 1706–1755 (2016). https://doi.org/10.1039/c6ja90034h
58. Clough, R., Harrington, C.F., Hill, S.J., Madrid, Y., Tyson, J.F.: Atomic spectrometry update: review of advances in elemental speciation. J. Anal. At. Spectrom. **35**(7), 1236–1278 (2020). https://doi.org/10.1039/d0ja90026e
59. Moradi, M., Yamini, Y., Kakehmam, J., Ahmadi, K.: A review in the sample preparation of aqueous solutions combined with X-ray fluorescence detection. J. Iran. Chem. Soc. **12**(5), 831–838 (2015). https://doi.org/10.1007/s13738-014-0545-0
60. Abe, W., Isaka, S., Koike, Y., Nakano, K., Nakamura, T.: X-ray fluorescence analysis of trace metals in environmental water using preconcentration with an iminodiacetate extraction disk. X-Ray Spectrom. **35**(3), 184–189 (2006). https://doi.org/10.1002/xrs.892
61. Vanhoof, C., Bacon, J.R., Fittschen, U.E.A., Vincze, L.: Atomic spectrometry update: review of advances in X-ray fluorescence spectrometry and its special applications. In J. Anal. At.Ic Spectrom. **37**(9), 1761–1775. Royal Society of Chemistry (2022). https://doi.org/10.1039/d2ja90035a
62. Evans, E.H., Pisonero, J., Smith, C.M.M., Taylor, R.N.: Atomic spectrometry update: Review of advances in atomic spectrometry and related techniques. J. Anal. At.Ic Spectrom. **35**(5), 830–851 (2020). Royal Society of Chemistry. https://doi.org/10.1039/d0ja90015j
63. Hagiwara, K., Inui, T., Koike, Y., Aizawa, M., Nakamura, T.: Speciation of inorganic arsenic in drinking water by wavelength-dispersive X-ray fluorescence spectrometry after in situ preconcentration with miniature solid-phase extraction disks. Talanta **134**, 739–744 (2015). https://doi.org/10.1016/j.talanta.2014.12.027
64. An, J., Kim, K.H., Kim, J.A., Jung, H., Yoon, H.O., Seo, J.: A simplified analysis of dimethylarsinic acid by wavelength dispersive X-ray fluorescence spectrometry combined with a strong cation exchange disk. J. Hazard. Mater. **260**, 24–31 (2013). https://doi.org/10.1016/j.jhazmat.2013.04.033
65. Inui, T., Abe, W., Kitano, M., Nakamura, T.: Determination of Cr(III) and Cr(VI) in water by wavelength-dispersive X-ray fluorescence spectrometry after preconcentration with an ion-exchange resin disk. X-Ray Spectrom. **40**(4), 301–305 (2011). https://doi.org/10.1002/xrs.1317
66. An, J., Jung, H., Bae, J.R., Yoon, H.O., Seo, J.: Feasibility of wavelength dispersive X-ray fluorescence spectrometry for a simplified analysis of bromine in water samples with the aid of a strong anion exchange disk. Spectrochim. Acta-Part B At.Ic Spectrosc. **91**, 1–4 (2014). https://doi.org/10.1016/j.sab.2013.10.005
67. Aranda, P.R., Colombo, L., Perino, E., de Vito, I.E., Raba, J.: Solid-phase preconcentration and determination of mercury(II) using activated carbon in drinking water by X-ray fluorescence spectrometry. X-Ray Spectrom. **42**(2), 100–104 (2013). https://doi.org/10.1002/xrs.2440
68. Marguí, E., Kregsamer, P., Hidalgo, M., Tapias, J., Queralt, I., Streli, C.: Analytical approaches for Hg determination in wastewater samples by means of total reflection X-ray fluorescence spectrometry. Talanta **82**(2), 821–827 (2010). https://doi.org/10.1016/j.talanta.2010.05.066
69. Zawisza, B., Skorek, R., Stankiewicz, G., Sitko, R.: Carbon nanotubes as a solid sorbent for the preconcentration of Cr, Mn, Fe Co, Ni, Cu, Zn and Pb prior to wavelength-dispersive X-ray

fluorescence spectrometry. Talanta **99**, 918–923 (2012). https://doi.org/10.1016/j.talanta.2012.07.059

70. de Carvalho, M.S., De, M., Domingues, L.F., Mantovano, J.L., da Cunha, J.W.S.D.: Preconcentration method for the determination of thorium in natural water by wavelength dispersive X-ray fluorescence spectrometry. J. Radioanal. Nucl. Chem. **253**(2) (2002)
71. Yamini, Y., Amiri, N., Karimi, M.: Determination of trace elements in natural water using X-ray fluorescence spectrometry after preconcentration with powdered silica gel. X-Ray Spectrom. **38**(6), 474–478 (2009). https://doi.org/10.1002/xrs.1201
72. Jeong, S., Kim, Y.T., Yoon, H.O.: Droplet method for the determination of sodium and potassium concentrations in highly saline water samples using X-ray fluorescence spectrometry. Spectrochim. Acta-Part B At.Ic Spectrosc. **153**, 28–33 (2019). https://doi.org/10.1016/j.sab.2019.01.007
73. Zawisza, B., Sitko, R.: Determination of trace elements in suspensions and filtrates of drinking and surface water by wavelength-dispersive X-ray fluorescence spectrometry. Anal. Bioanal. Chem. **384**(7–8), 1600–1604 (2006). https://doi.org/10.1007/s00216-006-0328-2
74. Shaltout, A.A., Welz, B., Ibrahim, M.A.: Influence of the grain size on the quality of standardless WDXRF analysis of river Nile sediments. Microchem. J. **99**(2), 356–363 (2011). https://doi.org/10.1016/j.microc.2011.06.006
75. Gonzalez-Fernandez, O., Margui, E., Queralt, I.: Multielemental analysis of dried residue from metal-bearing waters by wavelength dispersive X-ray fluorescence spectrometry. Spectrochim. Acta-Part B At.Ic Spectrosc. **64**(2), 184–190 (2009). https://doi.org/10.1016/j.sab.2008.11.011
76. Lee, J., An, J., Kim, J.A., Yoon, H.O.: Effectiveness of activated carbon disk for the analysis of iodine in water samples using wavelength dispersive X-ray fluorescence spectrometry. Chemosphere **142**, 72–76 (2016). https://doi.org/10.1016/j.chemosphere.2015.06.017
77. Pashkova, G.v., Revenko, A.G.: A review of application of total reflection X-Ray fluorescence spectrometry to water analysis. Appl. Spectrosc. Rev. **50**(6), 443–472 (2015). Bellwether Publishing, Ltd. https://doi.org/10.1080/05704928.2015.1010205
78. Musielak, M., Serda, M., Sitko, R.: Ultrasensitive and selective determination of mercury in water, beverages and food samples by EDXRF and TXRF using graphene oxide modified with thiosemicarbazide. Food Chem. **390** (2022). https://doi.org/10.1016/j.foodchem.2022.133136
79. Garg, V.K., Tanta, A., Lal Srivastav, A., Tiwari, M.K., Sharma, A., Kanwar, V.S.: Water quality assessment using synchrotron-based TXRF. Water Environ. Res. **94**(7) (2022). https://doi.org/10.1002/wer.10759
80. Borgese, L., Bilo, F., Tsuji, K., Fernández-Ruiz, R., Margui, E., Streli, C., Pepponi, G., Stosnach, H., Yamada, T., Vandenabeele, P., Maina, D.M., Gatari, M., Shepherd, K.D., Towett, E.K., Bennun, L., Custo, G., Vasquez, C., Depero, L.E.: First Total Reflection X-Ray Fluorescence round-robin test of water samples: preliminary results. Spectrochim. Acta-Part B At.Ic Spectrosc. **101**, 6–14 (2014). https://doi.org/10.1016/j.sab.2014.06.024
81. Borgese, L., Dalipi, R., Riboldi, A., Bilo, F., Zacco, A., Federici, S., Bettinelli, M., Bontempi, E., Depero, L.E.: Comprehensive approach to the validation of the standard method for total reflection X-ray fluorescence analysis of water. Talanta **181**, 165–171 (2018). https://doi.org/10.1016/j.talanta.2017.12.087
82. Barreiros, M.A., Carvalho, M.L., Costa, M.M., Marques, M.I., Ramos, M.T.: Application of total reflection XRF to elemental studies of drinking water. X-Ray Spectrom. **26**(4), 165–168 (1997). https://doi.org/10.1002/(SICI)1097-4539(199707)26:4%3c165::AID-XRS197%3e3.0.CO;2-P
83. Sanyal, K., Chappa, S., Bahadur, J., Pandey, A.K., Mishra, N.L.: Arsenic quantification and speciation at trace levels in natural water samples by total reflection X-ray fluorescence after pre-concentration with: N-methyl-d-glucamine functionalized quartz supports. J. Anal. At. Spectrom. **35**(11), 2770–2778 (2020). https://doi.org/10.1039/d0ja00385a
84. Bahadir, Z., Bulut, V.N., Hidalgo, M., Soylak, M., Marguí, E.: Cr speciation in water samples by dispersive liquid-liquid microextraction combined with total reflection X-ray fluorescence spectrometry. Spectrochim. Acta-Part B At.Ic Spectrosc. **115**, 46–51 (2016). https://doi.org/10.1016/j.sab.2015.11.001

85. Floor, G.H., Marguí, E., Hidalgo, M., Queralt, I., Kregsamer, P., Streli, C., Román-Ross, G.: Study of selenium sorption processes in volcanic ash using total reflection X-ray fluorescence (TXRF). Chem. Geol. **352**, 19–26 (2013). https://doi.org/10.1016/j.chemgeo.2013.05.034
86. Shukla, V.K., Dhara, S., Mishra, N.L.: Total reflection X-ray fluorescence spectrometric determination of ultra-trace uranium in natural water samples using a dispersive liquid-liquid microextraction method. J. Anal. At. Spectrom. **35**(8), 1632–1640 (2020). https://doi.org/10.1039/d0ja00039f
87. Oskolok, K.v., Monogarova, O.v., Alov, N.v.: Total reflection X-ray fluorescence determination of rare earth elements in mineral water. Mosc. Univ. Chem. Bull., **72**(1), 10–14 (2017). https://doi.org/10.3103/S0027131417010102
88. Majumder, S., Marguí, E., Roman-Ross, G., Chatterjee, D., Hidalgo, M.: Hollow fiber liquid phase microextraction combined with total reflection X-ray fluorescence spectrometry for the determination of trace level inorganic arsenic species in waters. Talanta **217** (2020). https://doi.org/10.1016/j.talanta.2020.121005
89. Marguí, E., Sagué, M., Queralt, I., Hidalgo, M.: Liquid phase microextraction strategies combined with total reflection X-ray spectrometry for the determination of low amounts of inorganic antimony species in waters. Anal. Chim. Acta **786**, 8–15 (2013). https://doi.org/10.1016/j.aca.2013.05.006
90. Musielak, M., Kocot, K., Zawisza, B., Talik, E., Margui, E., Queralt, I., Walczak, B., Sitko, R.: Ultratrace determination of metal ions using graphene oxide/carbon nanotubes loaded cellulose membranes and total-reflection X-ray fluorescence spectrometry: a green chemistry approach. Spectrochim. Acta-Part B At.Ic Spectrosc. **177** (2021). https://doi.org/10.1016/j.sab.2021.106069
91. Kocot, K., Zawisza, B., Marguí, E., Queralt, I., Hidalgo, M., Sitko, R.: Dispersive micro solid-phase extraction using multiwalled carbon nanotubes combined with portable total-reflection X-ray fluorescence spectrometry for the determination of trace amounts of Pb and Cd in water samples. J. Anal. At. Spectrom. **28**(5), 736–742 (2013). https://doi.org/10.1039/c3ja50047k
92. Bahadir, Z., Bulut, V.N., Hidalgo, M., Soylak, M., Marguí, E.: Determination of trace amounts of hexavalent chromium in drinking waters by dispersive microsolid-phase extraction using modified multiwalled carbon nanotubes combined with total reflection X-ray fluorescence spectrometry. Spectrochim. Acta-Part B At.Ic Spectrosc. **107**, 170–177 (2015). https://doi.org/10.1016/j.sab.2015.03.010
93. Romero, V., Costas-Mora, I., Lavilla, I., Bendicho, C.: Headspace thin-film microextraction onto graphene membranes for specific detection of methyl(cyclopentadienyl)-tricarbonyl manganese in water samples by total reflection X-ray fluorescence. Spectrochim. Acta-Part B At.Ic Spectrosc. **126**, 65–70 (2016). https://doi.org/10.1016/j.sab.2016.10.011
94. de la Calle, I., Romero, V., Lavilla, I., Bendicho, C.: Nanoparticle-assisted stabilization of metal species as an alternative to conventional approaches for avoiding volatilization errors in total reflection X-ray fluorescence: a review. Spectrochim. Acta-Part B At.Ic Spectrosc. **168** (2020). Elsevier B.V. https://doi.org/10.1016/j.sab.2020.105843
95. Okoro, H.K., Ayika, S.O., Ngila, J.C., Tella, A.C.: Rising profile on the use of metal–organic frameworks (MOFs) for the removal of heavy metals from the environment: an overview. Appl. Water Sci. **8**(6). Springer Verlag (2018). https://doi.org/10.1007/s13201-018-0818-3
96. Adam, A.M., Saad, H.A., Atta, A.A., Alsawat, M., Hegab, M.S., Altalhi, T.A., Refat, M.S.: An environmentally friendly method for removing Hg(II), Pb(II), Cd(II) and Sn(II) heavy metals from wastewater using novel metal–carbon-based composites. Crystals **11**(8) (2021). https://doi.org/10.3390/cryst11080882
97. Barkács, K., Varga, A., Gál-Solymosc, K., Záray, G.: Direct determination of metal concentrations in freshwater algae by total reflection X-ray fluorescence spectrometry † (1999)

X-Ray Fluorescence Analysis of Atmospheric Particulate Matters

16

Volkan Aylikci and Nuray Kup Aylikci

Abstract

The energy demand has increased with the growth of the world's population and the development of technology. Since most of the world's energy supplies from fossil fuels, the emission of greenhouse gases has not only forced the world to deal with the global climate crisis but has also led scientists to research how to improve the indoor and outdoor air quality of the habitat. Particulate matters are known as serious damaging pollutants affecting cardiovascular and neurologic systems. For the detection of atmospheric particulate matters that threaten the human health, elemental analysis technique will be suitable for the wide range elemental detection. For sensitive detection of hazardous chemicals in atmospheric particulate matters, X-ray fluorescence technique supplies direct analysis. Other detection methods such as inductively coupled plasma and atomic absorption spectroscopy are not practical in sample preparation, and these types of analysis methods are time-consuming and expensive. The detection of atmospheric particulate matters will be possible using suitable commercial membrane filters. This chapter provides to inform the reader about the recent studies and applications about the detection of atmospheric chemicals by X-ray fluorescence and will give information about possible future studies.

V. Aylikci
Department of Metallurgical and Material Science Engineering, Iskenderun Technical University–ISTE, Hatay, Turkey

N. K. Aylikci (✉)
Department of Engineering Sciences, Iskenderun Technical University–ISTE, Hatay, Turkey
e-mail: nuray.aylikci@iste.edu.tr

V. K. Singh (ed.), *X-Ray Fluorescence Spectroscopy and Chemometrics*,
https://doi.org/10.1007/978-3-031-98375-7_16

16.1 Introduction

Pioneering developments in industrial areas and the increase in population density around the world are threatening the resources of air, water and vegetation, and this situation is also leading to adverse effects on human health, in particular the respiratory organs. Especially for children, the exposure to different types of air pollutants results in reduced lung function and growth, allergic rhinitis and asthma. The term "particulate matter" is generally defined as a mixture of solid particles and liquid droplets that are suspended in the air. These substances can be emitted directly or to be formed later by reacting with chemicals in the atmosphere. Types of air particulates are categorized according to the particle size as summarized in Fig. 16.1.

Despite the fact that such particles are classified according to their particle size, their diameter are generally not spherical. The size of the particulate matter can be calculated from the equivalent diameter in accordance with the equation below;

$$R_a = R_o K \left(\frac{\rho_P}{\rho_0} \right)^{1/2} \tag{16.1}$$

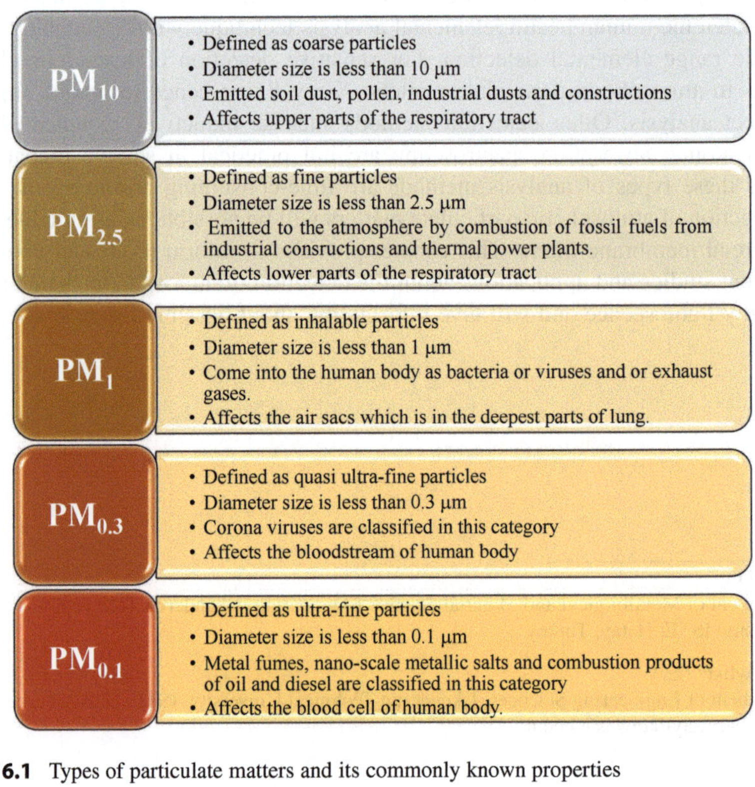

Fig. 16.1 Types of particulate matters and its commonly known properties

Here, R_0 is the geometric and R_a is the aerodynamic diameter. The density of particle is designated as ρ_P, , shape factor is K (equal to 1 for perfect sphere) and finally reference density is ρ_0.

The particulate matters can be emitted by natural and anthropogenic way. The particulates from volcanic activities, desert dust, post-earthquake rubble, pollen, fungal spores and sea spray are classified in naturally emitted particulate matter. The anthropogenic activities such as forest and grassland burnings, burning of fossil fuels release harmful gases contributing to climate change. Air pollution released into the atmosphere as a result of anthropogenic forest fires is shown in Fig. 16.2. In particular, as a result of the earthquakes that occurred on 6 February with the epicenter in Kahramanmaraş, Turkey, and affected 11 provinces, the contents of building rubble represented a threat to human health. Natural disasters, rather than man-made factors, can also lead air pollution.

Air pollution has detrimental effects on human-being cardiovascular system. Considering that the thickness of a hair fiber is in the range of 50–70 μm, it is certain that even the coarsest particulate matters will have adverse effects on human health. In urban areas, suspended particulate matters are emerged as the most serious environmental problems and their residence time in air can take a long time. The adverse effects of particulate matter on climate change, the environment and human health are motivating scientists to reduce chemical/biological hazards in the air. However, some researchers have improved fly ash based (collected from power plant) geo-polymer concretes [1–3] as structural materials for waste recycling but the air filtration system is the best choice for the purification of indoor/outdoor air increasing the inhalable air quality.

Fig. 16.2 An anthropogenic forest fires in Iskenderun/HATAY province

Monitoring air pollution levels is not possible with air filtration system alone. Detailed analytical methods are needed to understand how the air is polluted. In the scope of this chapter, a general description about the air filters and new studies about its production will be presented. Afterwards, the qualitative and quantitative analysis of particulate matters via X-ray fluorescence will be explained in detail giving information about sample preparation procedure, strategies and calibration process.

16.2 A General Insights about the Air Filter Production Studies

Technological breakthroughs and industrial revolutions have improved people's standard of living, but at the same time the quality of the air they breathe has deteriorated. The polluted air contains harmful gases, suspended chemical particles, spores, bacteria and viruses. All of these components are grouped together with a term defined particulate matters. According to the world health organization (WHO) data, annually 4–9 million fatalities were resulted from the airborne particulate matters [4, 5]. To overcome with particulate matter pollution, high-efficiency air filters are more favorable candidate. The capture mechanism can be explained by four mechanisms defined impaction, interception, diffusion and electrostatic interaction [5]. Coarse particles may be easy to adsorb by traditional filters but, the filter materials for fine and ultra-fine particles should manage the problems such as lower efficiency, high pressure drop and worse mechanical strength. On the other hand, several issues such as stabilities and physicochemical properties remain to be resolved or improved [6].

High efficiency particulate air filters are generally known as HEPA including the petro-chemical components. Since such filters are not environmentally friendly, scientists have focused on developing a new generation of stable and environmentally friendly filters with high efficiency and heat stability. The fibrous membrane type air filters were proposed for efficient particulate matter capture owing to their rich porous structure, finer diameters and better ability to incorporate active chemicals [7, 8]. However, these types of materials may be a good candidate for designing metal oxide, metallic nano particle, carbon, organic or MOF based electro-spun nanofiber air filters to remove particles with different aerodynamic radii. For instance, the ionic form of metallic species can easily interact with bacterial cell walls with electrostatic interaction to damage them. In addition, their finer fiber diameter can supply a protective surface to avoid from the fine and ultra-fine particulates.

The filtration process is generally designated with the characteristic of high filtration performance, low airflow resistance and quality factor [8]. The quality factor is calculated by the following equation written as [8];

$$\mathrm{QF} = -\mathrm{In}(1 - \eta)/\Delta P \tag{16.2}$$

In which the filtration efficiency is η, and the pressure drop is ΔP. The percentage of particulate matter removal is defined as filtration efficiency.

A bio-degradable poly-lactic acid based filtration materials was proposed as having high filtration efficiency, but most of them are known with their imbalanced characteristic between filtration efficiency and resistance. To prevail with imbalanced characteristic, double-layer fibers with bead like fiber structure were designed for increasing the efficient air particulate adsorption and the roughness of the material surfaces were attributed to the high filtration efficiency [9]. The other bio-degradable material is nanocellulose fibers which have hydrophilic character depending on their lignin content [10]. Due to the petroleum based content of polyvinylpyrrolidone (PVP), and polyacrylonitrile (PAN), a biodegradable nanofibrous membrane filter material was also developed using ethyl cellulose. Using the differences in viscosity forces at microscale, a bimodal structure was designed for these air filter material by electrospinning method. The filtration efficiency, pressure drop and quality factor were explained as 99.1%, 42.2 Pa and 0.112 Pa^{-1} respectively [11]. For better filtration performance, an antibacterial and bio-based zein@chitosanhydrochloride@phloretin fibrous filter material were synthesized. Phloretin has been used for the improvement of the electrospinning capability for green solution by means of the pulling mechanism of rigid polymer chains. A yield of 99.65%, a pressure drop of 57.7% and a quality factor of 0.098% was achieved [12]. In general chitosan has not a soluble property in water or ethanol solution but its carboxylated form is suitable for synthesizing air filter material by electrospinning.

In the last two decades, global climate change and the increase in greenhouse gas emissions have become a major concern for humanity. This is why the scientist have been focusing on the studies in which the energy consumption is low, green methodologies are used and waste disposal is a priority. It is well known that PVC (polyvinyl chloride) has been in widespread usage, particularly as a construction material. However, over time, the disposal of such materials as waste in the environment threatens many vital natural resources worldwide. Recycling of such wastes is possible by designing a new type of air filter. Zulfi et al. [13], utilized polymer waste materials for fabricating nanofiber membrane air filters with different concentrations of PVC between 10 and 30%. Air filter materials in particular are adversely affected by atmospheric humidity, and PVC material can have a successful filtration rate due to its hydrophobic nature. Indeed, the results of the study from Zulfi et al. [13], in which filtration efficiency of 94.35% for $PM_{2.5}$ and 94.60% for $PM_{0.3}$ were obtained, confirmed this prediction. High efficiency particulate air filters, known as HEPA, are defined as an effective particulate matter remover but in general, HEPA filtration systems are not sensitive to variation in particle contents resulting in high energy consumption. To reduce the energy consumption, an air filtration system for $PM_{2.5}$ was designed using thermoplastic polyurethane (TPU) filters and Internet of Things (IoT) [14]. A new type air filter material was developed by Nam and Park et al. [15] called super-absorbent polymer which has water-resistant feature. This type of air filter material was produced as super absorbent polymer beads as a wet air purifier. Especially for ammonia gas removal, an exceeded 80% filtration efficiency was reported and suggested for indoor and industrial usage due to its washable property. In addition, a different recyclable particulate air filter material is available by using aligned (uniform morphology) carbon nanotubes [16]. The fact that the increase in efficiency is in the same direction as the increase in ΔP is a common problem with air filter materials. The uniform structure of the air filter material based on carbon nanotubes sheets showed the best performance, with a higher filtration efficiency and a lower pressure drop being achieved. Different criteria for high performance filtration is the creating multi-layer morphology and ZIF-8 is the best choice for the given circumstances. A ZIF-8 functionalized wood composite material was reported with two 89.8% and 91.6%, removal efficiencies for $PM_{2.5}$ and PM_{10} respectively [17]. Cellulose is generally known as a raw material of wood and nano-cellulose fiber based air filters are capable of harmful gases and particulate matters due to the high porosity and low pressure drop [18]. A long life air filter was tried to study by Ye et al. [19]. The composites of polyacrylontrile and chitosan biguanide hydrochloride material were synthesized by coaxial electrospinning method. This type of filtering material was highlighted as a good absorber of both some harmful bacteria and also particulate matter due to the polarity dominated forces.

Plant based materials may be a good candidate which can compete with the other traditional filter materials. An integrated small-scale systems was proposed by Zhou et al. using plant-based filtration systems [20]. In addition, due to the rapid increase in carbon emission rates, carbon capture technologies are getting more importance. These technology groups are classified into five main different groups identified as absorption, adsorption, membrane systems (fibers and other microporous structures), biological methods and cold separation. In the context of this chapter, biological methods are the most relevant to our topic of removing particulate matter. Especially, algae and micro-algae are the most preferable plants for carbon capture mechanism. Not only the accumulation of carbon dioxide (CO_2), but also the simultaneous production of biodiesel from algae plants was performed by Alami et al. [21]. These technologies are crucial for reducing the carbon emission rates and also for the absorption of particulate matter in air. In addition, algal nanofibers can easily be synthesized by using electrospinning method [22]. Owing to its carbon content, a sustainable algal carbon nanoparticle composites have higher thermal resistance [23, 24] than that of high-performance silicon dioxide [25], but silicon dioxide nanofiber membranes were emphasized as the best removal of $PM_{0.3}$ particulates compared to HEPA filters. Nanofiber membrane materials are generally known with its effective particulate removal efficiency by the process of direct capture, Brownian motion, inertial impact, gravitational settlement and electrostatic interactions [25]. *Taihu blue algae* is an N-rich plant material that is efficient at removing or sequestering CO_2 [26]. In another study, which supported this result, it was emphasized that carbon and its derivatives [27–29] were effective in air filtration in which a visible improvement in CO_2 adsorption were observed according to the N containing functional groups.

Biochar is also a carbon based material capable of the adsorption copper ions [30]. In addition, different type of air purifier materials can be designed for the adsorption of particulate matters with different particulate matters. A water resistance ceramic filters with longer life time [31], modified halloysite nanotube based composites [32], metal organic framework composites [33–37], ferroelectric polyvinylidene fluoride (PVDF) nanofibers [38], ZIP8@graphene based aerogels [39] and finally Zeolitic-imidazolate-framework based porous fibers [40] were studied by different researchers to examine the optimum filtration conditions. Many different studies on air filter materials can be found in a more detailed literature search. The filter materials which have been presented so far have been introduced on the basis of priority criteria such as being bio-based, energy saving, reducing carbon emission and highly efficient. Depending on the existing conditions, different filter materials can be used. To the best of the authors' knowledge and in the light of the literature, a new type of filter material can be developed with using graphitic carbon nitride (g-C_3N_4). Due to the high N content of g-C_3N_4 based composites and the porous structure of carbon derivatives [41], the authors proposed that composites based on carbon and its derivatives can provide higher filtration efficiency and lower pressure drop.

16.3 X-Ray Fluorescence Studies about the Analysis of Particulate Matters

Human health especially in urban sites, cannot be maintained without improving the quality of the air they breathe and removing hazardous chemicals from the environment they live. However, as some of these fine and ultrafine particulate matters are too small, they can cause serious damage to living organisms. Although the studies on air filters in the previous section refer to removing harmful particles from the air, it's important to be informed about the extent of the damage and to draw up serious roadmaps to reduce the harmful effects. Understanding the extent of damage, monitoring technologies can be beneficial with comparing the results obtained from clean and polluted days [42]. In addition, the composition of particulate matters, its origin and seasonal variability [43], the relationship between rainwater elemental composition and aerosol [44], and size distribution of suspended particulate matters [45, 46] can be useful tools in the roadmaps to reduce the air pollution.

It is necessary to know about the filtering mechanism before deciding to study with a filter material in sampling process. The filtration mechanism is generally affected by fiber diameter, particle size, pore size, air flow and the other interaction mechanisms [47, 48]. The adsorbent ability is related with the filtration mechanism. In addition, material morphology, elemental composition and the layer structure are the another important parameters affecting the filtration mechanism. To optimize the filtration mechanism parameters, different air filter materials can be designed by modifying fiber morphologies, using hybrid structures, designing multiple layers and using different charging mechanisms. After the process of filtering mechanism, the adsorbed particulates can be analyzed by XRF systems.

However, the adsorbed particulates on any filter material should be analyzed as a thin film layer samples.

A dichotomous sampler in which the two filter materials were used eliminating the collection losses was used for the air particulate analysis using X-ray tube. The emitted X-rays from the samples were detected by lithium drifted silicon detector. Due to the negative effects on analyzing results of impurities, teflon membrane filters were used together with clean glass fiber filter. The unacceptable impurities such as Ca and Cl were kept with glass fiber filter material [49]. An elemental characterization of air particulates was performed using portable X-ray systems. The plot of calibration curve was achieved with weighted least-square regression from the measurements of several reference materials. The reference materials were prepared by using probable elemental materials in air particulates which is absorbed by polytetrafluoroethylene (PTFE) [50]. The unknown particulates were analyzed with the adsorption process on Teflon filters including PM_{10} and $PM_{2.5}$ particulates. Despite the applicability of portable-XRF systems for the broad-range elemental analysis from Na to Pb, some limitations were reported such as significant background contributions. These contributions were explained as the instrumental designing originating from the scattered tube radiation from the Teflon substrate. As it can be seen from this situation, filter materials are crucial for X-ray fluorescence analysis. Common commercial filter materials are cellulose, fiberglass, graphite, organic and metallic membranes. In addition, different types of samplers can be generally used for the air particulate matters defined as standard and size selective hivol system, dichotomous sampler, multiple stage impactor and cyclones [51]. In some circumstances, using filter may be not applicable due to the disadvantages explained as the hygroscopic properties of cellulose and fiberglass, low mechanical strength of organic membrane, and high cost of metallic membranes. To overcome problems caused by the filter material is available with the removing of particulate matters from the filters in an ultrasonic bath and then dissolved in an acidic solution. This transferring process of air particulates enables the analysis and using the Vaseline applied plexi-glass as an impaction plate, direct determination is possible for the analysis of air particulates with TXRF [52]. However, in acidic solution, some of elemental analysis may not be performed with an accurate manner. Considering the problems caused by both of the filter material and acid solution, a new methodology can be developed as explained in the study Bontempi et al. [53]. They patented a new approach for designing sandwich type filter material to overcome the negative effects of acidic solution.

In general, the sample preparation procedure was performed using digestion technique after the collection of air particulates on filter. Klockow et al. applied three different digestion techniques for TXRF sample preparation using the filtering and impactor collection systems. Open digestion with nitric acid and hydrogen peroxide, high pressure acid digestion, and cool plasma ashing procedures were applied but, cool plasma ashing process was explained as more reliable for the high detection with TXRF [54]. At the beginning of the analysis, 10 ml aliquot solution was prepared for each sample for then pipetting onto quartz glass which surface covered with hydroxyl bonding. Using the infrared light, the solution onto the carrier was dried and elemental analyze was performed using TXRF. Another cold plasma ashing technique was applied with polycarbonate filters due to their easy-to-decompose property and non-contamination by trace elements [55]. Before the process of ashing, air particulates collected polycarbonate filters were attached onto a quartz carrier and the aliquot solution made ready by adding Ga for internal standard and Rb for quality control. TXRF capability for the determination of fine and ultra-fine aerosol particles was performed by using May-type cascade impactor sampling. This combined new method was explained as short sampling times for major and trace elements which had different aerodynamic sizes with a high precision and wide range size distribution span [56]. The differences of detection limits between various studies may be attributed to the different solid angles of studied detectors and the mass collected for coarse particulates.

There are certain problems for each analytic determination methods that scientists use for environmental analysis. It is generally known that ICP analysis has a destructive effect on analyzed sample and there is a one chance for determining the elemental composition. TXRF method is a non-destructive and sensitive way to perform the analyses for different type of samples, however it also has some negative features. Even though, filters or impactors are part of the air analysis process, stripe-like deposition on a filter using a liquid droplet is not convenient for internal standardization. The external standardization may be proposed but we are confronted with the reproducibility of the measurement geometry [56]. The other problem is that the antimony (Sb) analysis due to the summarized reasons explained in the study of Smichowski [57]. The summarized reasons were the low concentrations, the extraction efficiencies, lack of stability, standards, certified reference materials of Sb and finally, the need for more appropriate matrix matched reference materials. As a result of these drawbacks, Sb analysis using the TXRF method may not give accurate results. A new filtering method was explained for lead determination, called SMART STORE procedure [58] due to the impractical applications of sandwiched filtering with a total thickness of 350 μm for wide range analysis. However, a sandwich filters were designed with two thin polypropylene sheets explained total thickness of 250 μm for directly placing onto the quartz carrier in TXRF instrument [59]. In filtering or impactor process for sampling, the collection of particulate matter may not be fully completed due the particle bounce-off from the filter surface. To overcome with this problem, a sticky thin film surface was proposed by Prost et al. [60, 61].

Air analysis will also possible by using EDXRF systems. Using a suitable filter paper, the adsorbed particulates can immediately be analyzed. The metallic contamination from glass fiber filters will give erroneous quantitative results, however Whatman 41 filter paper [62] and Teflon [63] were proposed for air particulate sampling in EDXRF analysis. Another adverse effect of using filter paper is air humidity. Due to the hydrophobic character of Teflon filter, it may be the best choice for air sampling. A study performed by Takahara et al. [64], the simultaneous sampling and analyzing processes were performed by using a virtual impactor. The device or monitoring system was designed for the separation of coarse (PM10) and fine particles (PM2.5) by making use of their different inertia. In monitoring system, PTFE filter tape and plastic scintillator were also included for hourly analyzes. This type of monitoring may have a unique advantage in the event of a sudden contamination of the air, such as a chemical attack. Unlike the filtering or impactor system, a collection gauge was designed by different scientist [65–67]. Mylar film and grease were coated on a PVC based wind vane to realize dry and wet deposition. However, this type of deposition obstructs the direct analysis in EDXRF systems. Different chemical solvent should be used to take the air particulates into the solution for subsequent drying and forming into pellets or thin films. The European Committee published a general standard method for air particulate sampling and Pb, Cd, As and Ni analysis described as sampling air particulates on quartz filters or cellulose membranes, digestion of them using different methods and analysis by using ICPMS other atomic absorption methods [68]. In general, two types of sampling and analysis procedure are known as online and offline measurement techniques. In online measurements, the collection of air particulates and analysis are performed simultaneously but, in offline measurements, there is a time delay between the sampling and analyzing processes.

Factors to consider in EDXRF analyzing for particulate analysis is the mass loss of samples resulting from the water soluble ionic compounds [69, 70]. In these circumstances, the combination of XRF and ICP analysis will be the best choice for the determination of mass loss origin. Another factor is the arsenic analysis with XRF equipment. The spectrum interference between As K_α line (10.53 keV) and Pb L_α line (10.55 keV) does not enable to accurate analysis of Pb and As together [71].

XRF analysis is generally defined as the practical, economic, non-destructive and so repeatable technique for the elemental analysis of particulate matters. This characteristic of XRF has made it the most preferred method for analyzing particulate matters [72–79] by using different filters or impactors. However, the filtering membranes are an exceptional candidate in sampling procedure in EDXRF systems owing to theirs high capturing efficiency of particulate matters and also, high collection probability of different particle sizes [80]. The sampling procedure is great of importance for better calibration curves and determining minimum detection limits. The generally used air samplers are categorized as explained in Fig. 16.3 [80].

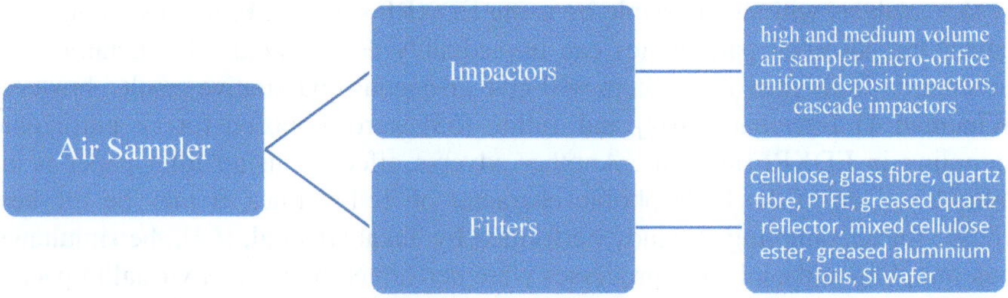

Fig. 16.3 Different types of air samplers

The calibration of the EDXRF systems is very practical. At first stage, the standards should be determined by using generally used chemical elements or compounds. For X-ray tube, K X-ray sensitivity for any element was generally calculated by the following formula;

$$K_j = \frac{A_j}{c_j \cdot I \cdot t} \tag{16.3}$$

In which A_j defines the peak area of element j resulting from the irradiation process, c_j is the area of the standard material, I symbol defines the current of X-ray tube and finally t is the measurement time [76]. After the sensitivity determination process, minimum detection limit can easily be determined in which the best sampling procedure applied.

16.4 Guidelines for Future Works

XRF based quantitative analysis is generally defined as a non-destructive method. However, the element of interest should not be present in the form of ions. For ionic compound analysis, different techniques should be applied. For elemental analysis of air particulates by XRF spectroscopy, the suitability of membranes or filters and different impactors in sample preparation process has been established in literature studies [81–87]. In addition, the surface of the leaf can be nominated as an effective material for the storage of airborne particles [88] in the sample preparation process. Furthermore, the artificial intelligence or deep learning can be a competitive candidate determining the ash content with a combination of XRF systems [89–91]. For both the reduction of micro plastics from the water and the reduction of particulate matter from the air is possible using porous composite membranes with click chemistry, in which a removal efficiency of 99.90% has been achieved [92]. The other material for removing air particulates is metal organic frameworks [93–96]. Owing to their porous structure, high specific surface area, high catalytic activity and adjustable pore sizes make the different range of particulate matters possible.

A new approach is investigated for air particulate mitigation using carbonaceous substances such carbon nanomaterials, graphene and carbon nanotubes [97]. The another approach will be the MOF/COF based nanomaterials for effective absorption air particulates using photocatalytic process. The designed hetero-structures give the material a special characteristic in the mitigation process of hazardous particulates [98]. In addition, nanofiber materials can supply an effective air purification for future prospects [99–102]. An in-depth analysis of the air filter studies has shown that high performance air filters can be generally classified as structure based approach and interaction based approaches [103]. The structure based approaches are related with the material morphologies, however, induction charging, triboelectric charging, corona charging, electrostatic precipitator and triboelectric nano-generator are classified in interaction based category.

As previously stated, the particulate matter sampling and its non-destructive XRF analysis most often performed by using glass fiber [104, 105], Teflon and quartz fiber [106], PTFE filter [107] and paper filters. Particle bounce off the filter surface is the most difficult problem to solve. Grease can be a solution to the problem, but polycarbonate substrates can minimize the negative impact on the particulate collection process [108].

References

1. Amran, M., Debbarma, S., Ozbakkaloglu, T.: Fly ash-based eco-friendly geopolymer concrete: a critical review of the long-term durability properties. In: Construction and Building Materials (Vol. 270) (2021). Elsevier Ltd. https://doi.org/10.1016/j.conbuildmat.2020.121857
2. Pavithra, P., Srinivasula Reddy, M., Dinakar, P., Hanumantha Rao, B., Satpathy, B.K., Mohanty, A.N.: A mix design procedure for geopolymer concrete with fly ash. J. Clean. Prod. **133**, 117–125 (2016). https://doi.org/10.1016/j.jclepro.2016.05.041
3. Meesala, C.R., Verma, N.K., Kumar, S.: Critical review on fly-ash based geopolymer concrete. Struct. Concr. **21**(3), 1013–1028 (2020). https://doi.org/10.1002/suco.201900326
4. Kundu, S., Stone, E.A.: Composition and sources of fine particulate matter across urban and rural sites in the Midwestern United States. Environ. Sci.: Process. Impacts **16**(6), 1360–1370 (2014). https://doi.org/10.1039/c3em00719g
5. Atalie, D., Chen, Z. X., Li, H., Liang, C. G., Gao, M. C., Cheng, X. X., & Ma, P. C. (2024). Eco-friendly and highly efficient PM0.3 air filter made from nonwoven basalt fiber and electrospun nanocellulose fiber. J. Hazard. Mater. **478**. https://doi.org/10.1016/j.jhazmat.2024.135608
6. Ali, S.B., Ghasemi, H., Ahmadi, R., Ghaffari, A.: Recent progresses in dry gas polymeric filters. J. Energy Chem. **62**, 103–119. Elsevier B.V. (2021). https://doi.org/10.1016/j.jechem.2021.03.008
7. Archer, B., Shaumbwa, V.R., Liu, D., Li, M., Iimaa, T., Surenjav, U.: Nanofibrous mats for particulate matter filtration. Ind. Eng. Chem. Res. **60**(20), 7517–7534 (2021). American Chemical Society. https://doi.org/10.1021/acs.iecr.1c00829
8. Wang, G., Xu, Z., Qi, Y., Fang, Y., Ning, G., Ye, J.: Electrospun nanofibrous membranes with antimicrobial activity for air filtration. Chin. Chem. Lett. **35**(10) (2024) Elsevier B.V. https://doi.org/10.1016/j.cclet.2024.109503
9. Ge, J., Lv, X., Zhou, J., Lv, Y., Sun, J., Guo, H., Wang, C., Hu, P., Spitalsky, Z., Liu, Y.: Multilevel structured polylactic acid electrospun fiber membrane based on green solvents for high-performance air filtration. Sep. Purif. Technol. **331** (2024). https://doi.org/10.1016/j.seppur.2023.125659
10. Gulipalli, P., Borle, S., Chivukula, K., Adusumalli, R.B.: Processing of nanocellulose sheet for capturing fine particulate matter. Mater. Today: Proc. **72**, 402–409 (2023). https://doi.org/10.1016/j.matpr.2022.08.189
11. Wang, Q., Shao, Z., Shen, R., Chen, R., Gui, Z., Li, G., Liu, Y., Zheng, G., Sui, J., Qi, Y., Song, W.: Preparation of ethyl cellulose bimodal nanofibrous membrane by green electrospinning based on molecular weight regulation for high-performance air filtration. Int. J. Biol. Macromol. **275** (2024). https://doi.org/10.1016/j.ijbiomac.2024.133411
12. Shen, R., Shao, Z., Chen, R., Wang, Q., Gui, Z., Qi, Y., Song, W., Liu, Y., Zheng, G.: Fully bio-based zein/chitosan hydrochloride/phloretin bimodal fibrous membrane for high-performance and antibacterial air filtration based on green electrospinning. Sep. Purif. Technol. **341** (2024). https://doi.org/10.1016/j.seppur.2024.126893

13. Amalia, R., Noviyanto, A., Rahma, L.A., Merita, Labanni, A., Fahroji, M., Purwajanti, S., Hapidin, D. A., Zulfi, A.: PVC waste-derived nanofiber: Simple fabrication with high potential performance for PM removal in air filtration. Sustain. Mater. Technol. **40** (2024). https://doi.org/10.1016/j.susmat.2024.e00928
14. Niu, Z., He, Q., Chen, C.: A PM2.5 pollution-level adaptive air filtration system based on elastic filters for reducing energy consumption. J. Hazard. Mater. **478** (2024). https://doi.org/10.1016/j.jhazmat.2024.135546
15. Yeo, Y.H., Gawk, M.A., Lee, S.Y., Nam, Y.S., Park, W.H.: Air purifier using super-absorbent polymer for removing air contaminants. J. Environ. Chem. Eng. **10**(3) (2022). https://doi.org/10.1016/j.jece.2022.107832
16. Song, S., Ham, W., Ahn, S.E.: Recyclable aligned carbon nanotube-sheet-based particulate air filter with high filtration efficiency and low pressure drop. Curr. Appl. Phys. **36**, 131–136 (2022). https://doi.org/10.1016/j.cap.2022.01.014
17. Wang, Z., Yin, F., Zhang, X. F., Zheng, T., Yao, J.: Delignified wood filter functionalized with metal-organic frameworks for high-efficiency air filtration. Sep. Purif. Technol. **293** (2022). https://doi.org/10.1016/j.seppur.2022.121095
18. Sepahvand, S., Ashori, A., Jonoobi, M.: Application of cellulose nanofiber as a promising air filter for adsorbing particulate matter and carbon dioxide. Int. J. Biol. Macromol. **244** (2023). Elsevier B.V. https://doi.org/10.1016/j.ijbiomac.2023.125344
19. Wang, G., Xiao, D., Fang, Y., Ning, G., Ye, J.: Polarity-dominated chitosan biguanide hydrochloride-based nanofibrous membrane with antibacterial activity for long-lasting air filtration. Int. J. Biol. Macromol. **254** (2024). https://doi.org/10.1016/j.ijbiomac.2023.127729
20. Wang, C., Zou, T., Zhou, T., Lyu, Y., Dawodu, A., Cheshmehzangi, A.: Advancing urban agriculture and air quality in edge computing environments through integrated small-scale plant-based filtration systems. Resour., Conserv. Recycl. **204** (2024). https://doi.org/10.1016/j.resconrec.2024.107522
21. Alami, A. H., Alasad, S., Ali, M., Alshamsi, M.: Investigating algae for CO2 capture and accumulation and simultaneous production of biomass for biodiesel production. Sci. Total. Environ. **759** (2021). https://doi.org/10.1016/j.scitotenv.2020.143529
22. Karaduman, F.R., Türk Çulha, S., Horzum, N.: Algal nanofibers: Current status and recent developments. Mater. Today Commun. **33** (2022). Elsevier Ltd. https://doi.org/10.1016/j.mtcomm.2022.104248
23. Mali, A., Agbo, P., Mantripragada, S., Jadhav, V.S., Wang, L., Zhang, L.: Sustainable algae-derived carbon particles from hydrothermal liquefaction: an innovative reinforcing agent for epoxy matrix composite. Sustainability (Switzerland) **16**(16) (2024). https://doi.org/10.3390/su16166870
24. Mathijsen, D.: Algae to sustainably produce carbon fiber and simultaneously take CO2 out of the atmosphere. Reinf. Plast. **64**(1), 50–53 (2020). https://doi.org/10.1016/j.repl.2019.04.070
25. Margarita Valencia-Osorio, L., Felipe Zapata-González, A., David Ojeda-Galeano, J., Lopes Aguiar, M., Lucía Álvarez-Láinez, M.: High-performance SiO2 nanofiber membrane applied for high-temperature air filtration. Sep. Purif. Technol. **328** (2024). https://doi.org/10.1016/j.seppur.2023.125030
26. Wang, H., Wang, H., Liu, G., Yan, Q.: In-situ pyrolysis of Taihu blue algae biomass as appealing porous carbon adsorbent for CO2 capture: role of the intrinsic N. Sci. Total. Environ. **771** (2021). https://doi.org/10.1016/j.scitotenv.2021.145424
27. Karanjikar, S.R., Sena, A.S., Manekar, P., Mudagi, S., Juneja, A.S.: Utilization of graphene and its derivatives for air & water filtration: a review. Mater. Today: Proc. **50**, 2007–2017 (2021). https://doi.org/10.1016/j.matpr.2021.09.346
28. M.B., Periyat, P.: Graphene and its derivatives for air purification: a mini review. Results Eng. **21** (2024). https://doi.org/10.1016/j.rineng.2024.101809
29. Khan, M.J., Karim, Z., Pongchaikul, P., Posoknistakul, P., Intra, P., Laosiripojana, N., Wu, K.C.W., Sakdaronnarong, C.: Nitrogen and sulfur doped carbon dots coupled cellulose nanofibers: A surface functionalized nanocellulose membranes for air filtration. J. Taiwan Inst. Chem. Eng. **160** (2024). https://doi.org/10.1016/j.jtice.2023.105324

30. Gao, Y., Zhu, X., Yue, Q., Gao, B.: Facile one-step synthesis of functionalized biochar from sustainable prolifera-green-tide source for enhanced adsorption of copper ions. J. Environ. Sci. (China) **73**, 185–194 (2018). https://doi.org/10.1016/j.jes.2018.02.012
31. Kwon, H.J., Yang, D.S., Koo, M.S., Ji, S.M., Jeong, J., Oh, S., Kuk, S.K., Heo, H. su, Ham, D.J., Kim, M., Choi, H., Lee, J.M., Shur, J.W., Lee, W.J., Bin, C.O., Timofeev, N., Wu, H., Wang, L., Lee, T, et.al.: Long-lifetime water-washable ceramic catalyst filter for air purification. Nat. Commun. **14**(1) (2023). https://doi.org/10.1038/s41467-023-36050-w
32. Ma, J., Jin, X., Lu, Y., Yang, M., Zhao, X., Guo, M., Zhang, H., Li, X., Wang, B.: Preparation of modified halloysite nanotubes@Ag3PO4/polyacrylonitrile electrospinning membranes for highly efficient air filtration and disinfection. Appl. Clay Sci. **253** (2024). https://doi.org/10.1016/j.clay.2024.107361
33. Wang, A., Fan, R., Zhou, X., Hao, S., Zheng, X., Yang, Y.: Hot-pressing method to prepare imidazole-Based Zn(II) metal-organic complexes coatings for highly efficient air filtration. ACS Appl. Mater. Interfaces. **10**(11), 9744–9755 (2018). https://doi.org/10.1021/acsami.8b01287
34. Wang, Z., He, Q., Liu, Y., Yu, C., Zhang, X.F., Kong, X.: Amino functionalized zeolitic imidazolate framework-8 coated cellulose aerogel for enhanced air purification. Sep. Purif. Technol. **355** (2025). https://doi.org/10.1016/j.seppur.2024.129741
35. Ji, X., Huang, J., Teng, L., Li, S., Li, X., Cai, W., Chen, Z., Lai, Y.: Advances in particulate matter filtration: Materials, performance, and application. Green Energy Environ. **8**(3), 673–697 (2023). KeAi Publishing Communications Ltd. https://doi.org/10.1016/j.gee.2022.03.012
36. Woo, H.C., Yoo, D.K., Jhung, S.H.: Particulate matters removal by using cotton coated with isomeric metal-organic frameworks (MOFs): Effect of voidage of MOFs on removal. J. Ind. Eng. Chem. **95**, 277–285 (2021). https://doi.org/10.1016/j.jiec.2020.12.035
37. Zhang, K., Huo, Q., Zhou, Y.Y., Wang, H.H., Li, G.P., Wang, Y.W., Wang, Y.Y.: Textiles/metal-organic frameworks composites as flexible air filters for efficient particulate matter removal. ACS Appl. Mater. Interfaces. **11**(19), 17368–17374 (2019). https://doi.org/10.1021/acsami.9b01734
38. Bui, T.T., Shin, M.K., Jee, S.Y., Long, D.X., Hong, J., Kim, M.G.: Ferroelectric PVDF nanofiber membrane for high-efficiency PM0.3 air filtration with low air flow resistance. Colloids Surf.S A: Phys.Chemical Eng. Asp. **640** (2022). https://doi.org/10.1016/j.colsurfa.2022.128418
39. Mao, J., Tang, Y., Wang, Y., Huang, J., Dong, X., Chen, Z., Lai, Y.: Particulate matter capturing via naturally dried ZIF-8/graphene aerogels under harsh conditions. IScience **16**, 133–144 (2019). https://doi.org/10.1016/j.isci.2019.05.024
40. Hu, M., Yin, L., Low, N., Ji, D., Liu, Y., Yao, J., Zhong, Z., Xing, W.: Zeolitic-imidazolate-framework filled hierarchical porous nanofiber membrane for air cleaning. J. Membr. Sci. **594** (2020). https://doi.org/10.1016/j.memsci.2019.117467
41. Attia, Y.A., Ezet, A.E., Saeed, S., Galmed, A.H.: Nano carbon-modified air purification filters for removal and detection of particulate matters from ambient air. Sci. Rep. **14**(1) (2024). https://doi.org/10.1038/s41598-023-50902-x
42. Malik, A., Aggarwal, S.G., Kunwar, B., Deshmukh, D.K., Shukla, K., Agarwal, R., Singh, K., Soni, D., Sinha, P.R., Ohata, S., Mori, T., Koike, M., Kawamura, K., Kondo, Y.: Physical and chemical properties of PM1 in Delhi: a comparison between clean and polluted days. Sci. Total. Environ. **892** (2023). https://doi.org/10.1016/j.scitotenv.2023.164266
43. Koçak, M., Mihalopoulos, N., Tutsak, E., Theodosi, C., Zarmpas, P., Kalegeri, P.: PM10 and PM2.5 composition over the central black sea: origin and seasonal variability. Environ. Sci. Pollut. Res. **22**(22), 18076–18092 (2015). https://doi.org/10.1007/s11356-015-4928-2
44. Türküm, A., Pekey, H., Pekey, B., Tuncel, G.: Investigating relationships between aerosol and rainwater compositions at different locations in Turkey. Atmos. Res. **89**(4), 315–323 (2008). https://doi.org/10.1016/j.atmosres.2008.03.010
45. Kuloglu, E., Tuncel, G.: Size distribution of trace elements and major ions in the eastern Mediterranean atmosphere. Water Air Soil Pollut. **167**, 221–241 (2005)

46. Herner, J.D., Green, P.G., Kleeman, M.J.: Measuring the trace elemental composition of size-resolved airborne particles. Environ. Sci. Technol. **40**(6), 1925–1933 (2006). https://doi.org/10.1021/es052315q
47. Mamun, A., Blachowicz, T., Sabantina, L.: Electrospun nanofiber mats for filtering applications technology, structure and materials. Polymers **13**(9) (2021). https://doi.org/10.3390/polym13091368
48. Han, S., Kim, J., Ko, S.H.: Advances in air filtration technologies: structure-based and interaction-based approaches. Mater. Today Adv. **9** (2021). Elsevier Ltd. https://doi.org/10.1016/j.mtadv.2021.100134
49. Dzubay, T.G., Stevens, R.K.: Ambient air analysis with dichotomous sampler and X-ray fluorescence spectrometer. Environ. Sci. Technol. **9**(7), 663–668 (1975). https://doi.org/10.1021/es60105a011
50. Chatoutsidou, S.E., Papagiannis, S., Anagnostopoulos, D.F., Eleftheriadis, K., Lazaridis, M., Karydas, A.G.: Application of a handheld X-ray fluorescence analyzer for the quantification of air particulate matter on Teflon filters. Spectrochim. Acta-Part B At.Ic Spectrosc. **196** (2022). https://doi.org/10.1016/j.sab.2022.106517
51. Schroeder, W.H., Dobson, M., Kane, D.M., Johnson, N.D.: Toxic trace elements associated with airborne Pariacnlaie matter: a review. J. Air Pollut. Control. Assoc. **37**(11), 1267–1285 (1987). https://doi.org/10.1080/08940630.1987.10466321
52. Klockenkämper, R., von Bohlen, A.: Elemental analysis of environmental samples by total reflection X-ray fluorescence: a review. In X-Ray Spectrom. **25**(4), 156–162 (1996). John Wiley and Sons Ltd. https://doi.org/10.1002/(SICI)1097-4539(199607)25:4<156::AID-XRS154>3.0.CO;2-3.
53. Bontempi, E., Zacco, A., Benedetti, D., Borgese, L., Colombi, P., Stosnach, H., Finzi, G., Apostoli, P., Buttini, P., Depero, L.E.: Total reflection X-ray fluorescence (TXRF) for direct analysis of aerosol particle samples. Environ. Technol. **31**(5), 467–477 (2010). https://doi.org/10.1080/09593330903513260
54. Schmeling, M., Klockow, D.: ANALYTICA CHIMICA ACTA ELSEVIER Sample collection and preparation for analysis of airborne particulate matter by total reflection X-ray fluorescence spectrometry. Analytica. Chimica. Acta. **346** (1997)
55. Wagner, A., Mages, M.: Total-Reflection X-ray fluorescence analysis of elements in size-fractionated particulate matter sampled on polycarbonate filters—Composition and sources of aerosol particles in Göteborg, Sweden. Spectrochim. Acta-Part B At.Ic Spectrosc. **65**(6), 471–477 (2010). https://doi.org/10.1016/j.sab.2010.02.007
56. Osán, J., Börcsök, E., Czömpöly, O., Dian, C., Groma, V., Stabile, L., Török, S.: Experimental evaluation of the in-the-field capabilities of total-reflection X-ray fluorescence analysis to trace fine and ultrafine aerosol particles in populated areas. Spectrochim. Acta-Part B At.Ic Spectrosc. **167** (2020). https://doi.org/10.1016/j.sab.2020.105852
57. Smichowski, P.: Antimony in the environment as a global pollutant: a review on analytical methodologies for its determination in atmospheric aerosols. Talanta **75**(1)2–14 (2008). Elsevier. https://doi.org/10.1016/j.talanta.2007.11.005
58. Borgese, L., Bilo, F., Zacco, A., Federici, S., Mutahi, A.W., Bontempi, E., Trzepla, K., Hyslop, N., Yatkin, S., Wobrauschek, P., Prost, J., Ingerle, D., Depero, L.E.: The assessment of a method for measurements and lead quantification in air particulate matter using total reflection X-ray fluorescence spectrometers. Spectrochim. Acta-Part B At.Ic Spectrosc. **167** (2020). https://doi.org/10.1016/j.sab.2020.105840
59. Borgese, L., Zacco, A., Pal, S., Bontempi, E., Lucchini, R., Zimmerman, N., Depero, L.E.: A new non-destructive method for chemical analysis of particulate matter filters: The case of manganese air pollution in Vallecamonica (Italy). Talanta **84**(1), 192–198 (2011). https://doi.org/10.1016/j.talanta.2010.12.048
60. Prost, J., Wobrauschek, P., Streli, C.: Quantitative total reflection X-ray fluorescence analysis of directly collected aerosol samples. X-Ray Spectrom. **46**(5), 454–460 (2017). https://doi.org/10.1002/xrs.2752

61. Prost, J., Zinkl, A., Ingerle, D., Wobrauschek, P., Streli, C.: Evaluation of a sample preparation procedure for total-reflection X-ray fluorescence analysis of directly collected airborne particulate matter samples. Spectrochim. Acta-Part B At.Ic Spectrosc. **147**, 13–20 (2018). https://doi.org/10.1016/j.sab.2018.05.005
62. Dittrich, T.R., Cothern, C.R.: Analysis of trace metal particulates in atmospheric samples using x-ray fluorescence. J. Air Pollut. Control Assoc. **21**(11), 716–719 (1971). https://doi.org/10.1080/00022470.1971.10469589
63. Sari, D.K., Lestiani, D.D., Kurniawati, S., Adventini, N., Atmodjo, D.P.D., Kusmartini, I.: Applicability of EDXRF for elemental analysis in airborne particulate matter (APM): assessment using APM reference material. J. Phys: Conf. Ser. **1436**(1), 012137 (2020). https://doi.org/10.1088/1742-6596/1436/1/012137
64. Takahara, H., Morikawa, A., Kitayama, S., Matsuyama, T., Tsuji, K.: Elemental analysis of hourly collected air filters with X-ray fluorescence under grazing incidence. Anal. Sci. **40**(3), 519–529 (2024). https://doi.org/10.1007/s44211-023-00483-6
65. Fakinle, B.S., Uzodinma, O.B., Odekanle, E.L., Sonibare, J.A.: Impact of elemental composition of particulate matter in the airshed of a University Farm on the local air quality. Heliyon **6**(1) (2020). https://doi.org/10.1016/j.heliyon.2020.e03216
66. Paode, R.D., Sofuoğlu, S.C., Sivadechathep, J., Noll, K.E., Holsen, T.M.: Dry deposition fluxes and mass size distributions of Pb, Cu, and Zn measured in southern lake Michigan during AEOLOS. Environ. Sci. & Technol. **32**(11), 1629–1635 (1998). https://pubs.acs.org/sharingguidelines.
67. Sekhavatjou, M.S., Alhashemi, A.H., Moaref, S., Alhashemi, H.: Determination of trace elements concentration in wet and dry atmospheric deposition and surface soil in the largest industrial city, Southwest of Iran. Artic. Int. J. Environ. Res. **8**(2), 335–346 (2014). https://www.researchgate.net/publication/261287647
68. Ogrizek, M., Kroflič, A., Šala, M.: Critical review on the development of analytical techniques for the elemental analysis of airborne particulate matter. Trends Environ. Anal. Chem. **33** (2022). Elsevier B.V. https://doi.org/10.1016/j.teac.2022.e00155.
69. Canepari, S., Perrino, C., Astolfi, M.L., Catrambone, M., Perret, D.: Determination of soluble ions and elements in ambient air suspended particulate matter: Inter-technique comparison of XRF, IC and ICP for sample-by-sample quality control. Talanta **77**(5), 1821–1829 (2009). https://doi.org/10.1016/j.talanta.2008.10.029
70. Yatkin, S., Gerboles, M.: Investigation of the mass loss of water soluble ionic compounds caused by EDXRF analysis of PM10 loaded filters. Atmos. Environ. **73**, 159–168 (2013). https://doi.org/10.1016/j.atmosenv.2013.03.022
71. Nakano, K., Oshiro, Y., Azechi, S., Somada, Y., Handa, D., Miyagi, Y., Tanahara, A., Arakaki, T., Itoh, A.: Preparation of standard materials of aerosol particles for X-ray fluorescence analysis using a small chamber sampling unit. X-Ray Spectrom. **47**(6), 450–458 (2018). https://doi.org/10.1002/xrs.2973
72. Öztürk, F., Zararsiz, A., Kirmaz, R., Tuncel, G.: An approach to measure trace elements in particles collected on fiber filters using EDXRF. Talanta **83**(3), 823–831 (2011). https://doi.org/10.1016/j.talanta.2010.10.038
73. Onat, B., Sahin, U.A., Akyuz, T.: Elemental characterization of PM2.5 and PM1 in dense traffic area in Istanbul, Turkey. Atmos. Pollut. Res. **4**(1), 101–105 (2013). https://doi.org/10.5094/APR.2013.010
74. Ryś, A., Wróbel, P., Samek, L.: Optimization of the energy dispersive x-ray fluorescence spectrometer for determination of elemental composition of air particulate matter samples. X-Ray Spectrom. **52**(6), 323–329 (2023). https://doi.org/10.1002/xrs.3328
75. Hyslop, N.P., Liu, Y., Yatkin, S., Trzepla, K.: Application of the U.S. EPA procedure for determining method detection limits to EDXRF measurement of filter-based aerosol samples. J. Air Waste Manag. Assoc. **72**(8), 905–913 (2022). https://doi.org/10.1080/10962247.2022.2064005
76. Giannoni, M., Calzolai, G., Chiari, M., Lucarelli, F., Mazzinghi, A., Nava, S., Ruberto, C.: Feasibility study of ED-XRF analysis of atmospheric particulate matter samples collected

with high time resolution. X-Ray Spectrom. **44**(4), 282–288 (2015). https://doi.org/10.1002/xrs.2620

77. Bernardoni, V., Cuccia, E., Calzolai, G., Chiari, M., Lucarelli, F., Massabò, D., Nava, S., Prati, P., Valli, G., Vecchi, R.: ED-XRF set-up for size-segregated aerosol samples analysis. X-Ray Spectrom. **40**(2), 79–87 (2011). https://doi.org/10.1002/xrs.1299
78. Mazzei, F., D'Alessandro, A., Lucarelli, F., Nava, S., Prati, P., Valli, G., Vecchi, R.: Characterization of particulate matter sources in an urban environment. Sci. Total Environ. **401**(1–3), 81–89 (2008). https://doi.org/10.1016/j.scitotenv.2008.03.008
79. Yatkin, S., Gerboles, M., Borowiak, A.: Evaluation of standardless EDXRF analysis for the determination of elements on PM 10 loaded filters. Atmos. Environ. **54**, 568–582 (2012). https://doi.org/10.1016/j.atmosenv.2012.02.062
80. Bilo, F., Cirelli, P., Borgese, L.: Elemental analysis of particulate matter by X-ray fluorescence methods: A green approach to air quality monitoring. TrAC—Trends Anal. Chem. **170** (2024). Elsevier B.V. https://doi.org/10.1016/j.trac.2023.117427
81. Yang, W., Pudasainee, D., Gupta, R., Li, W., Wang, B., Sun, L.: An overview of inorganic particulate matter emission from coal/biomass/MSW combustion: Sampling and measurement, formation, distribution, inorganic composition and influencing factors. Fuel Process. Technol. **213** (2021). Elsevier B.V. https://doi.org/10.1016/j.fuproc.2020.106657
82. Zhang, Y., Xu, Y., Peng, B., Chen, W., Cui, X., Zhang, T., Chen, X., Yao, Y., Wang, M., Liu, J., Zheng, M., Zhu, T.: Quantifying metallic components in aerosol filter samples using micro-synchrotron radiation X-ray fluorescence: With quartz filter as an example. Atmos. Environ. **318** (2024). https://doi.org/10.1016/j.atmosenv.2023.120205
83. Perrino, C., Tofful, L., Canepari, S.: Chemical characterization of indoor and outdoor fine particulate matter in an occupied apartment in Rome Italy. Indoor Air **26**(4), 558–570 (2016). https://doi.org/10.1111/ina.12235
84. Gokul, T., Kumar, K.R., Prema, P., Arun, A., Balaji, P., Faggio, C.: Particulate pollution and its toxicity to fish: An overview. Comp. Biochem. Physiol. Part-C: Toxicol. Pharmacol. **270** (2023). Elsevier Inc. https://doi.org/10.1016/j.cbpc.2023.109646
85. Milićević, T., Relić, D., Urošević, M.A., Castanheiro, A., Roganović, J., Samson, R., Popović, A.: Non-destructive techniques for the determination of magnetic particle and element contents in grapevine leaves and soil as an eco-sustainable tool for environmental pollution assessment in the agricultural areas. Environ. Monit. Assess. **195**(7) (2023). https://doi.org/10.1007/s10661-023-11402-7
86. Vanhoof, C., Chen, H., Berghmans, P., Corthouts, V., de Brucker, N., Tirez, K.: A risk assessment study of heavy metals in ambient air by WD-XRF spectrometry using aerosol-generated filter standards. X-Ray Spectrom. **32**(2), 129–138 (2003). https://doi.org/10.1002/xrs.627
87. Furger, M., Rai, P., Slowik, J.G., Cao, J., Visser, S., Baltensperger, U., Prévôt, A.S.H.: Automated alternating sampling of PM10 and PM2.5 with an online XRF spectrometer. Atmos. Environ. **X**(5) (2020). https://doi.org/10.1016/j.aeaoa.2020.100065
88. Zhou, S., Zhang, Z., Hipsey, M.R., Liu, J., Zhang, M.: Differences in mass concentration and elemental composition of leaf surface particulate matter: Plant species and particle size ranges. Process Saf. Environ. Prot. **175**, 599–610 (2023). https://doi.org/10.1016/j.psep.2023.05.040
89. Wen, Z., Liu, H., Zhou, M., Liu, C., Zhou, C.: Explainable machine learning rapid approach to evaluate coal ash content based on X-ray fluorescence. Fuel **332** (2023). https://doi.org/10.1016/j.fuel.2022.125991
90. Zhou, M., Wen, Z., Xu, G., Zhang, Z., Zhou, C.: Deep learning with TabNet: rapid coal ash content estimation via X-ray fluorescence. Int. J. Coal Prep. Util. (2024). https://doi.org/10.1080/19392699.2024.2341952
91. Rakotondrajoa, A., Radtke, M.: Machine learning based quantification of synchrotron radiation-induced x-ray fluorescence measurements—a case study. Mach. Learn.: Sci. Technol. **2**(2), 025004 (2021). https://doi.org/10.1088/2632-2153/abc9fb

92. Bai, W., Wang, F., Yan, L., Sun, H., Zhu, Z., Chen, L., Li, J., Liang, W., Li, A.: Porous organic polymers (POPs) membrane via thiol-yne click chemistry for efficient particulate matter capture and microplastics separation. Microporous Mesoporous Mater. **329** (2022). https://doi.org/10.1016/j.micromeso.2021.111509
93. Woellner, M., Hausdorf, S., Klein, N., Mueller, P., Smith, M. W., Kaskel, S.: Adsorption and detection of hazardous trace gases by metal–organic frameworks. Adv. Mater. **30**(37) (2018). Wiley-VCH Verlag. https://doi.org/10.1002/adma.201704679
94. Li, P., Li, J., Feng, X., Li, J., Hao, Y., Zhang, J., Wang, H., Yin, A., Zhou, J., Ma, X., Wang, B.: Metal-organic frameworks with photocatalytic bactericidal activity for integrated air cleaning. Nat. Commun. **10**(1) (2019). https://doi.org/10.1038/s41467-019-10218-9.
95. Duan, X., Chen, X., Li, C., Witherspoon, E., Burcar, E., Wang, Z., Peng, W., Pereira, A., Li, H.: Progress of MOFs and their derivatives for mitigating air pollution. Adv. Compos. Hybrid Mater. **7**(4) (2024). Springer Science and Business Media B.V. https://doi.org/10.1007/s42114-024-00930-6
96. Zhang, Y., Yuan, S., Feng, X., Li, H., Zhou, J., Wang, B.: Preparation of nanofibrous metal-organic framework filters for efficient air pollution control. J. Am. Chem. Soc. **138**(18), 5785–5788 (2016). https://doi.org/10.1021/jacs.6b02553
97. Islam, N., Saikia, B.K.: An overview on atmospheric carbonaceous particulate matter into carbon nanomaterials: a new approach for air pollution mitigation. Chemosphere **303** (2022). Elsevier Ltd. https://doi.org/10.1016/j.chemosphere.2022.135027
98. Li, Y., Liu, L., Meng, T., Wang, L., Xie, Z.: Structural engineering of ionic MOF@COF heterointerface for exciton-boosting sunlight-driven photocatalytic filter. ACS Nano **17**(3), 2932–2942 (2023). https://doi.org/10.1021/acsnano.2c11339
99. Zhu, M., Han, J., Wang, F., Shao, W., Xiong, R., Zhang, Q., Pan, H., Yang, Y., Samal, S.K., Zhang, F., Huang, C.: Electrospun nanofibers membranes for effective air filtration. Macromol. Mater. Eng. **302**(1) (2017). Wiley-VCH Verlag. https://doi.org/10.1002/mame.201600353
100. Lv, D., Zhu, M., Jiang, Z., Jiang, S., Zhang, Q., Xiong, R., Huang, C.: Green Electrospun nanofibers and their application in air filtration. Macromol. Mater. Eng. **303**(12) (2018). Wiley-VCH Verlag. https://doi.org/10.1002/mame.201800336
101. Lakshmanan, A., Gavali, D.S., Thapa, R., Sarkar, D.: Synthesis of CTAB-functionalized large-scale nanofibers air filter media for efficient PM2.5 capture capacity with low airflow resistance. ACS Appl. Polym. Mater. **3**(2), 937–948 (2021). https://doi.org/10.1021/acsapm.0c01203
102. Wang, L., Qiao, H., Xu, G., Liang, Y., Yang, J., Hu, J.: Preparation and properties of flame retardant air filter media by incorporation of melamine-formaldehyde fiber and phosphorous-containing poly-acrylate binder. J. Ind. Text.Es **52** (2022). https://doi.org/10.1177/15280837221121947
103. Han, S., Kim, J., Ko, S.H.: Advances in air filtration technologies: structure-based and interaction-based approaches. Mater. Today Adv. **9** (2021). Elsevier Ltd. https://doi.org/10.1016/j.mtadv.2021.100134
104. Fujiwara, F., dos Santos, M., Marrero, J., Polla, G., Gómez, D., Dawidowski, L., Smichowski, P.: Fractionation of eleven elements by chemical bonding from airborne particulate matter collected in an industrial city in Argentina. J. Environ. Monit. **8**(9), 913–922 (2006). https://doi.org/10.1039/b604307k
105. Li, X., Zhang, X., Qin, J., Zhang, Y., Chen, X., Jia, J., Feng, F., Wu, H., Bai, Y.: Facile and fast elemental analysis using glass fiber filter by X-ray fluorescence. Spectrochim. Acta-Part B At.Ic Spectrosc. **216** (2024). https://doi.org/10.1016/j.sab.2024.106951
106. Chiari, M., Yubero, E., Calzolai, G., Lucarelli, F., Crespo, J., Galindo, N., Nicolás, J.F., Giannoni, M., Nava, S.: Comparison of PIXE and XRF analysis of airborne particulate matter samples collected on Teflon and quartz fibre filters. Nucl. Instrum. Methods Phys. Res., Sect. B **417**, 128–132 (2018). https://doi.org/10.1016/j.nimb.2017.07.031

107. Bilo, F., Borgese, L., Wambui, A., Assi, A., Zacco, A., Federici, S., Eichert, D.M., Tsuji, K., Lucchini, R.G., Placidi, D., Bontempi, E., Depero, L.E.: Comparison of multiple X-ray fluorescence techniques for elemental analysis of particulate matter collected on air filters. J. Aerosol Sci. **122**, 1–10 (2018). https://doi.org/10.1016/j.jaerosci.2018.05.003
108. Ault, A.P., Peters, T.M., Sawvel, E.J., Casuccio, G.S., Willis, R.D., Norris, G.A., Grassian, V.H.: Single-particle SEM-EDX analysis of iron-containing coarse particulate matter in an urban environment: Sources and distribution of iron within Cleveland Ohio. Environ. Sci. Technol. **46**(8), 4331–4339 (2012). https://doi.org/10.1021/es204006k

17 Wavelength Dispersive X-Ray Fluorescence Spectroscopy for Diagnosis of Nematode Infested Plants

Neha Sharma and Vivek K. Singh

Abstract

Wavelength dispersive X-ray fluorescence (WDXRF) is emerging as a powerful and efficient tool for detecting chemical changes in crops affected by plant-parasitic nematodes. These microscopic pests target plant roots, causing nutrient imbalances, stunted growth, and ultimately plant death. Given the increasing threat these nematodes pose to global food security, early detection is essential. However, traditional diagnostic methods, such as DNA analysis and physical extraction techniques, can be time-consuming and complex. As the need for more rapid solutions grows, advanced spectroscopic methods, including WDXRF, LIBS, FTIR, UV-Vis, and ICP-OES, are becoming invaluable tools for monitoring plant health. These techniques allow for the identification of subtle chemical alterations in plants caused by nematode infestations, enabling earlier intervention and reducing the reliance on harmful chemical treatments. By offering faster, more accurate analysis, spectroscopic methods like WDXRF improve our understanding of plant health, support more efficient pest management strategies, and contribute to sustainable agricultural practices. This chapter explores the application of WDXRF in diagnosing nematode-induced damage in crops, demonstrating its potential to enhance food security while promoting environmentally friendly farming approaches.

N. Sharma (✉)
Department of Applied Sciences & Humanities, KCC Institute of Technology and Management, Greater Noida, Uttar Pradesh, India
e-mail: neha17.ns.sharma@gmail.com

V. K. Singh
Department of Physics, University of Lucknow, Lucknow, Uttar Pradesh, India

V. K. Singh (ed.), *X-Ray Fluorescence Spectroscopy and Chemometrics*,
https://doi.org/10.1007/978-3-031-98375-7_17

17.1 Introduction

The goal of plant science is to comprehend the biology of plants, their interactions with the environment, and how to apply this understanding to address significant issues in human health and food security. Since they make up the majority of our daily food, plants are vital to human nutrition. Twenty plant species provide over 85% of human calories, with three major grasses—rice (Oryza sativa), maize (Zea mays), and wheat (Triticum aestivum)—making up 60% of this total [1]. In addition to providing energy, plants also provide essential vitamins, minerals, lipids, carbohydrates, and amino acids for human health.

Water, hydrogen, oxygen, carbon dioxide, and about 14 other vital components are necessary for plant growth and development. Based on how much of each of these elements is found in plant tissues, they are divided into macronutrients and micronutrients. In order to maintain their health, plants need higher concentrations of macronutrients such nitrogen, phosphorus, potassium, calcium, magnesium, and sulfur (>1000 mg per kg dry weight). Conversely, micronutrients including boron, chlorine, copper, iron, manganese, molybdenum, nickel, and zinc are as important for healthy plant growth and development but are required in smaller amounts. In order to address the issues that contemporary society faces, especially those related to food security, sustainability, and human health, the study of plants is essential. The detection of trace elements is particularly crucial when dealing with plant parasitic nematodes, which seriously harm crops. Deficits or excessive buildup of trace mineral elements may result from variations in the concentration and transport of these elements in plants and soil over time and across different places [2]. Determining the concentrations of these trace elements is essential for both identifying diseases in agricultural crops and evaluating how well plants treat different health conditions. Issues including fast population increase, climate change, and nutrient stress from overuse of fertilizers and pesticides are posing a growing danger to agriculture's sustainability. Agricultural practices and industrial pollutants exacerbate nutritional problems that arise from either toxicities or deficiencies. The severity of plant diseases brought on by biotic agents like bacteria, nematodes, and fungi is largely determined by the availability of nutrients. According to research, the activation of defense mechanisms, such as the synthesis of antioxidant chemicals and the fortification of physical barriers, is intimately related to the ideal mineral nutrient levels in plants. According to Gupta et al. [3], these elements improve the plant's resistance to pathogen invasion. An estimated $220 billion is lost globally each year due to plant diseases, which significantly reduce yields [4]. Pathogen-induced crop losses in India surpass 35% annually [5]. Some examples are false smut disease, which reduces rice crop yields by 0.2 to 49% [6], and root-knot nematodes, which can cause as much as 29% losses in tomato and 28% in legumes [7]. In rice, bacterial leaf blight causes losses of 74 to 81% [8], while powdery mildew reduces chile yield by 42.82% [9]. These findings demonstrate how infections have a major effect on agricultural output. Thus, avoiding extensive crop damage and preserving food security depend heavily on early detection and surveillance of crop health and nutrient levels.

Plant parasitic nematodes are tiny, roundworms that range in size from 0.25 to 3 mm. They mostly attack roots, although they can also harm stems, leaves, flowers, and fruits. They attack on plant tissues with needle-like stylets and spread by soil movement caused by animals, wind, water, and equipment. Nutrient deficiencies, poor production, stunted development, and root galls are some of the symptoms. In order to release proteins and absorb nutrients, root-knot nematodes develop enormous cells that facilitate invasion. It takes tools like growth monitoring, DNA analysis, and extraction techniques to identify them. The soybean cyst nematode and foliar nematodes are two examples of the diseases that these nematodes can cause. The signs of these microscopic roundworms' attacks on plant roots include stunted growth, low yields, and nutrient deficits. Globally, nematode infestations are a major cause of production losses; in some cases, crop losses can reach 29% for tomatoes and 81% for rice [7, 8]. There is currently no sensor that can identify the health of plants in time to stop additional harm. When detected early, symptoms are the main markers of microbes and plant diseases. However, in certain instances, symptoms might not be identified in time, which could result in the spread of illnesses and significant harm over wide areas. As a result, early plant quality and disease monitoring is essential to maintaining plant science's sustainability. To tackle critical agricultural challenges such as heavy metal toxicity and mineral deficiencies caused by diseases, researchers globally are turning to advanced spectroscopic techniques like X-ray fluorescence spectrometry (XRF), fourier transform infra red spectroscopy (FTIR), laser induced breakdown spectroscopy (LIBS), UV-visible spectroscopy (UV-Vis) and inductively coupled optical emission spectroscopy (ICP-OES). These techniques enable the detection of subtle changes in the elemental composition of plants, which can indicate the presence of pathogens or nutrient imbalances. For instance, LIBS has been effectively applied to detect nutrient changes in citrus leaves infected by bacteria [10] and to map the migration of elements in sunflower leaves [11]. The elemental composition of various plant components has also been analyzed using LIBS in studies on rice plants, which helps to identify deficits and possible disease [12]. The importance of these spectroscopic methods in promoting agricultural sustainability is shown in the literature. Plant science may more effectively address issues related to plant longevity and sustainability by utilizing these cutting-edge instruments, which provide early detection techniques that reduce crop losses and improve the efficiency of food production. This is crucial to lowering the environmental impact of farming methods while meeting the world's expanding food demand.

This chapter provides a thorough analysis of how X-ray fluorescence (XRF) spectroscopy is used to analyze the elements and molecules of plant specimens that are both healthy and infected with nematodes. This work intends to identify compositional changes brought on by nematode infestations by utilizing cutting-edge spectroscopic techniques. It uses a synergistic method that combines XRF and laser-induced breakdown spectroscopy (LIBS) to accomplish a thorough evaluation of elemental distributions and molecular compositions, with a focus on papaya and okra plants.

17.2 Methodology

For this work, we gathered root samples from contaminated field sites in Uttar Pradesh, India, including papaya (Fig. 17.1) and both healthy and nematode-infested okra (Fig. 17.1). Following collection, the materials underwent meticulous cleaning, drying, and grinding in preparation for spectroscopic examination. We isolated the body parts of male, female, and juvenile nematodes using Cobb's screening and decanting techniques in order to determine which nematode species were causing the infestations in the papaya and okra roots. Using the procedures described by Goodey [13] and Eisenback et al. [14], the species was identified by looking at the female worms' perineal patterns. In order to examine the chemical alterations in the materials' internal composition, we used elemental and molecular spectroscopic methods like FTIR and WDXRF. The outcomes of FTIR and WDXRF investigations were also confirmed by supplementary techniques such as UV-Visible spectroscopy and LIBS. Sharma et al.'s studies provide further information on the diseased samples, nematode species, and thorough investigation [7, 8].

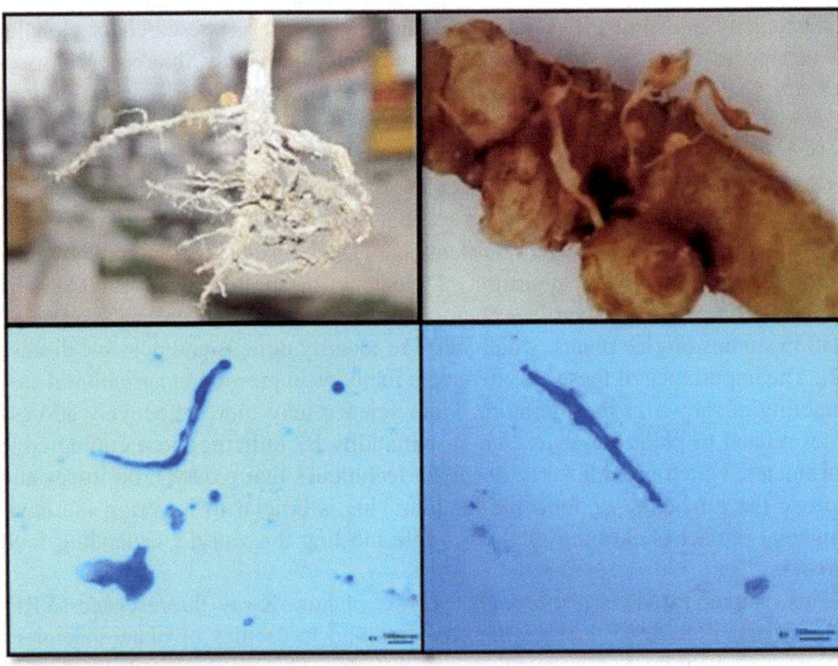

Fig. 17.1 Snapshot of Root-knot nematode infested okra and papaya plant root along with their microscopic observations

17.3 Application of WDXRF Spectroscopy in Agricultural Diseases

17.3.1 Study of Root-Knot Nematode Infected Okra Plant Roots

XRF has proven to be a useful tool for plant sample analysis. With an emphasis on agricultural crops, we investigated its use in this work for crop disease diagnosis. We analyzed both healthy and nematode-infested okra root samples using WDXRF spectroscopy [15]. The root samples contained trace levels of heavy metals and a variety of mineral components, according to our thorough spectral analysis. Crucially, when we contrasted healthy samples with those impacted by nematode infestation, we noticed noticeable variations in the composition of the roots. Elements like manganese, iron, nickel, zinc, and copper were found to be relatively abundant in the 5–10 keV energy range of the WDXRF spectra of both healthy and nematode-infested okra roots as shown in Fig. 17.2a–c [15]. According to our investigation, the elemental composition of the infected and healthy roots differed significantly. Specifically, the okra roots afflicted with nematodes had increased levels of calcium, sulfur, potassium, sodium, chlorine, and phosphorus. The diseased samples, however, had reduced levels of copper, phosphorus, manganese, silicon, and zinc. Infected roots had lower concentrations of toxic metals including chromium and aluminum. The outcomes of our experiment are in line with those of other research and published works. In particular, we discovered that nematode-infested okra root samples had reduced levels of components like iron, phosphorus, and zinc, which is consistent with research by Farahat and associates [17]. When comparing healthy and diseased tissues in nematode-infected root samples from a variety of plants, including citrus, eggplant, jasmine, cowpea, and papaya, they similarly found lower levels of these components [17]. Additionally, when comparing the data from infected and healthy roots, Sharma et al. [15] found that phosphorus concentrations decreased in okra roots infested with nematodes. The concentrations and patterns of different metabolites and mineral nutrients in the roots of various okra cultivars, both healthy and infected with M. incognita, were the main focus of their investigation [15]. These results are consistent with our investigation, which showed that the nematode-infested okra roots had lower phosphorus concentrations. The study by Sharma et al. [15] also observed same patterns, describing a similar pattern of falling phosphorus in okra roots infected with nematodes. Both research' findings show comparable patterns and are in agreement with one another.

Specifically, aluminum is a major hazard to crops because it prevents root development, upsets the nutritional equilibrium, and causes oxidative stress, which causes the cytoplasmic membrane to peroxide and eventually kill cells [18]. These alterations are linked to the development of gall nematodes on the okra roots' surface, which are brought on by nematodic and fungal infections. There is still uncertainty regarding the exact biological function of chromium in plant physiology [19], however high chromium levels have been connected to negative

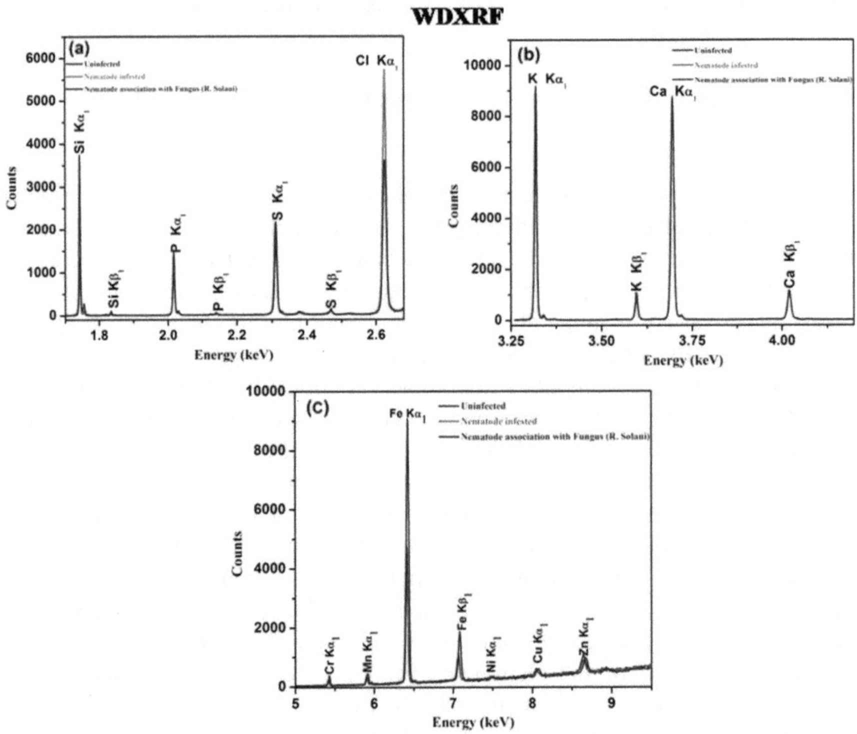

Fig. 17.2 **a–c** WD-XRF spectra of root-knot nematode-infected Okra plants showing different elemental ranges

consequences for plant health, such as morphological, physiological, and biochemical damage [19, 20]. The literature has extensively established the harmful effects of chromium on plants [19]. In this investigation, we found that okra root samples contaminated by nematodic and fungal diseases exhibited notable differences in the quantities of mineral components and are clearly shown in Fig. 17.3a, b [15]. The health of plants may suffer significantly as a result of these elemental alterations. They have the potential to hamper growth, interfere with photosynthesis, impede the flow of vital nutrients, alter water absorption, and ultimately result in a reduction in the okra plant's overall quality. Thus, it is possible to interpret the observed elemental changes as significant effects of nematode infection and the ensuing development of galls on plant roots.

17.3.2 Study of Root-Knot Nematode Infested Papaya Plant

To examine the compositional variations among healthy and infected papaya roots, we also used the WDXRF technique to examine samples of papaya roots afflicted with root-knot nematodes [16]. In both healthy and diseased papaya plant

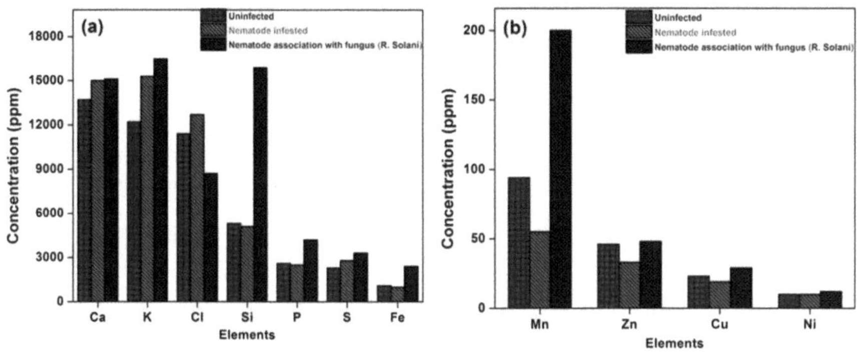

Fig. 17.3 **a, b** Bar diagrams showing the concentration of elements present in uninfected, root-knot infected Okra, and root-knot infected Okra associated with *R. Solani* as measured using WD-XRF

root samples, the study found a number of elements, including chlorine, copper, chromium, calcium, manganese, iron, potassium, phosphorus, nickel, silicon, sulfur, and zinc [16]. The WDXRF spectra of healthy and diseased papaya roots overlap in Fig. 17.4a–c, showing the presence of elements including potassium, calcium, and Sulfur throughout a variety of energy levels [16]. WDXRF spectra of both healthy and infected papaya root samples were analyzed, and the results showed that nematode infestation significantly altered the mineral element content. The importance of mineral components in plants has been highlighted in earlier studies, especially in boosting antioxidant activity, initiating defensive mechanisms, and fortifying physical barriers [21]. Our investigation found that the calcium level increased from 2.4% in healthy root samples to 3.14% in nematode-infected samples, which is consistent with these findings.

In our investigation, the silicon content of papaya root samples rose dramatically, rising from 0.23% in roots that were healthy to 0.53% in roots afflicted with nematodes. In their elemental and molecular examination of okra roots, Sharma et al. found similar patterns of higher calcium, magnesium, and silicon levels [16]. The integrity of cell membranes is maintained by these three elements—calcium, magnesium, and silicon—which improves the plant's resistance to pathogen incursions. Our study's observed rise in these components can be attributed to the plant's defensive mechanisms being activated in reaction to a pathogen attack. Potassium and phosphorus are vital macroelements that are involved in many plant functions, such as enzymatic system activation, protein synthesis induction, photosynthesis facilitation, stomatal movement regulation, cation and anion balance, and energy transfer [16]. In this investigation, WDXRF examination of samples of healthy and diseased papaya roots showed that when worm and fungal infections worsened in the plants, the levels of both potassium and phosphorus decreased. These components are known to encourage cell wall thickening, forming barriers that aid in stopping diseases from spreading further into the plant [16]. Along with potassium and phosphorus, the infected papaya root samples also showed decreased

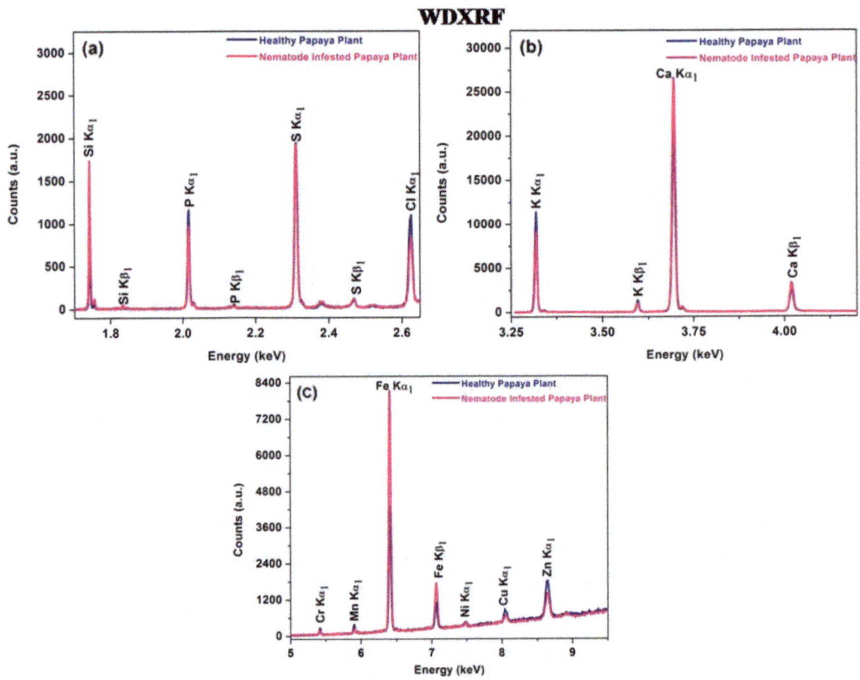

Fig. 17.4 **a–c** WDXRF spectra of healthy and diseased papaya plant samples in different energy ranges

levels of copper, manganese, sodium, and zinc. Sharma et al. [15] published comparable results, showing a similar pattern in okra roots afflicted with nematodes. Figure 17.4a–c shows the relative amounts of calcium, silica, potassium, and iron. Significant fluctuations in the amounts of vital components are caused by nematode invasions in papaya trees, which can negatively impact the longevity and general health of the plant.

The fluctuations in the concentrations of various elements due to nematode infestations can significantly impact the quality and health of papaya plants.

To validate our findings, we also employed the laser-induced breakdown spectroscopy (LIBS) technique, as presented in Fig. 17.5a–c [16]. It has been noted that when the LIBS method is free from self-absorption, the intensity of elemental characteristic lines correlates directly with the concentration of the analytes [16]. To ensure the accuracy of our LIBS results, we checked key conditions such as stoichiometric ablation, local thermal equilibrium, and the optical thickness of the plasma. Furthermore, to facilitate a precise comparative analysis, peak area intensity ratios were determined for key elements—calcium (393.36 nm), potassium (766.48 nm), iron (238.40 nm), and silicon (288.15 nm)—derived from the LIBS spectra of papaya root samples. Following the calculation of peak area intensities for specific elements using the LIBS technique, these values were utilized

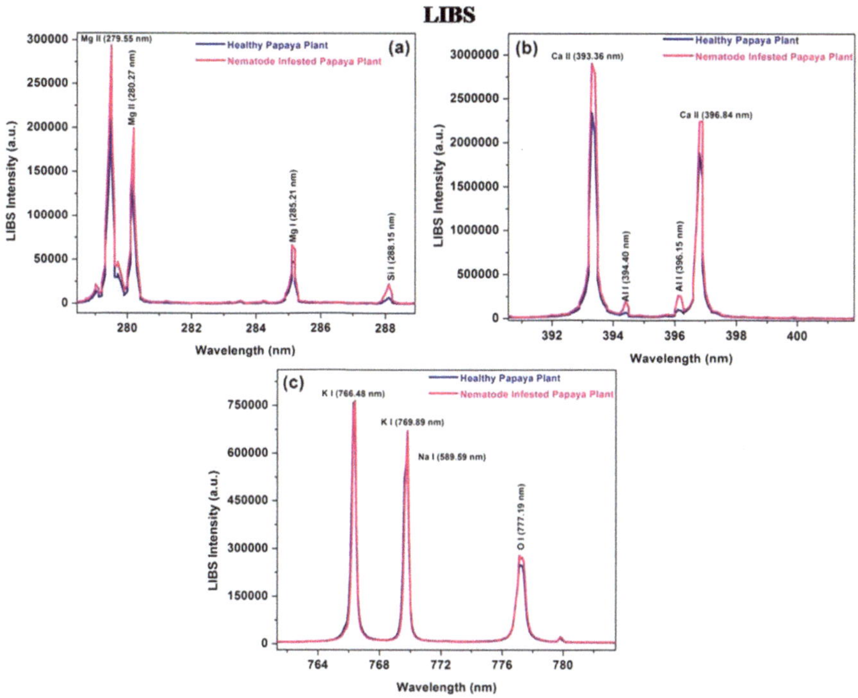

Fig. 17.5 **a–c** LIBS spectra of healthy papaya plant (blue line) and nematode-infested papaya plant (pink line) in different ranges and thereby shows the presence of the elements Mg, Si, Ca, Al, and K

as approximations of elemental concentrations and directly compared with those obtained from the WDXRF spectra of papaya plants. Given the fundamental differences between LIBS and WDXRF, our analysis focused on identifying and evaluating the elemental concentration trends across papaya plant samples rather than absolute values. These trends are distinctly represented in the bar graphs of Fig. 17.6 [16]. Specifically, Fig. 17.6a, b illustrate the concentration patterns of essential elements including calcium, iron, potassium, and silicon in both healthy and nematode-infected papaya root samples, as determined through LIBS and WDXRF analyses. The consistent alignment of trends between both techniques highlights the robustness and complementary potential of LIBS and WDXRF as effective methodologies for the qualitative and quantitative evaluation of mineral composition in both healthy and diseased plant samples.

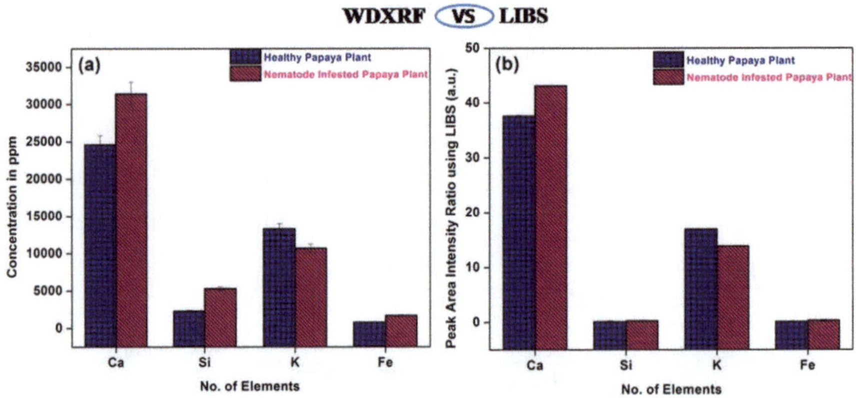

Fig. 17.6 **a, b** Comparison of the concentrations of elements present in papaya plant samples (healthy and nematode-infested papaya plant) measured by using LIBS and WDXRF methods

17.4 Conclusion

Agricultural crops exhibit immense biochemical and elemental diversity, which plays a crucial role in their growth, development, and defense mechanisms. Understanding these elemental compositions is essential, particularly when assessing the impact of plant diseases on crop health. This study underscores the significance of WDXRF as a powerful analytical tool for investigating the elemental profiles of both healthy and diseased plant samples. To the best of our knowledge, this marks the first time WDXRF has been employed within our research group to analyze disease-induced elemental variations, offering valuable insights into plant pathology. The ability of WDXRF to provide both qualitative and quantitative assessments of mineral elements in plant tissues makes it an indispensable technique for evaluating plant health. However, to achieve a more comprehensive understanding of disease progression, integrating advanced imaging modalities such as scanning electron microscopy (SEM) with energy dispersive X-ray analysis (EDX), field emission SEM, and micro-XRF would be highly beneficial. These techniques would allow for precise visualization and spatial distribution mapping of minerals and organic matter at the micrometer scale, enhancing our ability to detect disease-induced elemental fluctuations. Moreover, the incorporation of multivariate analysis alongside WDXRF data could further refine our understanding of plant diseases, such as nematode infestations, false smut, and various bacterial and fungal infections. By leveraging these integrative approaches, researchers can unravel complex interactions between elemental imbalances and disease progression, paving the way for more effective disease monitoring, early detection, and management strategies. The findings of this study reinforce the potential of WDXRF as a robust technique for plant disease diagnostics. When combined with complementary high-resolution analytical methods and advanced statistical modeling, it can contribute significantly to agricultural research by offering deeper insights into plant health, disease mechanisms, and potential mitigation strategies.

References

1. Bennett, B.C.: Encyclopaedia of Life Support Systems (EOLSS), pp. 30–34 (1998)
2. Dalcorso, G., Manara, A., Piasentin, S., Furini, A.: Metallom **6**, 1770–1788 (2014)
3. Gupta, N., Debnath, S., Sharma, S., Sharma, P., Purohit, J.: Agriculturally Important Microbes for Sustainable Agriculture. Springer, New Delhi (2017)
4. Agrius, G.N.: Plant pathology, 5th edn. Elsevier academic press, Burlington, MA USA (2005)
5. Mohhapatra, T.: ICAR News July–September 2018, Published in Monthly News Letter https://www.icar.org.in/sites.default/files/ICARNnewsjulyseptember2018.pdf (2018)
6. Dodan, D.S., Singh, R.: False smut of rice: present status. Agric. Rev. **17**(4), 227–240 (2013)
7. Sasser, J.N.: Root-Knot Nematodes (*Meloidogyne* spp.): Systematic, Biology & Control, F. Lamberti and C.E Taylor, Academic Press, New York (1979)
8. Rangarajan, S., Saleena, L.M., Vasudevan, P., Nair, S.: Biological suppression of rice diseases by Pseudomonas spp. under saline soil conditions. Plant Soil **251**(1), 73–82 (2003)
9. Ashtaputre, S.A.: Assessment of yield loss due to powdery mildew of chilli. Trend Biosci. **7**(11), 1138–1141 (2014)
10. Preira, F.M.V., Milori, D.M.B.P., Venancio, A.L., Russo, M.S., Martins, P.K., Astina, J.F.: Evaluation of the effects of Candidatus Liberibacter asiaticus on inoculated citrus plants using laser-induced breakdown spectroscopy (LIBS) and chemometrics tools. Talanta **83**, 351–356 (2010)
11. Kaiser, J., Galiova, M., Novotny, K., Cervenka, R., Reale, L., Novotny, J., Liska, M., Samek, O., Kanicky, V., Hrdlicka, A., Stejskal, K., Adam, V., Kizek, R.: Mapping of lead, magnesium and copper accumulation in plant tissues by laser induced breakdown spectroscopy and laser ablation inductively coupled plasma mass spectrometry. Spectrochimica Acat Part B **64**, 67–73 (2009)
12. Jabbar, A., Akhtar, M., Mehmood, S., Kurniawan, K.H., Hedwig, R., Baig, M.A.: Analytical approach of laser-induced breakdown spectroscopy to detect elemental profile of medicinal plant leaves. Indones. J. Chem. **19**(2), 430–440 (2019)
13. Goodey, J.B.: Laboratory Methods for Work with Soil and Plant Nematodes, Technical Bulletin Ministry of Agriculture, London, United Kingdom (1963)
14. Eisenback, J.D.: Biol. Control **1**, 95–122 (1985)
15. Sharma, N., Khajuria, Y., Sharma, J., Tripathi, D.K., Chauhan, D.K., Singh, V.K., Kumar, V., Singh, V.K.: Vacuum **158**, 126–135 (2018)
16. Sharma, N., Khajuria, Y., Singh, V.K., Kumar, S., Lee, Y., Rai, P.K., Singh, V.K.: At. Spectrosc.Spectrosc. **41**, 110–118 (2020)
17. Farahat, A.A., Al-Sayed, A., Mahfoud, N.A.: Egypt. J. Agronematol. **12**, 139–158 (2013)
18. Brhane, H., Fikru, E., Haileselassie, T., Tesfaye, K.: African J. Agric. Res. **12**, 1144–1148 (2017)
19. Reale, L., Ferranti, F., Mantilacci, S., Corboli, M., Aversa, S., Landucci, F., Baldisserotto, C., Ferroni, L., Pancaldi, S., Venanzoni, R.: Chemosphere **145**, 98–105 (2016)
20. UdDin, I., Bano, A., Masood, S.: Ecotoxicol. Env. Safety **113**, 271–278 (2014)
21. Dordas, C.: Agron. Sustain. Dev. Sustain. Dev. **28**, 33–46 (2008)

MIX
Papier aus verantwortungsvollen Quellen
Paper from responsible sources
FSC® C105338

If you have any concerns about our products,
you can contact us on
ProductSafety@springernature.com

In case Publisher is established outside the EU,
the EU authorized representative is:
Springer Nature Customer Service Center GmbH
Europaplatz 3, 69115 Heidelberg, Germany

Printed by Libri Plureos GmbH
in Hamburg, Germany